COVID-19 Prevention and Treatment

COVID-19 Prevention and Treatment

Editors

Silvia De Francia
Sarah Allegra

MDPI

Basel • Beijing • Wuhan • Barcelona • Belgrade • Novi Sad • Cluj • Manchester

Editors
Silvia De Francia
Clinical and Biological Sciences
University of Turin
Orbassano
Italy

Sarah Allegra
Clinical and Biological Sciences
University of Turin
Orbassano
Italy

Editorial Office
MDPI AG
Grosspeteranlage 5
4052 Basel, Switzerland

This is a reprint of articles from the Special Issue published online in the open access journal *Life* (ISSN 2075-1729) (available at: https://www.mdpi.com/journal/life/special_issues/COVID_19_prevention_treatment).

For citation purposes, cite each article independently as indicated on the article page online and as indicated below:

Lastname, A.A.; Lastname, B.B. Article Title. *Journal Name* **Year**, *Volume Number*, Page Range.

ISBN 978-3-7258-2205-8 (Hbk)
ISBN 978-3-7258-2206-5 (PDF)
doi.org/10.3390/books978-3-7258-2206-5

Contents

life

Editorial

COVID-19 Prevention and Treatment

Silvia De Francia *, Francesco Chiara and Sarah Allegra

Department of Clinical and Biological Sciences, University of Turin, Regione Gonzole 10, 10043 Orbassano, Italy
* Correspondence: silvia.defrancia@unito.it

1. Introduction

Coronavirus disease 2019 (COVID-19) has spread and become a substantial public health concern worldwide. COVID-19 is an infectious disease caused by the recently discovered SARS-CoV-2 virus. This new virus and the associated disease were unknown before the outbreak reported in Wuhan, China in December 2019. The incubation period for COVID-19 is 1–14 days, most commonly around 5–7 days. The disease causes respiratory illness with symptoms such as a cough, fever, tiredness, and, in more severe cases, difficulty breathing. The COVID-19 virus may persist on surfaces for a few hours up to several days and this may vary under different conditions, e.g., the type of surface, temperature, or humidity of the environment. COVID-19 was first described as a respiratory disease, but presently it is considered a systemic infection comprising multiple systems and causing chronic complications [1]. The pathology results not only from viral infection but from an aberrant inflammatory host immune response [2]. The immune response has been well described in acute COVID-19 patients, but the lasting consequences of the infection are still not well known. The principal mode by which people are infected with SARS-CoV-2 is through exposure to respiratory fluids carrying the infectious virus. Exposure occurs in three principal ways: (1) inhalation of respiratory droplets and aerosol particles; (2) deposition of these particles on exposed mucous membranes in the mouth, nose, or eye by direct splashes and sprays; and (3) touching mucous membranes with hands that have been soiled either directly by virus-containing respiratory fluids or indirectly by touching surfaces with virus on them [3]. The infectious dose of SARS-CoV-2 needed to transmit infection has not yet been established. Current evidence strongly suggests transmission from contaminated surfaces does not contribute substantially to new infections. The risk of SARS-CoV-2 transmission can be reduced by covering coughs and sneezes and maintaining distance from others. When consistent distancing is not possible, well-fitted masks may reduce the spread of infectious droplets from individuals with SARS-CoV-2 infection to others. Frequent hand washing also effectively reduces the risk of infection [4]. Health care providers should follow the Centers for Disease Control and Prevention (CDC)'s recommendations for infection control and the appropriate use of personal protective equipment [5]. At the end of 2021, COVID-19 vaccines received approvals for human use in several countries worldwide. Vaccination is the most effective way to prevent COVID-19. The COVID-19 Treatment Guidelines Panel recommends COVID-19 vaccination as soon as possible for everyone who is eligible according to the CDC's Advisory Committee on Immunization Practices. Four vaccines are authorized or approved for use in the United States to prevent COVID-19. For primary and booster vaccinations, the mRNA vaccines BNT162b2 (Pfizer-BioNTech) and mRNA-1273 (Moderna) and the recombinant spike protein with matrix-M1 adjuvant vaccine NVX-CoV2373 (Novavax) are preferable to the adenovirus vector vaccine Ad26.COV2.S (Johnson & Johnson/Janssen), because of the latter's risk of causing serious adverse events [6]. Reports have suggested that there is an increased risk of thrombosis with thrombocytopenia syndrome (TTS) in adults who received the Johnson & Johnson/Janssen vaccine [7] and, rarely, the Moderna vaccine [8]. TTS is a rare but serious condition that causes blood clots in large blood vessels and low

Citation: De Francia, S.; Chiara, F.; Allegra, S. COVID-19 Prevention and Treatment. *Life* **2023**, *13*, 834. https://doi.org/10.3390/life13030834

Received: 15 March 2023
Accepted: 16 March 2023
Published: 20 March 2023

platelet levels. The American Society of Hematology has published considerations relevant to the diagnosis and treatment of TTS that occurs in people who receive the Johnson & Johnson/Janssen vaccine [9]. These considerations include information on administering a non-heparin anticoagulant and intravenous immunoglobulin to these patients. Pregnant and lactating individuals were not included in the initial COVID-19 vaccine trials. However, the CDC, the American College of Obstetricians and Gynecologists, and the Society for Maternal-Fetal Medicine recommend vaccination for pregnant and lactating people based on the accumulated safety and efficacy data on the use of these vaccines in pregnant people [10,11]. These organizations also recommend vaccination for people who are trying to become pregnant or who may become pregnant in the future. As of March 2023, there are three Food and Drug administration (FDA)–authorized antiviral treatments available to treat people who have mild or moderate symptoms of COVID-19 infection and have risk factors for severe illness. These treatments have been shown to reduce the risk of hospitalization and death in this population: a combination of nirmatrelvir and ritonavir (Paxlovid), molnupiravir (Lagevrio), remdesivir (Veklury). These drugs work by targeting specific parts of the SARS-CoV-2 virus to stop it from multiplying in the body. Another treatment that may benefit immunocompromised people with COVID-19 infection is convalescent plasma. As of March 2023, tocilizumab (Actemra) is the only monoclonal antibody treatment with FDA emergency use authorization (EUA) to treat severe COVID-19 infection, helping to prevent complications. As of March 2023, in case of mild symptoms of COVID-19, to reduce fever and ease aches and pains, acetaminophen and ibuprofen are recommended [12].

The Special Issue "COVID-19 Prevention and Treatment" published in *Life* (ISSN 2075-1729), belonging to the section "Epidemiology", collects a series of research and review articles related to better design routes of prevention and treatment for COVID-19. This issue contains 21 articles. The following is an overview divided by themes addressed by the manuscripts.

2. Treatment and Prevention

Eleven of the papers focus on one of the most important topics related to the COVID-19 pandemic: how to prevent and how to treat the disease. Seven of them are focused on treatment and four of the papers focus on prevention. Three of the papers about treatment are related to the use of convalescent plasma, considered for its immunological mechanisms as benefit for patients in moderate and severe stages of COVID-19. The studies evaluated the safety and efficacy of its use [13–15]. As of December 2020, when the therapeutic agents approved for COVID-19 were limited all over the world, plasma from individuals recovered from COVID-19 was the first therapeutic tool adopted. Another four papers addressing treatment are related to the potential therapeutic implications of different substances or models [16–19]. The studies covered the use of Renessans, a product with iodine complexes and ascorbic acid (the study was designed to determine its efficacy for SARS-CoV-2 in *Rhesus macaque*), characterized by antimicrobial activity; the use of oxygen (the study clearly underlines how, in patients presenting with early dyspnea, the primary goal of therapy should be the reversal of brain hypoxia with a first approach of intermittent treatment with 100% oxygen using a tight oronasal mask or a hood); the use of 14 Mev neutron irradiation (neutron radiation is usually used to sterilize viruses because neutron radiation is 10 times more effective than gamma rays in inactivating viruses: the authors established a closed SARS-CoV-2 inactivation container model and simulated the inactivation performance by using several different neutron sources); and the application of lung imaging scores (the main goal of the paper was to propose a prediction model involving imaging methods, specifically ultrasound). Four of the eleven papers are focused on prevention, the use of vaccines, and their complications [20–23]: the VeroCell vaccine (in peruvian health workers), three different vaccination schedules administered in Chile until January 2022, the potential neurological complications of vaccines, and KD-414 as a booster vaccine for SARS-CoV-2 in healthy adults (KAPIVARA).

3. Mechanism of Action

Two of the papers of the Special Issue focus on potential mechanism of action involved in COVID-19's pathogenicity [24,25]: one paper explored the role of krebs von del Lungen-6 in severe-to-critical COVID-19 patients (KL-6 is a glycoprotein expressed mainly from type II alveolar cells with pro-fibrotic properties: serum KL-6 concentrations have been found in patients with COVID-19); and the second one showed potential therapeutic implications of how COVID-19 hijacks the cytoskeleton (after attaching to membrane receptors and entering cells, the SARS-CoV-2 virus co-opts the dynamic intra-cellular cytoskeletal network of microtubules, actin, and the microtubule-organizing center, enabling three factors that lead to clinical pathology: viral load due to intra-cellular trafficking, cell-to-cell spread by filopodia, and immune dysfunction).

4. Health System Organization

Three of the papers of the Special Issue focus on health system organization involved in COVID-19 management [26–28]: one paper explored the potential role of sex and gender differences in COVID-19 management (the virus mainly affected men with worse symptomatology due to a different immune system, which is stronger in women, and to the Angiotensin-converting enzyme 2 and Transmembrane protease serine 2 roles, which are differently expressed among the sexes; additionally, women are more inclined to maintain social distance and smoke less). The main objective of second study was to describe the measures taken to provide optimal medical care to patients who presented themselves in one of the large emergency hospitals of Romania in the fourth wave of the COVID-19 pandemic, and the third study was aimed at evaluating the government of Nepal's response to the COVID-19 pandemic.

5. General and Specific Condition of Pathology

Five of the papers of the Special Issue focus on the general and specific conditions of COVID-19 pathology [29–33]: one paper explored the potential role of inflammation as a prognostic hallmark of clinical outcome in infected patients. The main objective of a second study was to describe the role of host genetic variability in modulating COVID-19 clinical outcomes. The third study evaluated post-COVID-19 conditions in terms of complications, adverse events, and risk factors. Another paper explored the association between asthma and COVID-19 (interestingly, asthma characterized by type 2 inflammation displays a cellular and molecular profile that may confer protective effects against COVID-19) and the last one was a case report about a 69-year-old man with known seropositive generalized myasthenia gravis, hypertension, ischaemic heart disease, cerebrovascular disease, and recurrent urinary tract infections. He was admitted to the ICU for mixed acute respiratory failure, elevated serum lactate and liver function enzymes, and severe thrombocytopenia and a SARS-CoV-2 PCR test was positive.

Author Contributions: Writing—original draft preparation, S.D.F.; review and editing, S.D.F., F.C. and S.A. All authors have read and agreed to the published version of the manuscript.

Conflicts of Interest: The authors declare no conflict of interest.

References

1. Yong, S.J. Long COVID or post-COVID-19 syndrome: Putative pathophysiology, risk factors, and treatments. *Infect. Dis.* **2021**, *53*, 737–754. [CrossRef] [PubMed]
2. Shuwa, H.A.; Shaw, T.N.; Knight, S.B.; Wemyss, K.; McClure, F.A.; Pearmain, L.; Prise, I.; Jagger, C.; Morgan, D.J.; Khan, S.; et al. Alterations in T and B cell function persist in convalescent COVID-19 patients. *Med* **2021**, *2*, 720–735.e4. [CrossRef] [PubMed]
3. Centers for Disease Control and Prevention. Scientific Brief: SARS-CoV-2 Transmission. 2021. Available online: https://www.cdc.gov/coronavirus/2019-ncov/science/science-briefs/sars-cov-2-transmission.html (accessed on 1 March 2023).
4. Centers for Disease Control and Prevention. How to Protect Yourself and Others. 2023. Available online: https://www.cdc.gov/coronavirus/2019-ncov/prevent-getting-sick/prevention.html (accessed on 1 March 2023).

5. Centers for Disease Control and Prevention. Infection Control Guidance for Healthcare Professionals about Coronavirus (COVID-19). 2020. Available online: https://www.cdc.gov/coronavirus/2019-ncov/hcp/infection-control.html (accessed on 1 March 2023).
6. Centers for Disease Control and Prevention. Overview of COVID-19 Vaccines. 2022. Available online: https://www.cdc.gov/coronavirus/2019-ncov/vaccines/different-vaccines/overview-COVID-19-vaccines.html (accessed on 1 March 2023).
7. Food and Drug Administration. Fact Sheet for Healthcare Providers Administering Vaccine (Vaccination Providers): Emergency Use Authorization (EUA) of the Janssen COVID-19 Vaccine to Prevent Coronavirus Disease 2019 (COVID-19). 2022. Available online: https://www.fda.gov/media/146304/download (accessed on 1 March 2023).
8. See, I.; Lale, A.; Marquez, P.; Streiff, M.B.; Wheeler, A.P.; Tepper, N.K.; Woo, E.J.; Broder, K.R.; Edwards, K.M.; Gallego, R.; et al. Case series of thrombosis with thrombocytopenia syndrome after COVID-19 vaccination—United States, December 2020 to August 2021. *Ann. Intern. Med.* **2022**, *175*, 513–522. [CrossRef]
9. American Society of Hematology. Vaccine-Induced Immune Thrombotic Thrombocytopenia. 2022. Available online: https://www.hematology.org/covid-19/vaccine-induced-immune-thrombotic-thrombocytopenia (accessed on 1 March 2023).
10. Society for Maternal-Fetal Medicine. Publications and Clinical Guidance. 2022. Available online: https://www.smfm.org/covidclinical (accessed on 1 March 2023).
11. Shimabukuro, T.T.; Kim, S.Y.; Myers, T.R.; Moro, P.L.; Oduyebo, T.; Panagiotakopoulos, L.; Marquez, P.L.; Olson, C.K.; Liu, R.; Chang, K.T.; et al. Preliminary findings of mRNA COVID-19 vaccine safety in pregnant persons. *N. Engl. J. Med.* **2021**, *384*, 2273–2282. [CrossRef] [PubMed]
12. Harvard Medical School. Treatments for COVID-19. Available online: https://www.health.harvard.edu/diseases-and-conditions/treatments-for-covid-19 (accessed on 1 March 2023).
13. Ventura-Enríquez, Y.; Cabello-Gutiérrez, C.; Pérez-Calatayud, Á.A.; Cortina-De la Rosa, E.; Fareli-González, C.J.; Castillo-Juárez, P.; Carlos, A.P.-P.; Zavaleta-Martínez, E.O.; Diaz-Padilla, E.; Murrieta, S.; et al. Convalescent plasma to treat COVID-19: A two-center, randomized, double-blind clinical trial. *Life* **2022**, *12*, 1767. [CrossRef] [PubMed]
14. Franchini, M.; Glingani, C.; Donno, G.D.; Lucchini, G.; Beccaria, M.; Amato, M.; Castelli, G.P.; Bianciardi, L.; Pagani, M.; Ghirardini, M.; et al. Convalescent plasma for hospitalized patients: A single-center experience. *Life* **2022**, *12*, 420. [CrossRef]
15. Tomita, N.; Saito, S.; Terada-Hirashima, J.; Mikami, A.; Uemura, Y.; Kutsuna, S.; Nomoto, H.; Fujisawa, K.; Nagashima, M.; Terada, M.; et al. A multi-center, open-label, randomized controlled trial to evaluate the efficacy of convalescent plasma therapy for coronavirus disease 2019: A trial protocol (COVIPLA-RCT). *Life* **2022**, *12*, 856. [CrossRef]
16. Nawaz, M.; Ashraf, M.A.; Ali, M.A.; Shabbir, M.Z.; Shabbir, M.A.B.; Altaf, I.; Raza, S.; Rafique, S.; Hassan, S.; Sardar, N.; et al. Renessans helps in early clearance of SARS-CoV-2: In-vivo activity of the iodine complex in *Rhesus macaque*. *Life* **2022**, *12*, 1424. [CrossRef]
17. Ylikoski, J.; Lehtimäki, J.; Pääkkönen, R.; Mäkitie, A. Prevention and treatment of life-threatening COVID-19 may be possible with oxygen treatment. *Life* **2022**, *12*, 754. [CrossRef]
18. Liu, F.; Zhong, Z.; Liu, B.; Jiang, T.; Zhou, H.; Li, G.; Yuan, X.; Yan, P.; Niu, F.; Ouyang, X. SARS-CoV-2 inactivation simulation using 14 MeV neutron irradiation. *Life* **2022**, *11*, 1372. [CrossRef]
19. Skopljanac, I.; Pavicic Ivelja, M.; Budimir Mrsic, D.; Barcot, O.; Jelicic, I.; Domjanovic, J.; Dolic, K. Can lung imaging scores and clinical variables predict severe course and fatal outcome in COVID-19 pneumonia patients? A single-center observational study. *Life* **2022**, *12*, 735. [CrossRef] [PubMed]
20. Solis-Castro, M.E.; Jaramillo-Corrales, A.; Gonzalez Seminario, R.V.; Janampa Grados, N.; Mamani Pilco, I.E.; Vargas Quispe, K.E.; La Torre Rosillo, L.Y.; Vásquez Dominguez, M.N.; Enriquez Cusi, D.T.; Minaya, P.; et al. Effectiveness of the inactivated SARS-CoV-2 (Vero Cell) vaccine in peruvian health workers. *Life* **2022**, *12*, 1318. [CrossRef] [PubMed]
21. Tabilo Valenzuela, P.B.; Flores Balter, G.; Saint-Pierre Contreras, G.; Conei Valencia, D.; Moreno Calderón, C.; Bohle Venegas, C.; Guajardo Rivera, M.; Silva Ojeda, F.; Vial Covarrubias, M.J. Cellular immune response in patients immunized with three vaccine doses of different vaccination schemes authorized by the Chilean ministry of health in January 2022. *Life* **2022**, *12*, 534. [CrossRef] [PubMed]
22. Tondo, G.; Virgilio, E.; Naldi, A.; Bianchi, A.; Comi, C. Safety of COVID-19 vaccines: Spotlight in neurological complications. *Life* **2022**, *12*, 1338. [CrossRef]
23. Terayama, Y.; Tomita, N.; Terada-Hirashima, J.; Uemura, Y.; Shimizu, Y.; Takeuchi, J.S.; Takamatsu, Y.; Maeda, K.; Mikami, A.; Ujiie, M.; et al. Protocol of an exploratory single-arm study to evaluate the safety and immunogenicity of KD-414 as a booster vaccine for SARS-CoV-2 in healthy adults (KAPIVARA). *Life* **2022**, *12*, 966. [CrossRef]
24. D'Agnano, V.; Scialò, F.; Perna, F.; Atripaldi, L.; Sanduzzi, S.; Allocca, V.; Vitale, M.; Pastore, L.; Bianco, A.; Perrotta, F. Exploring the role of krebs von del Lungen-6 in severe to critical COVID-19 patients. *Life* **2022**, *12*, 1141. [CrossRef] [PubMed]
25. Aminpour, M.; Hameroff, S.; Tuszynski, J.A. How COVID-19 hijacks the cytoskeleton: Therapeutic implications. *Life* **2022**, *12*, 814. [CrossRef]
26. De Francia, S.; Ferretti, A.; Chiara, F.; Allegra, S.; Mancardi, D.; Allice, T.G.; Milia, M.G.; Gregori, G.; Burdino, E.; Avanzini, C.; et al. The influence of sex, gender, and age on COVID-19 data in the Piedmont region (northwest Italy): The virus prefers men. *Life* **2022**, *12*, 643. [CrossRef]
27. Oprita, B.; Davidoiu, A.; Dinu, A.B.; Oprita, R. The rescue of the Romanian health system by the emergency departments during the fourth wave of COVID-19 pandemic. *Life* **2022**, *12*, 1547. [CrossRef]

28. Pandey, B.D.; Ngwe Tun, M.M.; Pandey, K.; Dumre, S.P.; Nwe, K.M.; Shah, Y.; Culleton, R.; Takamatsu, Y.; Costello, A.; Morita, K. How an outbreak of COVID-19 circulated widely in Nepal: A chronological analysis of the national response to an unprecedented pandemic. *Life* **2022**, *12*, 1087. [CrossRef]
29. Fuzio, D.; Inchingolo, A.M.; Ruggieri, V.; Fasano, M.; Federico, M.; Mandorino, M.; Dirienzo, L.; Scacco, S.; Rizzello, A.; Delvecchio, M.; et al. Inflammation as prognostic hallmark of clinical outcome in patients with SARS-CoV-2 infection. *Life* **2022**, *12*, 322. [CrossRef] [PubMed]
30. Upadhyai, P.; Shenoy, P.U.; Banjan, B.; Albeshr, M.F.; Mahboob, S.; Manzoor, I.; Das, R. Exome-wide association study reveals host genetic variants likely associated with the severity of COVID-19 in patients of European ancestry. *Life* **2022**, *12*, 1300. [CrossRef] [PubMed]
31. Boaventura, P.; Macedo, S.; Ribeiro, F.; Jaconiano, S.; Soares, P. Post-COVID-19 condition: Where are we now? *Life* **2022**, *12*, 517. [CrossRef] [PubMed]
32. Wang, C.-J.; Cheng, S.-L.; Kuo, S.-H. Asthma and COVID-19 associations: Focus on IgE-related immune pathology. *Life* **2022**, *12*, 153. [CrossRef] [PubMed]
33. Buzatu, G.-C.; Bobirca, F.-T.; Isac, S.; Mihalache, O.A.; Cotorogea-Simion, M.; Tita, A.; Cobilinschi, C.; Tanasescu, M.D.; Bobirca, A.; Droc, G. The cumulative detrimental effect of COVID-19 pneumonia with myasthenic crisis: A case report and overview of the literature. *Life* **2022**, *12*, 1482. [CrossRef] [PubMed]

life

MDPI

Article

Inflammation as Prognostic Hallmark of Clinical Outcome in Patients with SARS-CoV-2 Infection

Diana Fuzio [1,*,†], Angelo Michele Inchingolo [2,†], Vitalba Ruggieri [1], Massimo Fasano [3], Maria Federico [3], Manuela Mandorino [2], Lavinia Dirienzo [4], Salvatore Scacco [5], Alessandro Rizzello [5], Maurizio Delvecchio [6], Massimiliano Parise [7], Roberto Rana [7], Nicola Faccilongo [7], Biagio Rapone [2], Francesco Inchingolo [2], Antonio Mancini [2], Maria Celeste Fatone [8], Antonio Gnoni [5,‡], Gianna Dipalma [2,‡] and Giovanni Dirienzo [1,‡]

1 Clinical Pathology Unit, Murgia Hospital, Asl Bari, 70022 Altamura, Italy
2 Department of Interdisciplinary Medicine, School of Medicine, University of Bari "Aldo Moro", 70124 Bari, Italy
3 Infectious Diseases Complex Unit, Murgia Hospital, Asl Bari, 70022 Altamura, Italy
4 Universitary Clinical Pathology Complex Unit, School of Medicine, University of Bari "Aldo Moro", 70124 Bari, Italy
5 Department of Basic Medical Sciences, Neuroscience and Sense Organs, University of Bari "Aldo Moro", 70124 Bari, Italy
6 Metabolic Disorders and Diabetes Unit, "Giovanni XXIII" Children's Hospital, School of Medicine, University of Bari "Aldo Moro", 70124 Bari, Italy
7 Department of Economics, University of Foggia, 71100 Foggia, Italy
8 PTA Trani, ASL BT, Internal Medicine Surgery, Viale Padre Pio, 76125 Trani, Italy
* Correspondence: diana.fuzio@asl.bari.it
† These authors contributed equally to this work.
‡ These authors also contributed equally to this work.

Abstract: Severe acute respiratory syndrome coronavirus 2 (SARS-CoV-2) is often characterized by a life-threatening interstitial pneumonia requiring hospitalization. The aim of this retrospective cohort study is to identify hallmarks of in-hospital mortality in patients affected by Coronavirus Disease 19 (COVID-19). A total of 150 patients admitted for COVID-19 from March to June 2021 to "F. Perinei" Murgia Hospital in Altamura, Italy, were divided into survivors (n = 100) and non-survivors groups (n = 50). Blood counts, inflammation-related biomarkers and lymphocyte subsets were analyzed into two groups in the first 24 h after admission and compared by Student's t-test. A multivariable logistic analysis was performed to identify independent risk factors associated with in-hospital mortality. Total lymphocyte count and $CD3^+$ and $CD4^+$ $CD8^+$ T lymphocyte subsets were significantly lower in non-survivors. Serum levels of interleukin-6 (IL-6), lactate dehydrogenase (LDH), C-reactive protein (CRP) and procalcitonin (PCT) were significantly higher in non-survivors. Age > 65 years and presence of comorbidities were identified as independent risk factors associated with in-hospital mortality, while IL-6 and LDH showed a borderline significance. According to our results, markers of inflammation and lymphocytopenia predict in-hospital mortality in COVID-19.

Keywords: biomarkers; COVID-19; C-reactive protein; interleukin-6; inflammation; in-hospital mortality; lactate dehydrogenase; lymphocytes; pneumonia; procalcitonin

Citation: Fuzio, D.; Inchingolo, A.M.; Ruggieri, V.; Fasano, M.; Federico, M.; Mandorino, M.; Dirienzo, L.; Scacco, S.; Rizzello, A.; Delvecchio, M.; et al. Inflammation as Prognostic Hallmark of Clinical Outcome in Patients with SARS-CoV-2 Infection. *Life* **2023**, *13*, 322. https://doi.org/10.3390/life13020322

Academic Editor: Fabrizio Montecucco

Received: 16 November 2022
Revised: 30 December 2022
Accepted: 19 January 2023
Published: 23 January 2023

1. Introduction

From March 2019 to the present, the pandemic caused by a novel betacoronavirus, named severe acute respiratory syndrome coronavirus 2 (SARS-CoV-2), has quickly spread from the city of Wuhan (China) to the entire world, with a variable trend characterized by flattening periods of the pandemic curve and periods of re-emergency [1]. Up-to-date information from the World Health Organization (WHO) reports 663,248,631 confirmed cases of Coronavirus Disease 19 (COVID-19) and 6,709,387 deaths (https://covid19.who.int/, accessed on 19 January 2023). The clinical course of COVID-19 is characterized by markedly divergent clinical manifestations. Aside from asymptomatic or pauci-symptomatic cases, nearly 20% of patients had severe bilateral interstitial pneumonia, which was associated with a rapid deterioration of clinical condition and necessitated hospitalization [2]. In COVID-19, multiple organ dysfunction syndrome (MODS) is triggered by an immune-mediated systemic inflammation, which often leads to death [3]. This pro-inflammatory state is reflected in specific laboratory parameters, such as blood count and markers of systemic inflammation. In severe COVID-19, some altered parameters are common to septic states, such as C-reactive protein (CRP), procalcitonin (PCT), ferritin, D-dimers and fibrinogen, while others indicate, more specifically, a hyper-activation of the immune system and are often present in systemic autoimmune diseases, such as pro-inflammatory cytokine levels, e.g., interleukin-6 (IL-6) and interleukin-10 (IL-10), and reduction of both total lymphocytes and specific lymphocyte subsets [4–6].

The understanding of the clinical and biochemical effects of COVID-19 is constantly expanding. Several strategies for dealing with this global emergency have been developed, ranging from preventive measures, from individual protection devices, isolation of positive cases and vaccines, to adjuvant therapies and specific antiviral treatments [7–9]. Due to the extreme variety of the spectrum of clinical manifestations of COVID-19, it is advisable to tailor preventive and therapeutic choices according to the characteristics of the patients. For instance, it is worth noting that elder patients and patients with pre-existing pathological conditions and comorbidities are at a higher risk of developing a severe outcome of COVID-19 [10]. Thus, it is crucial for physicians to distinguish, as soon as possible, the patients who will suffer from severe illness and are at risk of death, especially in the hospital setting during pandemic waves.

The aim of this retrospective study is to detail the clinical and laboratoristic features of hospitalized COVID-19 patients in a center of Southern Italy, identifying significant differences between the groups of survivors and non-survivors, and independent factors of death, in an attempt to define threshold values indicative of patient outcome.

2. Materials and Methods

2.1. Study Design and Participants

A total of 150 patients hospitalized from March to June 2021 in the public hospital "F. Perinei" Murgia Hospital in Altamura, Italy, were enrolled; patients were divided into survivors (n = 100, group 1) and non-survivors (n = 50, group 2) according to the clinical outcome. This study was approved by the Ethics Committee of "Azienda Ospedaliero Universitaria Consorziale Policlinico", Bari, Italy (0015987, 17 February 2022), and it was conducted in accordance with the Declaration of Helsinki for human studies.

2.2. COVID-19 RT-PCR Assay for Nasal and Pharyngeal Swab Specimens

Real-time reverse-transcriptase polymerase-chain reaction (RT-PCR) assay for nasal and pharyngeal swab specimens was employed to confirm COVID-19. Briefly, RNA extraction was performed using a NeoPlex COVID-19 Detection kit (GeneMatrix Inc., Temecula, CA, USA) on a KingFisher Extraction System (ThermoFisher Scientific, Waltham, MA, USA), according to the manufacturer's instructions. Amplification conditions included reverse transcription at 50 °C for 30 min, denaturation at 95 °C for 15 min and 40 cycles of 95 °C for 15 s and 60 °C for 60 s for fluoresce detection. A cycle threshold value (Ct-value) $\leq$ 38 was defined as a positive test, following Centers of Disease Control and Prevention (CDC) recommendations.

2.3. Laboratory Medicine Analyses and Clinical Data Collection

Laboratory medicine analyses (e.g., blood routine, lymphocyte subsets and inflammation-related biomarkers) were performed on patients' blood samples collected and analyzed in the first 24 h after admission to the Infective Diseases Department of "F. Perinei" Murgia Hospital. The total number of lymphocytes in peripheral blood was counted with an automated hematology analyzer (Pentra ABX, HORIBA, Kyoto, Japan). An Olympus

AU680 (Beckman Coulter Brea, CA, USA) was used to collect LDH and CRP data. PCT was determined by Liaison (DiaSorin S.p.A., Saluggia, Italy). IL-6 and ferritin were measured by ADVIA Centaur XP Immunoassay System (SIEMENS Health, Erlangen, Germany, GmbH). D-Dimers and Fibrinogen values were measured by ACL TOP 500 (Werfen, Bedford, MA USA). Peripheral blood lymphocyte typing was performed in patients' whole blood samples by cytofluorimetric analysis using AQUIOS CL Flow Cytometer (AQUIOS—Beckman Coulter CA, USA). Antibodies used for cell staining were TETRA-1 Panel (CD45, CD4, CD8 and CD3), and obtained data were analyzed using flow cytometry analysis software (Aquios system software, V2.2.0).

2.4. Statistical Analysis

Statistical analysis was performed by setting categorical variables as frequency rates and percentages, and continuous variables as means and 95% confidence intervals (95% CI) or median and interquartile range (IQR) values. The comparison of means for continuous variables that were normally distributed was performed with Student's *t*-test. The Mann–Whitney U test was used for continuous variables there were not normally distributed. Proportions for categorical variables were compared using the $\chi 2$ test, while receiver operating characteristic (ROC) curve analysis was performed using the Wilson/Brown method (95% confidence interval and standard error). Multinomial binary logistic regression analysis results were reported as Odds Ratios (OR) with 95% CI. Statistical analyses were performed by MedCalc® (Mariakerke, Belgium) and GraphPad Prism version 8.2 (GraphPad Software Inc., San Diego, CA, USA). Two-sided *p*-values lower than 0.05 were considered statistically significant.

3. Results

The median age of the patients was 70 years, showing a statistically significant difference between non-survivors and survivors (79 vs. 65 years, $p < 0.001$) (Table 1).

Table 1. Baseline characteristics of 150 enrolled patients with COVID-19. Values are reported as median (IQR) or number of patients (rate%), as appropriate.

Characteristic	Patients (n = 150)	Survivors (n = 100)	Non-Survivors (n = 50)	p-Values
Age (years)	70 (62–80)	65 (57–73)	79 (73–85)	<0.0001
Males	79 (52.7%)	64 (55.25–69.75)	78 (69–83)	<0.0001
Females	71 (47.3%)	68 (57.25–77.5)	83 (74–85)	<0.0001
Days from clinical onset	7 (7–10)	7 (3–20)	7 (2–30)	0.994
Males	7 (7–10)	7 (7–10.75)	7 (7–10)	0.310
Females	7 (7–10)	7 (7–10)	7 (7–10)	0.392

Male patients were significantly older than females (79 vs. 71 years), while the median period from the onset of the symptoms to hospital admission was the same in females and males (7 days). One hundred and twenty-six patients (84%) had comorbidities such as hypertension (55.3%), diabetes mellitus (24%), chronic cardiac disease (22.6%), malignancies (4.6%), obesity (35.3%), chronic pulmonary disease (8%), chronic kidney disease (7.3%) and chronic neurological disorders (16%) (Table 2).

All the patients with severe and moderate disease were given empirical antimicrobial treatment (cephalosporin, azithromycin and levofloxacin). Nineteen patients (12%) received antiviral therapy with remdesivir. In addition, all severe and moderate cases were administered corticosteroids (CTS) during hospitalization. Nine patients (6%) received hyperimmune plasma, and thirty-four patients (22%) required admission to the intensive care unit (ICU) (Table 3).

Table 2. Comorbidities of 150 enrolled patients with COVID-19. Values are reported as number of patients (rate%).

Comorbidities	Patients (n = 150)
Hypertension	83 (55.3%)
Obesity	53 (35.3%)
Diabetes	36 (24%)
Chronic cardiac disease	34 (22.6%)
Chronic neurological disorders	24 (16%)
Chronic pulmonary disease	12 (8%)
Chronic kidney disease	11 (7.3%)
Malignancies	7 (4.6%)
Autoimmune disorders	6 (4%)
HIV	0

Table 3. Medical treatment of 150 enrolled patients with COVID-19 pneumonia. Values are reported as median (IQR) or number of patients (rate%), as appropriate.

Treatments	Patients (n = 150)
Antiviral therapy	
Remdesivir, No (%)	19 (12)
Immune therapy	
Hyperimmune plasma, No (%)	9 (6)
CTS therapy	
Prednisone or equivalent > 1.5 mg/kg/day	148 (98)
ICU admission, No (%)	34 (22)
Length of hospital, days, median (range)	13 (1–87)

At baseline, 102 patients (68%) needed respiratory support by continuous positive airway pressure, 25 patients (16.7%) by a Venturi-type mask, 18 patients (12%) by a simple face mask. Compared with the reference range, significant differences in blood count, lymphocyte subsets and inflammatory-related biomarkers were observed between survivor and non-survivor groups (Table 4).

The median lymphocyte count was lower in the non-survivors group ($p < 0.0001$). When we tested different subsets of T cells, we found that even though both helper T cells ($CD3^+CD4^+$) and suppressor T cells ($CD3^+CD8^+$) in patients with COVID-19 were below reference range ($CD3^+CD4^+$: 500–1700 cells/μL, $CD3^+CD8^+$: 244–1100 cells/μL), the lowering of helper T cells was considerably pronounced in fatal cases (184 vs. 353 cells/uL; $p < 0.0001$). Suppressor T cells also showed a decreasing trend (83 vs. 172 cells/uL; $p < 0.0001$). Conversely, T-helper and T-suppressor ratio ($CD4^+/CD8^+$ ratio) remained in the normal range and showed no difference between the two subgroups. As shown in Figure 1, the area under the curve (AUC) derived from $CD8^+$ T cells was as large as that derived from $CD3^+$ cells or $CD4^+$ cells (AUC $CD8^+$ = 0.741 [0.655–0.827] vs. AUC $CD3^+$ = 0.769 [0.690–0.848] or AUC $CD4^+$ = 0.752 [0.670–0.833], $p < 0.001$).

Non-survivors had significantly higher serum levels of IL-6, LDH, CRP and PCT than survivors (Table 4). Conversely, except for fibrinogen, no significant differences were found in the levels of D-dimers and ferritin between the two groups. Figure 2 shows that the AUC of IL-6 was 0.735 [0.651–0.818] and LDH was 0.784 [0.703–0.864] ($p < 0.001$), and age and comorbidities had AUCs of 0.805 [0.736–0.873] and 0.709 [0.622–0.800], respectively.

By multivariable logistic regression analysis, two indicators were identified to be independent risk factors associated with in-hospital mortality: age > 65 years (OR = 1.14; 95% CI, 1.07–1.22, $p = 0.0001$) and number of comorbidities (OR = 1.84; 95% CI, 1.11–3.05; $p = 0.0178$). IL-6 levels > 20 pg/mL (OR = 1.03; 95% CI, 1.00–1.06) and LDH levels > 489 U/L (OR = 1.01; 95% CI, 1.00–1.01) showed a borderline 95% CI (Table 5).

Table 4. Laboratory data of SARS-CoV-2 patients on admission in the survivors and non-survivors groups and survival correlation [CRP: C-reactive protein, Hb: haemoglobin, IL-6: interleukin-6, LDH: lactate dehydrogenase, PCT: procalcitonin; PLT: platelets, WBC: white blood cells].

Blood Count	Patients ($n = 150$)	Survivors ($n = 100$)	Non-Survivors ($n = 50$)	p-Values
WBC ($\times 10^3/\mu L$)	8 (6–11)	7.8 (5.5–10)	8.2 (5.7–11)	0.3957
Lymphocytes ($\times 10^3/\mu L$)	0.9 (0.6–1.3)	1.0 (0.70–1.4)	0.70 (0.50–0.90)	<0.0001
Monocytes ($\times 10^3/\mu L$)	0.5 (0.3–0.6)	0.50 (0.30–0.60)	0.45 (0.30–0.60)	0.7255
Neutrophil ($\times 10^3/\mu L$)	6.2 (1–23.1)	6.0 (3.8–8.4)	6.7 (4.9–9.9)	0.1263
CD3 (cells/μL)	443 (281–707)	556 (358–836)	304 (194–491)	<0.000 1
CD4 (cells/μL)	272 (174–445)	353 (207–548)	184 (113–272)	<0.0001
CD8 (cells/μL)	135 (77–228)	172 (97–265)	83 (49–139)	<0.0001
CD4/CD8 ratio	1.98 (1.4–3.2)	1.97 (1.53–2.98)	2.00 (1.15–3.98)	0.9539
Hb (g/dL)	13.4 (12–14)	14(12–15)	13 (11–14)	0.0535
PLT ($\times 10^3/\mu L$)	236 (179–310)	254 (197–318)	200 (160–258)	0.0013
Inflammation-Related Biomarkers	**Patients ($n = 150$)**	**Survivors ($n = 100$)**	**Non-Survivors ($n = 50$)**	**p-Values**
D-Dimers (ng/mL)	1118 (696–2160)	1041 (652–1692)	1490 (997–3604)	0.0114
Ferritin (ng/mL)	738 (407–1293)	655 (349–1013)	1111 (618–1561)	0.0018
Fibrinogen (g/L)	533 (418–643)	523 (406–633)	544 (453–678)	0.4193
IL-6 (pg/mL)	13 (2–28)	9 (3–19)	20 (11–57)	<0.0001
LDH (U/L)	357 (272–464)	318 (257–398)	489 (359–557)	<0.0001
CRP (mg/L)	67 (26–124)	51 (16–80)	128 (68–170)	<0.0001
PCT (ng/mL)	0.08 (0.03–0.23)	0.05 (0.02–0.12)	0.21 (0.10–0.51)	<0.0001

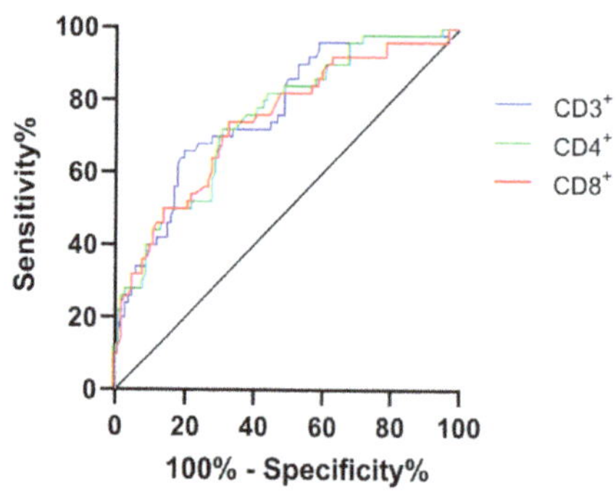

Figure 1. ROC curve of T lymphocyte subsets.

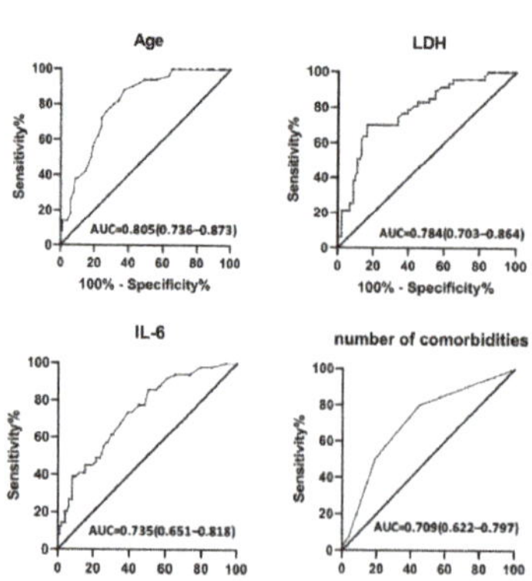

Figure 2. ROC curves of age, IL-6, LDH and number of comorbidities [AUC: area under the ROC curve].

Table 5. Variables independently associated with survival (odds ratios and 95% confidence intervals).

Variables	Odds Ratio	95% CI	p-Value
Age	1.14	1.07 to 1.22	0.0001
Number of comorbidities	1.84	1.11 to 3.05	0.0178
Variables not included in the model			
D-Dimers, Ferritin, Fibrinogen, CRP, PCT, IL-6, LDH			

4. Discussion

Infection by SARS-CoV-2 can cause sustained responses of pro-inflammatory cytokines and chemokines (namely, a "cytokine storm"), leading to a life-threatening immune-mediated MODS [3]. The identification of specific immunological and inflammatory profiles of patients, and their association with COVID-19 severity, is a challenge in order to promptly block systemic inflammation with targeted therapeutic interventions and, on the other hand, minimize unnecessary treatments, especially during pandemic waves.

In this study, we investigated the predictive values of markers of inflammation and lymphocytopenia in hospitalized severe COVID-19 patients, already assessed in other previous studies, with, in part, conflicting results. Our research showed that non-surviving hospitalized COVID-19 patients had significantly higher PCT, CRP, LDH and IL-6 levels, which are indicators of both in-hospital mortality and inflammation. Age and comorbidities—particularly hypertension, obesity, diabetes and chronic heart disease—were found to be independent risk factors linked to in-hospital mortality, while LDH and IL-6 showed a borderline significance.

PCT, the precursor of the hormone calcitonin, is an acute-phase glycoprotein produced by C-cells of thyroids and monocytes. PCT dramatically increases during bacterial and fungal infections while slightly increasing during viral infections, making it an important biomarker of sepsis. In our series of hospitalized COVID-19 patients, PCT levels correlated with disease severity. Data on the value of PCT as prognostic markers for COVID-19 are contradictory in the literature. When compared to moderate illness, PCT is four times higher in severe patients and eight times higher in critical patients, according to Hu R. et al. [11]. Likewise, several authors found that PCT levels are increased in patients with a fatal outcome of severe COVID-19 both at admission and during the course of hospitalization [10,12,13]. In addition, Sayah W. et al. assessed that PCT and the neutrophil-to-lymphocyte ratio are not influenced by the administration of CTS. For this reasons, they may constitute valid alternative markers to assess severe forms in patients already undergoing CTS [14]. The markedly increased levels of PCT in COVID-19 patients can be explained by several factors: a coexistent bacterial infection, prolonged invasive mechanical ventilation and the up-regulation of the signal transducer and activator of the transcription 3 (STAT3)-dependent pathway, which stimulates angiotensin-converting enzyme 2 (ACE2) and PCT production in monocytes [15–17]. Accordingly, other clinical trials refuted the negative prognostic value of PCT in severe COVID-19 [18,19].

CRP is a non-specific acute-phase glycoprotein produced by the liver in response to trauma, myocardial ischemia and infections. Bacterial infections usually determine a marked increase of CRP, while viral infections are associated with a mild increase in CRP levels. In our study, CRP was significantly higher in non-survivor COVID-19 patients compared with survivor groups. CRP proved to be one of the earliest negative prognostic markers in COVID-19 because its levels increased before the appearance of radiologic findings at chest computer tomography (CT) [20]. Furthermore, CRP levels increased both at the beginning and during the progression of COVID-19 disease, and correlated with severity and mortality [20,21]. Several authors tried to improve the predictive value of CRP by relating it to other parameters, such as the CRP/albumin ratio and the CRP/lymphocytes ratio [22,23]. In particular, CRP is associated with in-hospital mortality due to venous thromboembolism and acute kidney injury [24], and a value of CRP equal to or higher than 40 mg/L is considered life-threatening in COVID-19-hospitalized patients [25].

LDH is an enzyme of the oxidoreductase class produced by different cells that is released into the blood as a result of cell damage or high turnover, such as in cancer, trauma, inflammation or infection. Other authors have assessed that LDH is correlated with poor prognosis in hospitalized COVID-19 patients, also indicating lung and other tissue injuries [26,27]. Thus, COVID-19 may lead to inadequate tissue perfusion and MODS, causing LDH elevation [28]. Thus, high values of LDH could represent a valid biomarker of mortality due to widespread infection in COVID-19.

In our study, we also showed that lymphocyte counts in COVID-19 patients and the $CD3^+$ and $CD4^+CD8^+$ subsets were significantly lower, especially in the non-survivor group. Lymphocytopenia indicates a dysregulation of the immune system and has been observed in COVID-19 patients showing different spectrums of clinical disease [5,24]. Consisting with the literature data, we found a significant decrease of total peripheral lymphocyte counts and T cell main subsets ($CD3^+$ and $CD4^+CD8^+$) in both the survivor and non-survivor groups, even if these parameters were significantly lower in the latter group. The etiopathogenesis of lymphocytopenia in COVID-19 patients has been related to different causes. ACE2, identified as the main cell entry receptor for SARS-CoV-2, is low-expressed by lymphocytes, and the viral genome is rarely detectable in peripheral blood of infected patients [29]. Thus, it is reasonable to speculate that the decrease of peripheral lymphocytes is not ascribable to the direct damage of SARS-CoV-2 on lymphocytes, but rather to an exhaustion of them in terms both of number and function due to the persistent exposure to viral antigens [30,31]. An alternative explanation is that the decrease of peripheral lymphocytes is a result of activation-induced apoptosis or aggressive migration from peripheral blood to the lungs, where robust viral replication occurs [32]. Lymphocytopenia in COVID-19 may be related to hyper-activation of STING (stimulator of interferon genes) due to DNA damage following acute distress respiratory syndrome (ARDS). STING is able to activate the NF–κB pathway and also to determine a progressive $CD4^+CD8^+$ T lymphocytopenia, similar to what occurs in STING-associated vasculopathy with onset in infancy (SAVI) syndromes [33].

Several studies detected a significant reduction in total lymphocytes count and in $CD3^+$ and $CD4^+CD8^+$ T-cell subsets, both at the early stages and in severe forms of COVID-19-associated disease in deceased hospitalized patients compared to survivors [34–37]. Interestingly, a Brazilian study found that reduction of T-cell subtypes is a prognostic factor not only of death but also of need for intubation [38]. In addition to lymphocyte subsets, a high neutrophil-to-lymphocyte ratio has been considered a sensible parameter of negative outcome in COVID-19 [38,39].

The reduction of lymphocytes in COVID-19 also affects B cells and NK (natural killer) cells, as suggested by other studies, highlighting a significant reduction of $CD19^+$ and NK cell count in hospitalized patients, which correlated with progression of disease and death [34,40].

Cytokines have been thought to play an important role in immunity and immunopathology during virus infections. "Cytokines storm" is a phenomenon of excessive inflammatory reaction in which cytokines are rapidly produced in large amounts in response to microbial infection, as well as to therapies, pathogens, cancers, autoimmune conditions and monogenic disorders [41]. This phenomenon has been considered an important contributor to ARDS and MODS in COVID-19 patients [3]. It has been reported that the levels of IL-6, interleukin-2 (IL-2), interleukin-7 (IL-7), IL-10, tumor necrosis factor-alpha (TNF-α), granulocyte colony-stimulating factor (G-CSF), interferon gamma induced protein (IP-10), monocyte chemoattractant protein-1 (MCP-1) and macrophage inflammatory protein-1 alpha (MIP-1α) were significantly higher in COVID-19 patients [42–46].

IL-6 is a protein produced by various types of cells, particularly T lymphocytes, macrophages, mature adipocytes and myocytes, in response to tissue damage from trauma, infection or inflammation, exerting both pro-inflammatory and anti-inflammatory effects [47].

In several studies, IL-6 has been proven to be an independent factor predictive of in-hospital mortality, and its levels are not influenced by the administration of CTS [14,42,43].

In particular, high levels of IL-6 are indicative of lung involvement, acute kidney injury, brain damage, cardiovascular events, and, finally, intestinal permeability, which allows viruses to become widespread in the general circulation [48–52]. The value of IL-6 has been included in a score system that predicts the need for non-invasive ventilation by combining systemic inflammatory biomarkers and a chest CT severity score [53].

Besides IL-6 and LDH, other laboratory parameters are cited in the literature as independent risk factors of death in COVID-19, such as neutrophil count, platelet count, CRP, D-dimers, troponin-I and low total testosterone [54–57].

We found that age and comorbidities—primarily hypertension and obesity, followed by diabetes, chronic cardiac disease, chronic neurological disorders, chronic pulmonary disease, chronic kidney disease and malignancies—are independent risk factors of COVID-19 mortality [25,50]. Consistently, other studies reported that patients older than 65 suffering from comorbidities experienced more severe symptoms, MODS and death [27,58]. Hypertension and obesity are the most frequent comorbidities in patients with severe or fatal COVID-19, and the main cause of death after ARDS is a cardiovascular acute event, such as myocardial dysfunction, arrhythmia or shock [10,12,13,26,59,60]. Instead, according to other authors, COVID-19-deceased patients presented, at admission, more frequently with chronic kidney disease and neurological diseases [13]. Obesity, diabetes and chronic kidney disease are also independent risk factors for intubation, as well as death [61]. Other pre-existing pathological conditions, such as anemia, hypotension, dyslipidemia, hyperglycemia and use of CTS, have been reported as independent risk factors of severe COVID-19 [26,53,62,63].

5. Conclusions

SARS-CoV-2 induces serious infectious diseases and becomes a continuous threat to human health. A rapid and well-coordinated immune response is the first line of defense against viral infections. However, when the immune response is dysregulated, it will result in excessive inflammation, even causing death. The higher expression of proinflammatory cytokines in COVID-19 patients, especially in severe cases, with the consumption of $CD4^+CD8^+$ T cells might result in aggravated inflammatory responses, the production of cytokine storms and clinical conditions worsening.

Our findings demonstrated that lymphocyte counts and $CD3^+$ and $CD4^+CD8^+$ subsets were significantly lower in COVID-19 patients, particularly in the non-survivor group. IL-6, LDH, CRP and PCT levels were significantly higher in non-survivor hospitalized patients affected by COVID-19, representing not only markers of inflammation but also in-hospital mortality. Age and comorbidities, especially hypertension, obesity, diabetes and chronic heart disease, were identified as independent risk factors associated with in-hospital mortality. Elevated LDH and IL-6 levels may be substantial risk factors for in-hospital mortality even though they did not achieve significance as independent risk variables.

Even though our findings resulted from a monocentric study and the time frame for patient enrollment was limited, they highlighted the pivotal role of IL-6, LDH, CRP and PCT in predicting mortality for hospitalized COVID-19 patients.

Nevertheless, the results of this study should be confirmed in further larger controlled trials, while also developing scoring systems to assess the severity of COVID-19 disease, correlating every stage to a specific risk of mortality. On the other hand, it appears critical to establish prognostic survival factors for COVID-19 patients in order to screen the most vulnerable patients, organize targeted therapeutic interventions and reduce healthcare costs and waste.

Author Contributions: Conceptualization, D.F., A.M.I., F.I., L.D. and M.C.F.; methodology, A.G., D.F., M.D., M.M., S.S. and V.R.; software, A.M., F.I., G.D. (Gianna Dipalma), M.C.F. and N.F.; validation, A.R., D.F., F.I., G.D. (Gianna Dipalma), M.F. (Massimo Fasano) and M.P.; formal analysis, B.R., D.F., V.R., F.I., G.D. (Gianna Dipalma), L.D., M.F. (Maria Federico) and R.R.; resources, A.G., A.M.I., A.R., M.C.F., M.M. and M.P.; data curation, F.I., G.D. (Gianna Dipalma), D.F., M.F. (Massimo Fasano), M.F. (Maria Federico), N.F. and V.R.; writing—original draft preparation, D.F., V.R., A.M., A.M.I., B.R., F.I., G.D. (Giovanni Dirienzo) and R.R.; writing—review and editing, A.G., D.F., V.R., A.M.I., F.I., G.D. (Gianna Dipalma) and M.C.F.; visualization, A.G., A.M.I., F.I., G.D. (Gianna Dipalma), M.C.F., M.D. and S.S.; supervision, A.M.I., F.I., D.F., G.D. (Giovanni Dirienzo), G.D. (Gianna Dipalma) and S.S.; project administration, F.I. and G.D. (Giovanni Dirienzo). All authors have read and agreed to the published version of the manuscript.

Funding: This research received no external funding.

Institutional Review Board Statement: Not applicable.

Informed Consent Statement: Not applicable.

Data Availability Statement: Not applicable.

Conflicts of Interest: The authors declare no conflict of interest.

Abbreviations

ACE2	angiotensin-converting enzyme 2
ARDS	acute respiratory distress syndrome
CI	confidence intervals
COVID-19	Coronavirus Disease
CRP	C-reactive protein
G-CSF	granulocyte colony-stimulating factor
IL-2	interkeukine-2
IL-6	interleukin-6
IL-7	interleukin-7
IL-10	interleukin-10
IP-10	interferon gamma-induced protein 10 kDa
IQR	interquartile range
LDH	lactate dehydrogenase
MCP-1	monocyte chemoattractant protein-1
MIP-1α	macrophage inflammatory protein-1 alpha
MODS	multiple organ dysfunction syndrome
NK	natural killer
OR	odds ratio
PCT	procalcitonin
RT-PCR	real-time reverse-transcriptase polymerase-chain reaction
SARS-CoV-2	severe acute respiratory syndrome coronavirus 2
STAT3	signal transducer and activator of transcription 3
TNF-α	tumor necrosis factor-alpha

References

1. Khan, M.; Adil, S.F.; Alkhathlan, H.Z.; Tahir, M.N.; Saif, S.; Khan, M.; Khan, S.T. COVID-19: A Global Challenge with Old History, Epidemiology and Progress So Far. *Molecules* **2020**, *26*, 39. [CrossRef] [PubMed]
2. Jiang, S.; Du, L.; Shi, Z. An emerging coronavirus causing pneumonia outbreak in Wuhan, China: Calling for developing therapeutic and prophylactic strategies. *Emerg. Microbes Infect.* **2020**, *9*, 275–277. [CrossRef] [PubMed]
3. Montazersaheb, S.; Hosseiniyan Khatibi, S.M.; Hejazi, M.S.; Tarhriz, V.; Farjami, A.; Ghasemian Sorbeni, F.; Farahzadi, R.; Ghasemnejad, T. COVID-19 infection: An overview on cytokine storm and related interventions. *Virol. J.* **2022**, *19*, 92. [CrossRef] [PubMed]
4. Ji, P.; Zhu, J.; Zhong, Z.; Li, H.; Pang, J.; Li, B.; Zhang, J. Association of elevated inflammatory markers and severe COVID-19: A meta-analysis. *Medicine* **2020**, *99*, e23315. [CrossRef]
5. Tavakolpour, S.; Rakhshandehroo, T.; Wei, E.X.; Rashidian, M. Lymphopenia during the COVID-19 infection: What it shows and what can be learned. *Immunol. Lett.* **2020**, *225*, 31–32. [CrossRef]

6. Fatone, M.C. COVID-19: A Great Mime or a Trigger Event of Autoimmune Manifestations? *Curr. Rheumatol. Rev.* **2021**, *17*, 7–16.
7. Meng, J.; Li, R.; Zhang, Z.; Wang, J.; Huang, Q.; Nie, D.; Fan, K.; Guo, W.; Zhao, Z.; Han, Z. A Review of Potential Therapeutic Strategies for COVID-19. *Viruses* **2022**, *14*, 2346. [CrossRef]
8. Inchingolo, A.D.; Inchingolo, A.M.; Bordea, I.R.; Malcangi, G.; Xhajanka, E.; Scarano, A.; Lorusso, F.; Farronato, M.; Tartaglia, G.M.; Isacco, C.G.; et al. SARS-CoV-2 Disease Adjuvant Therapies and Supplements Breakthrough for the Infection Prevention. *Microorganisms* **2021**, *9*, 525. [CrossRef]
9. Inchingolo, A.D.; Dipalma, G.; Inchingolo, A.M.; Malcangi, G.; Santacroce, L.; D'Oria, M.T.; Isacco, C.G.; Bordea, I.R.; Candrea, S.; Scarano, A.; et al. The 15-Months Clinical Experience of SARS-CoV-2: A Literature Review of Therapies and Adjuvants. *Antioxidants* **2021**, *10*, 881. [CrossRef]
10. Pan, F.; Yang, L.; Li, Y.; Liang, B.; Li, L.; Ye, T.; Li, L.; Liu, D.; Gui, S.; Hu, Y.; et al. Factors associated with death outcome in patients with severe coronavirus disease-19 (COVID-19): A case-control study. *Int. J. Med. Sci.* **2020**, *17*, 1281–1292. [CrossRef]
11. Hu, R.; Han, C.; Pei, S.; Yin, M.; Chen, X. Procalcitonin levels in COVID-19 patients. *Int. J. Antimicrob. Agents* **2020**, *56*, 106051. [CrossRef]
12. Todor, S.B.; Bîrluțiu, V.; Topîrcean, D.; Mihăilă, R.G. Role of biological markers and CT severity score in predicting mortality in patients with COVID-19: An observational retrospective study. *Exp. Ther. Med.* **2022**, *24*, 698. [CrossRef]
13. Poskurica, M.; Stevanović, Đ.; Zdravković, V.; Čekerevac, I.; Ćupurdija, V.; Zdravković, N.; Nikolić, T.; Marković, M.; Jovanović, M.; Popović, M.; et al. Admission Predictors of Mortality in Hospitalized COVID-19 Patients-A Serbian Cohort Study. *J. Clin. Med.* **2022**, *11*, 6109. [CrossRef]
14. Sayah, W.; Berkane, I.; Guermache, I.; Sabri, M.; Lakhal, F.Z.; Yasmine Rahali, S.; Djidjeli, A.; Lamara Mahammed, L.; Merah, F.; Belaid, B.; et al. Interleukin-6, procalcitonin and neutrophil-to-lymphocyte ratio: Potential immune-inflammatory parameters to identify severe and fatal forms of COVID-19. *Cytokine* **2021**, *141*, 155428. [CrossRef]
15. Lugito, N.P.H. Is procalcitonin a part of human immunological response to SARS-CoV-2 infection or "just" a marker of bacterial coinfection? *Curr. Res. Transl. Med.* **2021**, *69*, 103289. [CrossRef]
16. Pink, I.; Raupach, D.; Fuge, J.; Vonberg, R.P.; Hoeper, M.M.; Welte, T.; Rademacher, J. C-reactive protein and procalcitonin for antimicrobial stewardship in COVID-19. *Infection* **2021**, *49*, 935–943. [CrossRef]
17. Heer, R.S.; Mandal, A.K.; Kho, J.; Szawarski, P.; Csabi, P.; Grenshaw, D.; Walker, I.A.; Missouris, C.G. Elevated procalcitonin concentrations in severe Covid-19 may not reflect bacterial co-infection. *Ann. Clin. Biochem.* **2021**, *58*, 520–527. [CrossRef]
18. Dolci, A.; Robbiano, C.; Aloisio, E.; Chibireva, M.; Serafini, L.; Falvella, F.S.; Pasqualetti, S.; Panteghini, M. Searching for a role of procalcitonin determination in COVID-19: A study on a selected cohort of hospitalized patients. *Clin. Chem. Lab. Med.* **2020**, *59*, 433–440. [CrossRef]
19. Liu, F.; Li, L.; Xu, M.; Wu, J.; Luo, D.; Zhu, Y.; Li, B.; Song, X.; Zhou, X. Prognostic value of interleukin-6, C-reactive protein, and procalcitonin in patients with COVID-19. *J. Clin. Virol.* **2020**, *127*, 104370. [CrossRef]
20. Tan, C.; Huang, Y.; Shi, F.; Tan, K.; Ma, Q.; Chen, Y.; Jiang, X.; Li, X. C-reactive protein correlates with computed tomographic findings and predicts severe COVID-19 early. *J. Med. Virol.* **2020**, *92*, 856–862. [CrossRef]
21. Mousavi-Nasab, S.D.; Mardani, R.; Nasr Azadani, H.; Zali, F.; Ahmadi Vasmehjani, A.; Sabeti, S.; Alavi Darazam, I.; Ahmadi, N. Neutrophil to lymphocyte ratio and C-reactive protein level as prognostic markers in mild versus severe COVID-19 patients. *Gastroenterol. Hepatol. Bed. Bench.* **2020**, *13*, 361–366. [PubMed]
22. Yang, M.; Chen, X.; Xu, Y. A Retrospective Study of the C-Reactive Protein to Lymphocyte Ratio and Disease Severity in 108 Patients with Early COVID-19 Pneumonia from January to March 2020 in Wuhan, China. *Med. Sci. Monit.* **2020**, *26*, e926393. [CrossRef] [PubMed]
23. Li, Y.; Li, H.; Song, C.; Lu, R.; Zhao, Y.; Lin, F.; Han, D.; Chen, L.; Pan, P.; Dai, M. Early Prediction of Disease Progression in Patients with Severe COVID-19 Using C-Reactive Protein to Albumin Ratio. *Dis. Markers* **2021**, *2021*, 6304189. [CrossRef] [PubMed]
24. Smilowitz, N.R.; Kunichoff, D.; Garshick, M.; Shah, B.; Pillinger, M.; Hochman, J.S.; Berger, J.S. C-reactive protein and clinical outcomes in patients with COVID-19. *Eur. Heart J.* **2021**, *42*, 2270–2279. [CrossRef]
25. Stringer, D.; Braude, P.; Myint, P.K.; Evans, L.; Collins, J.T.; Verduri, A.; Quinn, T.J.; Vilches-Moraga, A.; Stechman, M.J.; Pearce, L.; et al. The role of C-reactive protein as a prognostic marker in COVID-19. *Int. J. Epidemiol.* **2021**, *50*, 420–429. [CrossRef]
26. Turan, O.; Mirici, A.; Duru Akçalı, S.; Turan, P.A.; Batum, Ö.; Şengül, A.; Ekici Ünsal, Z.; Işık Kabakoğlu, N.; Ogan, N.; Torun, Ş.; et al. Characteristics of hospitalised COVID-19 patients and parameters associated with severe pneumonia. *Int. J. Clin. Pract.* **2021**, *75*, e14786. [CrossRef]
27. Guan, X.; Zhang, B.; Fu, M.; Li, M.; Yuan, X.; Zhu, Y.; Peng, J.; Guo, H.; Lu, Y. Clinical and inflammatory features based machine learning model for fatal risk prediction of hospitalized COVID-19 patients: Results from a retrospective cohort study. *Ann. Med.* **2021**, *53*, 257–266. [CrossRef]
28. Vincent, J.L.; Taccone, F.S. Understanding pathways to death in patients with COVID-19. *Lancet Respir. Med.* **2020**, *8*, 430–432. [CrossRef]
29. Inchingolo, A.D.; Inchingolo, A.M.; Bordea, I.R.; Malcangi, G.; Xhajanka, E.; Scarano, A.; Lorusso, F.; Farronato, M.; Tartaglia, G.M.; Isacco, C.G.; et al. SARS-CoV-2 Disease through Viral Genomic and Receptor Implications: An Overview of Diagnostic and Immunology Breakthroughs. *Microorganisms* **2021**, *9*, 793. [CrossRef]
30. Scialo, F.; Daniele, A.; Amato, F.; Pastore, L.; Matera, M.G.; Cazzola, M.; Castaldo, G.; Bianco, A. ACE2: The Major Cell Entry Receptor for SARS-CoV-2. *Lung* **2020**, *198*, 867–877. [CrossRef]

31. Signore, A.; Lauri, C.; Colandrea, M.; Di Girolamo, M.; Chiodo, E.; Grana, C.M.; Campagna, G.; Aceti, A. Lymphopenia in patients affected by SARS-CoV-2 infection is caused by margination of lymphocytes in large bowel: An [18F] FDG PET/CT study. *Eur. J. Nucl. Med. Mol. Imaging* **2022**, *49*, 3419–3429. [CrossRef]
32. Ashrafi, F.; Nematollahi, P.; Salmasi, M.; Hedayat, A.; Amra, B. Association of lymphocyte subsets with mortality in severe COVID-19 pneumonia patients. *J. Clin. Lab. Anal.* **2021**, *35*, e24046. [CrossRef]
33. Berthelot, J.M.; Lioté, F.; Maugars, Y.; Sibilia, J. Lymphocyte Changes in Severe COVID-19: Delayed Over-Activation of STING? *Front. Immunol.* **2020**, *11*, 607069. [CrossRef]
34. Cantenys-Molina, S.; Fernández-Cruz, E.; Francos, P.; de Quirós, J.C.L.B.; Muñoz, P.; Gil-Herrera, J. Lymphocyte subsets early predict mortality in a large series of hospitalized COVID-19 patients in Spain. *Clin. Exp. Immunol.* **2021**, *203*, 424–432. [CrossRef]
35. Zhao, C.; Hu, X.; Xue, Q.; Chen, L. Reduced Counts of Various Subsets of Peripheral Blood T Lymphocytes in Patients with Severe Course of COVID-19. *Bull. Exp. Biol. Med.* **2022**, *172*, 721–724. [CrossRef]
36. Iannetta, M.; Buccisano, F.; Fraboni, D.; Malagnino, V.; Campogiani, L.; Teti, E.; Spalliera, I.; Rossi, B.; Di Lorenzo, A.; Palmieri, R.; et al. Baseline T-lymphocyte subset absolute counts can predict both outcome and severity in SARS-CoV-2 infected patients: A single center study. *Sci. Rep.* **2021**, *11*, 12762. [CrossRef]
37. Balzanelli, M.G.; Distratis, P.; Dipalma, G.; Vimercati, L.; Catucci, O.; Amatulli, F.; Cefalo, A.; Lazzaro, R.; Palazzo, D.; Aityan, S.K.; et al. Immunity Profiling of COVID-19 Infection, Dynamic Variations of Lymphocyte Subsets, a Comparative Analysis on Four Different Groups. *Microorganisms* **2021**, *9*, 2036. [CrossRef]
38. de Oliveira, D.C.; Spiri, B.S.; Schluga, Y.C.; Justus, J.L.P.; Neto, F.D.N.L.; de Azambuja, A.P. Evaluation of lymphocyte count, T-cell subsets and neutrophil-to-lymphocyte ratio as early predictors for severity and outcome of COVID-19 disease-a report from a highly complex hospital in Brazil. *Hematol. Transfus. Cell Ther.* 2022; *in press*.
39. Montiel-Cervantes, L.A.; Medina, G.; Pilar Cruz-Domínguez, M.; Pérez-Tapia, S.M.; Jiménez-Martínez, M.C.; Arrieta-Oliva, H.I.; Carballo-Uicab, G.; López-Pelcastre, L.; Camacho-Sandoval, R. Poor Survival in COVID-19 Associated with Lymphopenia and Higher Neutrophile-Lymphocyte Ratio. *Isr. Med. Assoc. J.* **2021**, *23*, 153–159.
40. Li, M.; Guo, W.; Dong, Y.; Wang, X.; Dai, D.; Liu, X.; Wu, Y.; Li, M.; Zhang, W.; Zhou, H.; et al. Elevated Exhaustion Levels of NK and CD8+ T Cells as Indicators for Progression and Prognosis of COVID-19 Disease. *Front. Immunol.* **2020**, *11*, 580237. [CrossRef]
41. Yuan, S.; Jiang, S.C.; Zhang, Z.W.; Fu, Y.F.; Hu, J.; Li, Z.L. Quantification of Cytokine Storms During Virus Infections. *Front. Immunol.* **2021**, *12*, 659419. [CrossRef]
42. Balzanelli, M.G.; Distratis, P.; Lazzaro, R.; Cefalo, A.; Catucci, O.; Aityan, S.K.; Dipalma, G.; Vimercati, L.; Inchingolo, A.D.; Maggiore, M.E.; et al. The Vitamin D, IL-6 and the eGFR Markers a Possible Way to Elucidate the Lung–Heart–Kidney Cross-Talk in COVID-19 Disease: A Foregone Conclusion. *Microorganisms* **2021**, *9*, 1903. [CrossRef] [PubMed]
43. Dhar, S.K.; Damodar, S.; Gujar, S.; Das, M. IL-6 and IL-10 as predictors of disease severity in COVID-19 patients: Results from meta-analysis and regression. *Heliyon* **2021**, *7*, e06155. [CrossRef] [PubMed]
44. Mulla, S.; Molla, M.M.A.; Ahmed, S.M.A.; Akhtaruzzaman, A.K.M.; Saleh, A.A.; Anwar, S. Association of interferon gamma inducible protein-10, monocyte chemoattractant protein-1, macrophage inflammatory protein-1 alpha, interleukin-6, and rs12252 single nucleotide polymorphism of interferon-induced transmembrane protein-3 gene with the severity of COVID-19 infection. *Egypt J. Intern. Med.* **2022**, *34*, 53. [PubMed]
45. Ergoren, M.C.; Paolacci, S.; Manara, E.; Dautaj, A.; Dhuli, K.; Anpilogov, K.; Camilleri, G.; Suer, H.K.; Sayan, M.; Tuncel, G.; et al. A pilot study on the preventative potential of alpha-cyclodextrin and hydroxytyrosol against SARS-CoV-2 transmission. *Acta Biomed.* **2020**, *91*, e2020022.
46. Balzanelli, M.G.; Distratis, P.; Aityan, S.K.; Amatulli, F.; Catucci, O.; Cefalo, A.; De Michele, A.; Dipalma, G.; Inchingolo, F.; Lazzaro, R.; et al. An Alternative "Trojan Horse" Hypothesis for COVID-19: Immune Deficiency of IL-10 and SARS-CoV-2 Biology. *Endocr. Metab. Immune Disord. Drug Targets* **2022**, *22*, 1–5. [CrossRef]
47. Hirano, T. IL-6 in inflammation, autoimmunity and cancer. *Int. Immunol.* **2021**, *33*, 127–148. [CrossRef]
48. Basheer, M.; Saad, E.; Kananeh, M.; Asad, L.; Khayat, O.; Badarne, A.; Abdo, Z.; Arraf, N.; Milhem, F.; Bassal, T.; et al. Cytokine Patterns in COVID-19 Patients: Which Cytokines Predict Mortality and Which Protect Against? *Curr. Issues Mol. Biol.* **2022**, *44*, 4735–4747. [CrossRef]
49. Medeiros, T.; Guimarães, G.M.C.; Carvalho, F.R.; Alves, L.S.; Faustino, R.; Campi-Azevedo, A.C.; Peruhype-Magalhães, V.; Teixeira-Carvalho, A.; de Souza Gomes, M.; do Amaral, L.R.; et al. Acute kidney injury associated to COVID-19 leads to a strong unbalance of circulant immune mediators. *Cytokine* **2022**, *157*, 155974. [CrossRef]
50. Hernandez-Diaz, S.; Huybrechts, K.F.; Desai, R.J.; Cohen, J.M.; Mogun, H.; Pennell, P.B.; Bateman, B.T.; Patorno, E. Topiramate use early in pregnancy and the risk of oral clefts: A pregnancy cohort study. *Neurology* **2018**, *90*, e342–e351. [CrossRef]
51. Kokkoris, S.; Stamataki, E.; Emmanouil, G.; Psachoulia, C.; Ntaidou, T.; Maragouti, A.; Kanavou, A.; Malachias, S.; Christodouli, F.; Papachatzakis, I.; et al. Serum inflammatory and brain injury biomarkers in COVID-19 patients admitted to intensive care unit: A pilot study. *eNeurologicalSci* **2022**, *29*, 100434. [CrossRef]
52. Nguyen, N.; Nguyen, H.; Ukoha, C.; Hoang, L.; Patel, C.; Ikram, F.G.; Acharya, P.; Dhillon, A.; Sidhu, M. Relation of interleukin-6 levels in COVID-19 patients with major adverse cardiac events. *Proc. Bayl. Univ. Med. Cent.* **2022**, *35*, 6–9. [CrossRef]
53. Halmaciu, I.; Arbănași, E.M.; Kaller, R.; Mureșan, A.V.; Arbănași, E.M.; Bacalbasa, N.; Suciu, B.A.; Cojocaru, I.I.; Runcan, A.I.; Grosu, F.; et al. Chest CT Severity Score and Systemic Inflammatory Biomarkers as Predictors of the Need for Invasive Mechanical Ventilation and of COVID-19 Patients' Mortality. *Diagnostics* **2022**, *12*, 2089. [CrossRef]

54. Luo, X.; Zhou, W.; Yan, X.; Guo, T.; Wang, B.; Xia, H.; Ye, L.; Xiong, J.; Jiang, Z.; Liu, Y.; et al. Prognostic Value of C-Reactive Protein in Patients With Coronavirus 2019. *Clin. Infect. Dis.* **2020**, *71*, 2174–2179. [CrossRef]
55. Chapanduka, Z.C.; Abdullah, I.; Allwood, B.; Koegelenberg, C.F.; Irusen, E.; Lalla, U.; Zemlin, A.E.; Masha, T.E.; Erasmus, R.T.; Jalavu, T.P.; et al. Haematological predictors of poor outcome among COVID-19 patients admitted to an intensive care unit of a tertiary hospital in South Africa. *PLoS ONE* **2022**, *17*, e0275832. [CrossRef]
56. Yousfi, N.; Fathallah, I.; Attoini, A.; Jones, M.; Henchir, M.; Ben Hassine, Z.; Kouraichi, N.; Ben Salah, N. Prognostic Value of Routine Blood Parameters in Intensive Care Unit COVID-19 Patients. *EJIFCC* **2022**, *33*, 121–130.
57. Peralta-Amaro, A.L.; Pecero-García, E.I.; Valadez-Calderón, J.G.; Ramírez-Ventura, J.C.; Coria-Moctezuma, L.A.; Hernández-Utrera, J.E. Low testosterone levels as an independent risk factor for mortality in male patients with COVID-19: Report of a Single-Center Cohort Study in Mexico. *Rev. Int. Androl.* 2022; *in press*.
58. Chen, T.; Dai, Z.; Mo, P.; Li, X.; Ma, Z.; Song, S.; Chen, X.; Luo, M.; Liang, K.; Gao, S.; et al. Clinical Characteristics and Outcomes of Older Patients with Coronavirus Disease 2019 (COVID-19) in Wuhan, China: A Single-Centered, Retrospective Study. Newman A, editor. *J. Gerontol. Ser. A* **2020**, *75*, 1788–1795. [CrossRef]
59. Heubner, L.; Petrick, P.L.; Güldner, A.; Bartels, L.; Ragaller, M.; Mirus, M.; Rand, A.; Tiebel, O.; Beyer-Westendorf, J.; Rößler, M.; et al. Extreme obesity is a strong predictor for in-hospital mortality and the prevalence of long-COVID in severe COVID-19 patients with acute respiratory distress syndrome. *Sci. Rep.* **2022**, *12*, 18418. [CrossRef]
60. Vimercati, L.; De Maria, L.; Quarato, M.; Caputi, A.; Gesualdo, L.; Migliore, G.; Cavone, D.; Sponselli, S.; Pipoli, A.; Inchingolo, F.; et al. Association between Long COVID and Overweight/Obesity. *J. Clin. Med.* **2021**, *10*, 4143. [CrossRef]
61. Nader, M.; Zmerli, O.; Platt, D.E.; Hamdan, H.; Hamdash, S.; Tayeh, R.A.; Azar, J.; Kadi, D.; Sultan, Y.; Bazarbachi, T.; et al. Observational study of factors associated with morbidity and mortality from COVID-19 in Lebanon, 2020-2021. *PLoS ONE* **2022**, *17*, e0275101. [CrossRef]
62. Kurian, N.M.; BYadav, S.C.; Gurjar, M.; Mishra, P.; Sanjeev, O.P.; Azim, A.; Gupta, D.; Hashim, Z.; Verma, A.; Bhatia, E.; et al. Association of hyperglycaemia at-admission & diabetes mellitus with 28 day mortality in patients admitted with moderate-severe SARS-CoV-2 infection: A retrospective study. *Indian J. Med. Res.* **2022**, *155*, 554–564.
63. Caira-Chuquineyra, B.; Fernandez-Guzman, D.; Alvarez-Arias, P.M.; Zarate-Curi, Á.A.; Herrera-Añazco, P.; Benites-Zapata, V.A. Association between prehospital medication and fatal outcomes in a cohort of hospitalized patients due to coronavirus disease-2019 in a referral hospital in Peru. *Travel. Med. Infect. Dis.* **2022**, *50*, 102472. [CrossRef] [PubMed]

life

MDPI

Article

Convalescent Plasma to Treat COVID-19: A Two-Center, Randomized, Double-Blind Clinical Trial

Yanet Ventura-Enríquez [1,†], Carlos Cabello-Gutiérrez [2,†], Ángel Augusto Pérez-Calatayud [3,†], Evelyn Cortina-De la Rosa [1,4,†], Christian Javier Fareli-González [5], Paola Castillo-Juárez [2,6], Alberto Peña-Pérez Carlos [7], Elí Omar Zavaleta-Martínez [7], Elizabeth Diaz-Padilla [1,8], Sandra Murrieta [1], Violeta Deyanira Álvarez-Jiménez [9], Juan Alberto Díaz Ponce-Medrano [10], Catalina Casillas-Suárez [3], María Angelica Ocampo-Ocampo [3], Cruz Vargas-De-León [11,12,*] and Verónica Fernández-Sánchez [1,11,13,*]

1 Banco de Sangre, Centro Médico Naval (CEMENAV), Coyoacán, Ciudad de México 04470, Mexico
2 Departamento de Investigación en virología y micología, Instituto Nacional de Enfermedades Respiratorias (INER), Ciudad de México 14080, Mexico
3 División de Medicina Crítica, Hospital General de México "Dr. Eduardo Liceaga", Ciudad de México 06720, Mexico
4 Departamento de Hematología, Instituto Nacional de Cardiología Ignacio Chávez, Ciudad de México 14080, Mexico
5 Guías de Práctica Clínica, National Centre of Technological Excellence in Health, Ciudad de México 06600, Mexico
6 Departamento de Microbiología, Escuela Nacional de Ciencias Biológicas, Instituto Politécnico Nacional (IPN), Ciudad de México 11340, Mexico
7 Unidad de Cuidados Intensivos, Centro Médico Naval (CEMENAV), Coyoacán, Ciudad de México 04470, Mexico
8 Facultad de Química, UNAM, Ciudad de México 04510, Mexico
9 Biología Molecular y BSL III, Centro Médico Naval (CEMENAV), Coyoacán, Ciudad de México 04470, Mexico
10 Dirección General, Centro Médico Naval (CEMENAV), Coyoacán, Ciudad de México 04470, Mexico
11 División de Investigación, Hospital Juárez de México, Ciudad de México 07760, Mexico
12 Sección de Estudios de Investigación y Posgrado, Escuela Superior de Medicina, Instituto Politécnico Nacional (IPN), Ciudad de México 11340, Mexico
13 Facultad de Estudios Superiores Iztacala (FES-Iztacala), Universidad Nacional Autónoma de México (UNAM), Ciudad de México 54090, Mexico
* Correspondence: leoncruz82@yahoo.com.mx (C.V.-D.-L.); alf2228@yahoo.com.mx (V.F.-S.)
† These authors have shared first authorship.

Citation: Ventura-Enríquez, Y.; Cabello-Gutiérrez, C.; Pérez-Calatayud, Á.A.; Cortina-De la Rosa, E.; Fareli-González, C.J.; Castillo-Juárez, P.; Carlos, A.P.-P.; Zavaleta-Martínez, E.O.; Diaz-Padilla, E.; Murrieta, S.; et al. Convalescent Plasma to Treat COVID-19: A Two-Center, Randomized, Double-Blind Clinical Trial. *Life* **2022**, *12*, 1767. https://doi.org/10.3390/life12111767

Academic Editors: Silvia De Francia and Sarah Allegra

Received: 29 September 2022
Accepted: 26 October 2022
Published: 2 November 2022

Abstract: Background: The use of convalescent plasma (CP) has been considered for its immunological mechanisms that could benefit patients in moderate and severe stages of COVID-19. This study evaluated the safety and efficacy of the use of donor CP for COVID-19. Material and methods: A double-blind, randomized controlled clinical trial was conducted from May to October 2020. Thirty-nine participants with moderate (II) and severe (III) stages of COVID-19 confirmed by RT-PCR were included. The study randomization rate was set at 3:1. CPs were chosen for application with a neutralizing antibody titer of $\geq$1:32. Results: We observed a significantly lower 21-day post-transfusion mortality HR: 0.17 (95.0% CI [0.07–0.45, $p < 0.001$]) in the group receiving CP compared with the control group; protective units (PU) in the group receiving convalescent plasma after seven days were significantly higher (512 (32–16,384) vs. 96 (32–256), $p = 0.01$); the PAO_2/FIO_2 index showed a significant improvement in the group receiving CP (251.01 (109.4) vs. 109.2 (62.4), $p < 0.001$, in the control group). Conclusion: CP is safe and effective, as it decreased mortality in the CP group compared with the control group.

Keywords: convalescent plasma treatment; COVID-19; neutralizing antibodies; SARS-CoV-2

1. Introduction

SARS-CoV-2 is spread by droplets of saliva or nasal discharge that become suspended in the air or on surfaces. Symptoms can vary; among the most frequent are respiratory distress, headache, fever, cough, and fatigue. The virus is known to affect mainly type II alveolar epithelial cells in the lungs, inducing the activation of immune cells that trigger an inflammatory response to the extent of pulmonary infiltration. In immunocompromised, chronically ill, or elderly persons, where cellular regeneration capacity is impaired, SARS-CoV-2 infection produces severe irreversible cellular damage that can lead to death [1].

Treatment of the disease is a challenge because of the lack of clinical evidence regarding the effectiveness and safety of specific antiviral agents for the management of COVID-19. When facing an emergency, different treatment regimens have been used, such as the lopinavir/ritonavir combination, which did not result in a reduction in overall mortality. Another randomized controlled trial using hydroxychloroquine showed a reduction in body temperature and remission of cough in the intervention group compared with controls; however, these studies were not robust due to short periods and small sample sizes [2].

In the absence of evidence for specific treatment against COVID-19, classical and historical interventions have emerged as options for disease control. A passive immunization strategy with CP has been used to prevent and manage emerging infectious diseases since the early 20th century. In 2015, a protocol for the use of CP to treat Middle East respiratory syndrome (MERS) [3], and in 2009 for pandemic H1N1 influenza, was conducted. In this prospective cohort study by Hung et al., a significant reduction in the relative mortality risk odds ratio 0.20 (95% CI [0.06–0.69]) was observed in patients treated with CP (p = 0.01) [4].

CP is obtained by apheresis from survivors with previous infections caused by pathogens of interest, in which antibodies develop against the causative agent of the disease. Since its rapid acquisition, CP has been considered an emergency intervention in different pandemics, namely: Spanish flu, West Nile virus, and Ebola virus [5]. Therefore, the strategy of using CP has been proposed as a treatment option for SARS-CoV-2 infection. To date, several studies have reported controversial results regarding the use of CP [6]. Based on the scientific evidence obtained so far, CP was granted emergency use authorization to treat hospitalized patients with COVID-19 on 23 August 2020 by the Food and Drug Administration (FDA) [7]. Its application met the "maybe effective" standard for emergency use, and it is reasonable to consider the already known potential benefits of CP [8]. It has been described that the antibodies present in PC are highly specific, so that the viral neutralization response is faster and more effective, which favors the decrease in the viral load in patients and the apparent improvement. However, there is also an unfavorable side where these antibodies may bind to specific sites for virus recognition or inhibit the host immune response [1]. There have been a lot of different results, some of them show positive results, such as Lili et al. [9] or Yogiraj et al. whom, in a single center assay open label and randomized controlled by standard treatment (PICP19), reported 25.0% versus 35.0% mortality in a CP arm (RR= 0.71, CI 95.0%: [0.36–1.41]) [10]. In another study, Avendaño et al., in a multicenter randomized clinical trial (ConPlas-19), found 0.0% mortality versus 9.3% in a standard treatment arm (p = 0.07) [11]. Sahu et al. summarize six studies in which the administration of PC plasma was encouraging in different groups of patients, including those in critical condition with severe COVID-19 pneumonia. Although the patient groups are small, and the inclusion criteria for selecting PC patients were different, most of them do not mention adverse effects or complications with the use of PC [12].

This study aimed to evaluate the safety and efficacy of using plasma from convalescent donor COVID-19 to reduce mortality in patients with moderate (II) and severe (III) stages of SARS-CoV-2 disease.

2. Materials and Methods

The study was conducted by the Blood Bank of the Centro Médico Naval (CEMENAV) and the Hospital General de México "Dr Eduardo Liceaga" (HGM) in Mexico City, from 20 May 2020 to 10 October 2020. The study was approved by the Research Ethics Committee and the Research and Biosafety Committee of both institutions. Written informed consent was obtained from each patient or responsible family member, as well as from convalescent plasma donors. The protocol was authorized by the Federal Commission for the Protection against Health Risks (In Spanish, COFEPRIS) with the registration number: CE/168/20. Trial Registration. ClinicalTrials.gov Identifier: NCT04405310.

2.1. Study Design

A parallel, double-blind, randomized phase II clinical trial compared CP and placebo (albumin 0.5%) for patients with stage II and II SARS-CoV-2 pneumonia. Four treatment groups were formed: group A, patients with stage II pneumonia who received CP and conventional therapy; group B, patients with stage II pneumonia who received albumin solution and conventional therapy; group C: patients with stage III pneumonia who received convalescent CP and conventional therapy; and group D, patients with stage III pneumonia who received albumin solution and conventional therapy. Conventional therapy consisted of azithromycin and hydroxychloroquine.

2.2. CP Donor Selection Criteria

Diagnosis of COVID-19 confirmed by RT-PCR, complete resolution of symptoms, with subsequent RT-PCR test with a negative result, male, no history of transfusion in the last 15 days, age 18–55 years old, weight > 50 kg, and authorization of informed consent comply with the requirements for blood donors established in the Mexican Official Standard (NOM-253-SSA1-2012).

2.3. CP Donor Selection Criteria

Adults ≥18 years of age with a diagnosis of COVID-19 confirmed by RT-PCR stages II and III. An Hscore ≥169 points and the presence of severe acute hypoxemia with SpO2 <90% on room air and PaO_2/FiO_2 <300 mmHg. Meeting imaging criteria for SARS-CoV-2 stage II and III pneumonia (plain chest CT scan or plain chest X-ray), with the requirement for supplemental oxygen either via face mask plus reservoir bag, high-flow nasal prongs or advanced airway management, or invasive mechanical ventilatory support. C-reactive protein (CRP) value increased the baseline by 3.5 or greater than 18 mg/dL. Stage II was defined as evidence of lower airway disease, either by clinical assessment or imaging, and oxygen saturation (SpO2) ≥94%. Stage III was defined as a respiratory rate of >30 breaths per minute, SpO2 <94%, ratio between the arterial partial pressure of oxygen and the fraction of inspired oxygen (PaO_2/FiO_2) <300 mmHg, or pulmonary infiltrates >50%.

2.4. Interventions

Recipients assigned to CP received two units of 300 mL intravenously, one on day one and the second unit on day three, with a neutralizing antibody titer ≥ 1:32, equivalent to 256 UP/mL. The recipients in the control group received two units of 300 mL of 20% albumin in Hartman's solution intravenously on day one and day three.

2.5. Generation and Concealment of Random Allocation

Random sequence generation was performed by an external monitor who assigned a code to each patient who met the inclusion criteria. The staff who recruited the patients had no way of knowing which intervention each participant was assigned. A 3:1 allocation of patients was chosen because of the ethical issues involved in not providing a potential clinical benefit to a population with a high fatality rate. Figure 1 shows the process of patient selection and allocation to the two treatment arms.

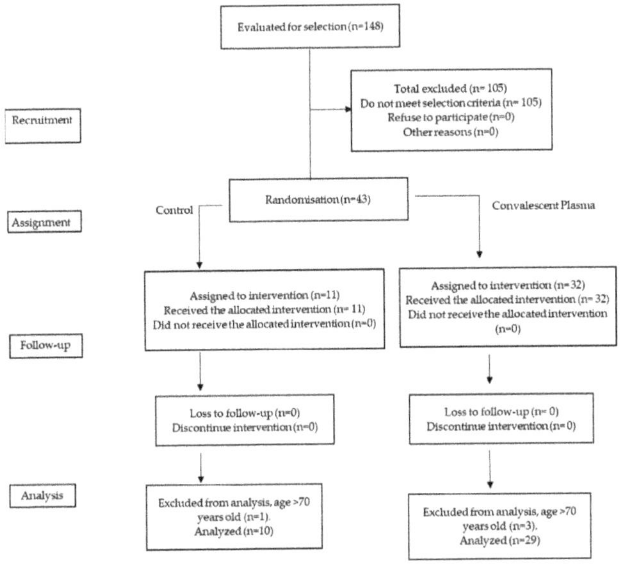

Figure 1. Selection and allocation of patients to receive convalescent plasma versus control group.

The CEMENAV and HGM blood banks provided the monitor with a list of convalescent plasma donors and potential recipients selected by the intensivist in charge of the intervention. Once the random allocation was performed, staff assigned by the Blood Bank, who were not involved in the randomization of treatments, placed the CP and albumin solution bags in a metal bag before taking them to the COVID-19 hospitalization area for administration. The staff involved in the infusion of plasma and solutions and the patients and staff who treated them were blinded to the treatment received. See Figure 1.

2.6. Inactivation of Covalescent Plasma

Inactivation of CP with riboflavin and UV light from each CP donor; 600 mL was collected by apheresis. The plasma was divided into two units (A and B) of approximately 300 mL for subsequent pathogen inactivation with riboflavin and UV light.

Two sterile disposable kits for plasma were used (ref. 10390 Terumo BCT); the plasma was transferred to the illumination bag provided in the kit using hose coupling equipment (T-SEAL II, Terumo Tube Sealing Device) and mixed with a riboflavin solution (500 μm in 0.9% sodium chloride solution, pH 3.5–5.1). After the riboflavin–plasma mixture was inoculated with the virus, the samples were placed into the Mirasol Illuminator (Terumo BCT) for UV treatment. The plasma units were exposed to 6.24 J/mL of energy.

2.7. Safety in the Use of PC

The safety of PC use was evaluated by the absence of adverse events associated with transfusion, according to the definition of the Mexican Official Standard on adverse reactions to blood transfusion: immediate reactions such as hemolysis, non-hemolytic fever, and allergies, among others (present before 24 h), as well as late reactions such as alloimmunization, hemolysis, and post-transfusion purpura, among others (present after 24 h).

2.8. Real-Time Reverse Transcriptase–Polymerase Chain Reaction (RT-PCR)

From nasopharyngeal samples, extraction of total genetic material from potential donors and recipients was performed using the QIAamp Viral RNA kit (Qiagen, Hilden, Germany), using the equipment (QIAcube-classics, Hilden, Germany). The amplification of specific genes (Rd, Rp, E, and N) for SARS-CoV-2 was performed using a GeneFinder

COVID-19 PLUS realAmp kit, and qRT-PCR was performed using the Applied Biosystems 7500 FAST kit (Applied Biosystems, Waltham, MA, USA). All samples were inactivated in a class A-II biosafety cabinet following the Biosafety and Good Laboratory Practice protocols issued by the WHO and the Institute of Diagnosis Epidemiological Reference (In Spanish, INDRE) in Mexico.

2.9. IgG Nucleocapsid Determination

Detection of IgG nucleocapsid antibodies from donor and recipient were analyzed in an Abbott Architect (Abbott Diagnostics®, Abbott Park, IL, USA) according to the manufacturer's instructions. The chemiluminescent assay detects IgG raised against the SARS-CoV-2 nucleocapsid protein. A signal to cut-off (S/CO) ratio of $\geq$1.4 was interpreted as reactive. Calibration was performed, and the positive quality control S/CO 1.65–8.40 and negative quality control S/CO $\leq$ 0.78 were met prior to performing the studies.

2.10. IgG S1/S2 Spike Protein Determinants

The LIAISON SARS-CoV-2 S1/S2 IgG antibody (quantitative assay) was performed using a Liaison-XL (DiaSorin®, Saluggia, Italy). The cut-off was >15.0 AU, which included a negative (<3.8 AU) and positive (>31.9 AU) control. The assay was performed according to the manufacturer's instructions.

2.11. Neutralizing Antibody Assays

An in vitro neutralization assay was developed similar to that reported by Beales et al. For this, serial dilutions were made with minimum essential medium and 50 μL of donor and recipient sera. A total of 5 μL of SARS-CoV-2 virus (MOI = 0.1) was added for each dilution, which was previously titrated using a lytic plate assay. The dilutions were incubated for one h at 37 °C. Each dilution was inoculated into a 96-well plate, seeded with VERO.E6 cells, and incubated at 37 °C and 5% CO_2 for two days. Plates were washed with PBS IX, fixed with an ethanol/acetone mixture (1:1) for 15 min, and stained with crystal violet for 20 min. The titer was obtained by considering the dilution at which the first lytic plaques appeared. The dilution titer was converted to protective units. The inverse of the dilution was divided by the final volume of each well (0.125 mL), and its titer per final volume of 1 mL was considered, which is equivalent to the protective units/mL (PU/mL). The lowest transfusion titer was 256 PUs.

2.12. Determination of Inflammatory Cytokines

Inflammatory cytokine levels were determined in plasma samples from recipients on the day before transfusion and on days 3 and 7 post-transfusion. A flow cytometry panel was analyzed, where 13 human cytokines were quantified as described in the test methodology (IL-1β, IFN-α2, IFN-γ, TNF-α, MCP-1, IL-6, hIL-8, IL-10, IL-12p70, IL-17A, IL-18, IL-23, and IL-33 Multi-analyte flow assay kit LEGENDplex (BioLegend®, San Diego, CA, USA), and samples were acquired using a FACS Canto II flow cytometer (BD®, Hackensack, NJ, USA), obtaining 4000–5000 events in a sample using the LEGENDplexv8 software to perform cytokine analysis.

2.13. Statistical Methods

As this was a phase II, and considering the health emergency caused by the pandemic, we chose to use a convenience sample size to evaluate the efficacy and safety of CP in patients with COVID-19 pneumonia.

Measures of central tendency and dispersion were calculated according to the type of variable. The t-test was performed for differences between standard variables and non-parametric tests for non-normal variables. Survival at 15 and 21 days was estimated by the relative risks (RR) and a 95% confidence interval (CI) was used to assess mortality. Kaplan–Meier method (95% confidence interval (CI)) and comparisons by group using the log-rank test were employed. Hazard ratio (HR) was estimated as a measure of effect size of the variables included in the Cox regression. Univariate and multivariate logistic analysis were performed to determine the contribution of some clinically essential variables to mortality.

3. Results

3.1. General

Thirty-nine patients were included in the trial: 16 in group A (moderate + CP), 3 in group B (moderate + albumin solution), 13 in group C (severe + CP), and 7 in group D (severe + albumin solution). Of the 39 patients (Figure 1), 8 (20.5%) were women, 28.2% had type II diabetes mellitus, 48.7% had systemic arterial hypertension, and only 10.25% were smokers; The distribution of these comorbidities did not differ significantly between treatment groups. There were no cases of liver disease or ischemic disease among the selected patients. Table 1 shows the comparison of comorbidities between treatments; there were no significative differences.

Table 1. Description of the comorbidities by treatment group.

	Groups		
	Convalescent Plasma	**Control**	***p*-Value**
Sex (Men)	23 (79.3%)	8 (80.0%)	1.000
DM	8 (27.6%)	3 (30.0%)	1.000
SAH	12 (41.4%)	7 (70.0%)	0.166
Smoker	4 (13.8%)	0 (0.0%)	0.556

DM: Diabetes mellitus; SAH: systemic arterial hypertension. Fisher exact test.

Table 2 shows the demographic characteristics of the patients included in the treatment group and the mean values of the variables assessing the clinical status and IgG anti-nucleocapsid antibody titers, anti-spicule proteins, and neutralizing antibodies at screening (baseline). In these values, only the PAO_2/FIO_2 index showed a significant difference between treated patients and controls; however, both results reflected the same clinical status in the two groups.

Table 2. Description of the population at the time of selection.

	Day before Transfusion		
	Convalescent Plasma	**Control**	***p*-Value**
Age (years old)	55.5 (10.3)	56.9 (7.5)	0.7 *
BMI (kg/m^2)	31.5 (5.5)	28.7 (5.24)	0.17 *
SOFA	4.8 (2.7)	6.1 (2.23)	0.2 *
PAO_2/FIO_2	160.6 (74.4)	107.8 (31.9)	**0.004** *
Ferritin (ng/mL)	855.6 (73.2–2607.5)	933.5 (321–1532)	0.56 #
CRP (mg/dL)	22.5 (1.7–372.0)	86.3 (0.68–377)	0.4 #
DD (ng/mL)	784 (154.8–5924.5)	956 (135–7285)	0.64 #
IgG (NC) (OJ)	7.4 (3.2–9.0)	6.6 (4.2–7.9)	0.09 #
IgG (Spike) OD)	190.7 (98.7)	181.8 (62.5)	0.804 *
PU	128 (16–2048)	128 (32–512)	0.94 #

Normally Distributed Variables (NDV): Mean (Standard deviation), Non-normal Distribution Variables (NnDV): Median (p5-p95%). BMI, body mass index; SOFA, Sequential Organ Failure Assessment; CRP, C-reactive protein; DD, D-Dimer; IgG (NC), IgG anti-nucleocapsid antibodies; OD, optical density; IgG (spike), IgG anti-spike antibodies; PU, protective units. *p*-value from a Student's *t*-test * and *p*-value from a Mann–Whitney test #. Significant *p*-values are bolded.

The changes in the variables studied in the patients at seven days post-transfusion are shown in Table 3, where the value obtained for the SOFA scale was significantly lower in the CP group, whereas in the case of the PAO_2/FIO_2 index and the PU titer, both results were significantly higher in the CP group than in the control group.

Table 3. Seven days post-treatment follow-up.

		Day 7 Post-Treatment		
	Severity	**Convalescent Plasma**	**Control**	***p*-Value**
SOFA	Moderate and severe	3.7 (2.02)	7.1 (2.8)	**0.001** *
	Moderate	3.0 (1.4)	5.5 (3.5)	0.373
	Severe	4.5 (2.4)	7.5 (2.4)	**0.005** **
PAO_2/FIO_2	Moderate and severe	251.01 (109.4)	109.2 (62.4)	**<0.001** *
	Moderate	316.8 (84.2)	175.6 (84.5)	0.031
	Severe	191.7 (96.7)	80.7 (17.2)	**0.005** *
Ferritin (ng/mL)	Moderate and severe	704.7 (182.9–2813)	1057.6 (237–3105)	0.34
	Moderate	665.6 (383.2)	665.6 (538.4)	1.0
	Severe	948.5 (682.2)	1364.7 (917.5)	0.274
CRP (mg/dL)	Moderate and severe	6.5 (0.51–157.4)	89 (0.65–382)	0.07
	Moderate	20.2 (48.2)	91.6 (97.7)	0.082
	Severe	44.2 (44.2)	111.9 (135.0)	0.32
DD (ng/mL)	Moderate and severe	1666.5 (278–9977.3)	1300.5 (142–5040)	0.61
	Moderate	1279.0 (883.9)	914.3 (711.8)	0.541
	Severe	1590.5 (2690.5)	907 (1697.1)	0.773
IgG (NC) (OD)	Moderate and severe	7.8 (4.8–9.14)	7.23 (5.10–8.4)	0.39
	Moderate	7.2 (0.9)	6.8 (1.3)	0.584
	Severe	7.3 (1.2)	6.9 (1.4)	0.519
IgG (Spike) (OD)	Moderate and severe	269.2 (100.5)	304.1 (94.3)	0.448
	Moderate	232.2 (111.9)	328.0 (5.6)	0.305
	Severe	287.7 (94.7)	294.6 (113.7)	0.903
PU	Moderate and severe	512 (32–16,384)	96 (32–256)	**0.01** **
	Moderate	790.8 (877.2)	128.0 (110.8)	0.094
	Severe	1536 (6671.4)	128 (92.1)	**0.40** **

Normally Distributed Variables (NDV): Mean (Standard deviation), Non-normal Distribution Variables (NnDV): Median (p5-p95%). SOFA, Sequential Organ Failure Assessment; CRP, C-reactive protein; DD, D-Dimer; IgG (NC), IgG anti-nucleocapsid antibodies; OD, optical density; IgG (spike), IgG anti-spike antibodies; PU, protective units. *p*-value from a Student's *t*-test* and *p*-value from a Mann–Whitney test **. Significant *p*-values are bolded.

Although CRP, ferritin, and DD values in the plasma group were lower at seven days post-transfusion, no significant differences were observed.

When comparing the change in SOFA and PAO_2/FIO_2 index values in the study groups between baseline and seven days after treatment, it was found that in the CP group vs. control group, the decrease in SOFA score was statistically more significant $p = 0.014$ in the CP group than in the control group ($p = 0.168$). In the case of the PAO_2/FIO_2 ratio, the increase observed at seven days in the CP group vs. the control group showed a significantly higher value, $p = 0.001$ vs. $p = 0.946$ (Tables 2 and 3).

When comparing IgG antibody titers against SARS-CoV-2 (anti-spike and anti-nucleocapsid) at day seven post-transfusion, no significant difference was found (Table 3); however, the UP/mL titer at day 7 in the CP group was higher than that in the control group: 512 (32–16,384) vs. 96 (32–256), $p = 0.01$.

3.2. *Quantification of Inflammatory Cytokines*

In the patients who received CP, an increase at day three post-transfusion was observed in 10 cytokines (IL-1β, INF-α2, INF-γ, MCP-1, IL-8, IL-10, IL-12p70, IL-18, IL-23, and IL-33), which in most cases was statistically significant. Nevertheless, there was a downward trend at day 7 for most cytokines compared with the baseline and day 3. The latter was expected, but it was not statistically significant because some patients could not be analyzed due to a lack of samples. In contrast, in the control group, we observed a different behavior with a decrease in some cytokines at day three but an increase at day seven, of which only IL-12p70 and IL-23 showed statistically significant differences (Table 4).

Table 4. Measurement of inflammatory cytokines by groups on the day before, day 3, and day 7 post-transfusion.

	Convalescent Plasma Group				Control Group			
	Basal	**Day-3**	**Day-7**	***p*-Value**	**Basal**	**Day-3**	**Day-7**	***p*-Value**
IL-1β	8.1 * (8.1–410)	17.5 * (8.1–1542)	8.1 (8.1–114.6)	0.059	8.1 (8.1–57)	8.1 (8.1–37.4)	29.4 (8.1–61.4)	NS
INF-α2	2.7 * (1.9–498)	5.1 * (2.2–3029)	5.1 (2.2–23.5)	**0.043**	11.1 * (1.9–19.6)	6.2 (2.2–16.4)	9.8 * (3.1–23.5)	NS
INF-γ	5.9 * (5.9–942)	19.9 * (5.9–4677)	7.2 (5.9–110)	**0.017**	5.9 (5.9–34)	7.2 (45.9–34)	23.4 (5.9–49)	NS
TNF- α	146.7 (14.3–20,513)	133.7 (16.0–8381)	107.0 (16.6–260)	NS	151 (14.3–11,654)	120.6 (54.7–198)	94.3 (55–160)	NS
MCP-1	988.8 * (25.4–6496)	6004 * (91.7–5636)	493 (33.5–1186)	**0.013**	908 (189–9427)	844 (339–5798)	733 (136–1685)	NS
IL-6	89.8 (12.0–722)	60 (12–18,835)	45 (12–823)	NS	104.7 (12.1–4162)	119.0 (45–1116)	60 (12.1–488)	NS
IL-8	80.0 * (69–706)	181 * (69–4402)	169 (69–1359)	**0.007**	326 (69–3672)	217 (90–3536)	267 69–665)	NS
IL-10	2.7 * (2.7–261)	27.9 * (1.7–479)	25 (2.7–142)	**0.006**	2.7 (2.7–217.0)	13.4 (2.7–6323)	64 (13.0–89)	NS
IL-12p70	3.2 (3.2–24)	11.3 (3.2–20.1)	8.0 (3.2–31.4)	NS	3.2 * (2.2–21.4)	9.6 (7–12.5)	14.9 * (8–27)	**0.028 ***
IL-17a	2.7 *° (2.7–20.5)	8.9 * (2.7–645)	8.9 ° (2.7–22.1)	**<0.001 *** **0.017**	2.7 (2.7–108)	8.3 (2.7–16.2)	11.9 (4–22)	NS
IL-18	427 * (27–1062)	549 * (55.4–2082)	572 (111–2436)	**0.036**	620 (250–2684)	476.0 (206–2868)	472 (45–1305)	NS
IL-23	7.1 *° (7.1–78.6)	26.6 * (7.1–125)	41.2° (7.1–95.8)	**0.002 *** **0.047°**	7.1 * (7.1–51.6)	36.2 (7.1–67.8)	46.4 * (26.6–90.0)	**0.018 ***
IL-33	3.8 * (3.8–154)	19 * (3.8–5967)	22.5 (3.8–95)	**0.007 ***	3.8 (23.8–40.7)	15.7 (3.8–37)	29.6 (3.8–60.1)	NS

Data expressed as: Median (p5-p95%). Superscripts °,* denote difference between groups, and their associated *p*-value with the Mann–Whitney test *. NS: Not significant. Significant *p*-values are bolded.

3.3. Mortality

In-hospital mortality of moderate and severe COVID-19 patients in the CP group was 13.8% (4/29), while in the control group, it was 80% (8/10), RR 0.17 (95% CI; 0.07–0.45). The mortality of the patients in the CP group was low compared with the mortality of COVID-19 patients in the CEMENAV and HGM centers, which was 55% and 45%, respectively.

In terms of the ability to prevent the development of severe disease or to prevent the use of Invasive Mechanical Ventilation, no significant differences were found between the plasma and control groups (Table 5), neither in the number of days after transfusion hospitalization. Considering severe (III) vs. moderate (II) patients, mortality was lower in both groups in those who received CP vs. those who received the albumin solution.

Table 5. Seven-day post-treatment follow-up.

Condition	Convalescent Plasma	Control	*p*-Value
Demographics related			
Sex (Men)	4 (17.4%)	7 (87.5%)	**0.001**
DM	1 (12.5%)	3 (66.7%)	0.152
HTA	1 (8.3%)	6 (85.7%)	**0.002**
Severity disease related			
Moderate disease state	0 (0.0%)	1 (33.3%)	N.S.
Severe disease state	4 (30.7%)	7 (100%)	**0.003**
IMV	1 (25%)	3 (100%)	0.143
Hospital discharge >15 days *	1 (11.1%)	3 (75%)	0.052

DM: diabetes mellitus, AH: arterial hypertension, IMV: Invasive Mechanical Ventilation, * Post transfusion. NS: Not significant. Significant *p*-values are bolded.

Of the 19 patients included in COVID-19 stage II, 16 received plasma. Mortality in the plasma group was 0.0% (0/16) and 33.33% (1/3) in the control and control groups, respectively. No statistically significant difference was found in the mortality of stage II patients receiving plasma compared with the control group RR 0.08 (95% CI; 0.00–1.59). Of the 20 patients included in stage III, 13 received plasma. Mortality in the plasma group was 30.76% (4/13), and in the control group it was 100% (7/7), RR 0.31 (95% CI; 0.14–0.7) (Table 4). Mortality analysis under different conditions is also shown. There was no association between sex and mortality, Diabetes Mellitus (DM), Arterial Hypertension (AHT), smoking, use of broad-spectrum antibiotics, or tocilizumab. The mortality rate of patients who received steroids was 42.3% compared with those who did not (8.3%) (p= 0.036). In the sub-analysis by severity, statistical significance was lost.

For the thrombosis prevention scheme, three grades of anticoagulation were used: prophylactic doses (40 mg/24 h), intermediate doses (60 mg/24 h), and total doses (1 mg/kg body weight every 12 h). Lower mortality was found patients with intermediate and full doses than in those who received only prophylaxis. (Table 6).

Table 6. Percentages of mortality between severity and thromboprophylaxis.

	Doses			*p*-Value
Patients	Prophylaxis	Intermediate	Anticoagulation	
Moderate and Severe	7.14%	50.0%	45.5%	**0.015**
Moderate	0.0%	0.0%	20.0%	0.097
Severe	100.0%	50.0%	52.9%	0.350

Prophylaxis: 40 mg/24 h, intermediate dose: 60 mg/24 h, and anticoagulation: mg/kg body weight. Significant p-values are bolded.

In the plasma group, the median survival at 21 days was 19.5 days (95% CI [17.5–21.5]), and in the control group, it was 16 days (95% CI 12.136–19.864), with a median of 17.0 days (95% CI [12.3–21.65]). There was an increased survival in patients receiving plasma (p = 0.001), regardless of the outcome (HR 0.129 (95% CI [0.039–0.432]). Figure 2 shows the cumulative 21-day mortality curves.

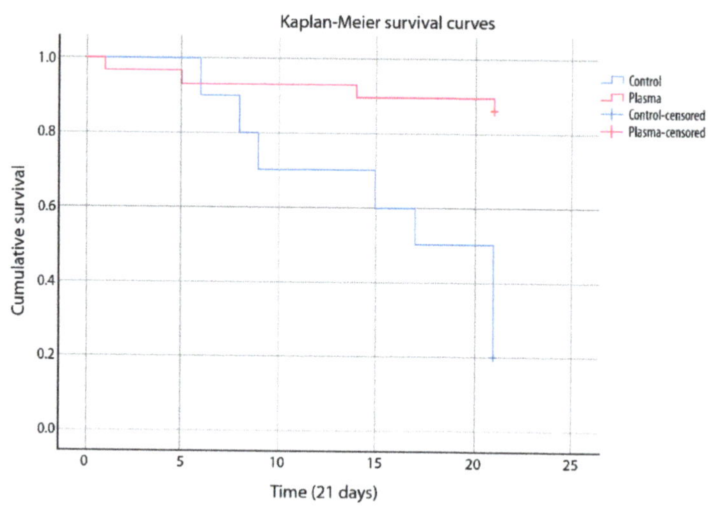

Figure 2. Twenty-one-day survival. Patients who received CP (n = 29) vs. control group (n = 10).

Univariate logistic analysis was performed to determine the contribution of some clinically essential variables to mortality; three variables were associated with the model: plasma application logistic (p < 0.001), disease severity (p = 0.006), and thromboprophylaxis (p = 0.028). However, in the multivariate analysis, no variables were independently associated with mortality.

3.4. Convalescent Plasma in Early Discharge

Early discharge was considered when patients were discharged up to 15 days after CP/albumin solution infusion; of the patients who received CP, 71.0% were discharged before 15 days, and 29.0% were discharged after 15 days. Of those who received albumin solution, 60.0% were discharged before 15 days and 40% after, which shows no significant difference in the number of days they remained in the hospital after either CP or albumin solution.

3.5. Convalescent Plasma as a Protector for Invasive Mechanical Ventilation

Of the 24 patients who were not treated with Invasive Mechanical Ventilation (IVM) at the time of CP or albumin application, only 4/19 who received CP received IVM, while in the group receiving albumin, 3/5 received IVM, showing no significant protective effect against IVM for CP application over albumin (RR = 0.35 [0.114–1.083], $p = 0.126$, by FET).

There were no adverse events related to the transfusion of convalescent plasma, not in the control group or treatment group.

4. Discussion

We reported the use of convalescent plasma together with standard treatment versus albumin in the control group plus standard treatment. We found a significant decrease in mortality in the CP group compared with that in the control group (13.8% vs. 80.0%, $p < 0.001$), as did Abolghasemi et al. [13], who found low mortality in the CP group: 14.8% versus 24.3% in control group ($p = 0.09$), with lower hospital stays, both partial and total.

Rasheed et al. reported a small study of 49 critically ill patients, 21/49 of whom received 400 mL of CP in addition to standard care. Patients who received PC reported shorter time to clinical improvement and low mortality (4.8% vs. 28%, $p = 0.03$) [14].

Libster et al. conducted a randomized, placebo-controlled, double-blind trial in which the plasma applied contained high titers of antibodies. Their results show a more significant effect of early infusion: RR = 0.4, 95% CI: [0.2–0.81]; however, these results were directly related to the number of antibodies in the transfused plasma [8]. In our study, the selection of plasmas for infusion was based on the specificity to neutralize SARS-CoV-2, which is reflected in the high PU concentration achieved at seven days in patients who received CP. In this study, we used a neutralization assay with a high degree of specificity, which also showed a significantly higher titer in the group that received CP at seven days than in those who received albumin (96 vs. 512, $p = 0.006$), a situation that may make a difference in the benefit of using convalescent plasma when it was chosen based on a neutralization assay such as ours. Conversely, several published studies did not perform neutralization assays, and this lack of standardization in the selection of donor plasmas could influence the difference in results between publications [9,10,15].

It is worth noting the change in the SOFA score value in the group that received plasma, which decreased by almost two units from baseline (SOFA baseline = 5.1, SOFA 7 days = 3.7, $n = 19$, $p = 0.014$). Although a 2-unit increase in this scale was used to predict mortality risk [16], we could consider the possible relationship with improvement when it decreased. Regarding the PAO_2/FIO_2 index, both groups presented clinically similar baseline values (Table 2). Seven days after the intervention, the group that received CP presented a significantly higher value than the control group. Our data are similar to those reported by Shen C et al., which in a series of five cases in which convalescent plasmas with titers equal to or higher than 1/1000, in patients with severe COVID-19, the Kirby index also showed a significant elevation in patients who evolved favorably [17]. Additionally, Sarkar S. et al., in a systematic review and meta-analysis, reported that convalescent plasma showed less mortality than traditional treatment OR = 0.44, CI 95%: [0.25–0.77] [18]. These results are related to increased viral clearance OR = 11.3, CI%: [4.9–25.9] and a better clinical evolution, even though the latter was not significant.

Our study found no significant differences in inflammation-related variables such as ferritin, CRP, and D-dimer levels, which may be due to the limited sample size.

During acute SARS-CoV-2 infection, pro-inflammatory cytokines may contribute to the pathology leading to acute respiratory distress syndrome, the life-threatening form of the infection. Several cytokines such as TNF-α, IL-6, IL-8, and IL-18, among others, have been associated; these proteins contribute to tissue damage [19]. However, when evaluating the different cytokines in this study, we only found an increase in IFN-γ and INF-α2 in the CP group at day three post-transfusion, and its maintenance at day 7, both of which are directly associated with viral elimination and the maintenance of cells of the innate immune response, promoting the activation of more cells and initiating a specific adaptive response. In previous studies, a good prognosis was observed in patients with increased type I interferon such as INF-α2, where patients treated with this cytosine improved and efficiently eliminated the virus in less time, demonstrating the direct effect on the clearance of the infection [20]. Cytokine levels have been reported as poor prognostic factors for patients with COVID-19 because they generate hyper-inflammation, including IL-2, IL-6, IL-7, IL-8, IL-10, G-SCF, IP10, MCP1, MCP1, MIP1A, and IL1-β [21]. In our study, we observed an increase in the cytokines as mentioned earlier in patients in the control group up to day 7, which indicates a state of continuous inflammation that can directly over-activate the innate immune response, such as macrophages and polymorphonuclears, among others, for their chemotaxis, perpetuating the arrival of more cells and promoting tissue damage, which could lead to clinical deterioration in the patient. In contrast, the levels of cytokines IL-17a, IL-23, and IL-33 were increased in patients with CP. These cytokines play an essential role in inflammation, since previous studies have shown that patients recovered from COVID-19 have persistent circulating cells that produce IL-33 in response to virus-specific T-cell activation. These cells were correlated with increased amounts of specific antibodies [22]. In our study, we observed a correlation between CP-treated patients and increased IL-33 expression, which could be attributed to the patient's clinical improvement.

Therefore, it is necessary to thoroughly evaluate immunological findings in these patients, from cytosines in supernatants to cell populations, as the difference in some treatments may be associated with immunological responses such as in CP.

One of the most extensive meta-analyses published included 430,781 participants with 115 fatal events without conclusive results. In this study, Klassen et al. reported an association with reduced mortality based on 13 non-randomized studies (OR: 0.50, 95% CI: [0.37–0.67]), and the authors allude to the lack of uniformity in the protocols for the use of CP, which allowed the application of plasmas in late stages or with very low antibody titers. Hence, they performed a sub-analysis excluding the results of the PLACID trial, where the donated plasmas showed very low titers of neutralizing antibodies 1:40 (interquartile range, 1.30–1.80). So, the sub-analysis is in favor of the use of plasma, and there is still a need for standardization in the processes involved in the use of convalescent plasma to treat COVID-19 [23,24].

The REMAP-CAP trial (NCT02735707), the RECOVERY trial (ISRCTN50189673), and the CONCOR-1 trial (NCT04348656) are large randomized controlled trials that issued the closure of recruitment for the PC intervention because their preliminary results do not show any benefit with its use and, in the case of mortality, they found no significant difference at 28 days. However, we must wait for their final results, and it is of great interest to know their experience with the use of PC [25–27].

Donor inclusion criteria in accordance with official national and international standards governing blood banks, antibody titer, AB0 blood group, and patient selection were crucial in this study. Specific antibodies in PCs may benefit the neutralization of viral load in patients or favor the binding of SARS-CoV-2 virus to target cells. In this study, measures were taken to ensure that the transfusion of PCs was safe and, to date, we have no reports of adverse effects.

Limitations of the Study

We recognize the limitations in terms of the number of participants that did not allow us to find significant differences in the variables studied on factors that could be independently associated with the use of plasma. It is also important to note that because the patients were in restricted areas, it was sometimes difficult to collect the samples. Finally, neutralization assays were also set up in our study, however, due to the stringency of using BSL-3 type areas for handling SARS-CoV-2 viruses, we were unable to conclude with the assays.

5. Conclusions

In this study, convalescent plasma was safe and patients did better, as it decreased mortality compared with the control group.

Author Contributions: Conceptualization, Y.V.-E., C.C.-G., Á.A.P.-C., E.C.-D.l.R., C.J.F.-G. and V.F.-S.; methodology, Y.V.-E., C.C.-G., Á.A.P.-C., E.C.-D.l.R., C.J.F.-G. and V.F.-S.; software, E.C.-D.l.R. and C.V.-D.-L.; validation, C.V.-D.-L. and V.F.-S.; formal analysis, E.C.-D.l.R. and C.V.-D.-L.; investigation, Y.V.-E., C.C.-G., Á.A.P.-C., E.C.-D.l.R., C.J.F.-G., P.C.-J., A.P.-P.C., E.O.Z.-M., E.D.-P., S.M., V.D.Á.-J., J.A.D.P.-M., C.C.-S., M.A.O.-O. and V.F.-S.; resources, V.F.-S.; data curation, E.C.-D.l.R.; writing—original draft preparation, Y.V.-E., C.C.-G., Á.A.P.-C., E.C.-D.l.R. and C.J.F.-G.; writing—review and editing, C.V.-D.-L. and V.F.-S.; supervision, V.F.-S.; project administration, V.F.-S.; funding acquisition, V.F.-S. All authors have read and agreed to the published version of the manuscript.

Funding: This research was supported by CEMENAV (Centro Medico Naval, Naval Medical Center).

Institutional Review Board Statement: This study was approved by the different committees of each hospital: Hospital General de México by Ethics in Investigation Committee with registration number CE/168/20 (23/04/2020) and Investigation Committee with registration number CI/140/20 (23/04/2020). Centro Medico Naval by Biosecurity Committee with registration number CB/01/20 (13/04/2020). The protocol was authorized by the Federal Commission for the Protection Against Health Risks (In Spanish, COFEPRIS) with the registration number: CE/168/20. The study is registered in ClinicalTrials.gov with identifier number: NCT04405310 (28/05/2020).

Informed Consent Statement: Informed consent was obtained from all the patients or responsible family member, as well as from convalescent plasma donors before their participation. Written informed consent was obtained from each patient or responsible family member, as well as from convalescent plasma donors.

Data Availability Statement: The data sets used to support the findings of this study are available from the corresponding author on reasonable request.

Acknowledgments: The authors thank the patients for their invaluable contribution to this study. We would like to acknowledge the enthusiasm and participation of the medical, nursing, and laboratory personnel from both Medical Centers.

Conflicts of Interest: The authors declare no conflict of interest.

References

1. Moubarak, M.; Kasozi, K.I.; Hetta, H.F.; Shaheen, H.M.; Rauf, A.; Al-Kuraishy, H.M.; Qusti, S.; Alshammari, E.M.; Ayikobua, E.T.; Ssempijja, F.; et al. The rise of SARS-CoV-2 variants and the role of convalescent plasma therapy for management of infections. *Life* **2021**, *11*, 734. [CrossRef] [PubMed]
2. Cao, B.; Wang, Y.; Wen, D.; Liu, W.; Wang, J.; Fan, G.; Ruan, L.; Song, B.; Cai, Y.; Wei, M.; et al. A trial of lopinavir-ritonavir in adults hospitalized with adults hospitalized with severe COVID-19. *N. Engl. J. Med.* **2020**, *382*, 1787–1799. [CrossRef]
3. Arabi, Y.; Balkhy, H.; Hajeer, A.H. Feasibility, safety, clinical, and laboratory effects of convalescent plasma therapy for patients with Middle East respiratory syndrome coronavirus infection: A study protocol. *Springerplus* **2015**, *4*, 709. [CrossRef]
4. Hung, I.F.N.; To, K.; Lee, C.-K.; Lee, K.-L.; Chan, K.K.C.; Yan, W.-W.; Liu, R.; Watt, C.-L.; Chan, W.-M.; Lai, K.-Y.; et al. Convalescent plasma treatment reduced mortality in patients with severe pandemic influenza A (H1N1) 2009 virus infection. *Clin. Infect. Dis.* **2011**, *52*, 447–456. [CrossRef] [PubMed]
5. Van Griensven, J.; Edwards, T.; de Lamballerie, X.; Semple, M.G.; Gallian, P.; Baize, S.; Horby, P.; Raoul, H.; Magassouba, N.; Antierens, A.; et al. Ebola-Tx Consortium Evaluation of Convalescent Plasma for Ebola Virus Disease in Guinea. *N. Engl. J. Med.* **2016**, *374*, 33–42. [CrossRef]
6. Agarwal, A.; Mukherjee, A.; Kumar, G.; Chatterjee, P.; Bhatnagar, T.; Malhotra, P. Convalescent plasma in the management of moderate COVID-19 in adults in India: Open-label phase II multicenter randomized controlled trial (PLACID Trial). *BMJ* **2020**, *371*, m4232. [CrossRef]

7. Food and Drug Administration. COVID-19 Convalescent Plasma EUA Decision Memo. Available online: www.fda.gov/media/141480 (accessed on 29 September 2022).
8. Libster, R.; Marc, G.P.; Wappner, D.; Coviello, S.; Bianchi, A.; Braem, V.; Esteban, I.; Caballero, M.T.; Wood, C.; Berrueta, M.; et al. Early High-titer plasma therapy to prevent severe COVID-19 in older adults. *N. Engl. J. Med.* **2021**, *384*, 610–618. [CrossRef]
9. Li, L. Effect of Convalescent Plasma Therapy on Time to Clinical Improvement in Patients with Severe and Life-threatening COVID-19. A randomized clinical trial. *JAMA* **2020**, *324*, 460–470. [CrossRef]
10. Ray, Y.; Paul, S.R.; Bandopadhyay, P.; D'Rozario, R.; Sarif, J.; Lahiri, A.; Bhowmik, D.; Vasudevan, J.S.; Maurya, R.; Kanakan, A.; et al. Clinical and immunological benefits of convalescent plasma therapy in severe COVID-19: Insights from a single center open label randomized control trial. *medRxiv*. **2020**. [CrossRef]
11. Avendaño-Solà, C.; Ramos-Martínez, A.; Muñez-Rubio, E.; Ruiz-Antorán, B.; de Molina, R.M.; Torres, F.; Fernández-Cruz, A.; Callejas-Díaz, A.; Calderón, J.; Payares-Herrera, C.; et al. Convalescent Plasma for COVID-19: A multicenter, randomized Clinical Trial. *medRxiv* **2020**, preprint. [CrossRef]
12. Sahu, K.K.; Mishra, A.K.; Raturi, M.; Lal, A. Current Perspectives of convalescent plasma therapy in COVID-19. *Acta Biomed.* **2020**, *91*, e2020175. [CrossRef] [PubMed]
13. Abolghasemi, H.; Eshghi, P.; Cheraghali, A.M.; Imani Fooladi, A.A.; Bolouki Moghaddam, F.; Imanizadeh, S.; Moeini Maleki, M.; Ranjkesh, M.; Rezapour, M.; Bahramifar, A.; et al. Clinical efficacy of convalescent plasma for treatment of COVID-19 infections: Results of a multicenter clinical study. *Transfus. Apher. Sci.* **2020**, *59*, 102875. [CrossRef] [PubMed]
14. Rasheed, A.M.; Fatak, D.F.; Hashim, H.A.; Maulood, F.; Kabah, K.K.; Almusawi, Y.; Abdulamir, A.S. The therapeutic potential of convalescent plasma therapy on treating critically-ill COVID-19 patients residing in respiratory care units in hospitals in Baghdad, Iraq. *Infez Med.* **2020**, *28*, 357–366. [PubMed]
15. Wang, Y.; Huo, P.; Dai, R.; Lv, X.; Yuan, S.; Zhang, Y.; Guo, Y.; Li, R.; Yu, Q.; Zhu, K. Convalescent plasma may be a possible treatment for COVID-19: A systematic review. *Int. Immunopharmacol.* **2020**, *91*, 107262. [CrossRef]
16. Raith, E.P.; Udy, A.A.; Bailey, M.; McGloughlin, S.; MacIsaac, C.; Bellomo, R.; Pilcher, D.V.; Australian and New Zealand Intensive Care Society (ANZICS) Centre for Outcomes and Resource Evaluation (CORE). Prognostic accuracy of the SOFA Score, SIRS Criteria, and SOFA Score for in-hospital mortality among adults with suspected infection admitted to the intensive care unit. *JAMA* **2017**, *3173*, 290–300. [CrossRef]
17. Shen, C.; Wang, Z.; Zhao, F.; Yang, Y.; Li, J.; Yuan, J.; Wang, F.; Li, D.; Yang, M.; Xing, L.; et al. Treatment of 5 critically ill patients with COVID-19 with convalescent plasma. *JAMA* **2020**, *323*, 1582–1589. [CrossRef]
18. Sarkar, S.; Dev Soni, K.; Khanna, P. Convalescent plasma a clutch at straws in COVID-19 management! A systematic review and meta-analysis. *J. Med. Virol.* **2020**, *93*, 1111–1118. [CrossRef]
19. Huang, C.; Wang, Y.; Li, X.; Ren, L.; Zhao, J.; Hu, Y.; Zhang, L.; Fan, G.; Xu, J.; Gu, X.; et al. Clinical features of patients infected with 2019 novel coronavirus in Wuhan, China. *Lancet* **2020**, *395*, 497–506. [CrossRef]
20. Wang, B.; Li, D.; Liu, T.; Wang, H.; Luo, F.; Liu, Y. Subcutaneous injection of IFN alpha-2b for COVID-19: An observational study. *BMC Infect Dis.* **2020**, *20*, 723. [CrossRef]
21. Qin, C.; Zhou, L.; Hu, Z.; Zhang, S.; Yang, S.; Tao, Y.; Xie, C.; Ma, K.; Shang, K.; Wang, W.; et al. Dysregulation of immune response in patients with coronavirus 2019 (COVID-19) in Wuhan, China. *Clin. Infect. Dis.* **2020**, *71*, 762–768. [CrossRef]
22. Stanczak, M.A.; Sanin, D.E.; Apostolova, P.; Nerz, G.; Lampaki, D.; Hofmann, M.; Steinmann, D.; Krohn-Grimberghe, M.; Thimme, R.; Mittler, G.; et al. IL-33 expression in response to SARS-CoV-2 correlates with seropositivity in COVID-19 convalescent individuals. *Nat. Commun.* **2021**, *12*, 2133. [CrossRef] [PubMed]
23. Joyner, M.J.; Klassen, S.A.; Senefeld, J.W.; Johnson, P.W.; Carter, R.E.; Wiggins, C.C.; Shoham, S.; Grossman, B.J.; Henderson, J.P.; Musser, J.M.; et al. Evidence favouring the efficacy of convalescent plasma for COVID-19 therapy. *medRxiv* **2020**. [CrossRef]
24. Simonovich, V.A.; Pratx, L.D.B.; Scibona, P.; Beruto, M.V.; Vallone, M.G.; Vázquez, C.; Savoy, N.; Giunta, D.H.; Pérez, L.G.; Sánchez, M.D.L.; et al. A Randomized Trial of Convalescent Plasma in COVID-19 Severe Pneumonia. *N. Engl. J. Med.* **2021**, *384*, 619–629. [CrossRef] [PubMed]
25. REMAP-CAP Statement 12 January 2021. Available online: https://www.sciencemediacentre.org/expert-reaction-to-remap-cap-recruitment-of-severely-ill-covid-19-patients-into-convalescent-plasma-trial-being-paused-after-initial-analysis-suggested-it-did-not-improve-outcomes (accessed on 5 February 2021).
26. RECOVERY Statement 15 Janurary 2021. Available online: https://www.recoverytrial.net/news/statement-from-the-recovery-trial-chief-investigators-15-january-2021-recovery-trial-closes-recruitment-to-convalescent-plasma-treatment-for-patients-hospitalised-with-covid-19 (accessed on 5 February 2021).
27. CONCOR-1: A Randomized, Open-Label Trial of Convalescents Plasma for Hospitalized Adults with Acute COVID-19 Respiratory Illness. Statement 3 February 2021. Available online: https://concor1.ca (accessed on 5 February 2021).

life

MDPI

Article

The Rescue of the Romanian Health System by the Emergency Departments during the Fourth Wave of COVID-19 Pandemic

Bogdan Oprita [1,2,*], Andrei Davidoiu [2], Alexandru Bogdan Dinu [2] and Ruxandra Oprita [1,3]

1 Faculty of Medicine, University of Medicine and Pharmacy Carol Davila Bucharest, 050474 Bucharest, Romania
2 Emergency Department, Clinical Emergency Hospital of Bucharest, 105402 Bucharest, Romania
3 Gastroenterology Department, Clinical Emergency Hospital of Bucharest, 105402 Bucharest, Romania
* Correspondence: bogdan.oprita@umfcd.ro; Tel.: +40-722-454-331

Simple Summary: The COVID-19 pandemic represents a public health problem that imposes a series of epidemiological measures with an important impact on the health system. Up to now, humanity has faced six waves of this pandemic, in each of which a certain strain of SARS-CoV-2 was dominant. In the fourth wave of the pandemic, the most presentations were recorded in emergency departments, which led to a need for special measures to allow the provision of medical care to as many patients as possible. We conducted a retrospective, observational study on a group of 1417 patients who presented themselves and received medical care at the Emergency Department of the Clinical Emergency Hospital of Bucharest, Romania, during the fourth wave of the COVID-19 pandemic. The average age of the patients included in our study (60 years) was higher than the average age reported for infection with the Omicron strains. Additionally, the severity of the cases was significant, with a rate of orotracheal intubation and mechanical ventilation of approximately 10%. In the condition of full occupancy of the hospital beds on the wards, all the necessary therapeutic measures for these patients were given in the emergency department by a multidisciplinary team.

Abstract: The COVID-19 pandemic has led to the confrontation of the health system with the need to identify solutions for providing medical care to a very large number of patients. The main objective of our study was to describe the measures taken to provide optimal medical care to patients who presented themselves in one of the large emergency hospitals of Romania in the fourth wave of the COVID-19 pandemic. Material and Methods: We conducted a retrospective, observational study on a group of 1417 patients. The statistical analysis was performed using R. Results: The average length of stay of patients in the emergency departments was approximately 2.6 h, increasing to up to 15 days in some more severe cases. For rapid antigen tests, the highest positivity rate for SARS-CoV-2 was identified in patients aged >75 years (53%). Among the identified risk factors associated with the need for mechanical ventilation were advanced age ($\alpha < 0.001$) and lack of vaccination against SARS-CoV-2 ($\alpha < 0.001$). Discussion and conclusions: A method of saving the Romanian health system in full hospital bed occupancy conditions in the wards proved to be the provision of medical care in emergency departments.

Keywords: COVID-19; emergency departments; fourth wave; mechanical ventilation; vaccination

Citation: Oprita, B.; Davidoiu, A.; Dinu, A.B.; Oprita, R. The Rescue of the Romanian Health System by the Emergency Departments during the Fourth Wave of COVID-19 Pandemic. *Life* **2022**, *12*, 1547. https://doi.org/10.3390/life12101547

Academic Editors: Silvia De Francia and Sarah Allegra

Received: 4 September 2022
Accepted: 3 October 2022
Published: 6 October 2022

1. Introduction

Up to now, humanity has faced six waves of the COVID-19 pandemic [1]. Each of these waves was characterized by a certain dominant variant of SARS-CoV-2:

- In the first wave, the ancestral strain with Asp614Gly mutation;
- In the second wave, the Beta variant (B.1.351);
- In the third wave, the Delta variant (B.1.617.2);
- In the fourth wave, the Omicron variant (B.1.1.529 and BA.1);
- In the fifth wave, the Omicron variant (BA.2);

- In the sixth wave, the Omicron variant (BA.4 and BA-5) [1–3].

For each of these waves, differences were reported regarding the symptoms of the patients, the severity of the disease, and the attitude of people or governments [4,5]. Anxiety about this new disease has decreased progressively, this mainly being explained by the approval of some vaccines for SARS-CoV-2 and the introduction of some alternative plans to control the pandemic [4]. However, the appearance of some mutant strains has increased the anxiety level of people again, with direct consequences on the burden of the health systems [4].

The COVID-19 pandemic imposed a series of epidemiological measures that led to consequences on the trends regarding presentation and hospitalization in medical care units. Thus, a decrease in the total number of hospitalizations was observed worldwide, but an increase in the percentage of patients that required hospitalization among those who presented themselves in emergency departments was also observed [6,7]. For example, Musseli M et al. reported for the March–May 2020 period, for Italy, a 60.4% decrease in overall visits to emergency departments, but an increase in the admission rates from 30% to 39% [6]. This phenomenon was explained by the reduced number of presentations for less severe medical conditions and the decreased ratio of non-urgent/urgent medical conditions [6].

Regarding hospitalization rates for patients diagnosed with COVID-19, this varies depending on the geographic region [8,9]. Broadly speaking, up to a quarter of these patients required hospital medical care, and approximately 5% required intensive therapy measures [8,9]. Under these conditions, the COVID-19 pandemic has put a lot of pressure on emergency and admission departments. According to data from the specialized literature, the negative impact of the pandemic on health care systems was felt more intensely in developing countries [10]. The higher mortality rates associated with SARS-CoV-2 infection in these regions were closely related to healthcare resources, such as the availability of staff or the number of intensive care beds [10]. Furthermore, the morbidity and mortality of patients with COVID-19 varied with the dominant SARS-CoV-2 variant in each wave but also with the immunity acquired through the disease or after vaccination [1].

In Romania, a developing European country, the evolution of the COVID-19 pandemic was unfavorable. During the first wave, the management of the pandemic was closely linked to the lockdown imposed by the government and the subsequent waves were more severe [11,12]. Thus, the number of patients who required hospitalization and intensive care measures increased progressively from one wave to the next [13]. In the fall of 2021, Romania faced the fourth wave of the COVID-19 pandemic [14]. At that time, the incidence of SARS-CoV-2 infection was 15 new daily cases per 100,000 inhabitants, and 20% of these patients required hospital medical care [14]. Under these conditions, the beds in the intensive care units were completely occupied quite quickly and the health system was faced with a lack of medical equipment and a lack of specialized medical assistance [15,16]. The rescue of the Romanian health system was offered at that time by the emergency departments. Thus, a significant number of patients diagnosed with COVID-19 or other medical conditions stayed and received specialized medical care in emergency departments.

The COVID-19 pandemic has put the health systems around the world in a position to find alternative measures for the care of a very large number of patients. Considering the geographical variations regarding the funding from the national state budgets for the health systems and, secondarily, the different levels of resources, the rescue measures showed variations from one region to another. The main objective of our study was to describe the measures taken to provide optimal medical care to the large number of patients who presented themselves in one of the large emergency hospitals of Romania in the fourth wave of the COVID-19 pandemic. The secondary objective was the description of some epidemiological parameters such as the positivity rate of the polymerase chain reaction (PCR) test for SARS-CoV-2, the positivity rate of SARS-CoV-2 rapid antigen test, and the vaccination rate for SARS-CoV-2 and its correlation with the severity of the case.

2. Materials and Methods

2.1. Ethical Statement

We conducted a retrospective, observational study on a group of 1417 patients who presented themselves and received medical care in the emergency department of the Emergency Clinical Hospital of Bucharest during the fourth wave of the COVID-19 pandemic (10.08.2021–30.12.2021). The study was approved by the Ethics Committee of the Clinical Emergency Hospital of Bucharest (no 34287/24.08.2022).

2.2. Data Collection

The inclusion criteria were represented by the patients who presented themselves and received medical care in the emergency department of this tertiary diagnostic and treatment center, under the conditions of full occupation of the beds on the wards and the impossibility of transfer to other hospitals for the same reasons. In conclusion, we selected the patients for whom the only medical care option could be offered by the emergency departments. A case was considered severe when endotracheal intubation was needed.

The exclusion criteria were represented by the absence of the informed consent of the patients for participation in clinical studies. None of the initially selected patients found themselves in this situation and were not excluded.

Patient data were obtained from the patient clinical observation sheets and were centralized in an Excel database. The statistical analysis was performed using R, with several integrated packages: effects, logisticRR, ggplot2, gtsummary, and ggpubr [17–23].

2.3. Statistical Analysis

We used a univariate binomial logistic regression with simple and multiple variants, the dependent variable representing the absence/existence of mechanical ventilation and the independent variables represent a series of clinical-demographic parameters followed in the study. Moreover, we conducted a descriptive statistical analysis of the variables. Thus, the means and standard deviations (SD) were reported for continuous variables, considering that we have a sufficient number of patients in the sample and the central limit theorem is applicable, while absolute and relative frequencies (in percentages) were reported for categorical variables. The level of significance α of the study was 0.05; therefore, p values below 0.05 were considered statistically significant.

3. Results

3.1. Demographical Parameters

The gender ratio was approximately equal between men and women (754 women, 663 men) (Table 1). The average age among women was 62.7 years, with a median of 66.0 years (15.0–99.00) and the average age among men was 58.9 years, with a median of 62.0 years (0–100). The ages of the patients were divided into four categories, depending on the values of the quartiles of the variable distribution: ≤47 years old (22.81% women, 30.17% men), 48–65 years old (24.54% women, 25.64% men), 66–75 years old (25.06% women, 24.59% men) and >75 years old (27.59% women, 19.60% men) (Table 1). The average period of hospitalization of patients in emergency departments was 2.6 h, increasing up to 15 days in some more severe cases. The mortality rate among women was 1.85% and among men, it was 1.50% (Table 1).

The vaccination against SARS-CoV-2 rate at that time turned out to be slightly higher among men than among women (32.12% among men vs. 29.58% among women) (Table 1). The highest vaccination rate was identified among patients aged between 66–75 years (40.06%), followed by the group of patients aged >75 years (34.32%), then the age category of 48–65 years (30.14%), and, finally, the age category <47 years (19.35%) (Table 2).

Table 1. Demographic data of the patients included in the study related to sex.

Sex	F (N = 754)	M (N = 663)
Age categories		
≤47 years old	172 (22.81%)	200 (30.17%)
48–65 years old	185 (24.54%)	170 (25.64%)
66–75 years old	189 (25.06%)	163 (24.59%)
>75 years old	208 (27.59%)	130 (19.60%)
Age		
Average age (SD)	62.7 (18.7)	58.9 (19.1)
Median (min, max)	66.0 (15.0, 99.0)	62.0 (0, 100)
Vaccine against SARS-CoV-2		
No	531 (70.42%)	450 (67.88%)
Yes	223 (29.58%)	213 (32.12%)
PCR		
Unfulfilled	546 (72.41%)	481 (72.55%)
Inconclusive	119 (15.78%)	102 (15.38%)
Positive	79 (10.48%)	71 (10.70%)
Negative	10 (1.33%)	9 (1.37%)
Rapid Antigen Test		
Unfulfilled	216 (28.65%)	205 (30.92%)
Positive	371 (49.20%)	318 (47.96%)
Negative	167 (22.15%)	140 (21.12%)
The need for MV with OTI		
No	674 (89.39%)	616 (92.91%)
Yes	80 (10.61%)	47 (7.09%)
Death		
Yes	14 (1.86%)	10 (1.51%)
No	740 (98.14%)	653 (98.49%)

MV = mechanical ventilation; OTI = orotracheal intubation

Table 2. Demographic data of the patients included in the study related to age categories.

Age Categories	≤47 Years Old (N = 372)	48–65 Years Old (N = 355)	66–75 Years Old (N = 352)	>75 Years Old (N = 338)
Sex				
F	172 (46.24%)	185 (52.11%)	189 (53.69%)	208 (61.54%)
M	200 (53.76%)	170 (47.89%)	163 (46.31%)	130 (38.46%)
Vaccine against SARS-CoV-2				
No	300 (80.65%)	248 (69.86%)	211 (59.94%)	222 (65.68%)
Yes	72 (19.35%)	107 (30.14%)	141 (40.06%)	116 (34.32%)
PCR				
Unfulfilled	278 (74.73%)	251 (70.70%)	249 (70.73%)	249 (73.67%)
Inconclusive	67 (18.01%)	54 (15.21%)	53 (15.06%)	47 (13.91%)
Positive	22 (5.91%)	48 (13.52%)	45 (12.78%)	35 (10.36%)
Negative	5 (1.34%)	2 (0.56%)	5 (1.42%)	7 (2.07%)
Rapid Antigen Test				
Unfulfilled	119 (31.99%)	106 (29.86%)	99 (28.13%)	97 (28.70%)
Positive	160 (43.01%)	170 (47.89%)	180 (51.14%)	179 (52.96%)
Negative	93 (25.00%)	79 (22.25%)	73 (20.74%)	62 (18.34%)
The need for OTI with MV				
No	368 (98.92%)	331 (93.24%)	311 (88.35%)	280 (82.84%)
Yes	4 (1.08%)	24 (6.76%)	41 (11.65%)	58 (17.16%)

MV = mechanical ventilation; OTI = orotracheal intubation.

3.2. Positivity Rate for SARS-CoV-2 Infection

All patients who presented to the emergency room underwent a PCR test for SARS-CoV-2 or a rapid SARS-CoV-2 antigen test. The type of test performed depended on the epidemiological and clinical suspicion regarding the probability of establishing the diagnosis of COVID-19, and the severity of the clinical picture. Therefore, among women, 79 positive PCR tests (10.48%) and 167 positive rapid antigen tests (49.20%) were identified. Among men, these percentages were approximately similar, identifying 71 positive PCR tests (10.70%) and 318 positive rapid antigen tests (47.96%) (Table 1). In regard to age, the highest rate of PCR test positivity for SARS-CoV-2 was identified among patients aged between 48–65 years (13.52%), followed by the age category of 66–75 years (12.78%), the category of age >75 years (10.36%), and the category of age <47 years (5.91%) (Table 2). Regarding the rapid antigen test, the highest rate of positivity was identified in patients aged >75 years (52.96%), and the lowest in patients aged <47 years (43.01%) (Table 2). As far as the SARS-CoV-2 vaccination status is concerned, the positivity of the PCR tests and rapid antigen tests for SARS-CoV-2 proved to be slightly higher in vaccinated patients than non-vaccinated ones. Hence, the PCR test was positive in 13.07% of the vaccinated patients and 9.48% of the non-vaccinated patients, and the rapid antigen test was positive in 61.93% of the vaccinated patients and 42.71% of the non-vaccinated ones (Table 3).

Table 3. Demographic data of the patients included in the study related to vaccination against SARS-CoV-2 status.

Vaccine Against SARS-CoV-2	No (N = 981)	Yes (N = 436)
Age Categories		
≤47 years old	300 (30.58%)	72 (16.51%)
48–65 years old	248 (25.28%)	107 (24.54%)
66–75 years old	211 (21.51%)	141 (32.34%)
>75 years old	222 (22.63%)	116 (26.60%)
Age		
Average (SD)	59.1 (19.7)	65.0 (16.4)
Median (min, max)	62.00 (1.00, 100)	68.50 (0, 95.0)
Sex		
F	531 (54.13%)	223 (51.15%)
M	450 (45.87%)	213 (48.85%)
PCR		
Unfulfilled	666 (67.98%)	361 (82.80%)
Inconclusive	204 (20.80%)	17 (3.90%)
Positive	93 (9.48%)	57 (13.07%)
Negative	18 (1.83%)	1 (0.23%)
Rapid Antigen Test		
Unfulfilled	282 (28.75%)	139 (31.88%)
Positive	419 (42.71%)	270 (61.93%)
Negative	280 (28.54%)	27 (6.19%)
The need for MV with OTI		
No	916 (93.37%)	371 (85.09%)
Yes	62 (6.33%)	65 (14.91%)

MV = mechanical ventilation; OTI = orotracheal intubation.

An important element was represented by the significant percentage of patients who required mechanical ventilation: 80 women (10.61%) and 47 men (7.09%) (Table 1). We mention that we only selected patients with a positive PCR test or rapid antigen test for SARS-CoV-2 who required orotracheal intubation and received medical care in the emergency department. The rest of the patients were transferred to the intensive care unit where there were more hospital beds available for patients without SARS-CoV-2 infection. The medical team that dealt with these patients was multidisciplinary, including emergency medicine doctors, intensive care doctors, internists, pulmonologists, cardiologists, gastroenterologists, nephrologists, and surgeons.

3.3. Risk Factors for Orotracheal Intubation

The statistical analysis indicates that there is a strong association between the age category and the need for mechanical ventilation. Thus, compared to patients younger than or equal to 47 years old, patients in the age category of 48–65 years have odds (probability) of mechanical ventilation 6.67 times greater, patients in the age category of 66–75 years have 12.1 times greater odds of mechanical ventilation, while patients over 75 years have 19.1 times greater odds of mechanical ventilation, all these associations being statistically significant (Table 4).

Table 4. The influence of age on the need for mechanical ventilation.

Predictor	N	OTI with MV	OR (95% CI) [1]	p Value
Age categories	1417			
≤47 years old	372	4	—	
48–65 years old	355	24	6.67 (2.55–22.9)	<0.001
66–75 years old	352	41	12.1 (4.83–40.7)	<0.001
>75 years old	338	58	19.1 (7.72–63.4)	<0.001

[1] OR = odds ratio, CI = confidence interval, MV = mechanical ventilation, OTI = orotracheal intubation.

In regard to gender, the statistical analysis indicates that men have odds of mechanical ventilation that are 46% lower than that of women, and the association is statistically significant ($p = 0.021$; CI 0.64 (0.44, 0.93)) (Table 5).

Table 5. The influence of sex on the need for mechanical ventilation.

Predictor	N	MV with OTI	OR (95% CI) [1]	p Value
Sex	1417			
F	754	80	—	
M	663	47	0.64 (0.44–0.93)	0.021

[1] OR = odds Ratio, CI = confidence Interval, MV = mechanical ventilation, OTI = orotracheal intubation.

Another factor associated with the need for mechanical ventilation proved to be vaccination status. As such, the statistical analysis indicates that non-vaccinated patients have 2.34 times higher odds of mechanical ventilation compared to vaccinated patients, and the effect is statistically significant ($p < 0.001$) (Table 6).

Table 6. The influence of the vaccination against SARS-CoV-2 status on the need for mechanical ventilation.

Predictor	N	OTI with MV	OR (95% CI) [1]	p Value
Vaccination against SARS-CoV-2	1417			
Yes	436	65	—	
No	981	62	2.34 (1.61–3.38)	<0.001

[1] OR = odds ratio, CI = confidence interval, MV = mechanical ventilation, OTI = orotracheal intubation.

Later, we performed a multiple univariate binomial logistic regression model, with all predictors for the need for mechanical ventilation. This statistical analysis excluded gender from the predictive factors for mechanical ventilation, but advanced age and a non-vaccinated status were considered risk factors (Table 7).

Table 7. Multiple univariate binominal logistic regression model, with all predictors for the need for mechanical ventilation.

Predictor	OR (95% CI) [1]	*p* Value
Age categories		
≤47 years old	—	
48–65 years old	6.07 (2.31–20.8)	<0.001
66–75 years old	10.2 (4.04–34.3)	<0.001
>75 years old	16.50 (6.64–54.90)	<0.001
Sex		
F	—	
M	0.72 (0.48–1.06)	0.10
Vaccination against SARS-CoV-2		
Yes	—	
No	1.99 (1.36–2.91)	<0.001

[1] OR = odds ratio, CI = confidence interval.

4. Discussion and Conclusions

The emergency department plays an important role in diagnostic and therapeutic management, as it is the first place where the patient comes in contact with the hospital [24]. In critical situations, when there is a significant increase in the number of patients who present themselves to the emergency room in a short time, the reorganization of emergency departments and an approach based on adaptive measures that allow the provision of medical care to as many patients as possible is required [25,26]. This kind of situation was seen during the COVID-19 pandemic which, through the epidemiological measures imposed, led to the reduction in the number of hospital beds available on the wards and the emergency departments facing a large number of patients. Particularly, the fourth wave of the COVID-19 pandemic recorded the highest number of presentations in emergency departments since the beginning of the pandemic [27]. This pandemic has strengthened the important role of emergency medicine in public health through one of its essential functions, namely the identification of patients at high risk of infection and their early isolation [24].

The emergency department of the Clinical Emergency Hospital of Bucharest was reorganized during the COVID-19 pandemic into specific areas arranged for patients infected with SARS-CoV-2 and uninfected patients. During the fourth wave of the COVID-19 pandemic, antigen tests for SARS-CoV-2 could be performed quickly for all patients who presented themselves to the emergency room. In case the clinical suspicion of COVID-19 was very high and the rapid antigen test was negative, the investigation continued with the SARS-CoV-2 PCR test. Additionally, patients who needed immediate surgical intervention or other invasive therapeutic intervention could benefit from the rapid PCR test. These measures led to an efficient triage of patients, with the isolation of those SARS-CoV-2 infected from those who did not present this infection. Another problem faced by the emergency department during this period was the large number of patients who presented themselves to the emergency room and the full occupancy of the hospital beds on the wards, associated with the impossibility of transferring these patients to other hospitals for the same reasons. Under these conditions, the solution adopted was the hospitalization of the patients and the provision of medical care in the emergency department. The severity of the cases was significant, with approximately 10% of the patients requiring mechanical ventilation. The medical care of these patients was provided by the doctors from the emergency department in collaboration with doctors from other departments. Even under these conditions, the mortality rate of these patients did not exceed 1.85%, agreeing with the data from the specialized literature. A study published in 2022 reported an increase in the in-hospital mortality rate in Romania from 2.1% in 2019 to 3.6% in 2020 [28]. However, in 2019, in-hospital deaths represented 33% of all deaths in Romania, while in 2020, they represented 31.1% [28].

The demographic analysis of the patients included in the study identified an approximately equal ratio between men and women. The average age among women was slightly higher than that among men (62.7 years vs. 58.9 years). Contrary to the data from the specialized literature, the average age of the patients included in our study was significantly higher [29,30]. The European Center for Disease Prevention and Control (ECDPC) reported the age of 30 years (20–33 years) as the average age of patients with Omicron infection [31]. Only 7% of the patients infected with the Omicron strain were >60 years old and 50% of them were men. Romania is a country with an aging population. Epidemiological studies claim that by 2050 the proportion of people aged >65 years in Romania will reach approximately 30% [32]. Another explanation for this result could also be found in the ECDC reports that identify Romania among the states where the Omicron strain was isolated but without dominating during the fourth wave (37.8%) [33]. The average period of hospitalization of these patients in the emergency department was 2.6 h, increasing up to 15 days in some cases.

One of the epidemiological parameters analyzed was the rate of vaccination against SARS-CoV-2 among the patients included in the study, which was approximately 30%. The differences in the vaccination rate between the sexes were relatively insignificant. With respect to age, however, the vaccination rate showed variations from 19.35% in patients aged $\leq$47 years to 40.06% in the 66–75 age group. Currently, the rate of vaccination against SARS-CoV-2 in Romania among the general population is approximately 42.1%, with variations between rural (29.69%) and urban areas (41.69%) [34,35]. These data indicate a slight increase in the vaccination rate in Romania from the moment our study was initiated until now. However, compared to the global vaccination rate of 62.9%, the vaccination against SARS-CoV-2 rate in Romania remains significantly lower [35].

Another analyzed parameter was the positivity rate of rapid antigen tests and PCR tests for SARS-CoV-2. Considering that only some of the patients underwent PCR tests, significant differences were identified between their positivity rate (approximately 10%) and the positivity rate of the rapid antigen test (approximately 50%). These percentages indicate the large number of patients who presented SARS-CoV-2 infection during the fourth wave of the pandemic and concur with the data from the specialized literature. Thus, Iuliano AD et al. report the fourth wave of the pandemic as the period with the highest number of presentations in emergency departments and the most hospitalizations for COVID-19 since the beginning of the pandemic [27]. However, the severity of the disease induced by the Omicron strain in wave four proved to be less compared to the severity of the disease related to the Delta strain [36–38]. This can also be explained by the higher rate of vaccination against SARS-CoV-2 rate in wave four [39]. Surprisingly, our study identified a higher incidence of detection of SARS-CoV-2 infection among vaccinated patients compared to non-vaccinated ones. However, we mention that in some of these patients the discovery of the SARS-CoV-2 infection was incidental, as they did not present symptoms specific to the disease. Singanayagam et al. explain this phenomenon by the fact that the vaccine reduces the risk of disease, accelerating viral clearance but not necessarily contacting the virus [40]. Another study that evaluated 565 nasopharyngeal exudates from patients infected with SARS-CoV-2 outlined the hypothesis that the infectious viral load (VL) in fully vaccinated individuals was lower in those infected with the Omicron strains compared to those infected with the Delta strain [41]. However, the number of infection cases was higher in wave four, in which the Omicron strains dominated, compared to wave three, in which the Delta strain dominated. This suggests that other mechanisms than VL growth are incriminated in the high infectivity of the Omicron strains [41].

Another important result of our study is the significant severity of the cases that received medical care in the emergency department, with approximately 10% of patients requiring orotracheal intubation and mechanical ventilation. A retrospective study carried out over 14 years in the United States highlighted an incidence of presentation in emergency departments at 371.9 per 1000 inhabitants (95% CI, 346.5–397.3) and an incidence of the need for mechanical ventilation in their rank of 0.85 per 1000 inhabitants (95%CI, 0.75–0.95) [42].

Thus, approximately 240,000 patients require mechanical ventilation annually in the USA, representing approximately 0.23% of all visits to emergency departments [42]. Regarding COVID-19, a study published in 2020 Montrief T et al. reported a rate of 10–20% regarding the need to provide specific measures to the intensive care unit, 3–10% for the need for mechanical ventilation [43]. Another study that comparatively evaluated the need for mechanical ventilation between the first four waves of the COVID-19 pandemic reported the following percentages: 16.3% in wave one, 8% in wave two, 12.4% in wave three, and 1.6% in wave four [40]. An important thing to mention is that the average age of patients from wave four included in the study was approximately 36 years [44]. In our study, the need for mechanical ventilation was higher (10%). This can be explained by the higher average age of the patients (60 years) and the increase in the number of comorbidities. Additionally, our study demonstrated an increase in the rate of mechanical ventilation with age in patients over 75 years old, this being approximately 19 times higher than patients aged ≤47 years old ($\alpha < 0.001$). Another explanation is offered by the ECDC, which reports that Romania is among the countries where the Omicron strain did not dominate during the fourth wave of the COVID-19 pandemic, along with Bulgaria, Croatia, Estonia, Latvia, Poland, and Slovakia [33]. Thus, some of these cases were most likely due to other strains of SARS-CoV-2. Feikin DR et al. outline the hypothesis that in many cases, the hospitalization of patients who associate infection with the Omicron variant is due to the exacerbation of their comorbidities induced by the infection rather than COVID-19 itself [45].

Another important conclusion of our study is the increase in the rate of mechanical ventilation by 2.34 times among unvaccinated patients as compared to vaccinated ones ($\alpha < 0.001$). This result validates the data from specialized literature, according to which there is an inversely proportional relationship between vaccination and the severity of Omicron infection [46–49]. The effectiveness of the vaccine for infection with the Omicron strain proved to be significantly higher in patients who had three doses of the vaccine, compared to those who had only two doses [47,49].

The limitations of our study are represented by the absence of data such as comorbidities and the number of vaccine doses given to each patient among the patients included in the study.

In conclusion, the emergency department offered the possibility of providing medical care to a significant number of patients in the fourth wave of the COVID-19 pandemic. The severity of these cases was variable, imposing a wide range of therapeutic measures, including orotracheal intubation. The possibility of working in multidisciplinary teams allowed the adoption of this adaptive measure, with important benefits for the patient. The volume of work in healthcare settings has increased, but it has allowed the saving of multiple human lives. All these patients received the therapeutic measures that they would have received in the hospital wards. The mortality rate among the patients included in our study did not exceed the in-hospital mortality rate reported in the specialized literature in the conditions of providing medical care in specific departments. This can be a model for reorganizing emergency departments in disaster situations when the number of patients exceeds the capacity of hospitals. We can conclude that the novelty and main impact of our study consists in the possibility of providing medical care in disaster situations for a significantly long period of time in emergency departments with good clinical results.

Author Contributions: Conceptualization, B.O. and R.O.; methodology, A.D.; software, A.B.D.; validation, B.O.; formal analysis, A.D.; investigation, A.B.D.; resources, A.D.; data curation, A.B.D.; writing—original draft preparation, B.O. and R.O.; writing—review and editing, B.O. and R.O.; visualization, B.O.; supervision, B.O. and R.O.; project administration, B.O.; funding acquisition, B.O. All authors have read and agreed to the published version of the manuscript.

Funding: This research received no external funding.

Institutional Review Board Statement: The study was conducted in accordance with the Declaration of Helsinki and approved by the Ethics Committee of CLINICAL EMERGENCY HOSPITAL OF BUCHAREST (no. 34287/24.08.2022).

Data Availability Statement: Not applicable.

Acknowledgments: In this section, you can acknowledge any support given which is not covered by the author's contribution or funding sections. This may include administrative and technical support, or donations in kind (e.g., materials used for experiments).

Conflicts of Interest: The authors declare no conflict of interest.

References

1. Jassat, W.; Karim, S.S.A.; Mudara, C.; Welch, R.; Ozougwu, L.; Groome, M.J.; Govender, N.; von Gottberg, A.; Wolter, N. Clinical severity of COVID-19 in patients admitted to hospital during the omicron wave in South Africa: A retrospective observational study. *Lancet* **2022**, *10*, E961–E969. [CrossRef]
2. Cheung, P.H.H.; Chan, C.P.; Jin, D.Y. Lessons learned from the fifth wave of COVID-19 in Hong-Kong in early 2022. *Emerg. Microbes Infect.* **2022**, *11*, 1072–1078. [CrossRef] [PubMed]
3. Implications of the Emergence and Spread of the SARS-CoV-2 Variants of Concern BA.4 and BA.5 for the EU/EEA. Available online: https://www.ecdc.europa.eu/en/news-events/implications-emergence-spread-sars-cov-2-variants-concern-ba4-and-ba5 (accessed on 20 July 2022).
4. El-Shabasy, R.M.; Nayel, M.A.; Taher, M.M.; Abdelmonem, R.; Shoueir, K.R.; Kenawy, E.R. Three waves changes, new variant strains, and vaccination effect against COVID-19 pandemic. *Int. J. Biol. Macromol.* **2022**, *204*, 161–168. [CrossRef] [PubMed]
5. Constantinescu, G.; Gheorghe, G.; Ionescu, V.I.; Plotogea, O.M.; Sandru, V.; Diaconu, C.C.; Varlas, V.N.; Bacalbasa, N.; Diaconu, C.C.; Stan-Ilie, M. COVID-19 pandemic and its impact on training programs of medical residency in Romania. *Gastroenterol. Insights* **2022**, *13*, 106–116. [CrossRef]
6. Muselli, M.; Cofini, V.; Mammarella, L.; Carmignani, C.; Fabiani, L.; Desideri, G.; Necozione, S. The impact of COVID-19 pandemic on emergency services. *Ann. Ig.* **2022**, *34*, 248–258. [PubMed]
7. Birkmeyer, J.D.; Barnato, A.; Birkmeyer, N.; Bessler, R.; Skinner, J. The Impact of the COVID-19 Pandemic on Hospital Admissions in The United States. *Health Aff.* **2020**, *39*, 2010–2017. [CrossRef] [PubMed]
8. Tsai, Y.; Vogt, T.M.; Zhou, F. Patient Characteristics and Costs Associated With COVID-19-Related Medical Care Among Medicare Fee-for-Service Beneficiaries. *Ann. Intern. Med.* **2021**, *174*, 1101–1109. [CrossRef]
9. Kim, L.; Garg, S.; O'Halloran, A.; Whitaker, M.; Pham, H.; Anderson, E.J.; Armistead, I.; Bennett, N.M.; Billing, L.; Como-Sabetti, K.; et al. Risk Factors for Intensive Care Unit Admission and In-hospital Mortality Among Hospitalized Adults Identified through the US Coronavirus Disease 2019 (COVID-19)-Associated Hospitalization Surveillance Network (COVID-NET). *Clin. Infect. Dis.* **2021**, *72*, e206–e214. [CrossRef] [PubMed]
10. Sen-Crowe, B.; Sutherland, M.; McKenney, M.; Elkbuli, A. A Closer Look into Global Hospital Beds Capacity and Resource Shortages During the COVID-19 Pandemic. *J. Surg. Res.* **2021**, *260*, 56–63. [CrossRef] [PubMed]
11. Vasile, M.C.; Arbune, A.A.; Lupasteanu, G.; Vlase, C.M.; Popovici, G.C.; Arbune, M. Epidemiologic and Clinic Characteristics of the First Wave of the COVID-19 Pandemic in Hospitalized Patients from Galati County. *J. Clin. Med.* **2021**, *10*, 4210. [CrossRef] [PubMed]
12. Plumper, T.; Neumayer, E. Lockdown policies and the dynamics of the first wave of the Sars-CoV-2 pandemic in Europe. *J. Eur. Public Policy* **2020**, *29*, 321–341. [CrossRef]
13. National Institute of Public Health. Available online: https://instnsp.maps.arcgis.com/apps/dashboards/5eced796595b4ee585bcdba03e30c127 (accessed on 10 July 2022).
14. Ministry of Internal Affairs. Available online: https://www.mai.gov.ro/category/comunicate-de-presa/ (accessed on 10 July 2022).
15. López-Valcárcel, B.G.; Vallejo-Torres, L. The costs of COVID-19 and the cost-effectiveness of testing. *App. Econ. J.* **2021**, *29*, 77–89.
16. McCabe, R.; Kont, M.D.; Schmit, N.; Whittaker, C.; Løchen, A.; Baguelin, M.; Knock, E.; Whittles, L.K.; Lees, J.; Brazeau, N.F.; et al. Modelling intensive care unit capacity under different epidemiological scenarios of the COVID-19 pandemic in three Western European countries. *Int. J. Epidemiol.* **2021**, *50*, 753–767. [CrossRef] [PubMed]
17. R Core Team. *R: A Language and Environment for Statistical Computing*; R Foundation for Statistical Computing: Vienna, Austria, 2020; Available online: https://www.R-project.org/ (accessed on 10 May 2022).
18. Fox, J.; Weisberg, S. *An R Companion to Applied Regression*, 3rd ed.; Sage Publications: Thousand Oaks, CA, USA, 2019.
19. Oaks, T.; Fox, J. Effect Displays in R for Generalised Linear Models. *J. Stat. Softw.* **2003**, *8*, 1–27. [CrossRef]
20. Wickham, H. *ggplot2: Elegant Graphics for Data Analysis*; Springer: New York, NY, USA, 2016.
21. Kassambara, A. ggpubr: 'ggplot2' Based Publication Ready Plots. R Package Version 0.4.0. 2020. Available online: https://CRAN.R-project.org/package=ggpubr (accessed on 10 May 2022).
22. Sjoberg, D.D.; Curry, M.; Hannum, M.; Larmarange, J.; Whiting, K.; Zabor, E.C. gtsummary: Presentation-Ready Data Summary and Analytic Result Tables. R Package Version 1.4.2. Available online: https://CRAN.R-project.org/package=gtsummary (accessed on 10 May 2022).
23. Lee, Y. logisticRR: Adjusted Relative Risk from Logistic Regression. R Package Version 0.3.0. 2020. Available online: https://CRAN.R-project.org/package=logisticRR (accessed on 10 May 2022).
24. Quah, L.J.J.; Tan, B.K.K.; Fua, T.P.; Wee, C.P.J.; Lim, C.S.; Nadarajan, G.; Zakaria, N.D.; Chan, S.J.; Wan, P.W.; Teo, L.T.; et al. Reorganising the emergency department to manage the COVID-19 outbreak. *Int. J. Emerg. Med.* **2020**, *13*, 32. [CrossRef] [PubMed]

25. Kayipmaz, A.E.; Demircan, A. Pandemic hospitals and reorganizing emergency departments. *Turk. J. Med. Sci.* **2021**, *51*, 3221–3228. [CrossRef] [PubMed]
26. Alquezar-Arbe, A.; Pinera, P.; Jacob, J.; Martin, A.; Jimenez, S.; Llorens, P.; Martin-Sanchez, F.J.; Burillo-Putze, G.; Garcia-Lamberechts, E.J.; Del Castillo, J.G.; et al. Impact of the COVID-19 pandemic on hospital emergency departments: Results of a survey of departments in 2020—The Spanish ENCOVUR study. *Emergencias* **2020**, *32*, 320–321.
27. Iuliano, A.D.; Brunkard, J.M.; Boehmer, T.K.; Peterson, E.; Adjei, S.; Binder, A.M.; Cobb, S.; Graff, P.; Hidalgo, P.; Panaggio, M.J.; et al. Trends in Disease Severity and Health Care Utilization During the Early Omicron Variant Period Compared with Previous SARS-CoV-2 High Transmission Periods—United States, December 2020–January 2022. *MMWR Morb. Mortal Wkly. Rep.* **2022**, *71*, 146–152. [CrossRef]
28. Vladescu, C.; Ciutan, M.; Rafila, A. In-hospital admission and deaths in the context of COVID-19 pandemic, in Romania. *Germs* **2022**, *12*, 169. [CrossRef]
29. Kim, M.K.; Lee, B.; Choi, Y.Y.; Um, J.; Lee, K.S.; Sung, H.K.; Kim, Y.; Park, J.S.; Lee, M.; Jang, H.C.; et al. Clinical Characteristics of 40 Patients Infected With the SARS-CoV-2 Omicron Variant in Korea. *J. Korean Med. Sci.* **2022**, *37*, e31. [CrossRef] [PubMed]
30. Yang, W.; Yang, S.; Wang, L.; Zhou, Y.; Xin, Y.; Li, H.; Mu, W.; Wu, Q.; Xu, L.; Zhao, M.; et al. Clinical characteristics of 310 SARS-CoV-2 Omicron variant patients and comparison with Delta and Beta variant patients in China. *Virol. Sin.* **2022**, *epub ahead of print*. [CrossRef] [PubMed]
31. Zhang, J.; Chen, N.; Zhao, D.; Zhang, J.; Hu, Z.; Tao, Z. Clinical Characteristics of COVID-19 Patients Infected by the Omicron Variant of SARS-CoV-2. *Front. Med.* **2022**, *9*, 912367. [CrossRef] [PubMed]
32. European Commission. Available online: https://ec.europa.eu/regional_policy/en/newsroom/news/2021/07/07-06-2021-romania-provides-social-infrastructure-for-its-elderly (accessed on 2 September 2022).
33. European Centre for Disease Prevention and Control. Available online: https://www.ecdc.europa.eu/en/news-events/weekly-epidemiological-update-omicron-variant-concern-voc-week-2-data-20-january-2022 (accessed on 26 August 2022).
34. The Vaccination Situation in Romania. Available online: https://vaccinare-covid.gov.ro/situatia-vaccinarii-in-romania (accessed on 25 August 2022).
35. Coronavirus (COVID-19) Vaccinations. Available online: https://ourworldindata.org/covid-vaccinations?country=OWID_WRL (accessed on 25 August 2022).
36. Wolter, N.; Jassat, W.; Walaza, S.; Welch, R.; Moultrie, H.; Groome, M.; Amoako, D.G.; Everatt, J.; Bhiman, J.N.; Scheepers, C.; et al. Early assessment of the clinical severity of the SARS-CoV-2 Omicron variant in South Africa. *Lancet* **2022**, *399*, 10323. [CrossRef]
37. Lewnard, J.A.; Hong, V.X.; Patel, M.M. Clinical outcomes among patients affected with Omicron (B.1.1.529) SARS-CoV-2 variant in southern California. *medRxiv* **2022**. [CrossRef]
38. Ferguson, N.; Ghani, A.; Hinsley, W.; Volz, E. Report 50—Hospitalisation Risk for Omicron Cases in England. Available online: https://www.imperial.ac.uk/mrc-global-infectious-disease-analysis/covid-19/report-50-severity-omicron/ (accessed on 25 August 2022).
39. Johnson, A.G.; Amin, A.B.; Ali, A.R. COVID-19 incidence and death rates among unvaccinated and fully vaccinated adults with and without booster doses during periods of Delta and Omicron variant emergence—25 U.S. Jurisdictions, April 4–December 25, 2021. *MMWR Morb. Mortal Wkly. Rep.* **2022**, *71*, 132–138. [CrossRef] [PubMed]
40. Singanayagam, A.; Hakki, S.; Dunning, J.; Madon, K.J.; Crone, M.A.; Koycheva, A.; Derqui-Fernandez, N.; Barnett, J.L.; Whitfield, M.G.; Varo, R.; et al. What is the vaccine effect on reducing transmission in the context of the SARS-CoV-2 delta variant? *Lancet Infect. Dis.* **2021**, *22*, 152.
41. Puhach, O.; Adea, K.; Hulo, N.; Sattonnet, P.; Genecand, C.; Iten, A.; Jacquerioz, F.; Kaiser, L.; Vetter, P.; Eckerle, I.; et al. Infectious viral load in unvaccinated and vaccinated individuals infected with ancestral, Delta or Omicron SARS-CoV-2. *Nat. Med.* **2022**, *28*, 1491–1500. [CrossRef] [PubMed]
42. Easter, B.D.; Fischer, C.; Fischer, J. The use of mechanical venilation in the ED. *Am. J. Emerg. Med.* **2012**, *30*, 1183–1188. [CrossRef] [PubMed]
43. Montrief, T.; Ramzy, M.; Long, B.; Gottlieb, M.; Hercz, D. COVID-19 respiratory support in the emergency department setting. *Am. J. Emerg. Med.* **2020**, *38*, 2160–2168. [CrossRef] [PubMed]
44. Maslo, C.; Friedland, R.; Toubkin, M.; Laubscher, A.; Akaloo, T.; Kama, B. Characteristics and Outcomes of Hospitalized Patients in South Africa During the COVID-19 Omicron Wave Compared with Previous Waves. *JAMA* **2021**, *327*, 583. [CrossRef] [PubMed]
45. Feikin, D.R.; Abu-Raddad, L.J.; Andrews, N.; Davies, M.A.; Higdon, M.M.; Orenstein, W.A.; Patel, M.K. Assessing vaccine effectiveness against severe COVID-19 disease caused by omicron variant. Report from a meeting of the World Health Organization. *Vaccine* **2022**, *40*, 3516–3527. [CrossRef] [PubMed]
46. Sheikh, A.; Kerr, S.; Woolhouse, M.; McMenamin, J.; Robertson, C. Severity of omicron variant of concern and effectiveness of vaccine boosters against symptomatic disease in Scotland (EAVE II): A national cohort study with nested test-negative design. *Lancet* **2022**, *22*, 959–966. [CrossRef]
47. Li, M.; Liu, Q.; Wu, D.; Tang, L.; Wang, X.; Yan, T.; An, Z.; Yin, Z.; Gao, G.F.; Wang, F.; et al. Association of COVID-19 Vaccination and Clinical Severity of Patients Infected with Delta or Omicron Variants—China, May 21, 2021–February 28, 2022. *China CDC Wkly.* **2022**, *4*, 293–297. [CrossRef] [PubMed]

48. Andrews, N.; Stowe, J.; Kirsebom, F.; Toffa, S.; Rickeard, T.; Gallagher, E.; Gower, C.; Kall, M.; Groves, N.; O'Connell, A.M.; et al. COVID-19 Vaccine Effectiveness against the Omicron (B.1.1.529) Variant. *N. Engl. J. Med.* **2022**, *386*, 1532–1546. [CrossRef] [PubMed]
49. Lauring, A.S.; Tenforde, M.W.; Chappel, J.D.; Gaglani, M.; Ginde, A.A.; NcNeal, T.; Ghamande, S.; Douin, D.J.; Talbot, H.K.; Casey, J.D.; et al. Clinical severity of, and effectiveness of mRNA vaccines against, COVID-19 from omicron, delta, and alpha SARS-CoV-2 variants in the United States: Prospective observational study. *BMJ* **2022**, *376*, e069761. [CrossRef] [PubMed]

life

MDPI

Article

Renessans Helps in Early Clearance of SARS-CoV-2: *In-Vivo* Activity of the Iodine Complex in *Rhesus macaque*

Muhammad Nawaz [1], Muhammad Adnan Ashraf [1], Muhammad Asad Ali [1], Muhammad Zubair Shabbir [1], Muhammad Abu Bakr Shabbir [1], Imran Altaf [1], Sohail Raza [1], Saira Rafique [1], Sohail Hassan [1], Nageen Sardar [1], Adnan Mehmood [1], Muhammad Waqar Aziz [1], Sehar Fazal [1], Muhammad Tahir Khan [1], Hafiz Muhammad Moavia Atique [1], Ali Asif [1], Zia Ullah [2], Mubashir Iqbal [1], Talha Imtiaz [1], Muhammad Anwar [1], Nadia Mukhtar [1] and Tahir Yaqub [1,*]

1 Institute of Microbiology, University of Veterinary and Animal Sciences, Lahore 54000, Pakistan
2 Pet Centre, University of Veterinary and Animal Sciences, Lahore 54000, Pakistan
* Correspondence: tahiryaqub@uvas.edu.pk

Simple Summary: Coronavirus has caused a wide range of mortality all over the world. The disease is being controlled by vaccines and antiviral agents. Nevertheless, the development of emerging variants keeps on spreading the disease. For the control of the disease, researchers are always finding new alternatives for the treatment of COVID-19. We explored the potential of repurposing iodine complexes for the treatment of COVID-19. In the first stage, we tested the iodine complexes for antiviral activity in the lab and found the atoxic anti-coronavirus doses to be effective. In the current study, we tested the complexes in animal trials on monkeys. The iodine complexes helped to clear coronavirus from monkeys earlier than the control group. This study provides a cheap alternative to be tested for human trials for COVID-19 that can be a good additive option for treatment.

Citation: Nawaz, M.; Ashraf, M.A.; Ali, M.A.; Shabbir, M.Z.; Shabbir, M.A.B.; Altaf, I.; Raza, S.; Rafique, S.; Hassan, S.; Sardar, N.; et al. Renessans Helps in Early Clearance of SARS-CoV-2: *In-Vivo* Activity of the Iodine Complex in *Rhesus macaque*. *Life* **2022**, *12*, 1424. https://doi.org/10.3390/life12091424

Academic Editors: Daniele Focosi and Christian Lehmann

Received: 12 August 2022
Accepted: 9 September 2022
Published: 13 September 2022

Abstract: Iodine complexes have known antimicrobial properties along with reported *in-vitro* antiviral activity for several viruses. Renessans is one such product with iodine complexes and ascorbic acid. The present study was designed to determine its efficacy for SARS-CoV-2 in *Rhesus macaque*. *Rhesus macaque* were assigned to: A) prophylactic group (n = 3), (B) treatment group (n = 3), (C) infection control group (n = 4), and (D) negative control group (n = 4). Groups A, B, and C were challenged with 2×10^6 TCID of SARS-CoV-2. The prophylactic group (A) was administered Renessans from 5 days before infection till 8 days postinfection (DPI). The treatment group (B) was administered Renessans from 3 till 8 DPI. Group C was administered water-insoluble fractions only. Nasal swabs from all monkeys of groups A, B, and C remained positive for SARS-CoV-2 till 2 and 7 DPI, while the swabs became negative for groups A and B at 14 DPI. Likewise, fecal matter of monkeys in group A returned negative results during the experiment, while that of group B had significantly decreased viral load ($10^{1.5}$ genome copies/mL) compared to group C (10^3 genome copies/mL). Hence, it is concluded that Renessans has *in-vivo* SARS-CoV-2 activity and may result in early clearance of SARS-CoV-2.

Keywords: iodine compound; antiviral; COVID-19; coronavirus; animal trial

1. Introduction

Severe acute respiratory syndrome-related coronavirus-2 (SARS-CoV-2) was first reported as the etiologic agent of coronavirus disease 2019 (COVID-19) in December 2019 at a wholesale seafood market in Wuhan, Hubei, China [1,2]. According to the World Health Organization (WHO), more than 613 million confirmed cases and more than 6.5 million fatalities have been reported worldwide since its first report [3].

SARS-CoV-2 belongs to the Coronaviridae family and *Nidovirales* order. SARS-CoV-2 is a single-stranded unsegmented positive-sense RNA with a size of 30 kb. The virus can spread from one human to another through coughing and sneezing. The predilection site of the virus is lung alveolar epithelial type 2 (AT2) cells. Several studies have reported that the spike proteins of SARS-CoV-2 bind to angiotensin-converting enzyme-2 (ACE-2) receptors present on AT2 cells [4,5]. It has been reported that ACE-2 receptors are also present in the tubular epithelia of the kidney, pancreas, heart, and endothelial cells [4,6,7]. Upon entering the host cell, the virus releases its positive-sense RNA, which dictates host-cell machinery and produces new virions [8]. SARS-CoV-2 infection can be asymptomatic, and in most cases may cause mild to severe complications [9]. Given the prevalence of asymptomatic individuals and the limited availability of molecular testing in different parts of the world, it is believed that the correct number of infections may be much higher than the estimates [10].

Notably, many variants of concern and variants of interest of the virus have been developed, which has kept on challenging the ongoing prevention and therapeutic strategies [11–13]. There is a constant need for improvement in the vaccines and repurposing of the already available antiviral compounds. Researchers are continuously working on providing alternative treatment options for COVID-19. Lopinavir–ritonavir, chloroquine, favipiravir, and remdesivir have shown good potential [14]. Alternatively, repurposing of other compounds with known antimicrobial properties in the inactivation of enveloped and nonenveloped viruses has been explored [15]. Antiviral activity of iodine formulations has been reported for herpes simplex virus [16], coronaviruses [17], influenza virus [18,19], adenoviral conjunctivitis [20], hepatitis C virus [21], and African swine fever virus [22]. During the pandemic of COVID-19, iodine formulations were repurposed for their antiviral potential as nasal sprays, mouthwash, eyewash, antiseptics, and overall anti-SARS-CoV-2 potential [23–26]. A pharmaceutical product of an iodine complex by the name of Renessans is already being used to treat polycystic fibrosis [27]. The antibacterial, antiulcer, and immunomodulatory effects have already been documented [28]. The closest formulation is of Balsam "Vozrozhdenie." It is an organic product with relatively less antimicrobial activity manufactured by MTI Medical based in Kazakhstan. The product is certified and is available in Kazakhstan, Kyrgyzstan, Russia, Belarus, Ukraine, and the UAE. The Renessans formulation has been approved by the Drug Regulatory Authority of Pakistan (DRAP registration 505620098). Renessans contains iodine (0.4–2.0%), potassium iodide (0.8–4.0%), starch (10.0–40.0%), ascorbic acid (0.4–2.0%), glucose (1.2–4.8%), and sodium chloride (0.3–1.8%) in weight. Our research group has previously reported the *in-vitro* antiviral potential of Renessans for SARS-CoV-2 [29]. Keeping in view the background, the present study was designed to further explore this alternative treatment for SARS-CoV-2 using monkeys (*Rhesus macaque*) as animal models.

2. Materials and Methods

2.1. Biosafety and Ethics Statement

High ethical standards were maintained during the conducting of experiments. Approval of the study was undertaken in compliance with the institutional guidelines of the ethics review committee (reference DR/317/7-7-202) of the University of Veterinary and Animal Sciences (UVAS), Lahore, Pakistan. Monkeys were housed in individual cages in a controlled comfortable environment in Animal Biosafety Laboratory-3 (ABSL-3) of the Institute of Microbiology (IM), UVAS. SARS-CoV-2-infected animals were housed under standard conditions of ambient temperature (22 ± 2 °C). Diagnostics of animal samples for the virus was performed in ABSL-3 for emerging pathogens (accreditation PHC/L&A/Lic/2020/COVID-13). A cycle of 12 h light and darkness was maintained throughout the experiment. Food comprising fruit and bread loaves was provided twice daily by trained staff and water provided to monkeys ad libitum throughout the experiment. Animals were kept under the supervision of a veterinarian.

2.2. Experimental Design

The monkeys were obtained from the wildlife department of Pakistan to determine the *in-vivo* efficacy of the antiviral drug (Renessans) for SARS-CoV-2. Before housekeeping, animals were tested for any communicable disease. Healthy ones were weighed and divided into four groups: (A) prophylactic (n = 3), (B) treatment group (n = 3), (C) infection control (n = 4), and (D) negative control (n = 4). SARS-CoV-2 (GenBank accession number MW031802) infection @ 2×10^6 TCID was given to groups A, B, and C through intranasal and oral routes (0.5 mL each) under anesthesia (mixture of ketamine and xylazine). Toxic (50 µg concentration) and effective (EC50 = 0.425 µg/mL) doses of the antiviral drug were calculated in our previous *in-vitro* study (Altaf et al., 2021). The antiviral drug (Renessans) was administered intravenously (IV) at 2.85 mg/7 kg (once daily) to group A from 5 days before the infection till 8 days postinfection (DPI). Group B was administered Renessans intravenously after the onset of clinical signs and symptoms from 3 to 8 DPI at 2.85 mg/7 kg (once daily). Groups C and D were administered IV with water-insoluble fractions only.

2.3. Fecal and Nasal Swab Sampling

Fecal and nasal swab sampling was performed to determine the shedding of SARS-CoV-2 through these routes. All monkeys of groups A, B, C and D were anesthetized for nasal sampling. Nasal sampling was performed five times during the whole experiment: firstly at 0 day before the infection and then at 2 DPI, 7 DPI, 14 DPI, and 21 DPI from all monkeys of groups A, B, and C. Fecal samples were collected daily from day 0 to 21 DPI. The experimental design for A, B and C groups is given in Figure 1.

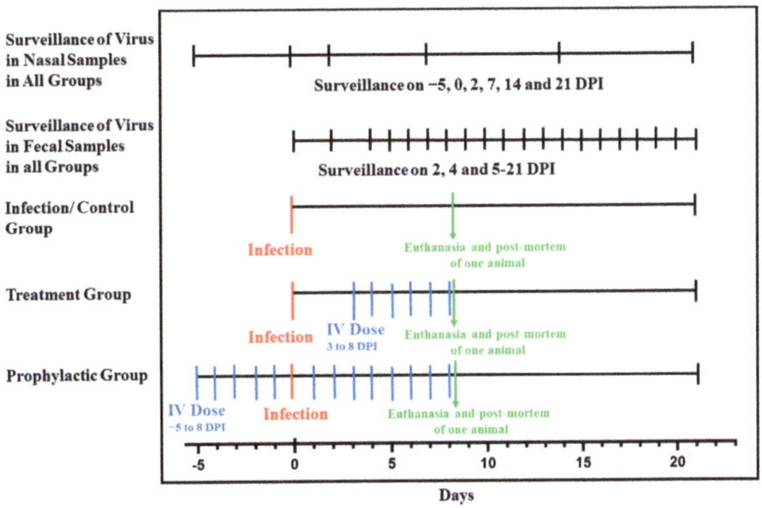

Figure 1. Detailed timeline of events for determining the *in-vivo* efficacy of the antiviral drug (Renessans) for SARS-CoV-2.

2.4. RNA Extraction from Fecal and Nasal Swab Samples

Fecal and swab samples were subjected to RNA extraction in ABSL-3. For RNA extraction, each sample was vortexed, 1 mL transferred to a microcentrifuge tube followed by centrifugation at 5000 rpm for 10 min at 4 °C, and supernatant was collected in a new microcentrifuge tube. The Hero-32 extraction system (Luoyang Ascend Biotechnology Co., Ltd., Luoyang, China) was used for RNA extraction. A total of 14 µL of proteinase K+ carrier RNA mixture was added to each well of the RNA extraction plate, followed by the addition of a supernatant (fecal and nasal swab samples) of 200 µL into the respective wells. The extraction plate was placed in an RNA extractor and RNA was extracted from the elution wells of plates and stored at −80 °C. RT-qPCR analysis was performed on the same day.

2.5. Detection of SARS-CoV-2 from Fecal and Nasal Swab Samples by Real-Time Quantitative PCR

A one-step RT-qPCR kit (Maccura Biotechnology, China) was used to synthesize the cDNA. The CFX96 touch real-time PCR (Bio-Rad) was used for amplification of the *ORF1ab* gene. The cyclic conditions were preincubation at 95 °C for 2 min, followed by 40 cycles of 15 s at 95 °C and 40 s at 95 °C. MS2 based pseudovirus containing exogenous RNA sequence serving as an internal process control (Maccura Biotechnology, Chengdu, China) was used as an internal codon for normalization. Reactions with a cycle threshold below 40 cycles were assumed positive. A standard curve was developed for the CT values of known concentration of $\log_{10}$ copies/mL of SARS-CoV-2. Genome copies of SARS-CoV-2 were calculated with the help of a standard curve [30].

2.6. SARS-CoV-2 Detection from Tissue by Real-Time Quantitative PCR

Three monkeys (one each from groups A, B and C) were euthanized for the determination of SARS-CoV-2 from different tissue samples by RT-qPCR. These monkeys were euthanized by intracardiac injection of potassium chloride 10 mL. Intestine, heart, lung, trachea, and ovary tissue samples were excised and stored at −80 °C till further use.

3. Results

3.1. SARS-CoV-2 Detection from Fecal Samples of Monkeys by Real-Time Quantitative PCR

Fecal samples were collected from A, B, C and D group monkeys at different times (Figure 1) for the detection of SARS-CoV-2 by real-time qualitative RT PCR. At 4 DPI and onwards, all three monkeys remained positive in C group while the virus was detected from fecal matter of one monkey of group A (P2) and B (T3). All the monkeys of D group remained negative. Noteworthily, at 21 DPI, P2 monkey from A group was found negative for SARS-CoV-2 and a group B (T3) monkey remained positive for SARS-CoV-2 (Supplementary Table S1). All monkeys of group C were still shedding the virus. Fecal matter of monkeys in the prophylactic group returned negative results at 20 DPI, while the monkeys of the treatment group revealed an average of less viral load ($10^{1.5}$ copies/mL) than the infection control group (10^3 copies/mL) (Figure 2, Supplementary Table S1). These findings suggested that Renessans has *in-vivo* SARS-CoV-2 activity and may result in early clearance of SARS-CoV-2.

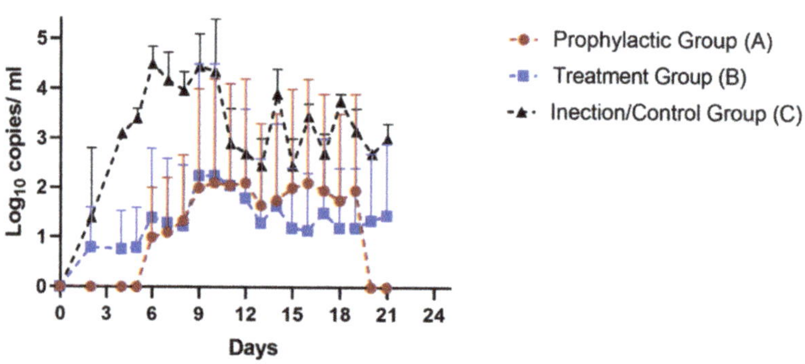

Figure 2. Longitudinal comparison of average fecal viral load of experimental and control groups.

3.2. SARS-CoV-2 Detection from Nasal Swab Samples of Monkeys by Real-Time Quantitative PCR

Nasal swabs were collected from A, B and C groups at different times for the detection of SARS-CoV-2 by real-time qualitative RT PCR. Preinfection nasal swab sampling was also performed to detect SARS-CoV-2 by real-time quantitative PCR from all the monkeys of groups A, B and C to rule out any previous exposure to SARS-CoV-2. However, after 48 h of infection, all the monkeys from groups A, B and C were found positive for SARS-CoV-2. At 2 and 7 DPI, there was no significant difference in the viral load of prophylactic and treatment group compared to group C. However, all (100%) nasal swabs from prophylactic and treatment groups were negative for SARS-CoV-2 at 14 and 21 DPI. Nevertheless, monkeys of the infection control (C) group were still found positive for SARS-CoV-2, and the viral load remained significantly high compared to the experimental groups (negative) at 14 (an average of $10^{3.6}$ copies/mL) and 21 (an average of $10^{3.4}$ copies/mL) DPI (Figure 3, Supplementary Table S2). All the monkeys of group D remained obviously negative in the negative-control group, therefore, results are not displayed in figures and tables. Based on these findings, it can be reported that Renessans (antiviral drug) helped in the early recovery of group A and B monkeys from SARS-CoV-2.

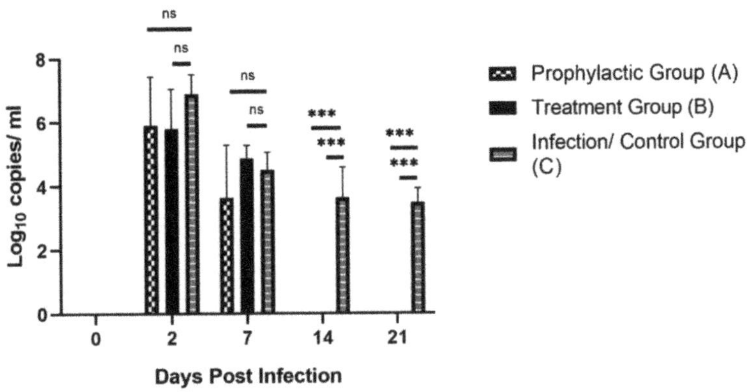

Figure 3. Graphical representation of the comparison of average nasal viral load of experimental and control groups. ***: $p < 0.01$, ns: non-significant.

3.3. *SARS-CoV-2 Detection from Tissue of Group A, B and C Monkeys by Real-Time Quantitative PCR*

For determining SARS-CoV-2 replication in body tissue, one monkey was euthanized at 2 DPI for postmortem from group C to confirm the infectivity potential. The animal was positive for SARS-CoV-2 in its lungs, trachea, and heart. Later, three monkeys (one from each of groups A, B and C) were euthanized at 8 DPI. SARS-CoV-2 detection from different tissue types was performed by real-time quantitative PCR. SARS-CoV-2 was detected in its intestine, lung, heart, ovary, and trachea tissue. At 8 DPI, the tissue of monkeys in the infection control group (group C) were positive for SARS-CoV-2, while the genome was not detected in groups A and B (Table 1). These findings also suggested that the antiviral drug Renessans protected against systemic SARS-CoV-2 infection.

Table 1. Log_{10} SARS-CoV-2 genome copies/mL from tissue of group A, B and C monkeys by real-time quantitative PCR at 8 DPI.

	Prophylactic (A Group)	Treatment (B Group)	Infection (C Group)
Organs	**P3**	**T2**	**I2**
Intestine	0	0	3
Lungs	0	0	5.4
Heart	0	0	3
Ovary	0	NA	2
Trachea	0	0	2.7

NA: not applicable.

4. Discussion

Morbidity and mortality due to COVID-19 are constantly on the rise, as many variants of SARS-CoV-2 have developed. Although different COVID-19 vaccines have reduced the transmission of the disease along with some of the antivirals, nevertheless, there is a need to explore more alternative therapeutics. Therefore, the already tested *invitro* potential of an iodine complex [29] was subjected to preclinical studies in the present study. Renessans, the iodine complex, proved its potential *in-vivo* by leading to early clearance in monkeys when used as prophylaxis and treatment compared to the control group. Therefore, iodine complexes can be further investigated and used in animal and human trials on a larger scale. Iodine compounds have been previously reported to bring about changes in the viral coat by modifying histidine and tyrosine amino acids leading to impairment of the attachment phase [31]. The present findings are consistent with the studies performed on MERS coronavirus and SARS-CoV-2 where researchers found iodine compounds to be clearing the viral load in cell-culture suspension assays, oral cavity, skin surface, and nasopharyngeal wash [23,24,32–34]. The present study's findings may pave the way for the repurposing of Renessans, which is already approved for the treatment of ulcerative conditions (European Patent Specification 2011).

Iodine compounds are known for their antimicrobial activity for topical treatment; however, different studies are exploring the possibilities of their use as oral and parenteral formulations. Systemic use can be associated with toxicity. Therefore, there was a need for evaluation of the toxicity potential of Renessans. The toxicity dose of Renessans was determined by MTT assay on Vero cell lines in the preliminary *in-vitro* study [29]. Researchers had reported the antiviral activity of iodine compounds for the family of coronaviruses even before the spread of SARS-CoV-2 [17]. One of the most studied compounds was povidone–iodine, which revealed its potential on cell lines *in-vitro* and on clinical trials *in-vivo*. Povidone–iodine formulations of antiseptic solution, skin cleanser, gargle, and mouthwash were effective in killing 99.9% of viral load when tested via suspension assays on Vero E6 cell lines [34], and cleared the viral presence *in-vivo* when used as a mouthwash in the oral cavity of four patients [23]. A concentration containing 1% active ingredient showed complete virucidal effect on salivary secretions [24]. The results of virucidal activity remained promising when the formulation was tested via nasal irrigation for nasopharyngeal wash as well [33]. Apart from other formulations of iodine, Renessans composition has also been tested for antiviral potential for influenza virus and hepatitis C virus [21,35]. As Renessans had shown potential for antiviral activity, a gap existed in the evaluation of iodine complexes in animal models, which the present study fills.

Baboons, marmosets and ferrets have been used as animal models for COVID-19 studies. Transgenic mice with hACE receptors and nonhuman primates have shown good potential for studies on the progression of SARS-CoV-2 infection. However, *Rhesus macaque* were selected as a more suitable nonhuman primate animal model for the present study, as there is variability of expression of hACE across different organs in transgenic murine models [36]. SARS-CoV-2 might not replicate in extrapulmonary organs in murine models [37]. Murine models have the limitation of not showing high SARS-CoV-2 viral loads as well. Furthermore, *Rhesus macaque* ACE has the highest receptor activity for SARS-CoV-2 of 14 mammalian species [38], and positivity or clearance of infection can be determined based on shedding of virus from nasal secretion, fecal matter, and molecular detection from body tissue after biopsy [39]. SARS-CoV-2 is excreted in fecal matter, which is a reliable source of early detection compared to nasopharyngeal samples. The results of detection from fecal matter demonstrated that Renessans showed better antiviral potential when used as prophylactic than as treatment, as monkeys in the A group returned negative in 20 days, while the average viral load for monkeys of B group decreased to $10^{1.5}$ copies/mL compared to group C's 10^3 copies/mL. A likely reason for clearance is perfused availability of iodine to systemic organs in the animals of prophylactic groups, as their dosage was started 8 days before the treatment group. As one animal from each group was euthanized for the possible detection and quantification of viral load in body tissue, the sampling results after death are termed "not applicable" (NA) in Supplementary tables.

Nasopharyngeal and nasal samples are also good alternative options to track the high or low titer of SARS-CoV-2. However, only nasal sampling was feasible due to the conventional size of swab buds and the short nostril passage of the monkeys. Notably, the highest viral load in C group reached $10^{6.9}$ copies/mL, which later decreased to $10^{3.4}$ copies/mL at 21 DPI. However, the viral genome copies reached $10^{5.8}$ copies/mL in A and B groups and later became negative at 14 DPI in nasal swab samples. One of the unusual findings was that animals from all the experimental groups became positive for SARS-CoV-2 on 2 DPI after challenge infection via nasal sampling; however, two animals from T1 and T2 appeared negative till 2 DPI via fecal sampling (Supplementary Table S1). Presumptively, onset of shedding of the virus in fecal matter was delayed compared to nasal sampling; however, shedding of virus stopped earlier in nasal samples than in fecal matter.

After the completion of the 21-day preclinical study, we kept on testing the animals for the virus in fecal matter and nasal sampling on a weekly basis. Becoming negative for nasal swab sampling is suggestive of clearance of pulmonary organs and overall progression to clearance of infection. Duration of positivity of SARS-CoV-2 remains less when detected by nasal sampling; however, duration of positivity remains relatively higher for fecal matter [40]. The treatment B group became negative at 28 DPI, while the infection control C group became negative for nasal swab and fecal matter at 28 DPI and 35 DPI, respectively. Alleviation of shedding of the virus was an interesting finding, in contrast to the study of Williamson et al., in which SARS-CoV-2 kept on shedding after the use of remdesivir [41]. Presumably prophylactic (A) and treatment (B) groups recovered early because of SARS-CoV-2 specific immune response as well [42,43]. Nasopharyngeal swab sampling is a standard method to detect SARS-CoV-2 [44], and hence was utilized in the study. Based on these findings, it can be reported that Renessans did have a positive effect and helped in the early recovery of infected monkeys.

The replication of virus in lungs and extrapulmonary body tissue is an important criterion for the progression of systemic viral infection and an important aspect of the study as well. The *Rhesus macaque* is a suitable animal model that can demonstrate the systemic infection of SARS-CoV-2 by having the potential of replication of the virus in extrapulmonary organs [36], as the ACE-2 receptors present in type II pneumocytes, nasal goblet secretory cells, and absorptive enterocytes share more than 91% homology with human ACE-2. Therefore, the experiment was designed to determine the susceptibility of body tissue to SARS-CoV-2. Upon real-time quantitative PCR of postmortem tissue samples, the virus was tested from lungs, trachea, ovary, intestine, and heart. The viral load in different tissue types of C group ranged from 10^2 to $10^{5.4}$ genome copies/mL. However, the virus was not detected on testing the tissue of A and B groups. Our time for viral detection in body tissue was justified as the viral load in lower respiratory tract becomes maximum at 9 DPI after intranasal inoculation [39]. The findings suggested that Renessans may have provided protection from systemic infection. Furthermore, SARS-CoV-2 can infect other organs apart from the lungs in C group. Similar findings were also observed in other studies where researchers detected SARS-CoV-2 in brain, lungs, trachea, spleen, kidney, eye, gastrointestinal tract, uterus, lymph nodes, heart and liver [45–47].

Considering the influence of iodine on functioning of thyroid glands, safety studies of Renessans have been completed in phase II clinical trials (license CT-0014, reference F 3-47/2020-DD (PS). One of the limitations of the study was that the disease progression and therapeutic response could not have been studied in severe or later stages of SARS-CoV-2 infection. Trials have already been initiated to test the efficacy of other iodine complexes [48]. It is recommended to extrapolate the trials to the next levels so clinicians can tackle the disease with multiple therapeutic options.

5. Conclusions

In light of the current study findings, it is concluded that Renessans has an *in-vivo* SARS-CoV-2 activity and may result in early clearance of SARS-CoV-2. Therefore, the current study may provide a basis for a clinical trial of the drug in SARS-CoV-2 patients and reveal its anti-SARS-CoV-2 potential.

Supplementary Materials: The following supporting information can be downloaded at: https://zenodo.org/record/6973267#.Yw90Q31BzIV. The following supporting information is available: Table S1: Log_{10} SARS-CoV-2 genome copies/mL from fecal samples of experimental and control groups by real-time quantitative PCR; Table S2: Log_{10} SARS-CoV-2 genome copies/mL from nasal swab samples of experimental and control groups by real-time quantitative PCR.

Author Contributions: Conceptualization: T.Y. and M.N.; supervision: T.Y. and M.N.; methodology: M.A.A. (Muhammad Adnan Ashraf), M.A.B.S., S.R. (Saira Rafique), S.H., N.S., A.M., M.W.A., M.T.K., A.A., Z.U. and S.F.; formal analysis: M.A.A. (Muhammad Asad Ali), I.A., H.M.M.A., T.I., S.R. (Sohail Raza), Z.U., M.A., M.I., N.M. and M.Z.S.; visualization: M.A.A. (Muhammad Adnan Ashraf), M.A.A. (Muhammad Asad Ali) and S.R. (Sohail Raza); writing—original draft: M.A.B.S.; writing—review and editing: M.A.A. (Muhammad Adnan Ashraf) and M.N. All authors have read and agreed to the published version of the manuscript.

Funding: This work was supported by a grant from MTI Medical, Pvt. Ltd. (Lahore, Pakistan). Tahir Yaqub has received research support funding.

Institutional Review Board Statement: Ethics approval was obtained from the ethics review committee with (reference DR/317/7-7-202) from the University of Veterinary and Animal Sciences (UVAS), Lahore, Pakistan.

Informed Consent Statement: Not applicable.

Data Availability Statement: Data are available in the manuscript.

Conflicts of Interest: The authors declare no conflict of interest.

References

1. Lake, M.A. What We Know so Far: COVID-19 Current Clinical Knowledge and Research. *Clin. Med.* **2020**, *20*, 124. [CrossRef] [PubMed]
2. WHO Statement on the Second Meeting of the International Health Regulations (2005) Emergency Committee Regarding the Outbreak of Novel Coronavirus (2019-NCoV). Available online: https://www.who.int/news-room/detail/30-01-2020-statement-on-the-second-meeting-of-the-international-health-regulations-(2005)-emergency-committee-regarding-the-outbreak-of-novel-coronavirus-(2019-ncov) (accessed on 18 August 2020).
3. Worldometer Coronavirus Cases. Available online: https://www.worldometers.info/coronavirus/ (accessed on 18 July 2020).
4. Wang, X.; Cao, R.; Zhang, H.; Liu, J.; Xu, M.; Hu, H.; Li, Y.; Zhao, L.; Li, W.; Sun, X.; et al. The Anti-Influenza Virus Drug, Arbidol Is an Efficient Inhibitor of SARS-CoV-2 in Vitro. *Cell Discov.* **2020**, *6*, 4–8. [CrossRef] [PubMed]
5. Zheng, Y.-Y.; Ma, Y.-T.; Zhang, J.-Y.; Xie, X. COVID-19 and the Cardiovascular System. *Nat. Rev. Cardiol.* **2020**, *17*, 259–260. [CrossRef] [PubMed]
6. Liu, F.; Long, X.; Zhang, B.; Zhang, W.; Chen, X.; Zhang, Z. ACE2 Expression in Pancreas May Cause Pancreatic Damage after SARS-CoV-2 Infection. *Clin. Gastroenterol. Hepatol.* **2020**, *18*, 2128–2130. [CrossRef]
7. Diao, B.; Wang, C.; Wang, R.; Feng, Z.; Zhang, J.; Yang, H.; Tan, Y.; Wang, H.; Wang, C.; Liu, L. Human Kidney Is a Target for Novel Severe Acute Respiratory Syndrome Coronavirus 2 Infection. *Nat. Commun.* **2021**, *12*, 2506. [CrossRef]
8. Sigrist, C.J.A.; Bridge, A.; Le Mercier, P. A Potential Role for Integrins in Host Cell Entry by SARS-CoV-2. *Antivir. Res.* **2020**, *177*, 104759. [CrossRef]
9. Guo, Y.-R.; Cao, Q.-D.; Hong, Z.-S.; Tan, Y.-Y.; Chen, S.-D.; Jin, H.-J.; Tan, K.-S.; Wang, D.-Y.; Yan, Y. The Origin, Transmission and Clinical Therapies on Coronavirus Disease 2019 (COVID-19) Outbreak—An Update on the Status. *Mil. Med. Res.* **2020**, *7*, 1–10. [CrossRef]
10. Pang, W.; Chehaitli, H.; Hurd, T.R. Impact of Asymptomatic COVID-19 Carriers on Pandemic Policy Outcomes. *Infect. Dis. Model.* **2022**, *7*, 16–29. [CrossRef]
11. Dhawan, M.; Emran, T.B.; Choudhary, O.P. Implications of COVID-19 Vaccine Boosters amid the Emergence of Novel Variants of SARS-CoV-2. *Ann. Med. Surg.* **2022**, *77*. [CrossRef]
12. Ciotti, M.; Ciccozzi, M.; Pieri, M.; Bernardini, S. The COVID-19 Pandemic: Viral Variants and Vaccine Efficacy. *Crit. Rev. Clin. Lab. Sci.* **2022**, *59*, 66–75. [CrossRef]
13. Hadj Hassine, I. Covid-19 Vaccines and Variants of Concern: A Review. *Rev. Med. Virol.* **2022**, *32*, e2313. [CrossRef]
14. Qomara, W.F.; Primanissa, D.N.; Amalia, S.H.; Purwadi, F.V.; Zakiyah, N. Effectiveness of Remdesivir, Lopinavir/Ritonavir, and Favipiravir for COVID-19 Treatment: A Systematic Review. *Int. J. Gen. Med.* **2021**, *14*, 8557. [CrossRef]
15. Cusinato, J.; Cau, Y.; Calvani, A.M.; Mori, M. Repurposing Drugs for the Management of COVID-19. *Expert Opin. Ther. Pat.* **2021**, *31*, 295–307. [CrossRef]
16. Hamuy, R.; Berman, B. Treatment of Herpes Simplex Virus Infections with Topical Antiviral Agents. *Eur. J. Dermatol.* **1998**, *8*, 310–319.

17. Kariwa, H.; Fujii, N.; Takashima, I. Inactivation of SARS Coronavirus by Means of Povidone-Iodine, Physical Conditions and Chemical Reagents. *Dermatology* **2006**, *212*, 119–123. [CrossRef]
18. Al-Horani, R.A.; Kar, S. Potential Anti-SARS-CoV-2 Therapeutics That Target the Post-Entry Stages of the Viral Life Cycle: A Comprehensive Review. *Viruses* **2020**, *12*, 1092. [CrossRef]
19. Derry, D. Iodine: The Forgotten Weapon against Influenza Viruses. *Thyroid Sci.* **2009**, *4*, 1–5.
20. Clement, C.; Capriotti, J.A.; Kumar, M.; Hobden, J.A.; Foster, T.P.; Bhattacharjee, P.S.; Thompson, H.W.; Mahmud, R.; Liang, B.; Hill, J.M. Clinical and Antiviral Efficacy of an Ophthalmic Formulation of Dexamethasone Povidone-Iodine in a Rabbit Model of Adenoviral Keratoconjunctivitis. *Investig. Ophthalmol. Vis. Sci.* **2011**, *52*, 339–344. [CrossRef]
21. Nabi, G.; Nasir, M.; Toor, I.; Zia, F.; Hassan, I.; Nadeem, M.F. An Open Labelled, Active Controlled, Three-Arm, Parallel-Group Study of the Safety and Efficacy of the Oral Formulation of Oral Iodine Complex (RENESSANS) Administered Alone and in Combination with Standard Interferon Therapy in Patients Suffering from Ch. *medRxiv* **2020**. [CrossRef]
22. Pan, L.; Luo, R.; Wang, T.; Qi, M.; Wang, B.; Sun, M.; Luo, Y.; Ji, C.; Sun, Y.; Qiu, H.-J. Efficient Inactivation of African Swine Fever Virus by a Highly Complexed Iodine. *Vet. Microbiol.* **2021**, *263*, 109245. [CrossRef]
23. Lamas, L.M.; Dios, P.D.; Rodríguez, M.T.P.; Del Campo, P.; Alvargonzalez, J.J.C.; Domínguez, A.M.L.; Feijoo, J.F.; Freitas, M.D.; Posse, J.L. Is Povidone-iodine Mouthwash Effective against SARS-CoV-2? First in Vivo Tests. *Oral Dis.* **2020**, *28*, 908–911. [CrossRef]
24. Pelletier, J.S.; Tessema, B.; Frank, S.; Westover, J.B.; Brown, S.M.; Capriotti, J.A. Efficacy of Povidone-Iodine Nasal and Oral Antiseptic Preparations against Severe Acute Respiratory Syndrome-Coronavirus 2 (SARS-CoV-2). *Ear Nose Throat J.* **2021**, *100*, 192S–196S. [CrossRef]
25. Shet, M.; Westover, J.; Hong, R.; Igo, D.; Cataldo, M.; Bhaskar, S. In Vitro Inactivation of SARS-CoV-2 Using a Povidone-Iodine Oral Rinse. *BMC Oral Health* **2022**, *22*, 1–6. [CrossRef]
26. Elzein, R.; Abdel-Sater, F.; Fakhreddine, S.; Abi Hanna, P.; Feghali, R.; Hamad, H.; Ayoub, F. In Vivo Evaluation of the Virucidal Efficacy of Chlorhexidine and Povidone-Iodine Mouthwashes against Salivary SARS-CoV-2. A Randomized-Controlled Clinical Trial. *J. Evid. Based Dent. Pract.* **2021**, *21*, 101584. [CrossRef]
27. Iftikhar, R. The Miracle of Iodine Complex in Treatment of Cases of Polycystic Ovarian Disease Induce by Subclinical Hypothyroidism. *Isra Med. J.* **2013**, *5*, 185–188.
28. Begaliev, S.S.; Nugerbekova, K.M. Medicinal Preparation "Renessans" Having an Antibacterial, Anti-Ulcerous and Immuno-Modulating Action—Google Patents. European Patent Specification EP2606885A1, 26 June 2013. Available online: https://patents.google.com/patent/EP2606885A1/en (accessed on 30 August 2022).
29. Altaf, I.; Nadeem, M.F.; Hussain, N.; Nawaz, M.; Raza, S.; Ashraf, M.A.; Ali, M.A.; Hassan, S.; Aziz, M.W.; Matti, N. An in Vitro Antiviral Activity of Iodine Complexes against SARS-CoV-2. *Arch. Microbiol.* **2021**, *203*, 4743–4749. [CrossRef]
30. Yaqub, T.; Nawaz, M.; Shabbir, M.Z.; Ali, M.A.; Altai, I.; Raza, S.; Shabbir, M.A.B.; Ashraf, M.A.; Aziz, S.Z.; Cheema, S.Q. A Longitudinal Survey for Genome-Based Identification of SARS-CoV-2 in Sewage Water in Selected Lockdown Areas of Lahore City, Pakistan: A Potential Approach for Future Smart Lockdown Strategy. *Biomed. Environ. Sci.* **2021**, *34*, 729. [CrossRef]
31. Gottardi, W. Iodine and Disinfection: Theoretical Study on Mode of Action, Efficiency, Stability, and Analytical Aspects in the Aqueous System. *Arch. Pharm. An Int. J. Pharm. Med. Chem.* **1999**, *332*, 151–157. [CrossRef]
32. Eggers, M.; Eickmann, M.; Zorn, J. Rapid and Effective Virucidal Activity of Povidone-Iodine Products against Middle East Respiratory Syndrome Coronavirus (MERS-CoV) and Modified Vaccinia Virus Ankara (MVA). *Infect. Dis. Ther.* **2015**, *4*, 491–501. [CrossRef] [PubMed]
33. Arefin, M.K.; Rumi, S.K.; Uddin, A.K.M.; Banu, S.S.; Khan, M.; Kaiser, A.; Chowdhury, J.A.; Khan, M.; Saeed, A.; Hasan, M.J. Virucidal Effect of Povidone Iodine on COVID-19 in the Nasopharynx: An Open-Label Randomized Clinical Trial. *Indian J. Otolaryngol. Head Neck Surg.* **2021**, 1–5. [CrossRef] [PubMed]
34. Anderson, D.E.; Sivalingam, V.; Kang, A.E.Z.; Ananthanarayanan, A.; Arumugam, H.; Jenkins, T.M.; Hadjiat, Y.; Eggers, M. Povidone-Iodine Demonstrates Rapid in Vitro Virucidal Activity against SARS-CoV-2, the Virus Causing COVID-19 Disease. *Infect. Dis. Ther.* **2020**, *9*, 669–675. [CrossRef]
35. Matti, N.; Ashraf, M.; Rasheed, M.A.; Altaf, I.; Carvalho, I.; Nadeem, M.F. Evaluation of Cytotoxic, Antiviral Effect and Mutagenic Potential of a Micronutrient Combination in Vitro Cell Culture. *BioRxiv* **2020**. [CrossRef]
36. Bi, Z.; Hong, W.; Yang, J.; Lu, S.; Peng, X. Animal Models for SARS-CoV-2 Infection and Pathology. *MedComm* **2021**, *2*, 548–568. [CrossRef]
37. Bao, L.; Deng, W.; Huang, B.; Gao, H.; Liu, J.; Ren, L.; Wei, Q.; Yu, P.; Xu, Y.; Qi, F. The Pathogenicity of SARS-CoV-2 in HACE2 Transgenic Mice. *Nature* **2020**, *583*, 830–833. [CrossRef]
38. Zhao, X.; Chen, D.; Szabla, R.; Zheng, M.; Li, G.; Du, P.; Zheng, S.; Li, X.; Song, C.; Li, R. Broad and Differential Animal Angiotensin-Converting Enzyme 2 Receptor Usage by SARS-CoV-2. *J. Virol.* **2020**, *94*, e00940-20. [CrossRef]
39. Zheng, H.; Li, H.; Guo, L.; Liang, Y.; Li, J.; Wang, X.; Hu, Y.; Wang, L.; Liao, Y.; Yang, F. Virulence and Pathogenesis of SARS-CoV-2 Infection in *Rhesus Macaques*: A Nonhuman Primate Model of COVID-19 Progression. *PLoS Pathog.* **2020**, *16*, e1008949. [CrossRef]
40. Zhang, N.; Gong, Y.; Meng, F.; Shi, Y.; Wang, J.; Mao, P.; Chuai, X.; Bi, Y.; Yang, P.; Wang, F. Comparative Study on Virus Shedding Patterns in Nasopharyngeal and Fecal Specimens of COVID-19 Patients. *Sci. China Life Sci.* **2021**, *64*, 486–488. [CrossRef]

41. Williamson, B.N.; Feldmann, F.; Schwarz, B.; Meade-White, K.; Porter, D.P.; Schulz, J.; Van Doremalen, N.; Leighton, I.; Yinda, C.K.; Pérez-Pérez, L. Clinical Benefit of Remdesivir in *Rhesus macaques* Infected with SARS-CoV-2. *Nature* **2020**, *585*, 273–276. [CrossRef]
42. Kuri-Cervantes, L.; Pampena, M.B.; Meng, W.; Rosenfeld, A.M.; Ittner, C.A.G.; Weisman, A.R.; Agyekum, R.; Mathew, D.; Baxter, A.E.; Vella, L. Immunologic Perturbations in Severe COVID-19/SARS-CoV-2 Infection. *bioRxiv* **2020**. [CrossRef]
43. Mathew, D.; Giles, J.R.; Baxter, A.E.; Greenplate, A.R.; Wu, J.E.; Alanio, C.; Oldridge, D.A.; Kuri-Cervantes, L.; Pampena, M.B.; D'Andrea, K. Deep Immune Profiling of COVID-19 Patients Reveals Patient Heterogeneity and Distinct Immunotypes with Implications for Therapeutic Interventions. *bioRxiv* **2020**. [CrossRef]
44. WHO. *WHO Laboratory Testing for 2019 Novel Coronavirus (2019-NCoV) in Suspected Human Cases*; Interim Guidance; 2 March 2020; WHO: Geneva, Switzerland, 2020.
45. Tian, S.; Xiong, Y.; Liu, H.; Niu, L.; Guo, J.; Liao, M.; Xiao, S.-Y. Pathological Study of the 2019 Novel Coronavirus Disease (COVID-19) through Postmortem Core Biopsies. *Mod. Pathol.* **2020**, *2020*, 1007–1014. [CrossRef] [PubMed]
46. Song, T.-Z.; Han, J.-B.; Yang, X.; Li, M.-H.; Zheng, Y.-T. Tissue Distribution of SARS-CoV-2 in Non-Human Primates after Combined Intratracheal and Intranasal Inoculation. *Sci. China Life Sci.* **2021**, *64*, 1792–1794. [CrossRef] [PubMed]
47. Böszörményi, K.P.; Stammes, M.A.; Fagrouch, Z.C.; Kiemenyi-Kayere, G.; Niphuis, H.; Mortier, D.; van Driel, N.; Nieuwenhuis, I.; Vervenne, R.A.W.; Haaksma, T. The Post-Acute Phase of SARS-CoV-2 Infection in Two Macaque Species Is Associated with Signs of Ongoing Virus Replication and Pathology in Pulmonary and Extrapulmonary Tissues. *Viruses* **2021**, *13*, 1673. [CrossRef] [PubMed]
48. Ashraf, S.; Ashraf, S.; Ashraf, M.; Imran, M.A.; Kalsoom, L.; Siddiqui, U.N.; Farooq, I.; Ghufran, M.; Khokhar, R.A.; Akram, M.K. A Quadruple Blinded Placebo Controlled Randomised Trial to Evaluate the Effectiveness of an Iodine Complex for Patients with Mild to Moderate COVID-19 in Pakistan (I-COVID-PK): A Structured Summary of a Study Protocol for a Randomised Controlled Trial. *Trials* **2021**, *22*, 1–3. [CrossRef]

life

MDPI

Article

Effectiveness of the Inactivated SARS-CoV-2 (Vero Cell) Vaccine in Peruvian Health Workers

Maria Edith Solis-Castro [1,*], Alex Jaramillo-Corrales [2], Rommell Veintimilla Gonzalez Seminario [1], Noemi Janampa Grados [3], Idania Edith Mamani Pilco [4], Karina Elizabeth Vargas Quispe [3], Lenin Yonel La Torre Rosillo [5], Mario Neyser Vásquez Dominguez [3], David Teodoro Enriquez Cusi [3], Percy Minaya [6], Karim Jacqueline Pardo Ruiz [7,8], Cristian Díaz-Vélez [9,10], Vannesa A. Pachas [11], Ricardo Wesley Alberca [12] and Paul E. Pachas [13,14]

1 Departamento Académico de Medicina Humana, Facultad de Ciencias de la Salud, Universidad Nacional de Tumbes, Av. Universitaria s/n, Pampa Grande, Tumbes 24001, Peru
2 Escuela Profesional de Medicina Humana, Facultad de Ciencias de la Salud, Universidad Nacional de San Antonio Abad del Cusco (UNSAAC), Av. de La Cultura 773, Cusco 08000, Peru
3 Centro Nacional de Epidemiología, Prevención y Control de Enfermedades CDC-Perú, Lima 15072, Peru
4 Hospital Regional Moquegua, Moquegua 18601, Peru
5 Red de Salud Bagua, Jr. Ayacucho s/n cdra N° 1, Bagua, Amazonas 01721, Peru
6 Instituto Nacional Materno Perinatal, Ministerio de Salud, Lima 15001, Peru
7 Facultad de Medicina, Universidad de Ciencias Aplicadas, Lima 15023, Peru
8 Ministerio de Salud, Lima 15072, Peru
9 Instituto de Evaluación de Tecnologías en Salud e Investigación, EsSalud, Lima 14072, Peru
10 Facultad de Medicina, Universidad Privada Antenor Orrego, Trujillo 13008, Peru
11 Escuela Profesional de Ingeniería Biomédica, Universidad Nacional Mayor de San Marcos de Lima, Lima 15081, Peru
12 Laboratorio de Dermatologia e Imunodeficiencias (LIM-56), Departamento de Dermatologia, Faculdade de Medicina FMUSP, Institute de Medicina Tropical da Universidade de Sao Paulo, Sao Paulo 05403-000, Brazil
13 Instituto Nacional de Salud-Perú, Lima 15072, Peru
14 Vice Ministerio de Salud Pública, Lima 15072, Peru
* Correspondence: esolisc@untumbes.edu.pe; Tel.: +51-986-172-813

Citation: Solis-Castro, M.E.; Jaramillo-Corrales, A.; Gonzalez Seminario, R.V.; Janampa Grados, N.; Mamani Pilco, I.E.; Vargas Quispe, K.E.; La Torre Rosillo, L.Y.; Vásquez Dominguez, M.N.; Enriquez Cusi, D.T.; Minaya, P.; et al. Effectiveness of the Inactivated SARS-CoV-2 (Vero Cell) Vaccine in Peruvian Health Workers. *Life* **2022**, *12*, 1318. https://doi.org/10.3390/life12091318

Academic Editors: Silvia De Francia and Sarah Allegra

Received: 27 July 2022
Accepted: 22 August 2022
Published: 26 August 2022

Abstract: Introduction: The COVID-19 pandemic has caused a global health crisis. Vaccines against this disease have demonstrated variable efficacy and safety, although effectiveness has not been evaluated. In February 2021, the Ministry of Health of Peru approved the emergency use of the inactivated SARS-CoV-2 (Vero Cell) vaccine and initiated vaccination with health personnel at the national level. The objective of the study is to determine the effectiveness of this vaccine to reduce infections, hospitalizations, and deaths due to COVID-19. Methodology: We performed a retrospective cohort study in the period from 23 February to 26 June 2021; data were obtained from the Ministry of Health (including demographic, epidemiologic, clinical, hospital, laboratory results, deaths, and both date and quantity of vaccine doses delivered). The exposed cohort were those who received one or two vaccine doses and the non-exposed were unvaccinated. The events studied were infections, hospitalizations and deaths in the cohorts. We consider a case confirmed for COVID-19 if the test result was positive for SARS-CoV-2, via PCR or antigen test. Effectiveness was measured with incidence density ratio and risk. Confounding factors were controlled using a Poisson model with robust variance. Results: We enlisted 520,733 health workers, of whom 415,212 had two vaccine doses and 105,521 were unvaccinated. The median age was 40 years (IQR: 32–50), and 65.6% were female. The effectiveness of two vaccine doses fourteen days after application adjusted by age, sex, hospitalization, and antecedent of having the infection was 90.9% (95% CI: 85.5–94.2%); effectiveness to avoid death from COVID-19; 67.7% (60.1–73.8%) effectiveness to avoid hospitalizations; and 26.3% (23.8–28.6%) effectiveness to reduce the risk of infection by SARS-CoV-2 relative to the unvaccinated cohort. Conclusions: The inactivated SARS-CoV-2 (Vero Cell) vaccine used in two doses has an acceptable effectiveness against death and risk of hospitalization, whereas it has less effectiveness in preventing COVID-19 infection.

Keywords: vaccine; effectiveness; SARS-CoV-2; COVID-19

1. Introduction

The COVID-19 disease pandemic caused by the severe acute respiratory syndrome coronavirus 2 (SARS-CoV-2) has caused a global health crisis [1], with high social and economic costs [2]. The COVID-19 mortality rate is higher among older persons, those with comorbidities and health workers, such as doctors, nurses and technical personnel, who are up to three times more likely to be infected compared to the general population, likely due to greater disease exposure within the healthcare sector [3].

From the first months of the pandemic, the development of vaccines against SARS-CoV-2 began, with acceptable efficacy and safety, and large-scale production and distribution worldwide [4]. The Vaccine Center at the London School of Hygiene & Tropical Medicine, through its COVID-19 vaccine tracker, registers 322 vaccine candidates developed against SARS-CoV-2; of these, 225 are in the pre-clinical phase, 97 in the clinical phase, and 17 have been approved for emergency use in multiple countries [5].

Phase III clinical trials have demonstrated that several vaccine candidates are safe and effective with acceptable immunogenicity, including BNT162, mRNA-1273, ChAdOx1 nCoV-19 (formerly AZD1222), BBIBP-CorV and Ad26.COV2.S. Consequently, numerous countries have started vaccination against COVID-19 [6]. Nonetheless, despite the efficacy demonstrated in phase III clinical trials, little is known about the vaccine effectiveness under real world conditions, [7,8] especially with the appearance of the variants of interest and concern with an increased ability to evade the immune response [9], creating further uncertainty.

In August 2020, Peru created the "Multisectoral Commission of temporary nature in charge of monitoring actions aimed at the development, production, procurement, donation and distribution of vaccines and/or treatments against COVID-19". During its tenure, the Commission formulated recommendations for the procurement of vaccines from Pfizer, Sinopharm, AstraZeneca, and Johnson & Johnson, as well as the COVAX Facility for vaccine administration.

In January 2021, Peru's Ministry of Health granted an exceptional authorization for the importation and use of the inactivated SARS-CoV-2 (Vero Cell) vaccine from the manufacturer Beijing Institute of Biologial Products Co., Ltd. (BIBP, Beijing, China)–China [10]. The implementation of the vaccine began on 9 February 2021, among healthcare workers in both the public and private sectors [11]. The phase I and phase II trials of the BBIBP-CorV vaccine demonstrated the vaccine to be both safe and well tolerated, with a humoral response against SARS-CoV-2 in all recipients of the vaccine on day 42 [12]; hence, its implementation was recommended on days 0 and 21–28 [13]. The phase III study undertaken in the United Arab Emirates showed an efficacy of 79.34% (preliminary results) [14].

Vaccination against COVID-19 in Peru began [15] with the increase in cases during the second wave of COVID-19 infections, with an attack rate of 4.8% in the general population [16]. Among medical doctors, there was an attack rate of 14.3% and a 3.5% fatality rate [14], the third highest rate in Latin America for doctors infected and killed by COVID- 19. To face this situation, the Ministry of Health of Peru approved the National Vaccination Plan and considered the healthcare workforce in Phase I [16]. The objective of the study is to determine the effectiveness of the inactivated SARS-CoV-2 (Vero Cell) vaccine in reducing infection, the severe forms of the disease that cause hospitalization, and death in health workers in Peru.

2. Materials and Methods

2.1. Study Design and Data Sources

A retrospective cohort study was conducted in health workers from 9 February to 26 June 2021, using secondary information from databases of the Ministry of Health that included demographic, epidemiological, clinical, hospital data, data from the laboratory (polymerase chain reaction test (PCR) and antigenic tests for the detection of SARS-CoV-2), death, as well as both the date and quantity of vaccination doses against SARS-CoV-2.

The integrated database was built from the amalgamation of the following data sources: the national vaccination registry of the General Office of Information Technology; the National Registry of Health Personnel; Smart Health System (ESSI) records of the Social Health Insurance of Peru; records of the Comprehensive Health Insurance (SIS); records from the Integrated System for COVID-19; clinical and laboratory data for SARS-CoV-2 from the laboratory information system (NetLab v2.0) of the National Institute of Health (NIH); and the information system of the epidemiological surveillance of COVID-19 of the National Center of Epidemiology, Prevention and Control of Diseases of Peru. The information on COVID 19-related deaths was obtained from the National Informatic System of Defunctions of the General Office of Information Technology. The type of profession was identified from the database of eleven professional associations in Peru.

The General Office of Information Technology of the Ministry of Health joined the different databases, providing the researchers with identification data of the anonymized subjects and the study's variables of interest. The quality control of the data was performed by the institution responsible for the database, the General Office of Information Technology, and the researchers.

2.2. Study Population

The study participants were health workers of both sexes, over 18 years of age, and comprised of both health professionals and technicians who deliver healthcare services as well as administrative and support services (including security, maintenance, etc.) across Peru's 24 departments and the Constitutional Province of Callao, in the three natural regions, at all the three levels (I, II and III) of the public and private health systems, providing services ranging from outpatient care to hospitalization and critical care for severe cases of COVID-19 [15].

The eligibility criteria for study inclusion were whether the health worker was vaccinated with the inactivated SARS-CoV-2 (Vero Cell) vaccine and registered in the database of vaccinated persons of the Ministry of Health of Peru. Workers who presented a severe event supposedly attributable to a second dose of vaccination administered after 25 May were excluded from the analysis; patients were also excluded if they presented one of the events of interest between the date of their first dose, and 14 days after their second vaccine dose. In the unvaccinated cohort, subjects were excluded who presented one of the events of interest from 9 February 2021 and 22 February 2021 (14 days from the start of vaccine deployment).

The retrospective monitoring for the two cohorts ran from 25 June 2021 to 14 days after the administration of the second dose for the vaccinated cohort, and up to 14 days after the start of the vaccination of health personnel (23 February) for the unvaccinated cohort. The monitoring period ended if the event of interest (infection, hospitalization or death) presented in both cohorts. The monitoring days were estimated through the difference between the day of the end of the study (26 June) or the presentation of the event of interest, and 14 days after the application of the 2nd dose, for the vaccinated cohort, or February 23 for the unvaccinated cohort.

In the vaccinated cohort, we considered infected cases if they had a positive PCR or antigenic test result 14 days after the 2nd dose was administered. In the unvaccinated cohort, we considered infected cases if they had a positive PCR or antigen result since 23 February 2021 (two weeks after starting the application of vaccines in Peru) to guarantee the monitoring of both cohorts throughout the same period of the pandemic. We considered an antecedent of previous infection to SARS-CoV-2 if the participant had a positive PCR, antigenic or serological test since the beginning of the pandemic, from March 2020 to 9 February 2021.

The PCR tests were performed in either the laboratories of the NIH, or in public or private laboratories accredited by the NIH. The antigenic tests were performed in each health establishment or in the house of the health worker as part of the epidemiological surveillance system. Samples for laboratory tests were collected in case the health worker had symptoms, or periodically as part of the regular surveillance of COVID-19 in health workers.

The inactivated vaccine (Vero Cell) from the manufacturer BEIJING INSTITUTE OF BIOLOGIAL PRODUCTS (BBIBP-CorV against SARS-CoV-2) was applied according to the suggestions in the laboratory insert, the first dose on day 0 and the second dose 21 days later.

2.3. Criteria for Analysis

We considered the following events for the study: infection, hospitalization and death related to the SARS-CoV-2 virus. Any subject participant with at least one positive test result was defined as infected. Subjects with a registered hospital admission date to a health facility and at least one recorded, positive test result were defined as hospitalized. Any subject registered in the National System of Deaths of Peru (SINADEF); Epidemiological Surveillance Notification System (NOTISP); Comprehensive Health Insurance (SIS); or Smart Health Service (EsSI) with a diagnosis related to COVID-19 and with a positive test result were defined as having died from COVID 19. The tests considered as a confirmatory diagnosis of SARS-CoV-2 in our study were PCR or antigenic tests.

2.4. Statistical Analysis

We used the statistical program STATA v16 (Serial number: 501606349486, StataCorp LLC, College Station, TX, USA). We carried out a descriptive analysis of the sociodemographic variables of the study population, including age, gender, professional association, occupational group and department of residence. Continuous variables were summarized as means and medians with standard deviation or interquartile ranges according to the Gaussian distribution. Categorical variables were summarized with frequencies and percentages.

We used the incidence density ratio as the measurement of events; this is the quotient between the total number of cases of a given event (infected, hospitalized or deceased) and the person-days of follow-up of each participant during the study period, multiplied by 1000.

Vaccine effectiveness was defined as the reduction in the incidence density ratio expressed in percentages. We considered the events presented 14 days after the second dose in the vaccinated cohort and 14 days after the start of vaccination in health workers for the cohort that did not receive any dose of vaccine, using the following formula:

$$\text{Effectiveness (VE)} = (1 - \text{Incident density ratio}) \times 100$$

We used a Poisson regression model with robust variance to adjust the results of the incidence ratio for risk factors or confounding variables such as age, sex, previous history of infection, hospitalization, and death. The 95% confidence intervals were estimated for each measure [17].

3. Results

We enrolled 520,733 health workers, of which 415,212 received two vaccine doses and 105,521 did not receive any dose; we excluded 60,114 health workers who were vaccinated with a single dose (Figure 1).

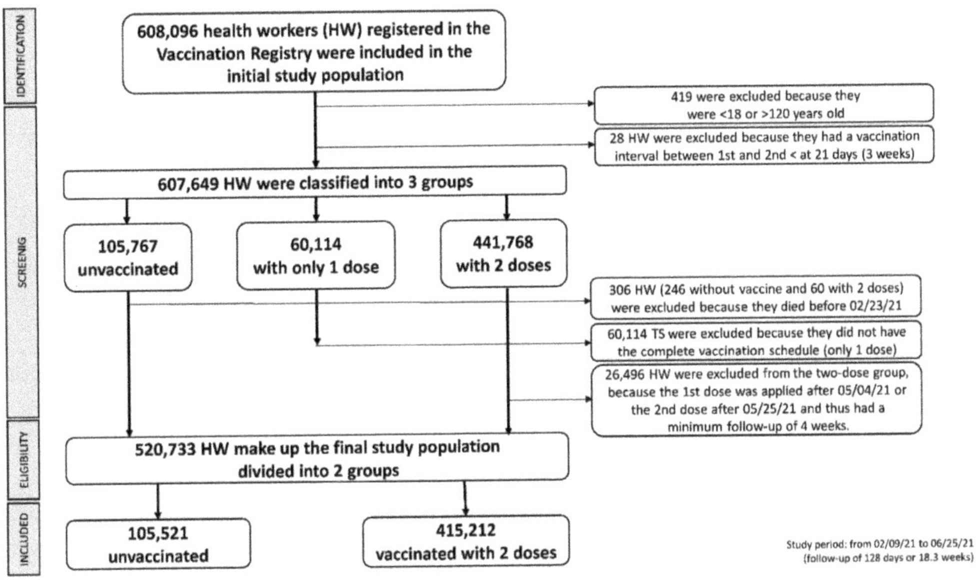

Figure 1. Study participants and eligibility.

The median age of the subjects was 40.0 years old (IQR: 31–51), and 65.6% were female. The median time interval between the administration of the first and second doses of the SARS-CoV-2 vaccine was 21 days (IQR: 21–22).

Table 1 shows the main characteristics of both the vaccinated and unvaccinated cohorts.

Table 1. Characteristics of the study cohort.

Variable	Unvaccinated Cohort		Cohort Vaccinated with 2 Doses		Total	
	n_0	%	n_2	%	N	%
Gender						
Female	72,746	68.9%	268,855	64.8%	341,601	65.6%
Male	32,775	31.1%	146,357	35.2%	179,132	34.4%
Median age (range)	37	(31–46) *	40	(32–51) *	40	(31–51) *
Mean age (SD)	39.75	11.89	42.66	12.98	42.04	12.79
Age group						
18–29	20,771	19.7%	65,914	15.9%	86,685	16.6%
30–59	77,569	73.5%	297,624	71.7%	375,193	72.1%
60 or more	7181	6.8%	51,674	12.4%	58,855	11.3%
Previous history of infection						
Present	19,802	18.8%	130,889	31.5%	150,691	28.9%
Absent	85,719	81.2%	284,323	68.5%	370,042	71.1%
Population	**105,521**	100%	**415,212**	100%	**520,733**	100%

* IQR: Updated 26 June 2021.

3.1. Vaccine Effectiveness in Preventing SARS-CoV-2 Infection

A total of 16,864 cases of SARS-CoV-2 were confirmed with PCR or antigenic tests. Of these, 10,560 received two doses of vaccine and 6304 received none.

The crude incidence density ratio for infection is 0.632 (95% CI 0.61–0.65); adjusted for age group (≥60 or <60 years), gender, previous history of infection, hospitalization and death, this is 0.737 (95% CI 0.714–0.762). The effectiveness of the vaccine in reducing SARS-CoV-2 infections is 26.3% (95% CI 23.8–28.6%) (Table 2).

Table 2. Incidence density ratio and effectiveness of the inactivated SARS-CoV-2 (Vero Cell) vaccine against COVID-19, according to gender, age and previous history *.

Categories, Outcomes, and Vaccination Status	Events	Persons-Days of Follow-Up	Incidence Density per 1000 Days of Follow-Up	Crude Incidence Density Ratio (CI 95%)	z	*p*-Value	Effectiveness of the 2 Doses of the Vaccine (CI 95%) **
Gender							
Female							
Infection							
Unvaccinated	4263	8,684,163	0.49089	0.667	−20.78	<0.0001	**33.3%**
Vaccinated with 2 doses	6975	21,290,825	0.32761	(0.64–0.69)			(30.7–35.8)
Hospitalization							
Unvaccinated	156	8,390,243	0.01859	0.224	−14.56	<0.0001	**77.6%**
Vaccinated with 2 doses	81	19,483,688	0.00416	(0.16–0.24)			(70.8–82.9)
Death							
Unvaccinated	68	8,972,580	0.00758	0.061	−8.08	<0.0001	**93.9%**
Vaccinated with 2 doses	10	21,657,884	0.00046	(0.034–0.075)			(89.9–97.7)
Male							
Infection							
Unvaccinated	2041	4,109,548	0.49665	0.628	−16.73	<0.0001	**37.2%**
Vaccinated with 2 doses	3585	11,492,768	0.31194	(0.60–0.66)			(33.6–40.5)
Hospitalization							
Unvaccinated	187	3,972,196	0.04708	0.180	−13.30	<0.0001	**82.0%**
Vaccinated with 2 doses	89	10,486,394	0.00849	(0.14–0.23)			(76.8–86.0)
Death							
Unvaccinated	137	4,241,419	0.03230	0.056	−12.17	<0.0001	94.4%
Vaccinated with 2 doses	21	11,679,587	0.00180	(0.031–0.082)			(91.8–96.9)
Age group							
18–59 years							
Infection							
Unvaccinated	5903	11,550,681	0.51105	0.658	−25.28	<0.0001	**34.2%**
Vaccinated with 2 doses	9698	28,831,895	0.33636	(0.64–0.68)			(32.0–36.2)
Hospitalization							
Unvaccinated	254	11,178,574	0.02272	0.203	−14.47	<0.0001	**79.7%**
Vaccinated with 2 doses	122	26,418,502	0.00462	(0.16–0.25)			(74.8–83.6)
Death							
Unvaccinated	101	11,962,139	0.00844	0.044	−9.50	<0.0001	**95.6%**
Vaccinated with 2 doses	11	29,340,246	0.00037	(0.019–0.073)			(92.7–98.1)
60 or more years							
Infection							
Unvaccinated	401	823,716	0.48682	0.449	−13.30	<0.0001	**55.1%**
Vaccinated with 2 doses	862	3,943,266	0.21860	(0.40–0.50)			(49.6–60.3)
Hospitalization							
Unvaccinated	89	792,047	0.11237	0.121	−11.82	<0.0001	**87.9%**
Vaccinated with 2 doses	48	3,544,451	0.01354	(0.08–0.17)			(82.9–91.5)
Death							
Unvaccinated	104	832,546	0.12492	0.040	−13.01	<0.0001	**96.0%**
Vaccinated with 2 doses	20	3,988,793	0.00501	(0.022–0.060)			(94.0–97.8)
Previous history of infection							
Absent							
Infection							
Unvaccinated	5636	10,025,416	0.56217	0.716	−19.57	<0.0001	**28.4%**
Vaccinated with 2 doses	8923	22,156,861	0.40272	(0.693–0.741)			(25.9–30.7)
Hospitalization							
Unvaccinated	319	2,242,740	0.14224	0.113	−13.23	<0.0001	**88.7%**
Vaccinated with 2 doses	155	9,653,245	0.01606	(0.093–0.137)			(86.3–90.7)

Table 2. *Cont.*

Categories, Outcomes, and Vaccination Status	Events	Persons-Days of Follow-Up	Incidence Density per 1000 Days of Follow-Up	Crude Incidence Density Ratio (CI 95%)	z	p-Value	Effectiveness of the 2 Doses of the Vaccine (CI 95%) **
Death							
Unvaccinated	191	10,398,828	0.01837	0.065	−13.19	<0.0001	**93.5%**
Vaccinated with 2 doses	27	22,621,886	0.00119	(0.040–0.092)			(90.8–96.0)
Present							
Infection							
Unvaccinated	668	2,348,981	0.28438	0.542	−13.32	<0.0001	**45.8%**
Vaccinated with 2 doses	1637	10,618,300	0.15417	(0.496–0.593)			(40.7–50.4)
Hospitalization							
Unvaccinated	24	2,242,740	0.01070	0.145	−5.86	<0.0001	**85.5%**
Vaccinated with 2 doses	15	9,653,245	0.00155	(0.076–0.277)			(72.3–92.4)
Death							
Unvaccinated	14	2,395,857	0.00584	0.032	−4.55	<0.0001	**96.8%**
Vaccinated with 2 doses	4	10,707,153	0.00037	(0.007–0.141)			(85.9–99.3)

* Updated 26 June 2021. ** There are significant differences ($p < 0.0001$) in the effectiveness of preventing infection, when comparing age groups and previous history of infection. There is no difference when comparing women and men in any of the 3 events.

3.2. *Vaccine Effectiveness in Hospitalized for SARS-CoV-2*

A total of 513 subjects were hospitalized for SARS-CoV-2, of which 170 had two doses of the vaccine and 343 received none. The incidence density ratio for hospitalization is 0.189 (95% CI 0.16–0.24), while the adjusted incidence density is 0.323 (95% CI 0.262–0.399). The effectiveness of two doses of vaccine in avoiding hospitalization for SARS-CoV-2 is 67.7% (95% CI 60.1–73.8%).

3.3. *Vaccine Effectiveness in Deaths from SARS-Co-V-2*

A total of 234 deaths from SARS-CoV-2 were identified, of which 29 had two doses of the vaccine and 205 were not vaccinated. The density ratio for the incidence of deaths is 0.054 (95% CI 0.034–0.075); the adjusted incidence density ratio is 0.092 (95% CI 0.058–0.145). The effectiveness of the vaccine to prevent deaths in the two-dose vaccine is 90.9% (95% CI 85.5–94.2%).

It was found that there is a difference in incidence density x per 100,000 person-days between the two cohorts of 1.6-fold for infection, 5.3-fold for hospitalization, and 17.8-fold for death. The risk reduction effect between those vaccinated with two doses and those not vaccinated is notable, with infection being the event with the highest probability of occurrence in both groups (see Figure 2).

In the follow-up of both cohorts for 15 weeks (Figure 3), it can be seen that the effect of the vaccine to prevent the events under study was considerable and sustained throughout the process. Thus, the risks decreased from 540.6 to 244.2 for infection, from 30.3 to 18.1 for hospitalization, and from 4.1 to 0.7 per 10,000 health workers for death. However, as the weeks go by, it seems that the protection decreases. A longer follow-up time is needed to determine the best time to receive a booster.

The greater effectiveness of the vaccine in the group of 60 years or older was for infection events (55.1% vs. 34.2% ($p < 0.001$)) and hospitalization (87.9% vs. 79.7%), compared to the group age from 18 to 59 years, while similar data were found for deaths (96.0% vs. 95.6%) (Figure 4).

When comparing the effectiveness of the vaccine to prevent the three events according to sex, no significant differences were found in any of them (infection, hospitalization and death). Therefore, the effectiveness is the same in women and men. These effectiveness data confirm what was analyzed in a systematic review and meta-analysis that concludes that, despite the significant biological and behavioral differences between both sexes, no significant differences were found in the efficacy of the COVID-19 vaccines (Figure 5).

A previous history of infection significantly increases the effectiveness of the vaccine against a new infection, increasing from 28.4% to 45.8% ($p < 0.0001$). A previous history of infection decreases the effectiveness against hospitalization, while it increases the effectiveness against death, but these variations are not significant (Figure 6).

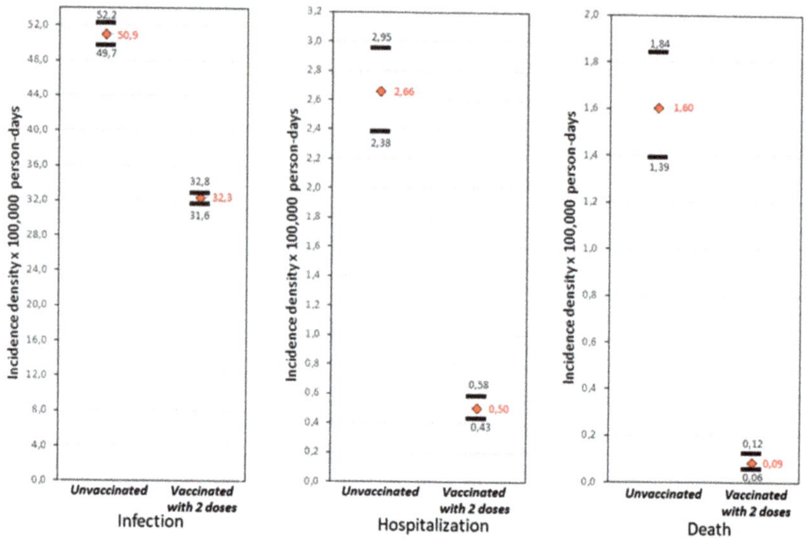

Figure 2. Crude incidence density rates for each event (CI 95%) in health workers, Peru Feb.–Jun. 2021.

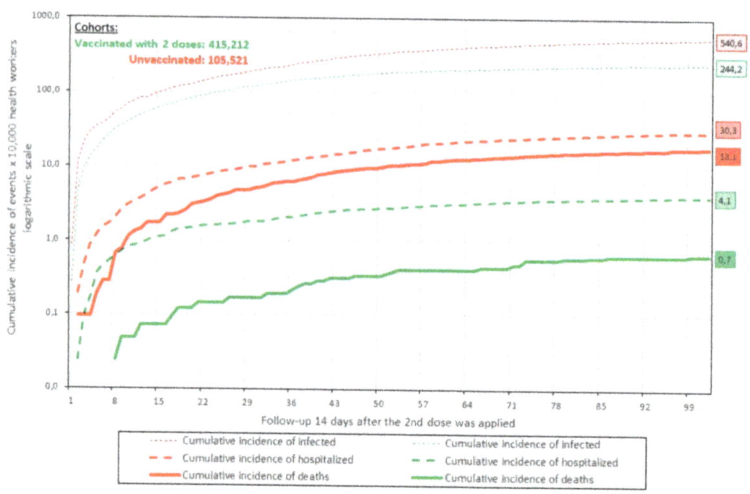

Figure 3. Follow-up and ridk of infection, hospitalization and death from COVID-19 per 10,000 vaccinated (inactivated SARS-CoV-2, Vero Cell) and unvaccinated health workers in Peru February–June 2021.

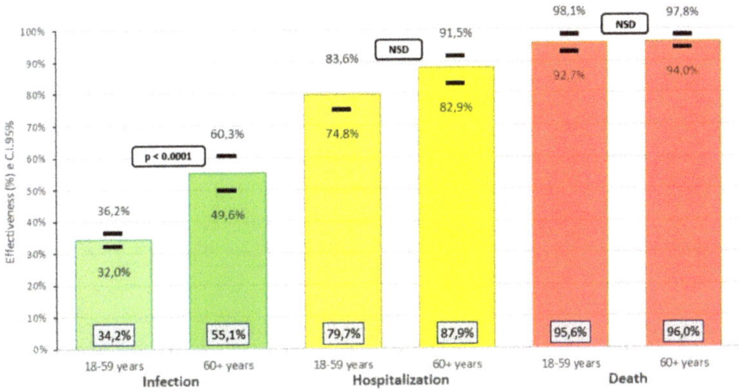

Figure 4. Effectiveness of the inactivated SARS-CoV-2 (Vero Cell) vaccine against COVID-19 according to age, in Peruvian health workers, February–June 2021.

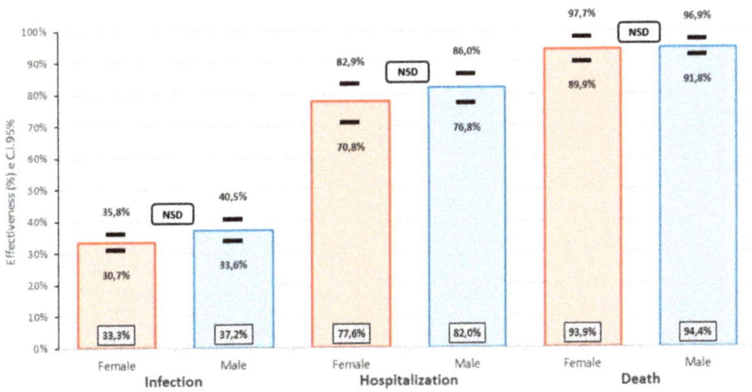

Figure 5. Effectiveness of the inactivated SARS-CoV-2 (Vero Cell) vaccine against SARS-CoV-2 according to sex, in Peruvian health workers, February–June 2021.

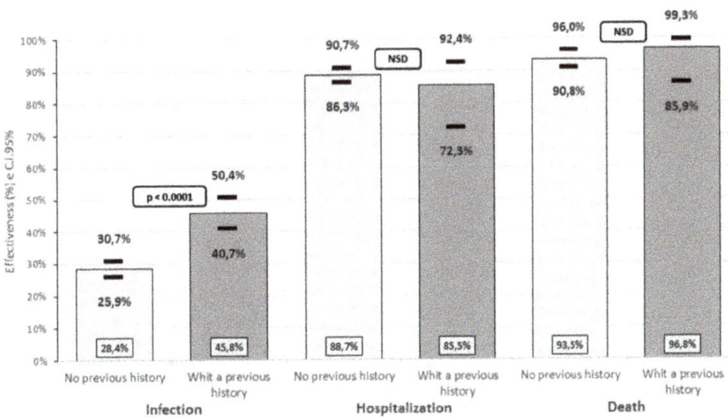

Figure 6. Effectiveness of the inactivated SARS-CoV-2 (Vero Cell) vaccine against COVID-19 according to previous history of infection, in Peruvian workers, February–June 2021.

In Figure 7, it can be seen that the effectiveness of the vaccine against infection decreases when adjusted for age, sex and previous history against the crude rate and when adjusted only for age and sex. We observed a similar situation with adjusted effectiveness against hospitalization; however, in the effectiveness of the vaccine against death, the rate adjusted for age, sex and previous history of infection decreases the effectiveness when compared with the rate adjusted for age and sex, but not when compared to the crude rate.

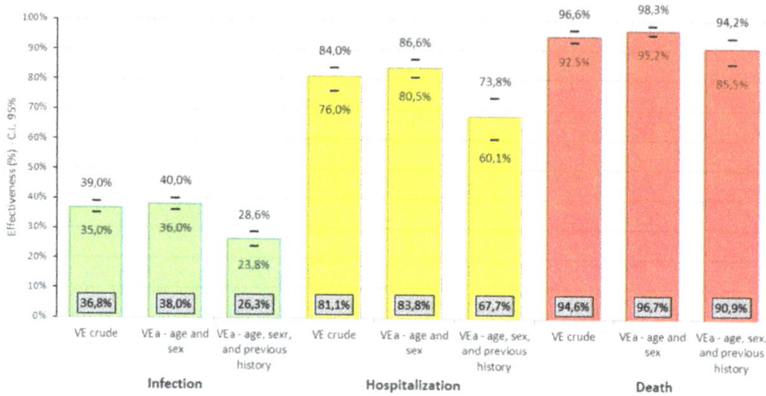

Figure 7. Crude and adjusted effectiveness of the inactivated SARS-CoV-2 (Vero Cell) vaccine against COVID-19, in Peruvian health workers, February–June 2021.

4. Discussion

The effectiveness of the inactivated SARS-CoV-2 (Vero Cell) vaccine applied in two doses to health workers is 90.9% to prevent death, 67.7% to prevent hospitalization and 26.3% to reduce the risk of infection by SARS-CoV-2. This is the first large cohort study on the inactivated SARS-CoV-2 (Vero Cell) vaccine effectiveness.

For the randomized phase 3 clinical trial on the effect of two inactivated vaccines against SARS-CoV-2 on symptomatic COVID-19 infection in adults, with an average age of 36.1 years, 84.4% of them men, where 94.6% received two doses, and an average monitoring of 77 days, the preliminary results reported a vaccine efficacy compared to placebo, of 72.8% (95% CI: 58.1–82.4%) for WIV04 and of 78.1% (95% CI: 64.8–86.3%) for HBO2. Two severe cases of COVID-19 occurred in the placebo-only group and none in the vaccine groups [18]. These results differ from this study, in the analysis period of 21 and 28 days recommended by the manufacturer, since this analysis also included people vaccinated with a second dose administered 28 days after the initial dose; this could alter the protective effect of the vaccine. In the same way, the study's monitoring period is up to 100 days following administration of the second vaccine dose. There were more than 500 thousand people included in the study, who, due to the work they perform, constitute a group with high exposure to the virus, representing the effectiveness of the study with the largest population followed.

On 7 May 2021, the Strategic Advisory Group of Experts on Immunization of the WHO, based on the evidence available, recommended the BBIBP-CorV vaccine for adults over 18 years of age over a two-dose schedule. With an interval of three to four weeks between doses, the reported efficacy to prevent symptomatic and hospitalized disease is 79% [19].

Worldwide, there are studies with similar results, while others have different methodology and findings. A study conducted in Israel, which included all vaccinated and unvaccinated people nationally, measured events (illness, infection, hospitalization, and death) at 14 and 20 days after the first dose, as well as 7 days after the second dose, between 20 December 2020 and 1 February 2021. For documented infection, the Israel study reported a 46% effectiveness following the first dose (95% CI, 40–51) and 92% following the second (95% CI, 88–95), respectively. Likewise, the study reported for symptomatic COVID-19:

57% (95% CI, 50 to 63) and 94% (95% CI, 87 to 98) for hospitalizations; 74% (95% CI, 56 to 86) and 87% (95% CI, 55 to 100) for severe disease; 62% (95% CI, 39 to 80) and 92% (95% CI, 75 to 100)to prevent death; and 72% (95% CI, 19 to 100) during days 14 to 20 after the first dose [20]. These results differ from those found in the present, probably because the study included the entire national population, with a different vaccine, and was also matched by clinical and demographic characteristics.

Compared with other vaccine types, the difference in effectiveness could be greater. A study conducted in Israel with the BNT162b2 vaccine (Pfizer Inc., New York, NY, USA), which used information from the national surveillance system corresponding to the first 4 months of the national vaccination campaign, reported that the effectiveness in reducing infection from the seventh day of the second dose was 95.3% (95% CI 94.9–95.7); 91.5% (95% CI 90.7–92.2) for asymptomatic infection; 97.0% (96.7% CI 96.7–97.2) for symptomatic infection; 97.2% (CI 95% 96.8–97.5) for hospitalization; 97.5% (95% CI 97.1–97.8) for severe COVID-19; and 96.7% (95% CI 96.0–97.3) for death from COVID-19 [21].

Another study conducted in Israel using the same BNT162b2 vaccine reported a 51.4% vaccine effectiveness against SARS-CoV-2 infection at 13–24 days after immunization with the first dose. These findings were similar in adults aged 60 years and over (44.5%); in those under 60 years of age (50.2%); men (52.1%); and women (50.0%) [22]. The reported values are also superior to the effectiveness found at present for the BBIBP-CorV vaccine.

On the other hand, a cohort study conducted in England with the BNT162b2 vaccine, also among health personnel, reported an incidence density of 14 infections per 10,000 person-days in the unvaccinated cohort; eight infections per 10,000 person-days at 21 days after the first dose; and four infections per 10,000 person-days 7 days after the second dose in the vaccinated cohort. In this way, it was determined that a single dose of the BNT162b2 vaccine was 70% (95% CI 55–85) effective against infection twenty-one days after the first dose, and 85% (95% CI 74–96) effective seven days after the second dose in the study population [23].

However, the BNT162b2 vaccine, when evaluated in effectiveness studies, showed different results. After the second dose, between weeks 1 and 2, the effectiveness in reducing positive cases was 73% and 85%, respectively. After 14 days of the second dose, the effectiveness was 89% to prevent hospitalizations and 97% to prevent severe disease [13].

If we consider other vaccines types such as Ad26.COV2.S (Janssen/Johnson & Johnson, New Brunswick, NJ, USA), which uses a viral vector, the protection may also be greater than that offered by BBIBP-CorV. In a phase III efficacy trial, the single-dose Ad26.COV2.S vaccine was 66.9% effective (95% CI 59.0–73.4) in preventing moderate to severe COVID-19 infections starting 14 days after vaccination. Efficacy was higher at 78% and 85%, 14- and 28-days following vaccination, respectively [24].

On the other hand, the ChAdOx1 nCoV-19/AZD1222 vaccine (University of Oxford, AstraZeneca, and the Serum Institute of India), in an interim report of a phase III study, showed an efficacy of 70.4% (95% CI 54.8–80.6) to prevent symptomatic COVID-19 at 14 days or more following the second dose. In this same study, a subgroup of participants was inadvertently given a lower dose of vaccine for the first of the two doses. In this subgroup, the efficacy reached 90.0% (95% CI 67.4 to 90.0), much higher than the 62.1% (95% CI 41.0 to 75.7) among those who received the full dose. Although the reasons for this difference are uncertain, the CI values suggest the difference in efficacy may not be statistically significant [25].

In a later report from this same study, the efficacy of the ChAdOx1 nCoV-19/AZD1222 vaccine for symptomatic COVID-19 was 76%, 21 days after receiving the first dose until receiving the second dose, or 90 days following the first dose, suggesting protection with a single dose [26]. Another report based on preliminary results from a placebo-controlled trial conducted in the United States, Chile, and Peru reported similar findings at two full doses four weeks apart. The reported values were 76% efficacy against symptomatic COVID-19 15 days following the second dose, 100% against serious or critical illnesses and hospitalization, and 85% efficacy for symptomatic COVID-19 in people aged 65 and older [27].

Even though these data mostly correspond to other types, and the methodology used is different (efficacy vs. effectiveness), the findings suggest less protection of the inactivated SARS-CoV-2 (Vero Cell) vaccine against SARS-CoV-2 to reduce infection, hospitalization and death associated with COVID-19, 14 days after the second dose. However, they are still acceptable, meeting the World Health Organization's (WHO) vaccine standard.

Similarly, our study results differ from the BBIBP-CorV vaccine's preliminary efficacy data for the prevention of symptomatic SARS-CoV-2 infections (78.1%), even with the post hoc results from the multicentered randomized clinical trial, which included asymptomatic cases (73.5%) and was later approved by the WHO for the use and marketing of the BBIBP-CorV vaccine [18,28].

These study results may differ because of the criteria for evaluating the effectiveness of the vaccine. The multicentered study evaluated effectiveness among symptomatic infections, while our study included both symptomatic and asymptomatic infections. Compliance with the definition of a suspected case of the standard epidemiological surveillance could not be 100% controlled. Despite the non-compliance at the hospital level of the periodic screenings of health personnel and the greater accessibility to diagnostic tests in case of work contacts, the suspected case definition generated an increase in the number of positive cases detected, underestimating the real effectiveness the vaccine may have against symptomatic disease.

Furthermore, the risk of exposure of health workers is greater, in terms of exposure time, viral load and in the likely exposure to multiple COVID-19 variants while delivering patient care, driving a decrease in vaccine effectiveness estimates.

Evaluating these results for public health in terms of infection and transmission is more challenging than for the results of the disease. Therefore, future research will include, in addition to the separate analysis of the effectiveness among symptomatic and asymptomatic patients, viral quantification, active surveillance in the professional and domestic environments and patient motivation for testing, whether due to COVID-19 exposure or routine practice, to refine estimates for vaccine effectiveness for disease infection and transmission.

Several studies of the early effectiveness of the vaccine against SARS-CoV-2 in medical settings for health workers have shown that vaccination in health personnel has a positive effect on the preservation of the workforce, through a marked reduction in disease incidence in medical personnel [29–31], including asymptomatic infections [32]. These studies rely on policies of weekly testing to identify asymptomatic infections, thus, facilitating the detection of asymptomatic SARS-CoV-2 infections after vaccination. In contrast, our analysis relies on regulations that only establish the testing of symptomatic personnel, overlooking asymptomatic infections. Such results are consistent with what we have observed, in such a way that, regardless of the brand or type of vaccine, immunizations have proven to be the most cost-effective public health intervention. It is highlighted that, despite this, the need to continue with complimentary interventions of social distancing, epidemiological surveillance, and laboratory and clinical monitoring, is emphasized. This is a crucial aspect that Peru must consider as new COVID-19 variants circulate and given the still limited knowledge of the effectiveness existing vaccines would have on them.

Another study has raised the need to evaluate vaccine effectiveness through the measurement of the antibody levels to determine the immune response to COVID-19 vaccination [33]. Although vaccines may provide long-term immunity via T cells and memory B cells [34], neutralizing antibody production after vaccination provide an immunological biomarker correlated with protection against SARS-CoV-2 in humans. Therefore, anti-SARS-CoV-2 antibodies are a valuable tool to assess the protective immunity after vaccination, and to aid in determining immunization schedules.

Additionally, our findings show that although the vaccine has a protective effect, this effect decreases five weeks after administration of the second dose for all groups observed (infections, hospitalizations, and deaths from COVID-19). Studies performed with mRNA vaccines indicate that the onset of protection was observed as early as 12 days after the administration of a single vaccine dose. The adaptive immune response that coincides with this onset of protection could represent the necessary elements of immunity against COVID-19 [35]. Preliminary data identified a reduced T and B cells response to inactivated vaccine in the elderly [36], suggesting the need to apply a third dose of an alternative vaccine, following the most common recommendation with mRNA vaccines.

The sensitivity of the diagnostic test used (65% for PCR tests and 56.2% for antigen tests [37] on average), directly influences the study results. This is due to the frequency of "false negative" test results. Additionally, the 53 different lineages of the SARS-CoV-2 virus identified to date also influence study results; this include variants identified in Lima, such as the B.1.1.7 (Alpha) lineage—the British variant, and the P1 (Gamma) lineage—the Manaus–Brazil variant (also identified in Huánuco and Loreto), which constituted 39.9% of the COVID-19 cases last sampled in Lima [38]. These variants increase viral transmission, the severity of COVID-19 symptoms, mortality rates, and importantly, they can also decrease the effectiveness of diagnostic methods, influencing the effectiveness of available vaccines [39,40].

Reports from Israel have linked vaccination with an increase in COVID-19 cases shortly following vaccination, driven by an increase in community and healthcare-related exposures. These same reports note that the coexistence of vaccination with the rapid spread of COVID-19 is a period in which vaccination-related symptoms pose a diagnostic dilemma, convoluting a vaccination-driven adverse reaction with a new COVID-19 infection. Considering these results, the COVID-19 cases identified in our study require additional monitoring and research to determine whether symptom onset began before vaccination. Likewise, the authors of another publication related to the same study conclude that both vaccinated subjects and unvaccinated controls must be sampled in order to determine a true decrease in infections, especially in asymptomatic cases among vaccinated persons [41].

Therefore, a second vaccine and the resulting benefit of immunization against SARS-CoV-2 with full vaccination should be emphasized.

The COVID-19 variants are emerging problems that may affect the effectiveness of existing vaccines. For this reason, the NIH has communicated the findings of genomic sequencing in Peru [42]. The NIH reports that, of 1714 samples extracted from March 2020 to May 2021, 71.6% (1228/1714) belong to the lambda or C.37 variant, which would be proportionally dominant at the time of evaluating the effectiveness of the BBIP-CorV vaccine. The same report shows that 14.1% (242/1714) of the sequenced samples correspond to the gamma or P1 variant (the "Brazilian variant"), present in 22 regions of Peru, except for Puno and Ancash. In the period of the study, neither the efficacy nor the effectiveness of the inactivated SARS-CoV-2 (Vero Cell) vaccine on the new SARS variants could be determined. CoV-2 (Delta and Omicron) appeared later.

To determine the effectiveness of the inactivated CoronaVac vaccine, a case-control study was conducted among people aged 70 years and over in Sao Paolo, Brazil. It was conducted in relation to the P1 variant (which represented 85% of the genotyped samples in the analysis period). Using surveillance and vaccination records, the case-control study found two doses of the CoronaVac vaccine to have an adjusted effectiveness of 41.6% (95% CI: 26.9 to 53.3) and 18.2% (95% CI 0.0 to 33.2) in the period $\geq$14 days and 0–13 days, respectively. When analyzed by age, the vaccine effectiveness decreases significantly after 14 days of the second dose, falling from 61.8% in the 70- to 74-year-old group to 28.0% in the over-80 year old group. These estimates approximate those identified in our study. [43].

This raises four major concerns, including: effects on viral transmissibility, the severity of the disease, the reinfection rates or escape from natural immunity, and the efficacy of the vaccine or escape from vaccine-induced immunity [44]. The evaluation of the effectiveness of vaccines on new variants could generate biases and confounding factors related to the variable risk of infection over time and the probability of receiving the vaccine or not [45].

The neutralizing activity of the serum for the B.1.351 (or the Beta, South African) variant among vaccinated people was lower by a factor of 1.6 to 8.6 for the BBIBP-CorV vaccine (Sinopharm), the BNT162b2 vaccine (BioNTech-Pfizer) and the mRNA-1273 (Modern) vaccine. It was also lower by a factor of up to 86, including complete immune escape, for the AZD1222 (Oxford-AstraZeneca) vaccine. This is why the administration of the latter vaccine was interrupted. The neutralizing activity for the P.1 variant among vaccinated persons was lower by a factor of 6.7 for the BNT162b2 vaccine and by a factor of 4.5 for the mRNA-1273 vaccine [44,46].

What has been described raises the hypothesis that the inactivated SARS-CoV-2 (Vero Cell) vaccine might be less effective due to mutations in the SARS-CoV-2 virus, specifically the predominant C.37 variant in the current epidemiological scenario. This should be corroborated with additional research. In the meantime, however, it puts Peru's health system on alert.

The strength of our study was the access to official information from the Ministry of Health, where the data were collected from the databases that served as a source of information for the study, which facilitated the capture and recapture of the data.

Despite the limitations of using national data for this analysis, it is necessary to incorporate hospitalization data from other parts of Peru's health system, including from the private sector, the armed forces, and the police force, for data integrity and quality control, in order to prevent omission in the registry of study information sources.

Further analysis should also include other relevant variables, such as exposure difference to the SARS-CoV-2 virus by types of activity carried out in workplaces (service provision, administrative tasks, etc.). Future analysis should also aim to complete the individual clinical information of infected, hospitalized, and deceased cases. Additionally, as it is an analysis of secondary sources, there may be information biases.

5. Conclusions

In conclusion, two doses of the inactivated SARS-CoV-2 (Vero Cell) vaccine applied to health workers from MINSA-Peru has acceptable effectiveness in reducing hospitalizations and death from COVID-19 fourteen days after the application of the second dose. The effectiveness in reducing SARS-CoV2 infection, however, is low.

It is necessary to continue surveillance studies for longer periods and to include other variables which were not included in the present analysis. Additionally, pre-vaccination periods should be compared and the effect of vaccines against the COVID-19 variants identified in Peru should be evaluated, as well as the need for additional vaccine doses.

Author Contributions: M.E.S.-C.: conceptualization, data curation, formal analysis, funding acquisition, investigation, methodology, project administration, resources, software, supervision, visualization, writing—original draft, writing—review and editing. A.J.-C.: conceptualization, data curation, formal analysis, funding acquisition, investigation, methodology, resources, software, validation, visualization, writing—original draft and writing—review and editing. R.V.G.S.: conceptualization, methodology, software, writing—original draft, writing—review and editing. N.J.G.: investigation, writing—original draft, Writing—review and editing. I.E.M.P.: conceptualization, investigation, writing—original draft, writing—review and editing. K.E.V.Q.: investigation, methodology, writing—original draft, writing—review and editing. L.Y.L.T.R.: formal analysis, methodology, writing—original draft, writing—review and editing. M.N.V.D.: investigation, methodology, resources, writing—original draft, writing—review and editing. D.T.E.C.: writing—original draft, writing—review and editing. P.M.: funding acquisition, resources, visualization, writing—original draft, writing—review and editing. K.J.P.R.: funding acquisition, resources, writing—original draft, writing—review and editing. C.D.-V.: resources, writing—original draft. V.A.P.: writing—original draft, writing—review and editing. R.W.A.: investigation, writing—original draft, writing—review and editing. P.E.P.: Conceptualization, data curation, formal analysis, software, Supervision, validation, writing—review and editing. All authors have read and agreed to the published version of the manuscript.

Funding: This research received no external funding. (R.W.A. receives a grant from Fundação de Amparo à Pesquisa do Estado de São Paulo (FAPESP) grant: 2019/02679-7).

Institutional Review Board Statement: The study was conducted in accordance with the Declaration of Helsinki and approved by the Ethics Committee of UNIVERSIDAD NACIONAL MAYOR DE SAN MARCOS-FACULTAD DE MEDICINA (protocol code 0115-15 July 2021).

Informed Consent Statement: Not applicable.

Data Availability Statement: Not applicable.

Acknowledgments: To the Peruvian health workers.

Conflicts of Interest: The authors declare no conflict of interest.

References

1. Santos-Sánchez, N.F.; Salas-Coronado, R. Origen, Características Estructurales, Medidas de Prevención, Diagnóstico y Fármacos Potenciales Para Prevenir y Controlar COVID-19. Medwave. 2020. Available online: /link.cgi/Medwave/Revisiones/RevisionClinica/8037.act (accessed on 28 March 2021).
2. Asahi, K.; Undurraga, E.A.; Valdés, R.; Wagner, R. The Effect of COVID-19 on the Economy: Evidence from an Early Adopter of Localized Lockdowns. *J. Glob. Health* **2021**, *11*, 05002. Available online: https://www.ncbi.nlm.nih.gov/pmc/articles/PMC7897430/ (accessed on 7 March 2021). [CrossRef]
3. Nguyen, L.H.; Drew, D.A.; Joshi, A.D.; Guo, C.-G.; Ma, W.; Mehta, R.S.; Sikavi, D.R.; Lo, C.H.; Kwon, S.; Song, M.; et al. Risk of COVID-19 among frontline healthcare workers and the general community: A prospective cohort study. *medRxiv* **2020**. medRxiv:2020.04.29.20084111.
4. Graham, B.S. Rapid COVID-19 vaccine development. *Science* **2020**, *368*, 945–946. [CrossRef]
5. Shrotri, S.; Kampmann, P. Rastreador de Vacunas COVID-19. 2021. Available online: https://vac-lshtm.shinyapps.io/ncov_vaccine_landscape/ (accessed on 28 March 2021).
6. Creech, C.B.; Walker, S.C.; Samuels, R.J. SARS-CoV-2 Vaccines. *JAMA* **2021**, *325*, 1318–1320. [CrossRef]
7. Abu Jabal, K.; Ben-Amram, H.; Beiruti, K.; Batheesh, Y.; Sussan, C.; Zarka, S.; Edelstein, M. Impact of age, ethnicity, sex and prior infection status on immunogenicity following a single dose of the BNT162b2 mRNA COVID-19 vaccine: Real-world evidence from healthcare workers, Israel, December 2020 to January 2021. *Eurosurveillance* **2021**, *26*, 5. [CrossRef]
8. Aran, D. Estimating real-world COVID-19 vaccine effectiveness in Israel using aggregated counts. *medRxiv* **2021**. [CrossRef]
9. Planas, D.; Veyer, D.; Baidaliuk, A.; Staropoli, I.; Guivel-Benhassine, F.; Rajah, M.M.; Planchais, C.; Porrot, F.; Robillard, N.; Puech, J.; et al. Reduced sensitivity of SARS-CoV-2 variant Delta to antibody neutralization. *Nature* **2021**, *596*, 276–280. [CrossRef]
10. DIGEMID/MINSA. Resolución Directoral No. 486–2021/digemid/dpf/ufpb/minsa. Resolución que Autoriza Excepcionalmente la Importación y uso por Situaciones de Salud Pública la Vacuna SARS-CoV-2 (Vero Cell) Inactivada × 0.5 mL del Fabricante Beijing Institute of Biological Products Co. Ltd. (BIBP), China. DIGEMID. 2021. Available online: http://www.digemid.minsa.gob.pe/Resoluciones/ResolucionesArchivos/DPF/2021/RD_DPF_0000486_2021.pdf (accessed on 9 March 2021).
11. Ministerio de Salud del Perú. Resolución Ministerial No. 848-2020-MINSA que aprueba el Documento Técnico «Plan Nacional de Vacunación contra la COVID-19. Ministerio de Salud. 2020. Available online: https://cdn.www.gob.pe/uploads/document/file/1394145/RM%20N%C2%B0848-2020-MINSA.PDF.PDF (accessed on 9 March 2021).
12. Xia, S.; Zhang, Y.; Wang, Y.; Wang, H.; Yang, Y.; Gao, G.F.; Tan, W.; Wu, G.; Xu, M.; Lou, Z.; et al. Safety and immunogenicity of an inactivated SARS-CoV-2 vaccine, BBIBP-CorV: A randomised, double-blind, placebo-controlled, phase 1/2 trial. *Lancet Infect Dis.* **2021**, *21*, 39–51. [CrossRef]
13. Organización Mundial de la Salud. Peru: WHO Coronavirus Disease (COVID-19) Dashboard. Global Peru. 2021. Available online: https://covid19.who.int/region/amro/country/pe (accessed on 7 March 2021).
14. Médicos Fallecidos por COVID-19 en Iberoamérica. Colegio Médico del Perú—Consejo Nacional. Available online: https://www.cmp.org.pe/medicos-fallecidos-por-covid-19-en-iberoamerica/ (accessed on 28 March 2021).
15. MINSA. Resolución Ministerial No. 848-2020-MINSA. 2020. Available online: https://www.gob.pe/institucion/minsa/normas-legales/1293043-848-2020-minsa (accessed on 9 March 2021).
16. Ministerio de Salud de Perú. Resolución Ministerial No. 161-2021-MINSA. 2021. Available online: https://www.gob.pe/institucion/minsa/normas-legales/1635499-161-2021-minsa (accessed on 27 March 2021).
17. Halloran, M.E.; Struchiner, C.J.; Longini, I.M. Study designs for evaluating different efficacy and effectiveness aspects of vaccines. *Am. J. Epidemiol.* **1997**, *146*, 789–803. [CrossRef]
18. Al Kaabi, N.; Zhang, Y.; Xia, S.; Yang, Y.; Al Qahtani, M.M.; Abdulrazzaq, N.; Al Nusair, M.; Hassany, M.; Jawad, J.S.; Abdalla, J.; et al. Effect of 2 Inactivated SARS-CoV-2 Vaccines on Symptomatic COVID-19 Infection in Adults: A Randomized Clinical Trial. *JAMA* **2021**, *326*, 35–45. Available online: https://jamanetwork.com/journals/jama/fullarticle/2780562 (accessed on 28 June 2021). [CrossRef]

19. WHO Lists Additional COVID-19 Vaccine for Emergency Use and Issues Interim Policy Recommendations. Available online: https://www.who.int/news/item/07-05-2021-who-lists-additional-covid-19-vaccine-for-emergency-use-and-issues-interim-policy-recommendations (accessed on 28 June 2021).
20. Dagan, N.; Barda, N.; Kepten, E.; Miron, O.; Perchik, S.; Katz, M.A.; Hernán, M.A.; Lipsitch, M.; Reis, B.; Balicer, R.D. BNT162b2 mRNA Covid-19 Vaccine in a Nationwide Mass Vaccination Setting. *N. Engl. J. Med.* **2021**, *384*, 1412–1423. [CrossRef]
21. Haas, E.J.; Angulo, F.J.; McLaughlin, J.M.; Anis, E.; Singer, S.R.; Khan, F.; Brooks, N.; Smaja, M.; Mircus, G.; Pan, K. Impact and effectiveness of mRNA BNT162b2 vaccine against SARS-CoV-2 infections and COVID-19 cases, hospitalisations, and deaths following a nationwide vaccination campaign in Israel: An observational study using national surveillance data. *Lancet* **2021**, *397*, 1819–1829. [CrossRef]
22. Chodick, G.; Tene, L.; Patalon, T.; Gazit, S.; Tov, A.B.; Cohen, D.; Muhsen, K. The effectiveness of the first dose of BNT162b2 vaccine in reducing SARS-CoV-2 infection 13–24 days after immunization: Real-world evidence. *medRxiv* **2021**. medRxiv:2021.01.27.21250612.
23. Hall, V.J.; Foulkes, S.; Saei, A.; Andrews, N.; Oguti, B.; Charlett, A.; Wellington, E.; Stowe, J.; Gillson, N.; Atti, A.; et al. COVID-19 vaccine coverage in health-care workers in England and effectiveness of BNT162b2 mRNA vaccine against infection (SIREN): A prospective, multicentre, cohort study. *Lancet* **2021**, *397*, 1725–1735. [CrossRef]
24. Sadoff, J.; Gray, G.; Vandebosch, A.; Cárdenas, V.; Shukarev, G.; Grinsztejn, B.; Goepfert, P.A.; Truyers, C.; Fennema, H.; Spiessens, B. Safety and Efficacy of Single-Dose Ad26.COV2.S Vaccine against Covid-19. *N. Engl. J. Med.* **2021**, *384*, 2187–2201. [CrossRef]
25. Voysey, M.; Clemens, S.A.C.; Madhi, S.A.; Weckx, L.Y.; Folegatti, P.M.; Aley, P.K.; Angus, B.; Baillie, V.L.; Barnabas, S.L.; Bhorat, Q.E.; et al. Safety and efficacy of the ChAdOx1 nCoV-19 vaccine (AZD1222) against SARS-CoV-2: An interim analysis of four randomised controlled trials in Brazil, South Africa, and the UK. *Lancet* **2021**, *397*, 99–111. [CrossRef]
26. Voysey, M.; Costa Clemens, S.A.; Madhi, S.A.; Weckx, L.Y.; Folegatti, P.M.; Aley, P.K.; Angus, B.; Baillie, V.L.; Barnabas, S.L.; Bhoratm, Q.E.; et al. Single-dose administration and the influence of the timing of the booster dose on immunogenicity and efficacy of ChAdOx1 nCoV-19 (AZD1222) vaccine: A pooled analysis of four randomised trials. *Lancet* **2021**, *397*, 881–891. [CrossRef]
27. AZD1222 Vaccine met Primary Efficacy Endpoint in Preventing COVID-19. 2021. Available online: https://www.astrazeneca.com/media-centre/press-releases/2020/azd1222hlr.html (accessed on 25 January 2021).
28. Evidence Assessment: Sinopharm/BBIBP COVID-19 Vaccine. Available online: https://cdn.who.int/media/docs/default-source/immunization/sage/2021/april/2_sage29apr2021_critical-evidence_sinopharm.pdf (accessed on 28 June 2021).
29. Keehner, J.; Horton, L.E.; Pfeffer, M.A.; Longhurst, C.A.; Schooley, R.T.; Currier, J.S.; Abeles, S.R.; Torriani, F.J. SARS-CoV-2 Infection after Vaccination in Health Care Workers in California. *N. Engl. J. Med.* **2021**, *384*, 1774–1775. [CrossRef]
30. Daniel, W.; Nivet, M.; Warner, J.; Podolsky, D.K. Early Evidence of the Effect of SARS-CoV-2 Vaccine at One Medical Center. *N. Engl. J. Med.* **2021**, *384*, 1962–1963. [CrossRef]
31. Benenson, S.; Oster, Y.; Cohen, M.J.; Nir-Paz, R. BNT162b2 mRNA Covid-19 Vaccine Effectiveness among Health Care Workers. *N. Engl. J. Med.* **2021**, *384*, 1775–1777. [CrossRef]
32. Angel, Y.; Spitzer, A.; Henig, O.; Saiag, E.; Sprecher, E.; Padova, H.; Ben-Ami, R. Association between Vaccination with BNT162b2 and Incidence of Symptomatic and Asymptomatic SARS-CoV-2 Infections among Health Care Workers. *JAMA* **2021**, *325*, 2457–2465. Available online: https://jamanetwork.com/journals/jama/fullarticle/2779853 (accessed on 16 June 2021). [CrossRef] [PubMed]
33. Khoury, D.S.; Cromer, D.; Reynaldi, A.; Schlub, T.E.; Wheatley, A.K.; Juno, J.A.; Subbarao, K.; Kent, S.J.; Triccas, J.A.; Davenport, M.P. Neutralizing Antibody Levels are Highly Predictive.pdf. 2021. Available online: https://www.nature.com/articles/s41591-021-01377-8.pdf (accessed on 24 February 2022).
34. Zollner, A.; Watschingera, C.; Rossler, A.; Farcet, M.R.; Pennerg, A.; Bohm, V.; Kiechlc, S.J.; Stampfel, G.; Hintenbergerg, T.; Tilg, H.; et al. B and T Cell Response to SARS-CoV-2 Vaccination in.pdf. 2021. Available online: https://www.thelancet.com/action/showPdf?pii=S2352-3964%2821%2900332-7 (accessed on 24 February 2022).
35. Kalimuddin, S.; Tham, C.Y.L.; Qui, M.; de Alwis, R.; Sim, J.X.Y.; Lim, J.M.E.; Tan, H.C.; Syenina, A.; Zhang, S.L.; Le Bert, N.; et al. Early T cell and binding antibody responses are associated with COVID-19 RNA vaccine efficacy onset. *Med* **2021**, 2, 682–688.e4. [CrossRef] [PubMed]
36. Reduced T Cell and Antibody Responses to Inactivated Coronavirus Vaccine among Males and Individuals above 55 Years Old | medRxiv. Available online: https://www.medrxiv.org/content/10.1101/2021.08.16.21262069v1 (accessed on 4 April 2022).
37. Dinnes, J.; Deeks, J.J.; Adriano, A.; Berhane, S.; Davenport, C.; Dittrich, S.; Emperador, D.; Takwoingi, Y.; Cunningham, J.; Beese, S. Rapid, point-of-care antigen and molecular-based tests for diagnosis of SARS-CoV-2 infection. *Cochrane Database Syst. Rev.* **2020**, *2021*, CD013705. Available online: http://doi.wiley.com/10.1002/14651858.CD013705 (accessed on 24 February 2021).
38. INS: Variante Brasilera Tiene una Amplia Circulación en Varios Distritos de Lima. Instituto Nacional de Salud. Available online: http://web.ins.gob.pe/es/prensa/noticia/ins-variante-brasilera-tiene-una-amplia-circulacion-en-varios-distritos-de-lima (accessed on 29 March 2022).
39. Lopez Bernal, J.; Andrews, N.; Gower, C.; Gallagher, E.; Simmons, R.; Thelwall, S.; Stowe, J.; Tessier, E.; Groves, N.; Dabrera, G.; et al. Effectiveness of Covid-19 Vaccines against the B.1.617.2 (Delta) Variant. *N. Engl. J. Med.* **2021**, *85*, 585–594. [CrossRef]

40. PubMed Central Full Text PDF. Available online: https://www.ncbi.nlm.nih.gov/pmc/articles/PMC8462339/pdf/21-1427.pdf (accessed on 24 February 2022).
41. Amit, S.; Beni, S.A.; Biber, A.; Grinberg, A.; Leshem, E.; Regev-Yochay, G. Post-Vaccination COVID-19 among Healthcare Workers, Israel. *Emerg. Infect. Dis.* **2021**, *27*, 1220. [CrossRef]
42. Secuenciación Genómica del virus SARS-CoV-2 en el Perú. Instituto Nacional de Salud. Available online: https://web.ins.gob.pe/es/covid19/secuenciamiento-sars-cov2 (accessed on 4 April 2022).
43. Ranzani, O.T.; Hitchings, M.; Dorion, M.; D'Agostini, T.L.; de Paula, R.C.; de Paula, O.F.P.; de Moura Villela, E.F.; Torres, M.S.; de Oliveira, S.B.; Schulz, W.; et al. Effectiveness of the CoronaVac vaccine in the elderly population during a P.1 variant-associated epidemic of COVID-19 in Brazil: A test-negative case-control study. *BMJ* **2021**, *374*, b2015. [CrossRef]
44. Abdool Karim, S.S.; de Oliveira, T. New SARS-CoV-2 Variants—Clinical, Public Health, and Vaccine Implications. *N. Engl. J. Med.* **2021**, *384*, 1866–1888. [CrossRef]
45. Patel, M.K.; Bergeri, I.; Bresee, J.S.; Cowling, B.J.; Crowcroft, N.S.; Fahmy, K.; Hirve, S.; Kang, G.; Katz, M.A.; Lanata, C.F.; et al. Evaluation of Post-Introduction COVID-19 Vaccine Effectiveness: Summary of Interim Guidance of the World Health Organization. *Vaccine* **2021**, *39*, 4013–4024. Available online: https://www.ncbi.nlm.nih.gov/pmc/articles/PMC8166525/ (accessed on 28 June 2021). [CrossRef]
46. Femeba Home | Fundación Femeba. Available online: https://www.fundacionfemeba.org.ar/home (accessed on 28 June 2021).

life

Article

Exome-Wide Association Study Reveals Host Genetic Variants Likely Associated with the Severity of COVID-19 in Patients of European Ancestry

Priyanka Upadhyai [1], Pooja U. Shenoy [2], Bhavya Banjan [3], Mohammed F. Albeshr [4], Shahid Mahboob [4], Irfan Manzoor [5] and Ranajit Das [2,*]

[1] Department of Medical Genetics, Kasturba Medical College, Manipal, Manipal Academy of Higher Education, Manipal 576104, India
[2] Yenepoya Research Centre, Yenepoya (Deemed to be University), Mangalore 575018, India
[3] Manipal School of Life Sciences, Manipal Academy of Higher Education, Manipal 576104, India
[4] Department of Zoology, College of Science, King Saud University, Riyadh 11451, Saudi Arabia
[5] Department of Biology, The College of Arts and Sciences, Indiana University, Bloomington, IN 47405, USA
* Correspondence: das.ranajit@gmail.com; Tel.: +91-8582802871

Abstract: Host genetic variability plays a pivotal role in modulating COVID-19 clinical outcomes. Despite the functional relevance of protein-coding regions, rare variants located here are less likely to completely explain the considerable numbers of acutely affected COVID-19 patients worldwide. Using an exome-wide association approach, with individuals of European descent, we sought to identify common coding variants linked with variation in COVID-19 severity. Herein, *cohort 1* compared non-hospitalized (controls) and hospitalized (cases) individuals, and in *cohort 2*, hospitalized subjects requiring respiratory support (cases) were compared to those not requiring it (controls). 229 and 111 variants differed significantly between cases and controls in *cohorts 1* and *2*, respectively. This included *FBXO34*, *CNTN2*, and *TMCC2* previously linked with COVID-19 severity using association studies. Overall, we report SNPs in 26 known and 12 novel candidate genes with strong molecular evidence implicating them in the pathophysiology of life-threatening COVID-19 and post-recovery sequelae. Of these few notable known genes include, *HLA-DQB1, AHSG, ALOX5AP, MUC5AC, SMPD1, SPG7, SPEG, GAS6*, and *SERPINA12*. These results enhance our understanding of the pathomechanisms underlying the COVID-19 clinical spectrum and may be exploited to prioritize biomarkers for predicting disease severity, as well as to improve treatment strategies in individuals of European ancestry.

Keywords: COVID-19 host genetics; genetic variation in COVID-19 patients; exome-wide association study for COVID-19 patients; common genetic variants

Citation: Upadhyai, P.; Shenoy, P.U.; Banjan, B.; Albeshr, M.F.; Mahboob, S.; Manzoor, I.; Das, R. Exome-Wide Association Study Reveals Host Genetic Variants Likely Associated with the Severity of COVID-19 in Patients of European Ancestry. *Life* **2022**, *12*, 1300. https://doi.org/10.3390/life12091300

Academic Editors: Silvia De Francia and Sarah Allegra

Received: 13 July 2022
Accepted: 23 August 2022
Published: 24 August 2022

1. Introduction

More than two years after its first outbreak, the coronavirus disease (COVID-19) caused by severe acute respiratory syndrome coronavirus 2 (SARS-CoV-2) remains a prominent healthcare challenge and disruptor of social and economic activities worldwide. It shows a complex array of manifestations ranging from the complete absence of symptoms to severe clinical outcomes, such as acute respiratory distress syndrome (ARDS) [1]. It is a multi-systemic disease often marked by cardiovascular, respiratory, neurological, gastrointestinal, renal, immunological, and hematological derangements [2–7]. The pathophysiological course of COVID-19 is modulated by the host immune response and progression to severe disease or death occurs due to an uncontrolled cytokine storm and hyperinflammation of the immune system [8]. The severity of clinical outcomes in COVID-19 patients are likely modified by multiple factors ranging from advanced age, male gender, and preexisting comorbidities, such as hypertension and obesity [9–12]. Moreover, ethnicity-specific differences are also noted in the clinical phenotype of SARS-CoV-2 infection [13,14]. However, these, as well as the discrepancies in socioeconomic attributes or access to healthcare and vaccination do not completely explain the striking individual and population specific disparities observed in the acuteness of COVID-19 clinical presentation worldwide [15–17]. For example, many SARS-CoV-2 infected individuals are known to remain asymptomatic despite significantly higher viral loads than their symptomatic counterparts [18]. Furthermore, middle-aged individuals (40–59 years) are known to develop severe disease even without the presence of underlying comorbidities [19]. The spectrum of COVID-19 clinical manifestation is multifactorial and plausibly explained, at least in part, by the variation in host genetic attributes.

This has been interrogated in a series of genome-wide association studies (GWASs) comparing COVID-19 patients and population controls. Ellinghaus et al., reported a significant association of variants in six candidate genes at chromosome 3p21.31, including *Solute Carrier Family 6 Member 20 (SLC6A20), Leucine Zipper Transcription Factor Like 1 (LZTFL1), C-C Motif Chemokine Receptor 9 (CCR9), C-X-C Motif Chemokine Receptor 6 (CXCR6), FYVE and Coiled-coil Domain Autophagy Adaptor 1 (FYCO1),* and *X-C Motif Chemokine Receptor 1 (XCR1)* and the *ABO* blood group locus at 9q34.2, with respiratory failure in COVID-19 patients from Italy and Spain [20]. These results were corroborated by an independent study in a cohort of SARS-CoV-2 infected individuals comprising of European, African–American and Latino ancestries [21]. Genetics of Mortality in Critical Care (GenOMICC) reported a robust association of single nucleotide polymorphisms (SNPs) within or proximal to genes, such as the *2′-5′ Oligoadenylate Synthetase* (*OAS*) cluster and *Interferon Alpha and Beta Receptor Subunit 2 (IFNAR2)* on chromosomes 12 and 21, respectively, that are involved in the antiviral type 1 interferon (IFN) response [22,23], *Tyrosine Kinase 2* (*TYK2*), required for IFN, interleukin 12 (IL-12), IL-23 and *T*-helper 1/*T*-helper 17 cell-dependent immunity [24], and *Dipeptidyl Peptidase 9* (*DPP9*), involved in antigen presentation [25], on chromosome 19 with severe outcomes in COVID-19 patients [26]. The COVID-19 Host Genetics Initiative (HG1) report concurred with the association of variants in *ABO, SLC6A20, TYK2, DPP9, IFNAR2,* and additionally reported a variant in *Protein Phosphatase 1 Regulatory Subunit 15A (PPP1R15A)* in influencing COVID-19 severity [27].

Since uninfected individuals or population controls may develop morbid COVID-19 subsequently upon contracting SARS-CoV-2 infection, we used asymptomatic COVID-19 patients as controls in a GWAS using a COVID-19 dataset generated by AncestryDNA [28], to uncover the genetic variants governing susceptibility to severe COVID-19 in individuals of European ancestry [29]. This study identified 621 SNPs that differed significantly between asymptomatic and acutely afflicted COVID-19 patients, and were associated with IFN and IL signaling, as well as obesity and cholesterol metabolism that are well-known COVID-19 comorbidities. In addition, it highlighted the putative ancestral genomic differences between patients with exacerbated COVID-19 phenotype versus those that were asymptomatic; the latter contained higher genomic fractions of Ancestral North Eurasian (ANE) and Eastern Hunter-Gatherer (EHG), and lower Western Hunter–Gatherer (WHG) fractions. A similar approach using asymptomatic individuals as controls was used in a GWAS on the Chinese population that revealed significant associations at the chromosome loci, 11q23.3 and 11q14.2 with severe COVID-19 and noted significant expression quantitative trail locus (eQTL) associations for RNA Exonuclease 2 (*REXO2), C11orf71, Nicotinamide N-Methyltransferase* (*NNMT1),* and *Cell Adhesion Molecule 1 Precursor* (*CADM1)* at 11q23.3 and *Cathepsin C (CTSC)* at 11q14.2 [30]. Another trans-ancestry GWAS of Europeans, South Asian, and East Asians from UAE used non-hospitalized COVID-19 patients as controls comparing them to those that were hospitalized [31]. This revealed prominent associations at *Von Willebrand Factor A Domain containing 8* (*VWA8;* 13p11.2), *Phosphodiesterase 8B* (*PDE8B*; 5q13.3), *CTSC* (11q14.2), *Thrombospondin Type 1 Domain containing 7B (THSD7B*; 2q22.1), *Serine–Threonine kinase 39* (*STK39*; 2q24.3), *F-box Protein 34* (*FBXO34*; 14q22.3), *Ribosomal Protein L6 Pseudogene 27 (RPL6P27*; 18p11.31) and *Methyltransferase 21C, AARS1 Lysine* (*METTL21C*; 13q33.3) that are expressed in the lung and associated with lung surface tension, airway obstruction, emphysema, T-cell mediated inflammation, and inflammatory cytokines [31].

Genetic determinants are known to modify the susceptibility to immune-mediated disorders and severe viral infections, such as from respiratory viruses, eg. Influenza virus A and SARS-CoV-1 [32–35]. As in case of other complex diseases, in addition to a large number of variants in regulatory elements and non-coding regions of the genome [36,37], variations in the coding regions may be involved in controlling the spectrum and severity of COVID-19 [38]. Already rare pathogenic single nucleotide variants in genes governing host immunity, such as *Toll Like Receptor 7 (TLR7), Toll Like Receptor Adaptor Molecule 1 (TICAM1), Interferon Regulatory Factor 3 (IRF3),* and *Interferon Alpha and Beta Receptor Subunit 1 (IFNAR1)* are reported in severe cases of COVID-19, including in young patients without pre-existing comorbidities [39,40]. Nevertheless, the monogenic inheritance of rare variants is unlikely to completely account for the large numbers of patients with life-threatening complications in COVID-19 worldwide.

Accordingly, we conducted a case–control based association study to query an exome dataset of 1,432,135 SNPs in 2692 COVID-19 patients of European ancestry from the GEN-COVID consortium, University of Sienna, Italy. We sought to identify common variations in the coding regions of the genome that may be associated with variable prognosis in SARS-CoV-2 infected patients. To this end we employed two strategies, first in *cohort 1* we compared non-hospitalized patients (controls; $N = 493$) to those that were hospitalized (cases; $N = 2199$) and in *cohort 2* we evaluated hospitalized COVID-19 patients on respiratory support (cases; $N = 1877$) to those not requiring the same (controls; $N = 815$).

2. Materials and Methods

2.1. Dataset

A novel whole-exome dataset of COVID-19 patients was obtained from the GEN-COVID consortium, University of Siena, Italy (Sienna_COVID). It comprised of the data from 2960 COVID-19 patients and assessed 1,432,135 SNPs. The patients were graded on a scale of 0–5 based on their hospitalization status and the requirement of respiratory support (Table 1).

Table 1. Gradation of COVID-19 patients employed in this study.

Grade	Hospitalization Status
0	Not hospitalized (NH)
1	Hospitalized without respiratory support (HWRS)
2	Hospitalized with O2 supplementation (HWOS)
3	Hospitalized with CPAP-biPAP (HWCB)
4	Hospitalized intubated (HI)
5	Dead (D)

The Sienna_COVID dataset was first merged with the genomic data of individuals from the 1000 Genomes Project (https://www.internationalgenome.org/data/, accessed on 22 April 2017). The merged dataset (Sienna_1K) comprised of 5464 individuals assessing 229,021 SNPs that were common between the two datasets. VCFtools v0.1.16 [41], and PLINK v1.90 [42], were used for all file conversions and manipulations.

Population structure within the Sienna_1K dataset was identified using Principal Component Analysis (PCA) implemented in PLINK v1.9 using—pca command. The PC1 and PC2 are plotted in RStudio v1.4.1717 (Figure 1). To control for population stratification and avoid genetic structure in the sample, we restricted our analysis to the COVID-19 patients that cluster with individuals of European ancestry from the 1000 Genomes Project [43]. Coordinates for the patients under analysis were delineated based on the clusters formed by GBR, FIN, IBS, CEU, and TSI genomes. 2692 COVID-19 patient exomes within the European cluster were selected (PC1 ranging from −0.0050 to 0.0050 and PC2 ranging from −0.0100 to 0) for downstream analysis (Figure 1 inset), and those outside this cluster were removed. Congruently, the data for 2692 COVID-19 patients of European descents were extracted from the original dataset (Sienna_COVID) and a new dataset assessing 1,432,135 SNPs was generated (European_Sienna_COVID).

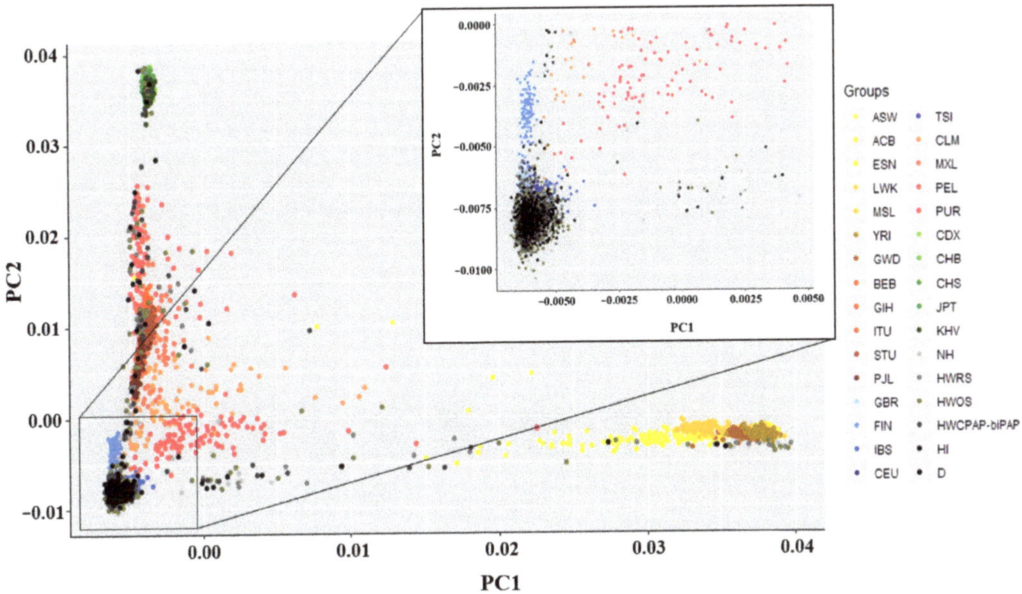

Figure 1. Principal Component Analysis (PCA) of COVID-19 patient genomes. PCA plot showing genetic differentiation among COVID-19 patient genomes. COVID-19 patients (NH, HWRS, HWOS, HWCB, HI, and D) were designated in various shades of grey and dead individuals were designated with black. African, South Asian, European, Latin American, and East Asian populations were designated with different shades of yellow, red, blue, pink, and green, respectively. We selected COVID-19 patients that cluster with European genomes (PC1 ranging from −0.0050 to 0.0050 and PC2 ranging from −0.0100 to 0) for downstream analysis. PCA was performed in PLINK v1.9. The PC1 and PC2 were plotted in RStudio v1.4.1717.

Among the 2692 patients considered, the number of individuals in each of the categories described here (Table 1) are Non-Hospitalized (NH; $N = 493$), Hospitalized Without Respiratory Support (HWRS; $N = 322$), Hospitalized Requiring Oxygen Supplementation (HWOS; $N = 882$), Hospitalized Intubated (HI; $N = 197$), Hospitalized With CPAP-biPAP patients (HWCB; $N = 611$), and Dead (D; $N = 187$).

2.2. Exome-Wide Association Studies

To identify genetic variants with significant frequency variation between hospitalized (HWRS + HWOS + HWCB + HI + D) versus non-hospitalized (NH) patients (*cohort 1*), and hospitalized patients on respiratory support (HWOS + HWCB + HI + D) versus those without it (NH + HWRS) (*cohort 2*), multiple regression-based case–control analyses were performed on the European_Sienna_COVID dataset in PLINK v1.9. Age, sex, and the first two principal components (PC1 and PC2) were employed as the covariates in the association analysis. To this end, COVID-19 patients were divided into two age groups: ≤50 years and >50 years. All covariates were included in a logistic regression model using –covar flag alongside the –logistic command. SNP-phenotype association was statistically delineated separately for all 1,432,135 SNPs to obtain Odd's Ratio and respective *p*-value for each SNP after controlling for the covariates. The p-value < 0.001 was considered statistically significant. The −log10 *p*-values of all assessed SNPs for both *cohort 1* and *cohort 2* were plotted as Manhattan plots using 'qqman' package in R v3.5.2 [44]. Significant SNPs were annotated using SNPnexus (https://www.snp-nexus.org/v4/, accessed on 15 August 2022) web-based server for GRCh38/hg38 [45].

3. Results

3.1. Exome-Wide Association Analyses

We assessed the genomes of non-hospitalized COVID-19 patients (N = 493; controls) against those hospitalized (N = 2199; cases) in *cohort 1*. And compared hospitalized COVID-19 patients on respiratory support (N = 1877; cases) with those not requiring it (N = 815; controls) in *cohort 2*. Out of 1,432,135 SNPs employed in exome-wide genetic analysis, 229 (*cohort 1*) and 111 (*cohort 2*) SNPs showed significant association with the severity of COVID-19 (p-value < 0.001) after controlling for the covariates (Figures 2A and 2B, respectively).

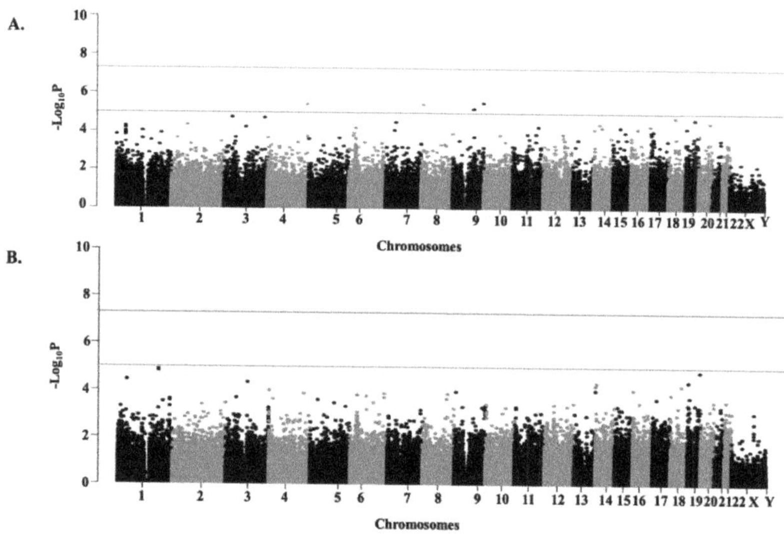

Figure 2. Manhattan Plots summarizing Exome-wide Association Study results. *X*-axis represents chromosomes (chr 1 to chr Y). SNPs present in the chromosomes are designated with dots. Negative log-transformed (−log10) covariate adjusted p-values are plotted in the *Y*-axis. The SNPs with p-value < 0.00001 are indicated with the blue line, and those with p-value < 0.0000001 are indicated with the red line. (**A**). *cohort 1*. Genomes of non-hospitalized COVID-19 patients (N = 493) were compared against the hospitalized patients (N = 2199). (**B**). *cohort 2*. Genomes of COVID-19 patients with respiratory support (N = 1877) were compared against those without respiratory support (N = 815).

The top 20 highly significant SNPs for *cohorts 1* and *2* are listed in Tables 2 and 3, respectively with their rsIDs, associated genes, Odd's Ratio (OR), and p-value (P).

SNPs that showed highly significant association (p-value < 0.001) in *cohort 1* were found to be associated with pathways (https://reactome.org/, accessed on 15 August 2022), such as infectious diseases, NOTCH signaling, immunoregulatory interactions between lymphoid and non-lymphoid cells, and extracellular matrix (ECM) organization (Figure 3A). SNPs with highly significant association (p-value < 0.001) in *cohort 2*, were linked with pathways involved in regulation of major histocompatibility complex (MHC) class II signaling, adaptive immune system modulation, carbohydrate metabolism, and RUNX transcription factor mediated regulation (Figure 3B).

Among the SNPs that varied significantly between cases and controls, 11 were found to be common between *cohorts 1* and *2* (Table 4).

Overall, we identified variants in 26 genes with strong links to COVID-19 severity via case–control or cellular studies (Table 5). In addition, we report 12 novel candidates with molecular function and supporting evidence highly suggestive of their contribution to COVID-19 pathology and post recovery sequelae.

Table 2. Top 20 highly significant SNPs identified in *cohort 1*.

SNP	rsID	Gene	OR	*P*
chr9:136451395:G:A	rs45519739	*SEC16A*	0.3082	2.99×10^{-6}
chr4:182680666:G:C	rs147269509	*TENM3*	0.2923	3.78×10^{-6}
chr9:91737515:A:G	rs16907720	*ROR2*	0.5227	6.16×10^{-6}
chr18:31542836:G:A	rs2278792	*DSG2*	1.623	1.78×10^{-5}
chr3:37481588:G:A	rs17227748	*ITGA9*	1.631	1.80×10^{-5}
chr3:186620636:A:C	rs1071592	*AHSG*	0.6521	1.91×10^{-5}
chr19:41426222:G:A	rs137913069	*B3GNT8*	0.2199	2.19×10^{-5}
chr7:44966456:ACAGCCCT:A	rs563691323	*MYO1G*	0.08036	3.14×10^{-5}
chr20:52165170:T:C	rs6126487	*ZFP64*	0.4896	3.24×10^{-5}
chr14:45247163:C:T	rs149551504	*MIS18BP1*	0.1085	3.92×10^{-5}
chr2:71434819:G:A	rs1398	*ZNF638*	0.1708	4.49×10^{-5}
chr16:89508537:G:A	rs187330648	*SPG7*	0.3642	5.00×10^{-5}
chr11:112961669:A:AT	rs782430131	*NCAM1*	0.3525	5.41×10^{-5}
chr1:43198448:A:G	rs603560	*CFAP57*	0.7111	5.45×10^{-5}
chr14:55351384:G:A	rs139920556	*FBXO34*	0.03484	5.69×10^{-5}
chr3:99830513:C:T	rs115071595	*FILIP1L*	0.2011	5.75×10^{-5}
chr15:52789603:G:C	rs61735385	*ONECUT1*	0.5803	5.77×10^{-5}
chr19:4497181:C:T	rs146793578	*HDGFL2*	0.4148	6.17×10^{-5}
chr6:32666522:A:ACC	rs749944694	*HLA-DQB1*	0.599	6.22×10^{-5}
chr1:43198547:C:T	rs603123	*CFAP57*	0.7139	6.64×10^{-5}

Table 3. Top 20 highly significant SNPs identified in *cohort 2*.

SNP	rsID	Gene	OR	*P*
chr1:184795690:C:T	rs35704242	*NIBAN1*	2.164	1.32×10^{-5}
chr1:184795547:G:A	rs35545276	*NIBAN1*	2.153	1.50×10^{-5}
chr19:55602871:G:A	rs310459	*ZNF524*	1.645	1.54×10^{-5}
chr1:40819499:C:A	rs33932028	*KCNQ4*	1.506	3.62×10^{-5}
chr19:2279020:A:T	-	*PEAK3*	2.914	4.11×10^{-5}
chr3:99830513:C:T	rs115071595	*CMSS1/FILIP1L*	0.2032	4.67×10^{-5}
chr14:23526719:C:T	rs45503996	*ZFHX2*	0.06768	4.72×10^{-5}
chr14:18967593:G:A	rs199622050	*POTEM*	0.4634	6.04×10^{-5}
chr18:50819654:G:T	rs3813089	*MRO*	0.7318	6.13×10^{-5}
chr16:1443487:G:C	rs112232284	*CCDC154*	3.285	9.29×10^{-5}
chr13:113828592:C:G	rs8191975	*GAS6*	0.6265	9.58×10^{-5}
chr4:987108:C:A	rs11248061	*SLC26A1/IDUA*	0.7571	1.01×10^{-4}
chr9:5892525:C:T	rs2233178	*MLANA*	1.725	1.10×10^{-4}
chr8:112313866:T:C	rs4308763	*CSMD3*	2.306	1.34×10^{-4}
chr6:155454846:G:T	rs12195525	*NOX3*	1.446	1.35×10^{-4}
chr4:159355878:G:GCCCCCCC	-	*RAPGEF2*	2.985	1.36×10^{-4}
chr18:3450457:C:A	rs238132	*TGIF1*	1.44	1.48×10^{-4}
chr6:32042795:G:T	rs7742632	*TNXB*	0.4841	1.57×10^{-4}
chr6:73111290:T:A	rs45536838	*KCNQ5*	0.3815	1.74×10^{-4}
chr19:8094455:G:C	rs34167077	*FBN3*	0.6772	1.96×10^{-4}

A.

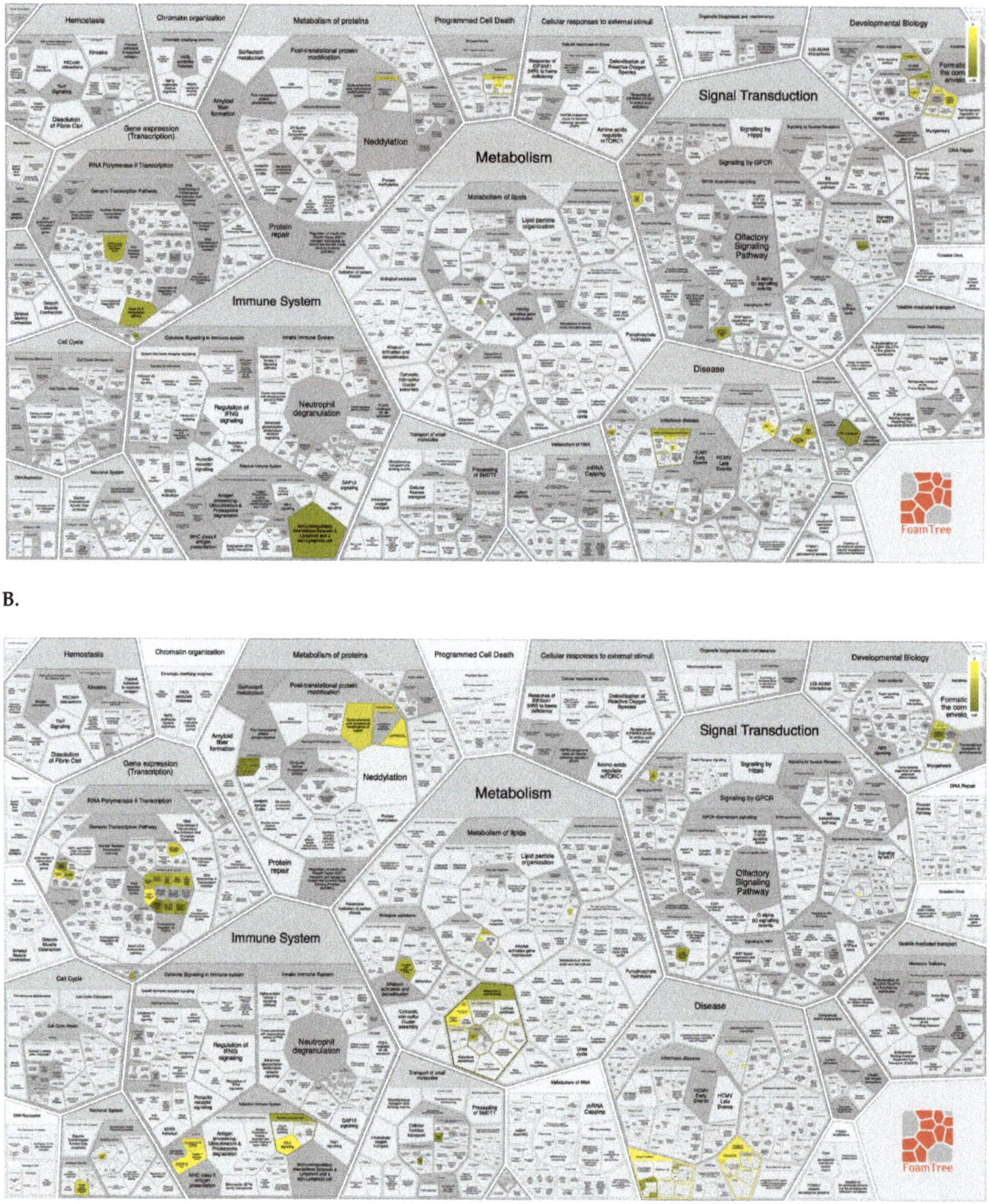

B.

Figure 3. FoamTree representing various Reactome pathways associated with the significant SNPs identified in (**A**). *cohort 1* and (**B**). *cohort 2*. Pathway map was generated using SNPnexus web-based server. Pathways associated with the submitted dataset are highlighted in various shades of yellow. The gray entities represent the pathways that are represented in the query dataset but absent in the submitted dataset.

Table 4. The SNPs identified in both *cohort 1* and *cohort 2*.

SNP	rsID	Gene	SNP Association	GO Annotation
chr20:62273292:C:T	rs2236526	*OSBPL2*	-	Cholesterol binding
chr16:1443487:G:C	rs112232284	*CCDC154*	-	Bone mineralization/Odontogenesis
chr14:104587865:G:A	rs144894622	*C14orf180*	-	Enables protein binding
chr19:1000718:C:A	rs139242998	*GRIN3B*	-	Calcium channel activity and ionotropic glutamate receptor activity
chr15:52789603:G:C	rs61735385	*ONECUT1*	-	DNA-binding transcription factor activity and RNA polymerase II-specific
chr8:107996952:G:A	rs716149	*RSPO2*	-	Signaling receptor binding and G protein-coupled receptor binding
chr8:112313866:T:C	rs4308763	*CSMD3*	-	ECM organization
chr4:17693320:A:G	rs3733579	*FAM184B*	-	-
chr20:44413724:C:T	rs1800961	*HNF4A*	-	DNA-binding transcription factor activity and sequence-specific DNA binding
chr6:73111290:T:A	rs45536838	*KCNQ5*	HDL cholesterol levels	Ion channel activity and delayed rectifier potassium channel activity
chr3:99830513:C:T	rs115071595	*CMSS1/FILIP1L*	-	Nucleic acid binding and RNA binding

Table 5. Genes identified in *cohorts 1* and *2* in this study that belong to various categories of cellular function and have been previously linked to the variation in severity of COVID-19. PubMed IDs (PMIDs) of these articles are also mentioned.

Molecular Pathway/Related Organelle	Genes	*Cohort 1*	*Cohort 2*	PMIDs
Class II HLA	*HLA-DQB1*	Yes	-	33972731
	HLA-DRB1	-	Yes	33972731
	HLA-DOB	Yes	-	34650566
Immunoglobulins	*KIR2DL1*	Yes	-	32612152
Complement cascade	*MASP2*	Yes	-	35511137
Inflammation modulatory	*AHSG*	Yes	-	34139154
	KLF6	-	Yes	33138195
	FUT4	-	Yes	33441124, 32810438
	GAS6	-	Yes	33810394
Lipid mediator pathway	*ALOX5AP*	Yes	-	34366882
	SMPD1	Yes	-	34987511
Mitochondria	*ATAD3A*	Yes	-	35233572
	SPG7	Yes		35005527
Endoplasmic reticulum	*TMCC2*	-	Yes	35368071
Cell surface glycans	*B3GNT8*	Yes	-	35737812
Extracellular matrix and cell adhesion	*CNTN2*	Yes	-	35368071
	MUC5AC	Yes	-	32776556
	MUC4	-	Yes	34448730
	ITGAL	-	Yes	32591762
	SPEG	-	Yes	33536081
Cell signaling	*GPRC5C*	Yes	-	35036860
	MAP2K7	Yes	-	32307550
	RIPK3	Yes	-	32753065
Ubiquitin-Proteasome system	*FBXO34*	Yes	-	34775353
	RNF213	Yes	-	34721400
	UBQLNL	-	Yes	34335605

3.2. Pharmacogenomic Annotation of Significant SNPs

SNPs that showed pronounced variation in our exome-wide association analysis were further annotated for their potential impact on drug responses using pharmacogenomic data obtained and curated from PharmGKB [46]. We found two notable SNPs, one on each cohort. According to PharmGKB, rs2228130 is associated with toxicity of gemcitabine, a chemotherapy drug and rs492602 is associated with alcoholism.

4. Discussion

Multiple large-scale case–control GWASs have investigated the genetic basis of the disparity in clinical severity in SARS-CoV-2 infected patients using unaffected individuals, population controls [20,26,27], or asymptomatic and mildly affected individuals as controls [29–31]. The exome or protein-coding region is a conserved component of the genome that may harbor potentially damaging variants, which may modulate the clinical phenotype in SARS-CoV-2 infection. Consistent with this, rare pathogenic variants in *TLR7* and eight candidate loci, including, *TLR3*, *IRF3*, and *IRF7* governing type I IFN response have been identified in patients with critical COVID-19 [39,40]. Further a meta-analysis of worldwide exome and genome data implicated rare disease-causing variants in *TLR7* in exacerbating the COVID-19 clinical course [47]. In contrast, largescale exome-wide association studies mostly using population controls did not find rare deleterious exonic variations to be significantly enriched in patients with life-threatening COVID-19 [48,49]. Given that COVID-19 is a genetically complex disorder with polygenic risk inheritance, we surmise that rare pathogenic exonic variants with large effect sizes are unlikely to completely explain the unfavorable disease course noted in large numbers of afflicted patients worldwide. Accordingly, we used an exome-wide association study to identify common genetic variants that are enriched in acutely affected COVID-19 patients of European ancestry. First in *cohort 1*, we compared non-hospitalized patients to those hospitalized and second, in *cohort 2* we evaluated hospitalized COVID-19 patients supported by ventilation to those not needing the same.

Our results included SNPs in the genes *FBXO34*, *Contactin 2* (*CNTN2*), and *Transmembrane And Coiled-Coil Domain Family 2* (*TMCC2*) which have been previously linked with COVID-19 severity [31,50]. *FBXO34* showed significant association with the acuteness of SARS-CoV-2 infection primarily in the European ancestry [31]. While its biological functions are not well understood, F-box proteins at large are a component of the Skip1-Cullin 1-F-box (SCF) E3 ubiquitin ligase complex that participate in proteasome-mediated protein turnover, which are manipulated during infection by many viruses including human immunodeficiency virus (HIV) [51]. Moreover, using GWAS, *FBXO34* has also been linked with plasma protein levels in cardiovascular disease risk among Europeans, as well as with blood cell count [52–55]. *CNTN2* and *TMCC2* occur in the same genomic region that is significantly associated with risk of poor COVID-19 outcomes in individuals with high European ancestry in Brazil [50].

Specifically, in *cohort 1* we identified SNPs in the following genes, *Killer Cell Immunoglobulin like Receptor, Two Ig Domains and Long Cytoplasmic Tail 1* (*KIR2DL1*) that is an inhibitory receptor upregulated in COVID-19 patients with acute respiratory distress reflecting reduced antiviral activity of natural killer cells in them [56]. *MBL Associated Serine Protease 2* (*MASP2*) that promotes complement cascade activation; variants in *MASP2* resulting in its reduced expression are noted in asymptomatic elderly COVID-19 patients [57]. *Alpha-2 HS glycoprotein* (*AHSG)* that modulates inflammation via attenuating macrophage activation and neutrophil degranulation and is significantly downregulated in severe COVID-19 [58]. The mitochondrial *ATPase family AAA domain containing 3A* (*ATAD3A*) that was upregulated in lymph nodes from COVID-19 autopsy cases [59]; pathogenic variants in *ATAD3A* have also been linked to type I interferonopathy [60]. *β-1,3-N-acetyl-glucosaminyltransferase 8* (*B3GNT8*) that encodes for a glycosyltransferase responsible for anchor point creation in poly-N-acetyl-lactosamine, glycan extensions that could modulate SARS-CoV-2 cellular invasion. Rare variants in *B3GNT8* are linked with milder COVID-19 presentations [61].

Arachidonate 5-Lipoxygenase Activating protein (*ALOX5AP*) encodes for 5-LOX activating protein (FLAP), an activating cofactor for the lipid mediator, Arachidonate 5-lipoxygenase (ALOX5) that generates leukotriene B_4 (LTB_4) associated with poor respiratory pathologies such as pneumonia, ARDS, and severe lung injury [62,63]. Congruently, increased *ALOX5* activity and *ALOX5AP* expression are observed in bronchoalveolar lavage (BAL) neutrophils in critical COVID-19 patients [64]. Increased expression of *ALOX5AP*, *ALOX5*, and plasma LTB_4 are also noted in diabetic COVID-19 cases requiring intensive care [65]. *G Protein-Coupled Receptor Class C Group 5 Member C* (*GPRC5C*) whose structural variants have been associated with poor prognosis in COVID-19 [66]. *Mitogen-activated Protein Kinase 7* (*MAP2K7*) encodes for an augmenter of the c-Jun kinase pathway during T-cell activation and is elevated in severe COVID-19 [67–69]. *Mucin 5 AC* (*MUC5AC*) that encodes for a gel forming secreted glycoprotein, which is upregulated in the airway mucus of seriously ill SARS-CoV-2 infected patients [70]. *Receptor-interacting kinase 3* (*RIPK3*), a serine–threonine kinase implicated in non-caspase dependent apoptosis termed as necroptosis that leads to ARDS after trauma and sepsis [71]; increased serum levels of RIPK3 are reported in morbid cases of COVID-19 [72]. *Ring Finger Protein 213* (*RNF213*), an E3 ubiquitin ligase and component of the proteasomal degradation machinery that is suppressed in monocytes from severe COVID-19 patients [73]. *Sphingomyelin phosphodiesterase 1* (*SMPD1*) that encodes for a sphinogomyelinase, which catalyzes the synthesis of ceramide, a bioactive lipid mediator involved in response to cellular damage. Upregulation of SMPD1 activity and ceramide have been noted in COVID-19 patients requiring intensive care [74]. *Spastic Paraplegia 7* (*SPG7*) shows increased expression upon SARS-CoV-2 infection [75]. It encodes for a component of the mitochondrial permeability transition pore (mPTP) that is strongly elevated in critical COVID-19 patients and correlated with increased levels of cardiac injury markers in them [76]. Further SARS-CoV-2 proteins localize to the mitochondria, and directly interact with SPG7 to disrupt mitochondrial morphology, energetics, and function in cardiomyocytes [76].

In *cohort 2* our results included SNPs in the following genes, *Kruppel Like Factor 6* (*KLF6*) that encodes for an inflammation regulator with high expression in the macrophages in BAL from grievously ill COVID-19 patients [77]. *Ubiquilin like* (*UBQLNL*) that encodes for a regulator of proteostasis and showed enhanced expression in seriously ill COVID-19 patients [78]. *Fucosyltransferase 4* (*FUT4*), a marker for premature/immature neutrophils that showed elevated expression in severe SARS-CoV-2 infection and is likely associated with poor outcomes in sepsis [79,80]. *Growth-Arrest Specific 6* (*GAS6*) that encodes for a ligand functional in the restorative program initiated to counterbalance pro-inflammatory immune response [81]. GAS6 levels were found to be elevated following SARS-CoV-2 infection and increased with exacerbation of severity [82]. Notably patients with high levels of GAS6 at initial stages were found to have the worst disease prognosis [82]. Interestingly, *GAS6-AS1* encodes for a long non-coding RNA that is downregulated in SARS-CoV-2 infection in vitro and likely interacts with immune modulatory and SARS-CoV-2 interacting proteins such as Adenosine Deaminase RNA Specific (ADAR) and A-Kinase Anchoring Protein 8 Like (AKAP8L) [83]. *Serpin Family A Member 12* (*SERPINA12*) that is involved in the inhibition of kallikrein-dependent inflammation and is downregulated in critical cases of COVID-19 promoting uncontrolled inflammation and worsening of disease outcome [84,85]. *Integrin subunit alpha L* (*ITGAL*), an integrin involved in monocyte migration across endothelium in anti-viral immune response was strongly activated in non-resident macrophages in severe COVID-19 [86]. *Striated Muscle Enriched Protein Kinase* (*SPEG*) that is associated with cardiomyopathy and COVID-19 mortality [87,88]. *Mucin 4* (*MUC4*) that is downregulated in the blood of critically ill COVID-19 patients [89].

The MHC class I and II performs pivotal roles in the host adaptive immunity by modulating the antigen presentation on the cell surface for T-cell recognition [90]. Genetic variation in *human leukocyte antigens* (*HLA*) at the MHC loci modifies the immune response to viral infections, including those caused by SARS-CoV-1 [91], influenza [92], and Middle East respiratory syndrome (MERS) [93]. We identified SNPs in Class II HLA genes, *HLA-*

DRB1, and HLA-DQB1 previously noted to be strongly repressed in a dominant population of dendritic cells [94], in acute cases of COVID-19 [95], in *cohort 1* and *2*, respectively. Reduced allele frequency of *HLA-DRB1* has also been observed in severe COVID-19 [96]. In addition, our results in *cohort 1* included *HLA-DOB*, involved in antigen processing and loading. Its alleles are overrepresented in symptomatic SARS-CoV-2 infected females in a Brazilian cohort [97].

Other findings of interest included variants in novel candidate genes, so far not demonstrated to be directly associated with variability in COVID-19 presentation but with strong accessory evidence for the same. These include *Lysine Acetyltransferase 2B* (*KAT2B*), an epigenetic regulator of TGFβ signaling that is a key pathway in cardiovascular development and disease [98]. Interestingly, SARS-CoV-2 infection is known to trigger aberrant TGFβ signaling that, in turn, mounts a chronic and sustained immune response likely exacerbating disease prognosis [99]. While the molecular mechanisms of KAT2B-dependent regulation of the TGFβ pathway in COVID-19 remain to be understood we surmise host variants in *KAT2B* may modify disease progression and severity following SARS-CoV-2 infection. *One Cut Homeobox 1* (*ONECUT1*) encodes for a transcription factor enriched in the liver and variants in it are associated with different forms of diabetes that is a known risk factor for severe outcomes in COVID-19 [100,101]. *Misshapen (Msn)/NIK related kinase 1* (*MINK1*) encodes for a component of the MAP kinase cascade that is involved in the phosphorylation and priming of the nucleotide-binding domain, leucine-rich-repeat containing family, pyrin-domain containing 3 (NLRP3) inflammasome [102]. Notably NLRP3 activation is central to inflammation and the pathogenesis of ARDS in severe COVID-19 [103]. *Spectrin alpha, Erythrocytic 1* (*SPTA1*) encodes for a structural protein in red blood cells (RBCs). In SARS-CoV-2 infected individuals increased levels and oxidation of SPTA1 peptides are noted reflecting structural aberration of RBCs [104], which may potentially contribute to the severity of hypoxemia, thromboembolism, and coagulation defects noted in the manifestation of COVID-19. *Transporter 2, ATP binding cassette Family B Member* (*TAP2*) encodes for a component of the immunoproteasome that replaces the proteasome in haematopoietic cells, as part of the first line of defense against pathogens following pro-inflammatory cytokine IFN-γ stimulation [105]. Furthermore, high viral load in SARS-CoV-2 infection activates the expression of *TAP2* and other immunoproteasome components in lungs of COVID-19 patients and may result in worsening of disease prognosis [106]. *Unc-93 Homolog B1, TLR Signaling Regulator* (*UNC93B1*) modulates activation of human plasmatoid predendritic cells, which play an important role in SARS-CoV-2 induced immune response [107]. A detrimental clinical course in COVID-19 pathology has been linked with an outburst of pro-inflammatory processes or a 'cytokine storm' [108]. Consistent with this we identified variants in *TGFβ Induced Factor Homeobox 1* (*TGIF1*), related to IFN signaling that is strongly upregulated following SARS-CoV-2 infection [109]. We also found variants in *Colony subunit factor 2 Receptor Subunit β* (*CSF2RB*) that encodes the common subunit of the receptor for IL-3, IL-5, and granulocyte-monocyte stimulating factor (GM-CSF). Increased levels of various cytokines including GM-CSF occur in later stages of SARS-CoV-2 infection producing a self-amplifying cytokine loop that leads to ARDS and mortality [110]. *Hepatocyte Nuclear Factor 4 alpha* (*HNF4A*) that encodes for an intestinal transcriptional regulator which promotes expression of *Angiotensin 1 Converting Enzyme 2* (*ACE2*) and suppresses *Transmembrane Serine Protease 2* (*TMPRSS2*), which are SARS-CoV-2 receptor and involved in viral protein priming, respectively, in the intestine [111]. Gastrointestinal symptoms of varying severity are noted in subsets of COVID-19 patients [112], and may be associated with variation in *HNF4A* in them.

Our analysis in *cohorts 1* and *2* identified SNPs in 17 and 9 genes, respectively, that are already associated with variability in SARS-CoV-2 infection severity (Table 5). We surmise that this is because *cohort 2* interrogates the entire pool of hospitalized patients distinguished only by the requirement of respiratory support, which has been used as a surrogate for more severe disease in the present study and may be regulated by variations at fewer genetic loci. In contrast, *cohort 1* compares individuals who were critically ill necessitating hospital care versus milder, non-hospitalized patients, which are likely distinguished by larger genetic differences modifying the disease severity. Polymorphisms in *AHSG, B3GNT8, DSG2, CFAP57, HLA-DQB1, ONECUT1,* and *SPG7* were among the top 20 SNPs delineated in *cohort 1*. In *cohort 2* variants in *TGIF1* and *GAS6* were among the most prominent 20 SNPs identified. Notable candidate genes common to both cohorts included *ONECUT1* and *HNF4A*.

Further, we also identified SNPs in genes, such as *FK506 binding protein* (*FKBP6*) that is a component of the synaptonemal complex with an indispensable role in meiotic chromosome pairing and fertility in males [113]. *Dipeptidase 3* (*DPEP3*) that encodes a membrane-bound protein in testicular germ cells, which may be important for testicular function and is downregulated in convalescent male COVID-19 patients, likely contributing to compromised fertility in them [114]. *Desmoglein 2* (*DSG2*) that occurs at the intercalated discs coupling adjacent cardiomyocytes and is implicated in arrhythmogenic cardiomyopathy [115]. High levels of circulating anti-DSG2 autoantibodies have been reported in recuperating COVID-19 male subjects, compared to healthy individuals, as well as a subgroup showed elevated levels compared to those in arrhythmogenic right ventricular cardiomyopathy [116]. These data further suggest that host genetic variability not only modifies the course of COVID-19 severity but may modulate features such as cardiovascular disease and male infertility that are also noted in post-COVID-19 sequelae in some patients [117,118].

Finally, we note the absence of information on the SARS-CoV-2 strains for the enrolled patients as a limitation of this approach. Different SARS-CoV-2 genetic strains are associated with distinct clinical outcomes [119]. Since 'hospitalization' has been used as a proxy for severe COVID-19 in this study, variance in hospitalization rates due to different COVID-19 strains can potentially obfuscate results obtained here.

Together with existing literature, our results improve the current understanding of genetic factors modulating the spectrum and gravity of the COVID-19 phenotype in individuals of European descent. The novel findings from this work warrant further validatory analysis and studies to facilitate a comprehensive molecular understanding of the COVID-19 pathology. Together the notable candidates from this work may be useful as biomarkers to inform on adverse prognosis and may facilitate the development and deployment of effective therapies, among COVID-19 patients of European ancestry.

5. Conclusions

Using exome-wide genetic analysis with data from individuals of European descent we identified common variants in candidate genes that are significantly associated with the severity of COVID-19. Our findings include *FBXO34, CNTN2,* and *TMCC2* discovered earlier in other case–control analyses studies. Overall we report SNPs in 26 genes with existing molecular links and 12 novel candidates with biological function or strong supporting evidences suggestive of a high possibility of involvement in modifying the COVID-19 clinical phenotype. These results not only broaden the current knowledge of molecular mechanisms underlying the COVID-19 pathophysiology but may be utilized in delineating a battery of biomarkers predictive of disease outlook in individuals with European ancestry.

Author Contributions: Conceptualization, R.D. and P.U.; methodology, P.U., P.U.S. and B.B.; software, P.U.S. and B.B.; validation, P.U., P.U.S., B.B., R.D., M.F.A., S.M. and I.M.; formal analysis, P.U., P.U.S. and B.B.; investigation, P.U., P.U.S., B.B., R.D., M.F.A., S.M. and I.M.; resources, P.U., R.D., M.F.A., S.M. and I.M.; data curation, P.U., P.U.S. and B.B.; writing—original draft preparation, P.U.; writing—review and editing, P.U., R.D., M.F.A., S.M. and I.M.; visualization, R.D. and B.B.; supervision, R.D.; project administration, R.D., M.F.A., S.M. and I.M.; funding acquisition, M.F.A., S.M. and I.M. All authors have read and agreed to the published version of the manuscript.

Funding: This research received no external funding, and The APC was funded by Researchers Supporting Project, Grant number RSP2022R436, King Saud University, Riyadh, Saudi Arabia.

Institutional Review Board Statement: Not applicable.

Informed Consent Statement: Not applicable.

Data Availability Statement: The data employed in this study can be obtained upon request from the GEN-COVID consortium, University of Siena, Italy.

Acknowledgments: We thank the GEN-COVID consortium, University of Siena, Italy for kind donation of the COVID-19 exome data. The authors also thank the Researchers Supporting Project, King Saud University, Riyadh, Saudi Arabia for the APC support.

Conflicts of Interest: The authors declare no conflict of interest.

References

1. Wiersinga, W.J.; Rhodes, A.; Cheng, A.C.; Peacock, S.J.; Prescott, H.C. Pathophysiology, Transmission, Diagnosis, and Treatment of Coronavirus Disease 2019 (COVID-19): A Review. *JAMA* **2020**, *324*, 782–793. [CrossRef] [PubMed]
2. Terpos, E.; Ntanasis-Stathopoulos, I.; Elalamy, I.; Kastritis, E.; Sergentanis, T.N.; Politou, M.; Psaltopoulou, T.; Gerotziafas, G.; Dimopoulos, M.A. Hematological findings and complications of COVID-19. *Am. J. Hematol.* **2020**, *95*, 834–847. [CrossRef] [PubMed]
3. Mehta, P.; McAuley, D.F.; Brown, M.; Sanchez, E.; Tattersall, R.S.; Manson, J.J.; Hlh Across Speciality Collaboration, U.K. COVID-19: Consider cytokine storm syndromes and immunosuppression. *Lancet* **2020**, *395*, 1033–1034. [CrossRef]
4. Mahdizade Ari, M.; Mohamadi, M.H.; Shadab Mehr, N.; Abbasimoghaddam, S.; Shekartabar, A.; Heidary, M.; Khoshnood, S. Neurological manifestations in patients with COVID-19: A systematic review and meta-analysis. *J. Clin. Lab. Anal.* **2022**, *36*, e24403. [CrossRef] [PubMed]
5. Driggin, E.; Madhavan, M.V.; Bikdeli, B.; Chuich, T.; Laracy, J.; Biondi-Zoccai, G.; Brown, T.S.; Der Nigoghossian, C.; Zidar, D.A.; Haythe, J.; et al. Cardiovascular Considerations for Patients, Health Care Workers, and Health Systems During the COVID-19 Pandemic. *J. Am. Coll. Cardiol.* **2020**, *75*, 2352–2371. [CrossRef] [PubMed]
6. Armaly, Z.; Kinaneh, S.; Skorecki, K. Renal Manifestations of COVID-19: Physiology and Pathophysiology. *J. Clin. Med.* **2021**, *10*, 1216. [CrossRef] [PubMed]
7. Aljaroudi, M.E.; Alghamdi, S.K.; Al Salman, B.A.; Alabdulghani, M.J. Liver and Gastrointestinal Involvement in Patients With COVID-19: A Retrospective Study. *Cureus* **2022**, *14*, e24580. [CrossRef]
8. Chen, R.; Lan, Z.; Ye, J.; Pang, L.; Liu, Y.; Wu, W.; Qin, X.; Guo, Y.; Zhang, P. Cytokine Storm: The Primary Determinant for the Pathophysiological Evolution of COVID-19 Deterioration. *Front. Immunol.* **2021**, *12*, 589095. [CrossRef]
9. Pijls, B.G.; Jolani, S.; Atherley, A.; Derckx, R.T.; Dijkstra, J.I.R.; Franssen, G.H.L.; Hendriks, S.; Richters, A.; Venemans-Jellema, A.; Zalpuri, S.; et al. Demographic risk factors for COVID-19 infection, severity, ICU admission and death: A meta-analysis of 59 studies. *BMJ Open* **2021**, *11*, e044640. [CrossRef]
10. Richardson, S.; Hirsch, J.S.; Narasimhan, M.; Crawford, J.M.; McGinn, T.; Davidson, K.W.; the Northwell COVID-19 Research Consortium; Barnaby, D.P.; Becker, L.B.; Chelico, J.D.; et al. Presenting Characteristics, Comorbidities, and Outcomes Among 5700 Patients Hospitalized With COVID-19 in the New York City Area. *JAMA* **2020**, *323*, 2052–2059. [CrossRef]
11. Du, Y.; Zhou, N.; Zha, W.; Lv, Y. Hypertension is a clinically important risk factor for critical illness and mortality in COVID-19: A meta-analysis. *Nutr. Metab. Cardiovasc. Dis.* **2021**, *31*, 745–755. [CrossRef] [PubMed]
12. McMichael, T.M.; Currie, D.W.; Clark, S.; Pogosjans, S.; Kay, M.; Schwartz, N.G.; Lewis, J.; Baer, A.; Kawakami, V.; Lukoff, M.D.; et al. Epidemiology of COVID-19 in a Long-Term Care Facility in King County, Washington. *N. Engl. J. Med.* **2020**, *382*, 2005–2011. [CrossRef] [PubMed]
13. Garg, S.; Kim, L.; Whitaker, M.; O'Halloran, A.; Cummings, C.; Holstein, R.; Prill, M.; Chai, S.J.; Kirley, P.D.; Alden, N.B.; et al. Hospitalization Rates and Characteristics of Patients Hospitalized with Laboratory-Confirmed Coronavirus Disease 2019-COVID-NET, 14 States, 1–30 March 2020. *MMWR Morb. Mortal. Wkly. Rep.* **2020**, *69*, 458–464. [CrossRef] [PubMed]
14. Niedzwiedz, C.L.; O'Donnell, C.A.; Jani, B.D.; Demou, E.; Ho, F.K.; Celis-Morales, C.; Nicholl, B.I.; Mair, F.S.; Welsh, P.; Sattar, N.; et al. Ethnic and socioeconomic differences in SARS-CoV-2 infection: Prospective cohort study using UK Biobank. *BMC Med.* **2020**, *18*, 160. [CrossRef] [PubMed]
15. Mathur, R.; Rentsch, C.T.; Morton, C.E.; Hulme, W.J.; Schultze, A.; MacKenna, B.; Eggo, R.M.; Bhaskaran, K.; Wong, A.Y.S.; Williamson, E.J.; et al. Ethnic differences in SARS-CoV-2 infection and COVID-19-related hospitalisation, intensive care unit admission, and death in 17 million adults in England: An observational cohort study using the OpenSAFELY platform. *Lancet* **2021**, *397*, 1711–1724. [CrossRef]
16. Muhsen, K.; Na'aminh, W.; Lapidot, Y.; Goren, S.; Amir, Y.; Perlman, S.; Green, M.S.; Chodick, G.; Cohen, D. A nationwide analysis of population group differences in the COVID-19 epidemic in Israel, February 2020-February 2021. *Lancet Reg. Health Eur* **2021**, *7*, 100130. [CrossRef]
17. Badwe, R.A.; Dikshit, R.; Chaturvedi, P.; Gupta, S. Geographical & seasonal variation in COVID-19 related mortality. *Indian J. Med. Res.* **2020**, *152*, 6–8. [CrossRef]
18. Glenet, M.; Lebreil, A.L.; Heng, L.; N'Guyen, Y.; Meyer, I.; Andreoletti, L. Asymptomatic COVID-19 Adult Outpatients identified as Significant Viable SARS-CoV-2 Shedders. *Sci. Rep.* **2021**, *11*, 20615. [CrossRef]

19. Wang, P.; Sha, J.; Meng, M.; Wang, C.; Yao, Q.; Zhang, Z.; Sun, W.; Wang, X.; Qie, G.; Bai, X.; et al. Risk factors for severe COVID-19 in middle-aged patients without comorbidities: A multicentre retrospective study. *J. Transl. Med.* **2020**, *18*, 461. [CrossRef]
20. Severe Covid, G.G.; Ellinghaus, D.; Degenhardt, F.; Bujanda, L.; Buti, M.; Albillos, A.; Invernizzi, P.; Fernandez, J.; Prati, D.; Baselli, G.; et al. Genomewide Association Study of Severe COVID-19 with Respiratory Failure. *N. Engl. J. Med.* **2020**, *383*, 1522–1534. [CrossRef]
21. Shelton, J.F.; Shastri, A.J.; Ye, C.; Weldon, C.H.; Filshtein-Sonmez, T.; Coker, D.; Symons, A.; Esparza-Gordillo, J.; The 23andMe COVID-19 Team; Aslibekyan, S.; et al. Trans-ancestry analysis reveals genetic and nongenetic associations with COVID-19 susceptibility and severity. *Nat. Genet.* **2021**, *53*, 801–808. [CrossRef] [PubMed]
22. Ivashkiv, L.B.; Donlin, L.T. Regulation of type I interferon responses. *Nat. Rev. Immunol.* **2014**, *14*, 36–49. [CrossRef] [PubMed]
23. Murira, A.; Lamarre, A. Type-I Interferon Responses: From Friend to Foe in the Battle against Chronic Viral Infection. *Front Immunol.* **2016**, *7*, 609. [CrossRef] [PubMed]
24. Velavan, T.P.; Pallerla, S.R.; Ruter, J.; Augustin, Y.; Kremsner, P.G.; Krishna, S.; Meyer, C.G. Host genetic factors determining COVID-19 susceptibility and severity. *EBioMedicine* **2021**, *72*, 103629. [CrossRef] [PubMed]
25. Geiss-Friedlander, R.; Parmentier, N.; Moller, U.; Urlaub, H.; Van den Eynde, B.J.; Melchior, F. The cytoplasmic peptidase DPP9 is rate-limiting for degradation of proline-containing peptides. *J. Biol. Chem.* **2009**, *284*, 27211–27219. [CrossRef] [PubMed]
26. Pairo-Castineira, E.; Clohisey, S.; Klaric, L.; Bretherick, A.D.; Rawlik, K.; Pasko, D.; Walker, S.; Parkinson, N.; Fourman, M.H.; Russell, C.D.; et al. Genetic mechanisms of critical illness in COVID-19. *Nature* **2021**, *591*, 92–98. [CrossRef]
27. Initiative, C.-H.G. Mapping the human genetic architecture of COVID-19. *Nature* **2021**, *600*, 472–477. [CrossRef]
28. Roberts, G.H.L.; Park, D.S.; Coignet, M.V.; McCurdy, S.R.; Knight, S.C.; Partha, R.; Rhead, B.; Zhang, M.; Berkowitz, N.; AncestryDNA Science Team; et al. AncestryDNA COVID-19 Host Genetic Study Identifies Three Novel Loci. *medRxiv* **2020**, *2020*, 2010.2006.20205864. [CrossRef]
29. Upadhyai, P.; Suresh, G.; Parit, R.; Das, R. Genomic and Ancestral Variation Underlies the Severity of COVID-19 Clinical Manifestation in Individuals of European Descent. *Life* **2021**, *11*, 921. [CrossRef]
30. Li, Y.; Ke, Y.; Xia, X.; Wang, Y.; Cheng, F.; Liu, X.; Jin, X.; Li, B.; Xie, C.; Liu, S.; et al. Genome-wide association study of COVID-19 severity among the Chinese population. *Cell Discov.* **2021**, *7*, 76. [CrossRef]
31. Mousa, M.; Vurivi, H.; Kannout, H.; Uddin, M.; Alkaabi, N.; Mahboub, B.; Tay, G.K.; Alsafar, H.S.; Partnership, U.C.-C. Genome-wide association study of hospitalized COVID-19 patients in the United Arab Emirates. *EBioMedicine* **2021**, *74*, 103695. [CrossRef] [PubMed]
32. Matzaraki, V.; Kumar, V.; Wijmenga, C.; Zhernakova, A. The MHC locus and genetic susceptibility to autoimmune and infectious diseases. *Genome Biol.* **2017**, *18*, 76. [CrossRef] [PubMed]
33. Hsieh, Y.H.; Chen, C.W.; Schmitz, S.F.; King, C.C.; Chen, W.J.; Wu, Y.C.; Ho, M.S. Candidate genes associated with susceptibility for SARS-coronavirus. *Bull. Math Biol.* **2010**, *72*, 122–132. [CrossRef] [PubMed]
34. Clohisey, S.; Baillie, J.K. Host susceptibility to severe influenza A virus infection. *Crit. Care* **2019**, *23*, 303. [CrossRef] [PubMed]
35. Ota, M.; Nagafuchi, Y.; Hatano, H.; Ishigaki, K.; Terao, C.; Takeshima, Y.; Yanaoka, H.; Kobayashi, S.; Okubo, M.; Shirai, H.; et al. Dynamic landscape of immune cell-specific gene regulation in immune-mediated diseases. *Cell* **2021**, *184*, 3006–3021.e17. [CrossRef]
36. Maurano, M.T.; Humbert, R.; Rynes, E.; Thurman, R.E.; Haugen, E.; Wang, H.; Reynolds, A.P.; Sandstrom, R.; Qu, H.; Brody, J.; et al. Systematic localization of common disease-associated variation in regulatory DNA. *Science* **2012**, *337*, 1190–1195. [CrossRef]
37. Farh, K.K.; Marson, A.; Zhu, J.; Kleinewietfeld, M.; Housley, W.J.; Beik, S.; Shoresh, N.; Whitton, H.; Ryan, R.J.; Shishkin, A.A.; et al. Genetic and epigenetic fine mapping of causal autoimmune disease variants. *Nature* **2015**, *518*, 337–343. [CrossRef]
38. Gusev, A.; Lee, S.H.; Trynka, G.; Finucane, H.; Vilhjalmsson, B.J.; Xu, H.; Zang, C.; Ripke, S.; Bulik-Sullivan, B.; Stahl, E.; et al. Partitioning heritability of regulatory and cell-type-specific variants across 11 common diseases. *Am. J. Hum. Genet.* **2014**, *95*, 535–552. [CrossRef]
39. van der Made, C.I.; Simons, A.; Schuurs-Hoeijmakers, J.; van den Heuvel, G.; Mantere, T.; Kersten, S.; van Deuren, R.C.; Steehouwer, M.; van Reijmersdal, S.V.; Jaeger, M.; et al. Presence of Genetic Variants Among Young Men With Severe COVID-19. *JAMA* **2020**, *324*, 663–673. [CrossRef]
40. Zhang, Q.; Bastard, P.; Liu, Z.; Le Pen, J.; Moncada-Velez, M.; Chen, J.; Ogishi, M.; Sabli, I.K.D.; Hodeib, S.; Korol, C.; et al. Inborn errors of type I IFN immunity in patients with life-threatening COVID-19. *Science* **2020**, *370*, eabd4570. [CrossRef]
41. Danecek, P.; Auton, A.; Abecasis, G.; Albers, C.A.; Banks, E.; DePristo, M.A.; Handsaker, R.E.; Lunter, G.; Marth, G.T.; Sherry, S.T.; et al. The variant call format and VCFtools. *Bioinformatics* **2011**, *27*, 2156–2158. [CrossRef] [PubMed]
42. Purcell, S.; Neale, B.; Todd-Brown, K.; Thomas, L.; Ferreira, M.A.; Bender, D.; Maller, J.; Sklar, P.; de Bakker, P.I.; Daly, M.J.; et al. PLINK: A tool set for whole-genome association and population-based linkage analyses. *Am. J. Hum. Genet.* **2007**, *81*, 559–575. [CrossRef] [PubMed]
43. The 1000 Genomes Project Consortium; Auton, A.; Brooks, L.D.; Durbin, R.M.; Garrison, E.P.; Kang, H.M.; Korbel, J.O.; Marchini, J.L.; McCarthy, S.; McVean, G.A.; et al. A global reference for human genetic variation. *Nature* **2015**, *526*, 68–74. [CrossRef] [PubMed]
44. Turner, S. qqman: An R package for visualizing GWAS results using Q-Q and manhattan plots. *J. Open Source Softw.* **2018**, *3*, 731. [CrossRef]

45. Dayem Ullah, A.Z.; Oscanoa, J.; Wang, J.; Nagano, A.; Lemoine, N.R.; Chelala, C. SNPnexus: Assessing the functional relevance of genetic variation to facilitate the promise of precision medicine. *Nucleic Acids Res.* **2018**, *46*, W109–W113. [CrossRef]
46. Whirl-Carrillo, M.; Huddart, R.; Gong, L.; Sangkuhl, K.; Thorn, C.F.; Whaley, R.; Klein, T.E. An Evidence-Based Framework for Evaluating Pharmacogenomics Knowledge for Personalized Medicine. *Clin. Pharmacol. Ther.* **2021**, *110*, 563–572. [CrossRef]
47. Butler-Laporte, G.; Povysil, G.; Kosmicki, J.; Cirulli, E.T.; Drivas, T.; Furini, S.; Saad, C.; Schmidt, A.; Olszewski, P.; Korotko, U.; et al. Exome-wide association study to identify rare variants influencing COVID-19 outcomes: Results from the Host Genetics Initiative. *medRxiv* **2022**. [CrossRef]
48. Povysil, G.; Butler-Laporte, G.; Shang, N.; Wang, C.; Khan, A.; Alaamery, M.; Nakanishi, T.; Zhou, S.; Forgetta, V.; Eveleigh, R.J.; et al. Rare loss-of-function variants in type I IFN immunity genes are not associated with severe COVID-19. *J. Clin. Investig.* **2021**, *131*, e147834. [CrossRef]
49. Kosmicki, J.A.; Horowitz, J.E.; Banerjee, N.; Lanche, R.; Marcketta, A.; Maxwell, E.; Bai, X.; Sun, D.; Backman, J.D.; Sharma, D.; et al. Pan-ancestry exome-wide association analyses of COVID-19 outcomes in 586,157 individuals. *Am. J. Hum. Genet.* **2021**, *108*, 1350–1355. [CrossRef]
50. Pereira, A.C.; Bes, T.M.; Velho, M.; Marques, E.; Jannes, C.E.; Valino, K.R.; Dinardo, C.L.; Costa, S.F.; Duarte, A.J.S.; Santos, A.R.; et al. Genetic risk factors and COVID-19 severity in Brazil: Results from BRACOVID study. *Hum. Mol. Genet.* **2022**. [CrossRef]
51. Correa, R.L.; Bruckner, F.P.; de Souza Cascardo, R.; Alfenas-Zerbini, P. The Role of F-Box Proteins during Viral Infection. *Int. J. Mol. Sci.* **2013**, *14*, 4030–4049. [CrossRef] [PubMed]
52. Folkersen, L.; Fauman, E.; Sabater-Lleal, M.; Strawbridge, R.J.; Franberg, M.; Sennblad, B.; Baldassarre, D.; Veglia, F.; Humphries, S.E.; Rauramaa, R.; et al. Mapping of 79 loci for 83 plasma protein biomarkers in cardiovascular disease. *PLoS Genet.* **2017**, *13*, e1006706. [CrossRef] [PubMed]
53. Vuckovic, D.; Bao, E.L.; Akbari, P.; Lareau, C.A.; Mousas, A.; Jiang, T.; Chen, M.H.; Raffield, L.M.; Tardaguila, M.; Huffman, J.E.; et al. The Polygenic and Monogenic Basis of Blood Traits and Diseases. *Cell* **2020**, *182*, 1214–1231.e11. [CrossRef] [PubMed]
54. Chen, M.H.; Raffield, L.M.; Mousas, A.; Sakaue, S.; Huffman, J.E.; Moscati, A.; Trivedi, B.; Jiang, T.; Akbari, P.; Vuckovic, D.; et al. Trans-ethnic and Ancestry-Specific Blood-Cell Genetics in 746,667 Individuals from 5 Global Populations. *Cell* **2020**, *182*, 1198–1213. [CrossRef] [PubMed]
55. Sakaue, S.; Kanai, M.; Tanigawa, Y.; Karjalainen, J.; Kurki, M.; Koshiba, S.; Narita, A.; Konuma, T.; Yamamoto, K.; Akiyama, M.; et al. A cross-population atlas of genetic associations for 220 human phenotypes. *Nat. Genet.* **2021**, *53*, 1415–1424. [CrossRef]
56. Demaria, O.; Carvelli, J.; Batista, L.; Thibult, M.L.; Morel, A.; Andre, P.; Morel, Y.; Vely, F.; Vivier, E. Identification of druggable inhibitory immune checkpoints on Natural Killer cells in COVID-19. *Cell Mol. Immunol.* **2020**, *17*, 995–997. [CrossRef] [PubMed]
57. D'Alterio, G.; Lasorsa, V.A.; Bonfiglio, F.; Cantalupo, S.; Rosato, B.E.; Andolfo, I.; Russo, R.; Esposito, U.; Frisso, G.; Abete, P.; et al. Germline rare variants of lectin pathway genes predispose to asymptomatic SARS-CoV-2 infection in elderly individuals. *Genet. Med.* **2022**, *24*, 1653–1663. [CrossRef]
58. Demichev, V.; Tober-Lau, P.; Lemke, O.; Nazarenko, T.; Thibeault, C.; Whitwell, H.; Rohl, A.; Freiwald, A.; Szyrwiel, L.; Ludwig, D.; et al. A time-resolved proteomic and prognostic map of COVID-19. *Cell Syst.* **2021**, *12*, 780–794. [CrossRef]
59. Guarnieri, J.W.; Dybas, J.M.; Fazelinia, H.; Kim, M.S.; Frere, J.; Zhang, Y.; Albrecht, Y.S.; Murdock, D.G.; Angelin, A.; Singh, L.N.; et al. Targeted Down Regulation Of Core Mitochondrial Genes During SARS-CoV-2 Infection. *bioRxiv* **2022**. [CrossRef]
60. Lepelley, A.; Della Mina, E.; Van Nieuwenhove, E.; Waumans, L.; Fraitag, S.; Rice, G.I.; Dhir, A.; Fremond, M.L.; Rodero, M.P.; Seabra, L.; et al. Enhanced cGAS-STING-dependent interferon signaling associated with mutations in ATAD3A. *J. Exp. Med.* **2021**, *218*, e20201560. [CrossRef]
61. Buchanan, C.J.; Gaunt, B.; Harrison, P.J.; Yang, Y.; Liu, J.; Khan, A.; Giltrap, A.M.; Le Bas, A.; Ward, P.N.; Gupta, K.; et al. Pathogen-sugar interactions revealed by universal saturation transfer analysis. *Science* **2022**, *377*, eabm3125. [CrossRef] [PubMed]
62. Masclans, J.R.; Sabater, J.; Sacanell, J.; Chacon, P.; Sabin, P.; Roca, O.; Planas, M. Possible prognostic value of leukotriene B(4) in acute respiratory distress syndrome. *Respir. Care* **2007**, *52*, 1695–1700. [PubMed]
63. Auner, B.; Geiger, E.V.; Henrich, D.; Lehnert, M.; Marzi, I.; Relja, B. Circulating leukotriene B4 identifies respiratory complications after trauma. *Mediators Inflamm.* **2012**, *2012*, 536156. [CrossRef] [PubMed]
64. Sahanic, S.; Loffler-Ragg, J.; Tymoszuk, P.; Hilbe, R.; Demetz, E.; Masanetz, R.K.; Theurl, M.; Holfeld, J.; Gollmann-Tepekoylu, C.; Tzankov, A.; et al. The Role of Innate Immunity and Bioactive Lipid Mediators in COVID-19 and Influenza. *Front. Physiol.* **2021**, *12*, 688946. [CrossRef] [PubMed]
65. Bonyek-Silva, I.; Machado, A.F.A.; Cerqueira-Silva, T.; Nunes, S.; Silva Cruz, M.R.; Silva, J.; Santos, R.L.; Barral, A.; Oliveira, P.R.S.; Khouri, R.; et al. LTB4-Driven Inflammation and Increased Expression of ALOX5/ACE2 During Severe COVID-19 in Individuals With Diabetes. *Diabetes* **2021**, *70*, 2120–2130. [CrossRef] [PubMed]
66. Sahajpal, N.S.; Jill Lai, C.Y.; Hastie, A.; Mondal, A.K.; Dehkordi, S.R.; van der Made, C.I.; Fedrigo, O.; Al-Ajli, F.; Jalnapurkar, S.; Byrska-Bishop, M.; et al. Optical genome mapping identifies rare structural variations as predisposition factors associated with severe COVID-19. *iScience* **2022**, *25*, 103760. [CrossRef]
67. Ouyang, Y.; Yin, J.; Wang, W.; Shi, H.; Shi, Y.; Xu, B.; Qiao, L.; Feng, Y.; Pang, L.; Wei, F.; et al. Downregulated Gene Expression Spectrum and Immune Responses Changed During the Disease Progression in Patients With COVID-19. *Clin. Infect. Dis.* **2020**, *71*, 2052–2060. [CrossRef]
68. Tournier, C.; Whitmarsh, A.J.; Cavanagh, J.; Barrett, T.; Davis, R.J. Mitogen-activated protein kinase kinase 7 is an activator of the c-Jun NH2-terminal kinase. *Proc. Natl. Acad. Sci. USA* **1997**, *94*, 7337–7342. [CrossRef]

69. Matsuda, S.; Moriguchi, T.; Koyasu, S.; Nishida, E. T lymphocyte activation signals for interleukin-2 production involve activation of MKK6-p38 and MKK7-SAPK/JNK signaling pathways sensitive to cyclosporin A. *J. Biol. Chem.* **1998**, *273*, 12378–12382. [CrossRef]
70. Lu, W.; Liu, X.; Wang, T.; Liu, F.; Zhu, A.; Lin, Y.; Luo, J.; Ye, F.; He, J.; Zhao, J.; et al. Elevated MUC1 and MUC5AC mucin protein levels in airway mucus of critical ill COVID-19 patients. *J. Med. Virol.* **2021**, *93*, 582–584. [CrossRef]
71. Shashaty, M.G.S.; Reilly, J.P.; Faust, H.E.; Forker, C.M.; Ittner, C.A.G.; Zhang, P.X.; Hotz, M.J.; Fitzgerald, D.; Yang, W.; Anderson, B.J.; et al. Plasma receptor interacting protein kinase-3 levels are associated with acute respiratory distress syndrome in sepsis and trauma: A cohort study. *Crit. Care* **2019**, *23*, 235. [CrossRef] [PubMed]
72. Nakamura, H.; Kinjo, T.; Arakaki, W.; Miyagi, K.; Tateyama, M.; Fujita, J. Serum levels of receptor-interacting protein kinase-3 in patients with COVID-19. *Crit. Care* **2020**, *24*, 484. [CrossRef] [PubMed]
73. Mukund, K.; Nayak, P.; Ashokkumar, C.; Rao, S.; Almeda, J.; Betancourt-Garcia, M.M.; Sindhi, R.; Subramaniam, S. Immune Response in Severe and Non-Severe Coronavirus Disease 2019 (COVID-19) Infection: A Mechanistic Landscape. *Front. Immunol.* **2021**, *12*, 738073. [CrossRef] [PubMed]
74. Abusukhun, M.; Winkler, M.S.; Pohlmann, S.; Moerer, O.; Meissner, K.; Tampe, B.; Hofmann-Winkler, H.; Bauer, M.; Graler, M.H.; Claus, R.A. Activation of Sphingomyelinase-Ceramide-Pathway in COVID-19 Purposes Its Inhibition for Therapeutic Strategies. *Front. Immunol.* **2021**, *12*, 784989. [CrossRef] [PubMed]
75. Bojkova, D.; Klann, K.; Koch, B.; Widera, M.; Krause, D.; Ciesek, S.; Cinatl, J.; Munch, C. Proteomics of SARS-CoV-2-infected host cells reveals therapy targets. *Nature* **2020**, *583*, 469–472. [CrossRef]
76. Ramachandran, K.; Maity, S.; Muthukumar, A.R.; Kandala, S.; Tomar, D.; Abd El-Aziz, T.M.; Allen, C.; Sun, Y.; Venkatesan, M.; Madaris, T.R.; et al. SARS-CoV-2 infection enhances mitochondrial PTP complex activity to perturb cardiac energetics. *iScience* **2022**, *25*, 103722. [CrossRef]
77. Shaath, H.; Vishnubalaji, R.; Elkord, E.; Alajez, N.M. Single-Cell Transcriptome Analysis Highlights a Role for Neutrophils and Inflammatory Macrophages in the Pathogenesis of Severe COVID-19. *Cells* **2020**, *9*, 2374. [CrossRef]
78. Prokop, J.W.; Hartog, N.L.; Chesla, D.; Faber, W.; Love, C.P.; Karam, R.; Abualkheir, N.; Feldmann, B.; Teng, L.; McBride, T.; et al. High-Density Blood Transcriptomics Reveals Precision Immune Signatures of SARS-CoV-2 Infection in Hospitalized Individuals. *Front. Immunol.* **2021**, *12*, 694243. [CrossRef]
79. Schulte-Schrepping, J.; Reusch, N.; Paclik, D.; Bassler, K.; Schlickeiser, S.; Zhang, B.; Kramer, B.; Krammer, T.; Brumhard, S.; Bonaguro, L.; et al. Severe COVID-19 Is Marked by a Dysregulated Myeloid Cell Compartment. *Cell* **2020**, *182*, 1419–1440. [CrossRef]
80. Aschenbrenner, A.C.; Mouktaroudi, M.; Kramer, B.; Oestreich, M.; Antonakos, N.; Nuesch-Germano, M.; Gkizeli, K.; Bonaguro, L.; Reusch, N.; Bassler, K.; et al. Disease severity-specific neutrophil signatures in blood transcriptomes stratify COVID-19 patients. *Genome Med.* **2021**, *13*, 7. [CrossRef]
81. Peng, C.K.; Wu, C.P.; Lin, J.Y.; Peng, S.C.; Lee, C.H.; Huang, K.L.; Shen, C.H. Gas6/Axl signaling attenuates alveolar inflammation in ischemia-reperfusion-induced acute lung injury by up-regulating SOCS3-mediated pathway. *PLoS ONE* **2019**, *14*, e0219788. [CrossRef] [PubMed]
82. Morales, A.; Rojo Rello, S.; Cristobal, H.; Fiz-Lopez, A.; Arribas, E.; Mari, M.; Tutusaus, A.; de la Cal-Sabater, P.; Nicolaes, G.A.F.; Ortiz-Perez, J.T.; et al. Growth Arrest-Specific Factor 6 (GAS6) Is Increased in COVID-19 Patients and Predicts Clinical Outcome. *Biomedicines* **2021**, *9*, 335. [CrossRef] [PubMed]
83. Turjya, R.R.; Khan, M.A.; Mir Md Khademul Islam, A.B. Perversely expressed long noncoding RNAs can alter host response and viral proliferation in SARS-CoV-2 infection. *Future Virol.* **2020**, *15*, 577–593. [CrossRef] [PubMed]
84. Buszko, M.; Park, J.H.; Verthelyi, D.; Sen, R.; Young, H.A.; Rosenberg, A.S. The dynamic changes in cytokine responses in COVID-19: A snapshot of the current state of knowledge. *Nat. Immunol.* **2020**, *21*, 1146–1151. [CrossRef]
85. Wilczynski, S.A.; Wenceslau, C.F.; McCarthy, C.G.; Webb, R.C. A Cytokine/Bradykinin Storm Comparison: What Is the Relationship Between Hypertension and COVID-19? *Am. J. Hypertens* **2021**, *34*, 304–306. [CrossRef] [PubMed]
86. Chua, R.L.; Lukassen, S.; Trump, S.; Hennig, B.P.; Wendisch, D.; Pott, F.; Debnath, O.; Thurmann, L.; Kurth, F.; Volker, M.T.; et al. COVID-19 severity correlates with airway epithelium-immune cell interactions identified by single-cell analysis. *Nat. Biotechnol.* **2020**, *38*, 970–979. [CrossRef] [PubMed]
87. Hu, J.; Li, C.; Wang, S.; Li, T.; Zhang, H. Genetic variants are identified to increase risk of COVID-19 related mortality from UK Biobank data. *Hum. Genom.* **2021**, *15*, 10. [CrossRef] [PubMed]
88. Agrawal, P.B.; Pierson, C.R.; Joshi, M.; Liu, X.; Ravenscroft, G.; Moghadaszadeh, B.; Talabere, T.; Viola, M.; Swanson, L.C.; Haliloglu, G.; et al. SPEG interacts with myotubularin, and its deficiency causes centronuclear myopathy with dilated cardiomyopathy. *Am. J. Hum. Genet.* **2014**, *95*, 218–226. [CrossRef]
89. Smet, A.; Breugelmans, T.; Michiels, J.; Lamote, K.; Arras, W.; De Man, J.G.; Heyndrickx, L.; Hauner, A.; Huizing, M.; Malhotra-Kumar, S.; et al. A dynamic mucin mRNA signature associates with COVID-19 disease presentation and severity. *JCI Insight* **2021**, *6*, e151777. [CrossRef]
90. Wieczorek, M.; Abualrous, E.T.; Sticht, J.; Alvaro-Benito, M.; Stolzenberg, S.; Noe, F.; Freund, C. Major Histocompatibility Complex (MHC) Class I and MHC Class II Proteins: Conformational Plasticity in Antigen Presentation. *Front. Immunol.* **2017**, *8*, 292. [CrossRef]

91. Lin, M.; Tseng, H.K.; Trejaut, J.A.; Lee, H.L.; Loo, J.H.; Chu, C.C.; Chen, P.J.; Su, Y.W.; Lim, K.H.; Tsai, Z.U.; et al. Association of HLA class I with severe acute respiratory syndrome coronavirus infection. *BMC Med. Genet.* **2003**, *4*, 9. [CrossRef] [PubMed]
92. Luckey, D.; Weaver, E.A.; Osborne, D.G.; Billadeau, D.D.; Taneja, V. Immunity to Influenza is dependent on MHC II polymorphism: Study with 2 HLA transgenic strains. *Sci. Rep.* **2019**, *9*, 19061. [CrossRef] [PubMed]
93. Hajeer, A.H.; Balkhy, H.; Johani, S.; Yousef, M.Z.; Arabi, Y. Association of human leukocyte antigen class II alleles with severe Middle East respiratory syndrome-coronavirus infection. *Ann. Thorac. Med.* **2016**, *11*, 211–213. [CrossRef] [PubMed]
94. Heger, L.; Hofer, T.P.; Bigley, V.; de Vries, I.J.M.; Dalod, M.; Dudziak, D.; Ziegler-Heitbrock, L. Subsets of CD1c(+) DCs: Dendritic Cell Versus Monocyte Lineage. *Front. Immunol.* **2020**, *11*, 559166. [CrossRef]
95. Saichi, M.; Ladjemi, M.Z.; Korniotis, S.; Rousseau, C.; Ait Hamou, Z.; Massenet-Regad, L.; Amblard, E.; Noel, F.; Marie, Y.; Bouteiller, D.; et al. Single-cell RNA sequencing of blood antigen-presenting cells in severe COVID-19 reveals multi-process defects in antiviral immunity. *Nat. Cell Biol.* **2021**, *23*, 538–551. [CrossRef]
96. Langton, D.J.; Bourke, S.C.; Lie, B.A.; Reiff, G.; Natu, S.; Darlay, R.; Burn, J.; Echevarria, C. The influence of HLA genotype on the severity of COVID-19 infection. *HLA* **2021**, *98*, 14–22. [CrossRef]
97. Castelli, E.C.; de Castro, M.V.; Naslavsky, M.S.; Scliar, M.O.; Silva, N.S.B.; Andrade, H.S.; Souza, A.S.; Pereira, R.N.; Castro, C.F.B.; Mendes-Junior, C.T.; et al. MHC Variants Associated With Symptomatic Versus Asymptomatic SARS-CoV-2 Infection in Highly Exposed Individuals. *Front. Immunol.* **2021**, *12*, 742881. [CrossRef]
98. Hou, Y.S.; Wang, J.Z.; Shi, S.; Han, Y.; Zhang, Y.; Zhi, J.X.; Xu, C.; Li, F.F.; Wang, G.Y.; Liu, S.L. Identification of epigenetic factor KAT2B gene variants for possible roles in congenital heart diseases. *Biosci. Rep.* **2020**, *40*, BSR20191779. [CrossRef]
99. Ferreira-Gomes, M.; Kruglov, A.; Durek, P.; Heinrich, F.; Tizian, C.; Heinz, G.A.; Pascual-Reguant, A.; Du, W.; Mothes, R.; Fan, C.; et al. SARS-CoV-2 in severe COVID-19 induces a TGF-beta-dominated chronic immune response that does not target itself. *Nat. Commun.* **2021**, *12*, 1961. [CrossRef]
100. Philippi, A.; Heller, S.; Costa, I.G.; Senee, V.; Breunig, M.; Li, Z.; Kwon, G.; Russell, R.; Illing, A.; Lin, Q.; et al. Mutations and variants of ONECUT1 in diabetes. *Nat. Med.* **2021**, *27*, 1928–1940. [CrossRef]
101. Norouzi, M.; Norouzi, S.; Ruggiero, A.; Khan, M.S.; Myers, S.; Kavanagh, K.; Vemuri, R. Type-2 Diabetes as a Risk Factor for Severe COVID-19 Infection. *Microorganisms* **2021**, *9*, 1211. [CrossRef] [PubMed]
102. Zhu, K.; Jin, X.; Chi, Z.; Chen, S.; Wu, S.; Sloan, R.D.; Lin, X.; Neculai, D.; Wang, D.; Hu, H.; et al. Priming of NLRP3 inflammasome activation by Msn kinase MINK1 in macrophages. *Cell Mol. Immunol.* **2021**, *18*, 2372–2382. [CrossRef] [PubMed]
103. Freeman, T.L.; Swartz, T.H. Targeting the NLRP3 Inflammasome in Severe COVID-19. *Front. Immunol.* **2020**, *11*, 1518. [CrossRef] [PubMed]
104. Thomas, T.; Stefanoni, D.; Dzieciatkowska, M.; Issaian, A.; Nemkov, T.; Hill, R.C.; Francis, R.O.; Hudson, K.E.; Buehler, P.W.; Zimring, J.C.; et al. Evidence of Structural Protein Damage and Membrane Lipid Remodeling in Red Blood Cells from COVID-19 Patients. *J. Proteome Res.* **2020**, *19*, 4455–4469. [CrossRef] [PubMed]
105. Schmidt, C.; Berger, T.; Groettrup, M.; Basler, M. Immunoproteasome Inhibition Impairs T and B Cell Activation by Restraining ERK Signaling and Proteostasis. *Front. Immunol.* **2018**, *9*, 2386. [CrossRef]
106. Desterke, C.; Turhan, A.G.; Bennaceur-Griscelli, A.; Griscelli, F. HLA-dependent heterogeneity and macrophage immunoproteasome activation during lung COVID-19 disease. *J. Transl. Med.* **2021**, *19*, 290. [CrossRef]
107. Onodi, F.; Bonnet-Madin, L.; Meertens, L.; Karpf, L.; Poirot, J.; Zhang, S.Y.; Picard, C.; Puel, A.; Jouanguy, E.; Zhang, Q.; et al. SARS-CoV-2 induces human plasmacytoid predendritic cell diversification via UNC93B and IRAK4. *J. Exp. Med.* **2021**, *218*, e20201387. [CrossRef]
108. Ragab, D.; Salah Eldin, H.; Taeimah, M.; Khattab, R.; Salem, R. The COVID-19 Cytokine Storm; What We Know So Far. *Front. Immunol.* **2020**, *11*, 1446. [CrossRef]
109. Schmidt, N.; Lareau, C.A.; Keshishian, H.; Ganskih, S.; Schneider, C.; Hennig, T.; Melanson, R.; Werner, S.; Wei, Y.; Zimmer, M.; et al. The SARS-CoV-2 RNA-protein interactome in infected human cells. *Nat. Microbiol.* **2021**, *6*, 339–353. [CrossRef]
110. Hue, S.; Beldi-Ferchiou, A.; Bendib, I.; Surenaud, M.; Fourati, S.; Frapard, T.; Rivoal, S.; Razazi, K.; Carteaux, G.; Delfau-Larue, M.H.; et al. Uncontrolled Innate and Impaired Adaptive Immune Responses in Patients with COVID-19 Acute Respiratory Distress Syndrome. *Am. J. Respir. Crit. Care Med.* **2020**, *202*, 1509–1519. [CrossRef]
111. Chen, L.; Marishta, A.; Ellison, C.E.; Verzi, M.P. Identification of Transcription Factors Regulating SARS-CoV-2 Entry Genes in the Intestine. *Cell Mol. Gastroenterol. Hepatol.* **2021**, *11*, 181–184. [CrossRef] [PubMed]
112. Zhang, J.; Garrett, S.; Sun, J. Gastrointestinal symptoms, pathophysiology, and treatment in COVID-19. *Genes Dis.* **2021**, *8*, 385–400. [CrossRef] [PubMed]
113. Crackower, M.A.; Kolas, N.K.; Noguchi, J.; Sarao, R.; Kikuchi, K.; Kaneko, H.; Kobayashi, E.; Kawai, Y.; Kozieradzki, I.; Landers, R.; et al. Essential role of Fkbp6 in male fertility and homologous chromosome pairing in meiosis. *Science* **2003**, *300*, 1291–1295. [CrossRef] [PubMed]
114. Ghosh, S.; Parikh, S.; Nissa, M.U.; Acharjee, A.; Singh, A.; Patwa, D.; Makwana, P.; Athalye, A.; Barpanda, A.; Laloraya, M.; et al. Semen Proteomics of COVID-19 Convalescent Men Reveals Disruption of Key Biological Pathways Relevant to Male Reproductive Function. *ACS Omega* **2022**, *7*, 8601–8612. [CrossRef] [PubMed]
115. Kant, S.; Holthofer, B.; Magin, T.M.; Krusche, C.A.; Leube, R.E. Desmoglein 2-Dependent Arrhythmogenic Cardiomyopathy Is Caused by a Loss of Adhesive Function. *Circ. Cardiovasc. Genet.* **2015**, *8*, 553–563. [CrossRef]

116. Lee, E.C.Y.; Tyler, R.E.; Johnson, D.; Koh, N.; Ong, B.C.; Foo, S.Y.; Tan, J.W.C. High frequency of anti-DSG 2 antibodies in post COVID-19 serum samples. *J. Mol. Cell Cardiol.* **2022**, *170*, 121–123. [CrossRef]
117. Ingul, C.B.; Grimsmo, J.; Mecinaj, A.; Trebinjac, D.; Berger Nossen, M.; Andrup, S.; Grenne, B.; Dalen, H.; Einvik, G.; Stavem, K.; et al. Cardiac Dysfunction and Arrhythmias 3 Months After Hospitalization for COVID-19. *J. Am. Heart Assoc.* **2022**, *11*, e023473. [CrossRef]
118. Best, J.C.; Kuchakulla, M.; Khodamoradi, K.; Lima, T.F.N.; Frech, F.S.; Achua, J.; Rosete, O.; Mora, B.; Arora, H.; Ibrahim, E.; et al. Evaluation of SARS-CoV-2 in Human Semen and Effect on Total Sperm Number: A Prospective Observational Study. *World J. Mens Health* **2021**, *39*, 489–495. [CrossRef]
119. Nakamichi, K.; Shen, J.Z.; Lee, C.S.; Lee, A.; Roberts, E.A.; Simonson, P.D.; Roychoudhury, P.; Andriesen, J.; Randhawa, A.K.; Mathias, P.C.; et al. Hospitalization and mortality associated with SARS-CoV-2 viral clades in COVID-19. *Sci. Rep.* **2021**, *11*, 4802. [CrossRef]

life

MDPI

Article

Exploring the Role of Krebs von den Lungen-6 in Severe to Critical COVID-19 Patients

Vito D'Agnano [1,†], Filippo Scialò [1,2,†], Francesco Perna [3,4], Lidia Atripaldi [1], Stefano Sanduzzi [1], Valentino Allocca [1], Maria Vitale [2,5], Lucio Pastore [2,5], Andrea Bianco [1] and Fabio Perrotta [1,*]

1 Department of Translational Medical Sciences, University of Campania "L. Vanvitelli", 80131 Naples, Italy; vito.dagnano@studenti.unicampania.it (V.D.); filippo.scialo@unicampania.it (F.S.); lidia.atripaldi@studenti.unicampania.it (L.A.); stefano.sanduzzi@studenti.unicampania.it (S.S.); valentino.allocca@unicampania.it (V.A.); andrea.bianco@unicampania.it (A.B.)

2 CEINGE-Biotecnologie Avanzate, 80131 Naples, Italy; vitalema@ceinge.unina.it (M.V.); lucio.pastore@unina.it (L.P.)

3 Section of Respiratory Diseases, Department of Clinical Medicine and Surgery, Monaldi Hospital, University of Naples Federico II, 80131 Naples, Italy; francesco.perna@unina.it

4 Section of Biochemical Clinical Laboratory, Department of Clinical Medicine and Surgery, University of Naples Federico II, 80131 Naples, Italy

5 Dipartimento di Medicina Molecolare e Biotecnologie Mediche, Università di Napoli Federico II, 80131 Naples, Italy

* Correspondence: fabio.perrotta@unicampania.it

† These authors contributed equally to this work.

Citation: D'Agnano, V.; Scialò, F.; Perna, F.; Atripaldi, L.; Sanduzzi, S.; Allocca, V.; Vitale, M.; Pastore, L.; Bianco, A.; Perrotta, F. Exploring the Role of Krebs von den Lungen-6 in Severe to Critical COVID-19 Patients. *Life* **2022**, *12*, 1141. https://doi.org/10.3390/life12081141

Academic Editors: Silvia De Francia and Sarah Allegra

Received: 24 June 2022
Accepted: 26 July 2022
Published: 28 July 2022

Abstract: COVID-19 encompasses a broad spectrum of clinical conditions caused by SARS-CoV-2 infection. More severe cases experience acute respiratory and/or multiorgan failure. KL-6 is a glycoprotein expressed mainly from type II alveolar cells with pro-fibrotic properties. Serum KL-6 concentrations have been found in patients with COVID-19. However, the relevance of KL-6 in patients with severe and critical COVID-19 has not been fully elucidated. Methods: Retrospective data from consecutive severe to critical COVID-19 patients were collected at UOC Clinica Pnuemologica "Vanvitelli", A.O. dei Colli, Naples, Italy. The study included patients with a positive rhinopharyngeal swab for SARS-CoV-2 RNA with severe or critical COVID-19. Results: Among 87 patients, 24 had poor outcomes. The median KL-6 value in survivors was significantly lower when compared with dead or intubated patients (530 U/mL versus 1069 U/mL $p < 0.001$). KL-6 was correlated with body mass index (BMI) (r: 0.279, p: 0.009), lung ultrasound score (LUS) (r: 0.429, $p < 0.001$), Chung Score (r: 0.390, $p < 0.001$). KL-6 was associated with the risk of death or oro-tracheal intubation (IOT) after adjusting for gender, BMI, Charlson Index, Chung Score, and PaO_2/FIO_2 (OR 1.003 95% CI 1.001–1.004, $p < 0.001$). Serum KL-6 value of 968 has a sensitivity of 79.2%, specificity of 87.1%, PPV 70.4%, NPV 91.5%, AUC: O.85 for risk of death or IOT. Conclusions: The presented research highlights the relevance of serum KL-6 in severe to critical COVID-19 patients in predicting the risk of death or IOT.

Keywords: COVID-19; SARS-CoV-2; Krebs von den Lungen-6; lung ultrasound score

1. Introduction

The COVID-19 syndrome caused by the new coronavirus SARS-CoV-2 is characterized by a broad range of clinical manifestations with different degrees of severity ranging from mild, moderate, severe, and critical. In the most severe and critical cases, symptoms include respiratory distress syndrome, a hyperinflammatory status caused by the cytokine storm syndrome, and a hypercoagulable state leading to disseminated intravascular coagulation (DIC), multiorgan failure, and death [1–3]. Therefore, since the beginning of this pandemic, many efforts have been adopted to find reliable biomarkers that in these patients could predict the need for intubation or mortality. Many studies have already shown that an increasing trend of biochemical parameters, such as CRP, IL-6, D-dimer, neutrophil, and lymphocyte count [4,5] gives only a limited picture of a patient's clinical condition and the search for new biomarkers that positively correlate with the need for intubation and mortality is still of great importance. For this reason, in our study, we sought to characterize the role of Krebs von den Lungen-6 (KL-6) protein in a subset of patients admitted to our hospital that progressed to severe conditions. We aimed to understand if a correlation would exist between its level and lung function in severe and critical COVID-19 patients and therefore be used to predict the need for intubation and if it would correlate with mortality.

KL-6 is produced by the shedding of the extracellular domain of MUCIN 1, a glycoprotein mainly expressed by damaged alveolar type II cells and has been shown to have an anti-inflammatory action through the inhibition of Toll-like receptor signaling but also participates in lung cancer progression and development of fibrotic processes [6]. KL-6 is formed by a stretch of 20 amino acids rich in glycosylated serine and threonine residue forming a rigid structure that protrudes into the extracellular space and provides a barrier against microbial and virus attacks [6]. Interestingly, the shedding of MUCIN 1 and release of KL-6 in the serum is operated by ADAM17 [7], which is also responsible for the cleavage of ACE2, the main receptor of SARS-CoV-2 [8,9]. KL-6 has initially been identified as a tumor marker [10] but has successively been found to play a role in inducing fibroblast migration [11] and inhibiting cell–cell adhesion [12], and could be used as a marker of interstitial lung fibrosis (ILF) [13]. Few studies have begun to describe an increase in KL-6 level in COVID-19 patients [14–19] although its physiological role in the context of SARS-CoV-2 infection needs further elucidation. As discussed previously, to characterize this biomarker further, here we add to the existing evidence the study of KL-6 specifically in the context of COVID-19 patients with more severe phases of respiratory involvement. Here, to the best of our knowledge, we clearly describe for the first time a strong correlation between KL-6 levels and lung function assessed by LUS. Moreover, based on our data, we suggest that KL-6 could be used as a marker to predict the need for intubation, and it strongly correlates with mortality.

2. Materials and Methods

A total of 87 consecutive patients admitted to U.O.C. Clinica Pnuemologica "Vanvitelli", A.O. dei Colli, Naples, Italy, have been included in the analysis. Demographic, clinical, laboratory, and treatment data were extracted from electronic medical records. To be included in the study, patients were required to have a positive rhinopharyngeal swab for SARS-CoV-2 RNA. Criteria for severe COVID-19 were clinical signs of pneumonia (fever, cough, dyspnoea) plus one of the following: respiratory rate > 30 breaths/min; severe respiratory distress; or $SpO_2 < 90\%$ on room air [20]. Criteria for critical COVID-19 were (1) respiratory symptoms onset within 1 week; (2) chest imaging showing bilateral opacities, not fully explained by volume overload, and (3) $PaO_2/FIO_2 < 300$ mmHg with PEEP or CPAP > 5 cm H_2O. [20] Peripheral blood samples were centrifuged at $500\times g$ for 10 min. Chemiluminescence enzyme immunoassay (CLEIA) using a KL-6 antibody kit (LUMIPULSE G1200, Fujirebio, Tokyo, Japan), in line with the manufacturer's protocol, has been employed to assess serum KL-6. The kit had a detection range between 50–10,000 U/mL, where the level in healthy donors has been specified to be between 118–627 U/mL.

The patients were followed up until discharge or in-hospital death. The respiratory support needed during hospitalization included a venturi mask (VM) or non-rebreathing mask (NRM), high-flow nasal cannula (HFNC), continuous positive airway pressure (CPAP), and/or pressure support non-invasive ventilation (NIV). Acute respiratory failure management was provided according to international recommendations [21]. PaO_2/FIO_2 was calculated based on the ratio between arterial O_2 pressure and oxygen inspiratory fraction administered.

LUS was performed as the patient entered the ward according to the 12 field 12-region model, 6 on each side as otherwise reported. Artifacts have been categorized as follows: A-line, horizontal artifacts observed in normal lungs; B-lines: vertical artifacts in a variety of patterns including focal and confluent; consolidations, including small peripheral subpleural or multifocal, translobar consolidation with occasional mobile air bronchograms and white lung. Finally, the state of pleural line was assessed. The findings were classified according to the following scoring method with scores ranging from 0 to 3: Score 0: normal A-lines with a continuous and regular pleural line.

Score 1: multiple separated B-lines. Score 2: coalescent B-lines pattern with alterations of the pleural line. Score 3: consolidation area and possibly a large white lung artifact. The total score was computed as the sum, which could range from 0 to 36.

3. Statistical Analysis

Categorical data were expressed as numbers and percentages, while continuous variables as either median and interquartile range or mean and standard deviation, according to the distribution assessed graphically and by the Shapiro–Wilk test. The presence of missing data has been reported. The study variable KL-6 was further described with an appropriate graph to analyze its distribution, and the possible association with other laboratory and clinical respiratory variables was analyzed using the Spearman correlation test. The endpoint was in-hospital IOT or mortality, assessed either from data at discharge, IOT, or death certificate. Univariable and multivariable logistic regression models were performed to evaluate the association between IOT or mortality with exposure variables. Odds ratios and 95% confidence intervals (OR—95% CI) have been calculated for all models. Logistic regression analysis was performed to evaluate the presence of risk factors for the mentioned endpoint. A multivariate model was built to include co-variates with significant interference. The *p*-value for statistical significance was set at <0.05 for all the tests. ROC curve and AUC test were calculated to assess KL-6 value with the best sensitivity, specificity, positive predicted value (PPV), and negative predicted value (NPV). All analyses were performed using statistical software STATA v16 (StataCorp. 2019. StataCorp LLC: College Station, TX, USA).

4. Results

In the present study, we analyzed data obtained from eighty-seven consecutive patients admitted to our hospital and that had tested positive for SARS-CoV-2 by real-time PCR analysis on rhinopharyngeal swabs. The classification of severe and critical COVID-19 patients was done based on specific parameters described in the material and methods. The clinical characteristics reported in Table 1 show that our patient population was almost equally distributed between males (52%) and females (48%). Twenty-four patients (27.6%) experienced worse outcomes, including IOT or death, whilst 63 subjects were discharged. As demonstrated in many studies since the beginning of this pandemic, also in our patient population the median age of those that were discharged was lower (67 years) compared to the patient that progressed to a critical phase (75.5 years; $p = 0.011$), confirming that age is a determining factor and positively correlates with the disease progression and exacerbation. Among the discharged patients ($n = 63$), the majority received Continuous Positive Airway Pressure (CPAP) (52.4%) or High-Flow Nasal Cannula (HFNC) (37%) support for the management of acute respiratory failure while six of them (9.5%) were treated with non-invasive ventilation (NIV). On the other side, NIV treatment has been adopted for 9 out of 24 patients (37.5%) who experienced worse outcomes. In this group of patients, LUS score was significantly higher (36 versus 29, $p < 0.001$) in comparison with discharged patients, reflecting a more extended lung involvement. Likewise, Chung CT Score was higher in patients who experienced worse outcomes (median value 15 versus 13, $p = 0.04$). Blood tests revealed a significant increase in white blood cells (WBC) and neutrophils among patients with more severe COVID-19 ($p = 0.008$ and $p = 0.014$, respectively). Furthermore, based on data that emerged from the analysis of inflammation markers, IL-6 and C-Reactive Protein (CRP) were significantly lower in survivors (25.3 pg/mL versus 115 pg/mL; $p = 0.003$ and 4.25 mg/L versus 6.5 mg/L; $p = 0.002$, respectively). When the KL-6 level was measured, we reported a higher median value (1969 U/mL) in patients who had undergone IOT or were deceased compared to discharged patients (530 U/mL; $p < 0.001$).

Table 1. Study population characteristics.

	Discharged (n = 63)	Death/IOT (n = 24)	p
Age	67 (60–73)	75.5 (67–80.3)	0.011
Gender (Male)	37 (58.7)	8 (33.3)	
BMI (Kg/m^2)	27.7 (25–31.5)	31.2 (27.7–35.2)	0.02
Charlson Index	3 (3–4)	4 (3–4)	
LUS	29 (23.8–32.3)	36 (36–36)	<0.001
Chung Score	13 (12–15)	15 (12–16)	0.04
WBC (10^3 cell/µL)	8.65 (6.86–11.1)	12.3 (9.5–12.9)	0.008
Neutrophils (10^3 cell/µL)	7.46 (6.31–10.2)	11 (7.66–11.5)	0.014
Lymphocytes (10^3 cell/µL)	0.69 (0.497–1.02)	0.685 (0.44–1.07)	0.851
Eosinophils (10^3 cell/µL)	0 (0.00–0.01)	0 (0.00–0.01)	0.859
NLR	11.8 (6.51–16.3)	13.1 (9.88–15.6)	0.113
RBC (10^6 cell/µL)	4.72 (4.56–5.41)	4.5 (3.89–4.92)	0.003
HGB (g/dL)	13.7 (12.3–14.3)	12.6 (9.6–14)	0.058
PLT (10^3 cell/µL)	221 (183–272)	203 (159–287)	0.382
CRP (mg/L)	4.25 (2.27–9.4)	6.5 (4–9.85)	0.002
D-Dimer (µg/L)	281(162–519)	499(435–903)	0.307
IL2R (U/mL)	1121 (809–1507)	973 (906–1737)	0.987
IL-6 (pg/mL)	25.3 (16.6–55.9)	115 (42.2–160)	0.003
KL-6 (U/mL)	530 (469–787)	1969 (1036–3669)	<0.001
PaO_2/FiO_2	119 (88–155)	100 (91.3–110)	0.211
Respiratory Support			
Nasal Cannula, face mask, or non-rebreathing mask	*4 (6.4)*	*0 (0)*	
HFNC	*20 (31.7)*	*2 (8.3)*	
CPAP	*33 (52.4)*	*13 (54.2)*	
NIV	*6 (9.5)*	*9 (37.5)*	

KL6 predicts negative outcomes in COVID-19 severe patients.

Therefore, to understand if KL-6 could be used as a biomarker of severity and suggest the need for intubation, we sought to perform a regression analysis. As shown in Table 2, we could demonstrate that KL-6 significantly correlated with BMI (r: 0.279, p: 0.009), LUS Score (r: 0.429, $p < 0.001$), and Chung Score (r: 0.390, $p < 0.001$), confirming our hypothesis.

Table 2. Correlation Matrix (KL-6 Dependent Variable).

		KL6
Age	Pearson's r	0.174
	p-value	0.110
BMI	Pearson's r	0.279
	p-value	0.009
Charlson index	Pearson's r	0.194
	p-value	0.073
LUS SCORE	Pearson's r	0.429
	p-value	<0.001
CHUNG-SCORE	Pearson's r	0.390
	p-value	<0.001
NLR	Pearson's r	0.236
	p-value	0.030
D-DIMERO	Pearson's r	0.005
	p-value	0.966
P/F	Pearson's r	0.180
	p-value	0.101

By performing multivariate analysis, KL-6 was associated with risk of death or IOT after adjusting for gender, BMI, Charlson Index, Chung Score, and PaO_2/FIO_2 (OR 1.003 95% CI 1.001–1.004, $p < 0.001$) (Table 3). Finally, we performed a ROC curve to define the KL-6 value with the best performance in predicting negative outcomes in patients with severe to critical COVID-19. We found that serum KL-6 value of 968 has a sensitivity of 79.2%, specificity of 87.1%, PPV 70.4%, NPV 91.5%, and AUC: O.85 for risk of death or IOT (Figure 1).

Table 3. Multivariable analysis of the risk of IOT or death in severe COVID-19 patients.

		95% Confidence Interval		
	Odds Ratio	**Lower**	**Upper**	p
Gender	0.46967	0.101	2.18	0.334
BMI	1.09941	0.952	1.27	0.196
Charlson index	1.14269	0.631	2.07	0.66
CHUNG-Score	0.97959	0.712	1.35	0.899
PaO_2/FiO_2	0.99005	0.97	1.01	0.328
KL-6	1.00266	1.001	1.004	<0.001

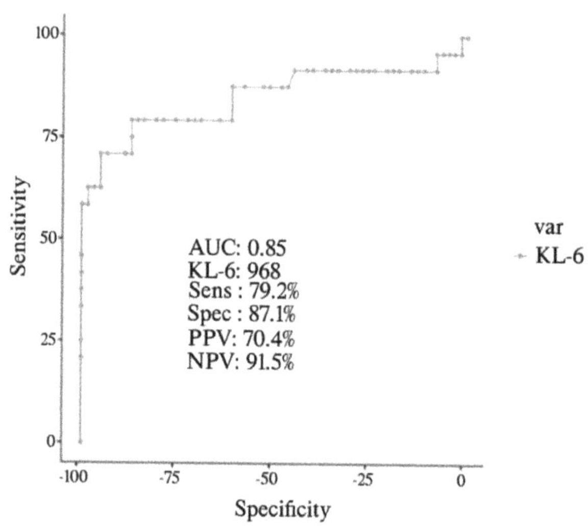

Figure 1. ROC Curve for KL-6 as a predictor of in-hospital mortality or IOT.

5. Discussion

Since the beginning of the SARS-CoV-2 pandemic, the clinical scenario related to COVID-19 has emerged as far as to be homogeneous, with an extremely broad spectrum of severity. Elevated levels of KL-6, a mucin-like glycoprotein mainly expressed by respiratory bronchiolar epithelial cells as well as type II alveolar epithelial cells, are thought to denote a disruption of alveolar epithelial cells. Indeed, abnormal values of serum KL-6 have been reported in a number of lung conditions, including acute lung injury, acute respiratory distress syndrome (ARDS), and interstitial lung diseases [22].

Although several studies have investigated multiple prognostic biomarkers in patients with COVID-19, the paucity of validated tools for predicting outcomes in these subjects still remains.

In this respect, the utility of KL-6 in COVID-19 patients has been lately explored as either a marker of severity or a prognostic biomarker. [23–25]. We have therefore completed a monocentric retrospective study aiming at assessing the role of KL-6 in hospitalized patients with more pronounced respiratory failure secondary to SARS-CoV-2 infection. In our study, among patients deceased or intubated, the level of serum KL-6 is significantly higher compared with patients discharged, with a median serum value of 1069 U/mL versus 530 U/mL ($p < 0.001$). According to our data, a significantly greater risk for worse outcomes should be considered for patients having a serum KL-6 value of 968 U/mL or more, with a sensitivity and specificity of 79.2% and 87.1%, respectively, and a positive predictive value of 70.4%. These findings fully match data obtained from previous studies in which elevated levels of serum KL-6 are a significant predictor of poor outcomes in patients with COVID-19 [23–27].

Interestingly, it has been reported that patients infected with SARS-CoV-2 with a KL-6 value persistently above 505 U/mL may be prone to develop pulmonary fibrosis, which may be irreversible when KL-6 raises above 674 U/mL [28].

Chest imaging has played a key role during the SARS-CoV-2 pandemic in detecting alterations in lung and severity quantification [29]. Severity scores have been developed for both chest computed tomography (CT) and lung ultrasound, such as Chung severity score [30] and lung ultrasound score (LUS) [31], respectively. Our study showed a significant correlation between KL-6 serum value and both Chung severity score and LUS (Table 2).

To the best of our knowledge, this is the first study showing a significant correlation between KL-6 and LUS score, highlighting the relevant role of point-of-care lung ultrasound as a ready and non-invasive tool to assess lung damage in these patients. Recent research aiming to investigate the relationship between KL-6 and chest CT scan documented a significant correlation between KL-6 and the extension of parenchymal lesions using a CT semiquantitative score. In particular, the authors found a higher frequency of crazy paving patterns and consolidations involving the right upper and middle lobe in COVID-19 patients with KL-6 > 400 U/mL [32].

Another main finding of our study was the identification of a strong correlation has been found between the serum level of KL-6 and BMI, reflecting a higher risk of lung damage in obese patients. The dramatic impact of obesity on COVID-19 severity and outcomes has been already proved [33,34]. In our study, a strong correlation has been found between the serum level of KL-6 and BMI, reflecting a higher risk of lung damage in obese patients. Accordingly, we found a significantly higher BMI in the group of patients who experienced worse outcomes, in line with existing data emerging from the literature.

A dysregulated immune system represents the hallmark of severe COVID-19. Interleukin-6 is considered a key factor in the inflammatory soup, and its inhibition has been assessed in the treatment algorithm [35]. A significant relationship between adverse clinical outcomes and abnormal levels of IL-6 has been reported [36]. IL-6 was significantly higher in patients experiencing worse outcomes compared to survivors (115 pg/mL versus 25.3, $p = 0.003$). White blood cells and neutrophils were also significantly higher in the former group compared to controls, with a mean value of 12.3×10^3 cells/μL versus 8.65×10^3 cells/μL ($p = 0.008$) and 11.0×10^3 cells/μL versus 7.46×10^3 cells/μL ($p = 0.014$), respectively (Table 1). Although the neutrophils-to-lymphocytes ratio was not found significantly different between the two groups, it shows a positive correlation with KL-6 ($p = 0.03$) (Table 2), suggesting its potential role as a surrogate marker of major risk of lung damage in these patients. The data at our disposal support the role of KL-6 as a key marker in the management of patients infected with SARS-CoV-2 in terms of severity quantification and prognosis. However, we cannot ignore that the study suffers from some limitations, such as the monocentric nature of the study and the use of retrospective data. In conclusion, our study shows that KL-6 represents a strategic tool of great utility in the diagnostic and therapeutic algorithm of more severe patients infected with SARS-CoV-2, as both a key parameter in the severity quantification of COVID-19 and a prognostic biomarker.

6. Conclusions

Severe and critical COVID-19 patients represent a major burden for health care systems. Prognostic stratification of patients with acute respiratory failure secondary to COVID-19 is complex because of the paucity of reliable biomarkers able to predict clinical behavior. KL-6 is a mucin-like glycoprotein mainly expressed by respiratory bronchiolar epithelial cells as well as type II alveolar epithelial cells highly expressed during COVID-19. In this study, we documented that it could be considered a valuable biomarker for predicting the risk of oro-tracheal intubation or death among hospitalized severe or critical COVID-19 patients. The relationship with imaging radiological severity score and other laboratory parameters suggests a robust association with severe prognosis among individuals during severe phases of COVID-19.

Author Contributions: Conceptualization, A.B., F.P. (Fabio Perrotta) and L.P.; methodology, F.S., V.D., F.P. (Francesco Pernaand), M.V., S.S., V.A. and L.A.; validation, F.S., V.D., M.V., L.A., V.A. and S.S.; formal analysis, F.S., V.D. and F.P. (Francesco Pernaand); writing—original draft preparation, A.B., F.P. (Fabio Perrotta), V.D. and F.S.; writing—review and editing, A.B., F.P. (Fabio Perrotta), L.P., V.D. and F.S.; supervision, A.B. and F.P. (Fabio Perrotta); funding acquisition, A.B. All authors have read and agreed to the published version of the manuscript.

Funding: The present research has been funded from Minister of Education, University and Research with PRIN n. 20209TB4AX_001.

Institutional Review Board Statement: The study was approved by the local Ethics Committee (n. 16223/2020, 1 June 2020) and is in accordance with 1976 Declaration of Helsinki and its later amendments. Written consent was waived based on the observational study design and to limit written form contamination.

Informed Consent Statement: Informed consent was obtained from all subjects involved in the study.

Data Availability Statement: The data will be available upon request.

Acknowledgments: The authors would like to acknowledge Dalila Manna for her work as laboratory data manager.

Conflicts of Interest: The authors declare no conflict of interest.

References

1. Carfora, V.; Spiniello, G.; Ricciolino, R.; Di Mauro, M.; Migliaccio, M.G.; Mottola, F.F.; Verde, N.; Coppola, N.; Sagnelli, C.; De Pascalis, S.; et al. Anticoagulant treatment in COVID-19: A narrative review. *J. Thromb. Thrombolysis* **2021**, *51*, 642–648. [CrossRef] [PubMed]
2. Nigro, E.; Perrotta, F.; Polito, R.; D'Agnano, V.; Scialò, F.; Bianco, A.; Daniele, A. Metabolic Perturbations and Severe COVID-19 Disease: Implication of Molecular Pathways. *Int. J. Endocrinol.* **2020**, *2020*, 8896536. [CrossRef] [PubMed]
3. Słomka, A.; Kowalewski, M.; Żekanowska, E. Coronavirus Disease 2019 (COVID–19): A Short Review on Hematological Manifestations. *Pathogens* **2020**, *9*, 49. [CrossRef] [PubMed]
4. Mueller, A.A.; Tamura, T.; Crowley, C.P.; DeGrado, J.R.; Haider, H.; Jezmir, J.L.; Keras, G.; Penn, E.H.; Massaro, A.F.; Kim, E.Y. Inflammatory Biomarker Trends Predict Respiratory Decline in COVID-19 Patients. *Cell Rep. Med.* **2020**, *1*, 100144. [CrossRef]
5. Boccia, M.; Aronne, L.; Celia, B.; Mazzeo, G.; Ceparano, M.; D'Agnano, V.; Parrella, R.; Valente, T.; Bianco, A.; Perrotta, F. COVID-19 and coagulative axis: Review of emerging aspects in a novel disease. *Monaldi Arch. Chest Dis.* **2020**, *90*, 271–276. [CrossRef]
6. Ballester, B.; Milara, J.; Cortijo, J. The role of mucin 1 in respiratory diseases. *Eur. Respir. Rev.* **2021**, *30*, 200149. [CrossRef]
7. Thathiah, A.; Blobel, C.P.; Carson, D.D. Tumor necrosis factor-α converting enzyme/ADAM 17 mediates MUC1 shedding. *J. Biol. Chem.* **2003**, *278*, 3386–3394. [CrossRef]
8. Scialo, F.; Vitale, M.; Daniele, A.; Nigro, E.; Perrotta, F.; Gelzo, M.; Iadevaia, C.; Cerqua, F.S.; Costigliola, A.; Allocca, V.; et al. SARS-CoV-2: One year in the pandemic. What have we learned, the new vaccine era and the threat of SARS-CoV-2 variants. *Biomedicines* **2021**, *9*, 611. [CrossRef]
9. Perrotta, F.; Matera, M.G.; Cazzola, M.; Bianco, A. Severe respiratory SARS-CoV2 infection: Does ACE2 receptor matter? *Respir. Med.* **2020**, *168*, 105996. [CrossRef]
10. Stahel, R.A.; Gilks, W.R.; Lehmann, H.-P.; Schenker, T. Third international workshop on lung tumor and differentiation antigens: Overview of the results of the central data analysis. *Int. J. Cancer* **1994**, *57* (Suppl. S8), 6–26. [CrossRef]
11. Ligtenberg, M.J.L.; Buijs, F.; Vos, H.L.; Hilkens, J. Suppression of Cellular Aggregation by High Levels of Episialin. *Cancer Res.* **1992**, *52*, 2318–2324.
12. Hirasawa, Y.; Kohno, N.; Yokoyama, A.; Inoue, Y.; Abe, M.; Hiwada, K. KL-6, a Human MUC1 Mucin, Is Chemotactic for Human Fibroblasts. *Am. J. Respir. Cell Mol. Biol.* **1997**, *17*, 501–507. [CrossRef]

13. Lee, J.S.; Lee, E.Y.; Ha, Y.J.; Kang, E.H.; Lee, Y.J.; Song, Y.W. Serum kl-6 levels reflect the severity of interstitial lung disease associated with connective tissue disease. *Arthritis Res. Ther.* **2019**, *21*, 58. [CrossRef]
14. Arnold, D.T.; Donald, C.; Lyon, M.; Hamilton, F.W.; Morley, A.J.; Attwood, M.; Dipper, A.; Barratt, S.L. Krebs von den Lungen 6 (KL-6) as a marker for disease severity and persistent radiological abnormalities following COVID-19 infection at 12 weeks. *PLoS ONE* **2021**, *16*, e0249607. [CrossRef]
15. Naderi, N.; Rahimzadeh, M. Krebs von den Lungen-6 (KL-6) as a clinical marker for severe COVID-19: A systematic review and meta-analyses. *Virology* **2020**, *566*, 106–113. [CrossRef]
16. Deng, K.; Fan, Q.; Yang, Y.; Deng, X.; He, R.; Tan, Y.; Lan, Y.; Deng, X.; Pan, Y.; Wang, Y.; et al. Prognostic roles of KL-6 in disease severity and lung injury in COVID-19 patients: A longitudinal retrospective analysis. *J. Med Virol.* **2021**, *93*, 2505–2512. [CrossRef]
17. d'Alessandro, M.; Bergantini, L.; Cameli, P.; Curatola, G.; Remediani, L.; Sestini, P.; Bargagli, E.; Bennett, D.; Bianchi, F.; Perillo, F.; et al. Peripheral biomarkers' panel for severe COVID-19 patients. *J. Med. Virol.* **2021**, *93*, 1230–1232. [CrossRef]
18. Awano, N.; Inomata, M.; Kuse, N.; Tone, M.; Takada, K.; Muto, Y.; Fujimoto, K.; Akagi, Y.; Mawatari, M.; Ueda, A.; et al. Serum KL-6 level is a useful biomarker for evaluating the severity of coronavirus disease 2019. *Respir. Investig.* **2020**, *58*, 440–447. [CrossRef]
19. WHO COVID-19 Clinical Management. *Living Guidance for Clinical Management of COVID-19*; World Health Organization: Geneva, Switzerland, 2021; p. 63.
20. Sato, H.; Callister, M.; Mumby, S.; Quinlan, G.; Welsh, K.; Dubois, R.; Evans, T. KL-6 levels are elevated in plasma from patients with acute respiratory distress syndrome. *Eur. Respir. J.* **2004**, *23*, 142–145. [CrossRef]
21. Alhazzani, W.; Møller, M.H.; Arabi, Y.M.; Loeb, M.; Ng Gong, M.; Fan, E.; Oczkowski, S.; Levy, M.M.; Derde, L.; Dzierba, A.; et al. Surviving Sepsis Campaign: Guidelines on the management of critically ill adults with Coronavirus Disease 2019 (COVID-19). *Intensive Care Med.* **2020**, *46*, 854–887. [CrossRef]
22. Oguz, E.O.; Kucuksahin, O.; Turgay, M.; Yildizgoren, M.T.; Ates, A.; Demir, N.; Kumbasar, O.O.; Kinikli, G.; Duzgun, N. Association of serum KL-6 levels with interstitial lung disease in patients with connective tissue disease: A cross-sectional study. *Clin. Rheumatol.* **2016**, *35*, 663–666. [CrossRef]
23. Ishizaka, A.; Matsuda, T.; Albertine, K.H.; Koh, H.; Tasaka, S.; Hasegawa, N.; Kohno, N.; Kotani, T.; Morisaki, H.; Takeda, J.; et al. Elevation of KL-6, a lung epithelial cell marker, in plasma and epithelial lining fluid in acute respiratory distress syndrome. *Am. J. Physiol. Lung Cell. Mol. Physiol.* **2004**, *286*, L1088–L1094. [CrossRef]
24. Scotto, R.; Pinchera, B.; Perna, F.; Atripaldi, L.; Giaccone, A.; Sequino, D.; Zappulo, E.; Sardanelli, A.; Moriello, N.S.; Stanziola, A.; et al. Serum KL-6 Could Represent a Reliable Indicator of Unfavourable Outcome in Patients with COVID-19 Pneumonia. *Int. J. Environ. Res. Public Health* **2021**, *18*, 2078. [CrossRef]
25. Peng, D.-H.; Luo, Y.; Huang, L.-J.; Liao, F.-L.; Liu, Y.-Y.; Tang, P.; Hu, H.-N.; Chen, W. Correlation of Krebs von den Lungen-6 and fibronectin with pulmonary fibrosis in coronavirus disease 2019. *Clin. Chim. Acta* **2021**, *517*, 48–53. [CrossRef]
26. Yamaya, T.; Hagiwara, E.; Baba, T.; Kitayama, T.; Murohashi, K.; Higa, K.; Sato, Y.; Otoshi, R.; Tabata, E.; Shintani, R.; et al. Serum Krebs von den Lungen-6 levels are associated with mortality and severity in patients with coronavirus disease 2019. *Respir. Investig.* **2021**, *59*, 596–601. [CrossRef]
27. Scarpati, G.; Baldassarre, D.; Lacava, G.; Oliva, F.; Pascale, G.; Boffardi, M.; Pagliano, P.; Calabrese, V.; Tripepi, G.L.; Piazza, O. Krebs von den Lungen 6 levels in COVID-19 ICU Patients are Associated with Mortality. *medRxiv* **2021**, 2021.11.17.21266464.
28. Xue, M.; Zhang, T.; Chen, H.; Zeng, Y.; Lin, R.; Zhen, Y.; Li, N.; Huang, Z.; Hu, H.; Zhou, L.; et al. Krebs Von den Lungen-6 as a predictive indicator for the risk of secondary pulmonary fibrosis and its reversibility in COVID-19 patients. *Int. J. Biol. Sci.* **2021**, *17*, 1565–1573. [CrossRef]
29. Bianco, A.; Valente, T.; Perrotta, F.; Stellato, E.; Brunese, L.; Wood, B.J.; Carrafiello, G.; Parrella, R.; Aronne, L.; Boccia, M.; et al. Remarkable vessel enlargement within lung consolidation in COVID-19 compared to AH1N1 pneumonia: A retrospective study in Italy. *Heliyon* **2021**, *7*, e07112. [CrossRef]
30. Chung, M.; Bernheim, A.; Mei, X.; Zhang, N.; Huang, M.; Zeng, X.; Cui, J.; Xu, W.; Yang, Y.; Fayad, Z.A.; et al. CT imaging features of 2019 novel coronavirus (2019–nCoV). *Radiology* **2020**, *295*, 202–207. [CrossRef]
31. Dargent, A.; Chatelain, E.; Kreitmann, L.; Quenot, J.-P.; Cour, M.; Argaud, L.; The COVID-LUS Study Group. Lung ultrasound score to monitor COVID-19 pneumonia progression in patients with ARDS. *PLoS ONE* **2020**, *15*, e0236312. [CrossRef]
32. Anastasi, E.; Manganaro, L.; Guiducci, E.; Ciaglia, S.; Dolciami, M. Association of serum Krebs von den Lungen-6 and chest CT as potential prognostic factors in severe acute respiratory syndrome SARS-CoV-2: A preliminary experience. *Radiol. Med.* **2022**, *127*, 725–732. [CrossRef] [PubMed]
33. Sjögren, L.; Stenberg, E.; Thuccani, M.; Martikainen, J.; Rylander, C.; Wallenius, V.; Olbers, T.; Kindblom, J.M. Impact of obesity on intensive care outcomes in patients with COVID-19 in Sweden—A cohort study. *PLoS ONE* **2021**, *16*, e0257891. [CrossRef] [PubMed]
34. Zhang, X.; Lewis, A.M.; Moley, J.R.; Brestoff, J.R. A systematic review and meta-analysis of obesity and COVID-19 outcomes. *Sci. Rep.* **2021**, *11*, 7193. [CrossRef] [PubMed]
35. Rubin, E.J.; Longo, D.L.; Baden, L. R Interleukin-6 Receptor Inhibition in COVID-19—Cooling the Inflammatory Soup. *N. Engl. J. Med.* **2021**, *384*, 1564–1565. [CrossRef] [PubMed]
36. Coomes, E.A.; Haghbayan, H. Interleukin-6 in COVID-19: A systematic review and meta-analysis. *Rev. Med. Virol.* **2020**, *30*, e2141. [CrossRef]

life

MDPI

Article

How COVID-19 Hijacks the Cytoskeleton: Therapeutic Implications

Maral Aminpour [1], Stuart Hameroff [2,3,4] and Jack A. Tuszynski [5,6,*]

1 Department of Biomedical Engineering, University of Alberta, Edmonton, AB T6G 1Z2, Canada; aminpour@ualberta.ca
2 Department of Anesthesiology, The University of Arizona Health Sciences Center, Tucson, AZ 85621, USA; hameroff@u.arizona.edu
3 Department of Psychology, University of Arizona, Tucson, AZ 85621, USA
4 Center for Consciousness Studies, The University of Arizona, Tucson, AZ 85621, USA
5 Department of Physics, University of Alberta, Edmonton, AB T6G 1Z2, Canada
6 Department of Mechanical and Aerospace Engineering (DIMEAS), Politecnico di Torino, 10129 Torino, Italy
* Correspondence: jackt@ualberta.ca

Abstract: The SARS-CoV-2 virus invades and replicates within host cells by "hijacking" biomolecular machinery, gaining control of the microtubule cytoskeleton. After attaching to membrane receptors and entering cells, the SARS-CoV-2 virus co-opts the dynamic intra-cellular cytoskeletal network of microtubules, actin, and the microtubule-organizing center, enabling three factors that lead to clinical pathology: (1) viral load due to intra-cellular trafficking, (2) cell-to-cell spread by filopodia, and (3) immune dysfunction, ranging from hyper-inflammatory cytokine storm to ineffective or absent response. These factors all depend directly on microtubules and the microtubule-organizing center, as do cell functions such as mitosis and immune cell movement. Here we consider how the SARS-CoV-2 virus may "hijack" cytoskeletal functions by docking inside the microtubule-organizing center's centriole "barrels", enabling certain interactions between the virus's positively charged spike ("S") proteins and negatively charged C-termini of the microtubules that the centriole comprises, somewhat like fingers on a keyboard. This points to the potential benefit of therapies aimed not directly at the virus but at the microtubules and microtubule-organizing center of the host cell on which the virus depends. These therapies could range from anti-microtubule drugs to low-intensity ultrasound (megahertz mechanical vibrations) externally applied to the vagus nerve at the neck and/or to the spleen (since both are involved in mediating inflammatory response). Given that ultrasound imaging machines suitable for vagal/splenic ultrasound are available for clinical trials in every hospital, we recommend an alternative therapeutic approach for COVID-19 based on addressing and normalizing the host cell microtubules and microtubule-organizing centers co-opted by the SARS-CoV-2 virus.

Keywords: SARS-CoV-2; COVID-19; microtubules; MTOC; low-intensity ultrasound

Citation: Aminpour, M.; Hameroff, S.; Tuszynski, J.A. How COVID-19 Hijacks the Cytoskeleton: Therapeutic Implications. *Life* **2022**, *12*, 814. https://doi.org/10.3390/life12060814

Academic Editors: Silvia De Francia, Sarah Allegra and Daniele Focosi

Received: 23 March 2022
Accepted: 25 May 2022
Published: 30 May 2022

Publisher's Note: MDPI stays neutral with regard to jurisdictional claims in published maps and institutional affiliations.

1. Introduction

The novel coronavirus, severe acute respiratory syndrome coronavirus 2 (SARS-CoV-2), is the causative agent of COVID-19, or coronavirus-19 disease, one of the most challenging pandemics of all time. Although vaccination is a proven strategy for encompassing coronavirus disease 2019 (COVID-19), the emergence of SARS-CoV-2 variants has led to waning protection against the virus. Significant efforts are underway to repurpose existing drugs, circumventing the lengthy wait time needed for testing de novo design drugs. Nevertheless, a specific treatment for COVID-19 has yet to be identified, and innovative strategies must be devised and implemented to address this challenge. In the ongoing battle against COVID-19, though, little attention has been given to the role of microtubules and of the microtubule-organizing center (MTOC) or to non-pharmacological therapeutic strategies targeting inflammatory and immunological operations that may be beneficial for decreasing COVID-19-induced complications and improving patient outcomes (33–35). In this paper, we recommend, as a complement to anti-microtubule drugs, alternative therapeutic approaches that treat COVID-19 by addressing host cell microtubules and MTOCs co-opted by the SARS-CoV-2 virus. In this section we provide background information concerning the MTOC and cytokine storm (Section 1.1) and the role of microtubules and the cytoskeleton in COVID-19 (Section 1.2) as well as the molecular architecture of the centriole and cartwheel and the relation of them with microtubules (Section 1.2).

1.1. "Brain" of the Cytoskeleton and Cytokine Storm

Although SARS-CoV-2 belongs to a family of infectious viruses that also includes SARS and Middle East Respiratory Syndrome (MERS), SARS and MERS never reached pandemic levels, and this raises questions as to why SARS-CoV-2 is so difficult to contain and treat. To some extent, all infectious viruses invade and "hijack" host cell machinery to proliferate and complete their life cycle, but SARS-CoV-2 seems to have extraordinary adaptive abilities in this regard. There have been several models [1] and hypotheses [2] about the life cycle of SARS-CoV-2 [3–5]. The early phases that are typically necessary in order for the coronavirus to gain cell entry—e.g., binding and viral entry via membrane fusion [6] or endocytosis, priming of spike protein, uncoating of RNA, RNA replication in double-walled vesicles (DMV), and transcription—are relatively well understood. The evidence on the basis of which to correctly or completely describe the late phases such as the assembly of the N-RNA complexes with M, E, and S proteins at the endoplasmic reticulum-Golgi intermediate compartment (ERGIC) toward several possible exit routes in small exocytic single-layer membrane vesicles (SMV), large virus-containing vesicle (LVCP) secretory pathways [7], or deacidified lysosomal vesicles [8] before plasma membrane exocytosis, however, is still incomplete. There are still uncertainties and unknowns concerning the process by which virions are formed and enveloped during exocytosis. Indeed, a wide field of research remains to be explored in clarifying the mechanisms underlying the particle formation, budding, transportation, and egress of SARS-CoV-2. Correspondingly, there are many more prospective pathways (models and hypotheses) that need to be considered and addressed across the life cycle of the virus with respect to its interactions with the cytoskeleton and microtubules as an essential component of egress [9–11].

The particular cell machinery hijacked by SARS-CoV-2 and other viruses includes the actin and microtubule cytoskeleton. The cytoskeleton is a dynamic network of filamentous protein polymers, including microtubules, actin filaments, and centrioles, that functions as both the cell's scaffolding and, apparently, an information processing and signaling system. The cytoskeleton forms part of the MTOC. Other microtubule-associated proteins (MAPs) connect cytoskeletal polymers into networks, and still others link them with membrane proteins. The focal point is the centrosome, or MTOC, which functions as the "brain" of the cytoskeleton, composed of a pair of microtubule-based cylinders, mutually perpendicular and embedded in an electronegative "pericentriolar material" (Figures 1 and 2).

SARS-CoV-2 appears to use and depend upon co-opted activities of the microtubules and MTOC for infection, proliferation, and host cell damage. "Cytokine storm", a hyper-inflammatory immune response causing extensive cellular damage, is organized and mediated through the MTOC (Figure 3).

1.2. Microtubules and the Cytoskeleton—Roles in COVID-19

The SARS-Cov-2 virus takes over human bodies by viral attachment, internalization, transport, transcription, replication, assembly, exocytosis, and cell-to-cell spread, causing further damage by inducing an excessive immune response on the part of the host, including "cytokine storm", as well as by disrupting normal cytoskeletal functions. SARS-CoV-2 seems to be particularly adept at subverting and co-opting microtubules.

Microtubules, it should be noted, are cylindrical lattice polymers of the protein tubulin that are involved in critical cellular processes, including mitosis, cellular motility, and morphology, and the function of many cell membrane receptors. Trafficking of viruses and other material through cells is often facilitated by active transport by "motor proteins", which convey cargo by efficiently moving it along these microtubules.

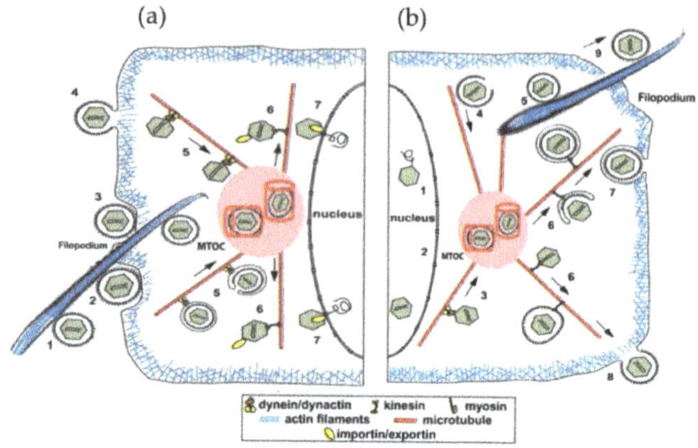

Figure 1. (**a**) Entry of SARS-CoV-2 into cells can take several routes: (1) virus delivered by filopodia extending from nearby infected cell; (2–4) fusion, invagination into cell; (5) transport by motor proteins toward MTOC; (6) transport from MTOC to different cell regions; and (7) entry into nucleus for reverse transcription and replication. The MTOC includes two perpendicular cylinders embedded in a dense electronegative "pericentriolar material". The MTOC appears to enable optimal traffic along microtubules to distribute and replicate SARS-CoV-2 virus. (**b**) After reverse transcription and replication, virion is transported to MTOC (3,4,5) and from there along microtubules to other cellular regions and, ultimately, egress from the cell (6–9).

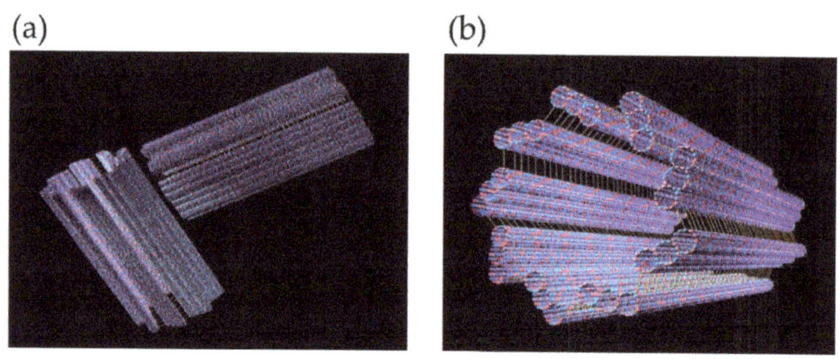

Figure 2. (**a**) A MTOC/centriole/centrosome comprising two cylinders, each consisting of nine longitudinally fused microtubules in a perpendicular array. (**b**) A single cylinder, i.e., half of a centriole/centrosome.

Microtubules self-assemble from tubulin, protein dimers composed of two similar 55 kDa monomer proteins, known as α-tubulin and β-tubulin, that self-assemble end-to-end to form long chains of tubulin dimers known as protofilaments. These then associate laterally into lattice sheets that roll into cylindrical microtubules. Adjacent protofilaments can associate in either of two ways corresponding to two types of lattices—the asymmetrical B-lattice with a seam and the seamless and hexagonal A-lattice with Fibonacci geometry and symmetry. Evidence indicates that microtubules have self-similar patterns of vibrational resonances and conductivity features at terahertz, gigahertz, megahertz, and kilohertz frequencies [12–14].

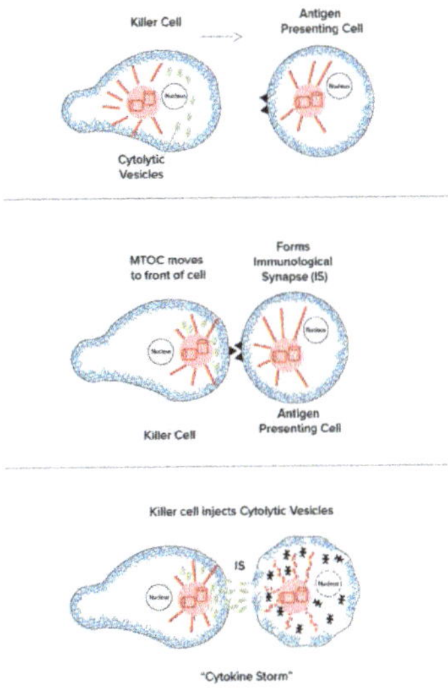

Figure 3. (**Left**): The effector cell (immune cell) makes contact and "immunological synapse" as the centrosome/MTOC moves to the "front" of the cell near the synapse. (**Right**): The centrosomes/MTOCs manage the release of cytolytic vesicles containing cytokines, tumor necrosing factor, interleukins, and other reactive agents (cytokine storm).

Microtubules' organizational abilities, lattice structure, and vibrational properties have prompted speculation that they process information, i.e., that they function as the cell's nervous system and "on-board computer". Each tubulin in a microtubule lattice can be modified ("encoded") by post-translational modification and phosphorylated by kinase enzymes, making them ideal for information and vibrational microtubule regulation of cellular activities. It has been proposed in this regard that phosphorylation of specific tubulins in a microtubule lattice by synaptic CaMKII (calcium-calmodulin kinase 2) serves to encode memory in microtubules inside neurons (Figure 4) [15], and specific locations of the MAP tau on microtubules appear to regulate trafficking of motor proteins and their cargo. Other regulatory kinases may phosphorylate specific patterns of tubulins in microtubule lattices to initiate specific cytoskeletal activities, suggesting the occurrence of some manner of encoding.

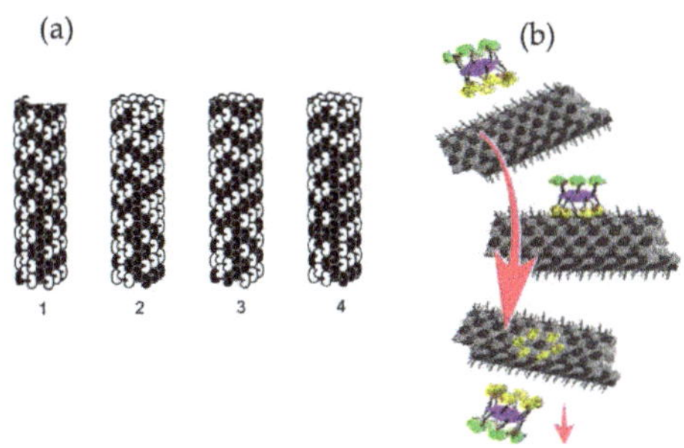

Figure 4. (**a**) Four steps in microtubule automata theoretical simulation. The state of each peanut-shaped tubulin protein is influenced by neighbor states in the microtubule lattice, resulting in moving patterns (1,2,3, and 4) capable of organizing cellular activities [16]. (**b**) In another theoretical simulation, a kinase (CaMKII) phosphorylates six tubulins in a microtubule lattice, encoding information. The hexagonal CaMKII precisely matches the microtubule A and B lattices [15].

Could SARS-CoV-2 hijack the cytoskeleton by hacking microtubule kinase codes, either by activating particular kinases or through direct effects on microtubules? If so, therapies aimed at disrupting microtubules may prevent virus spread and minimize damage to the host cell. Anti-microtubule drugs used in cancer chemotherapy to block mitosis, and in gout to immobilize immune cells, have already been shown to reduce viral load and to help in treating COVID-19 symptoms [17] and are currently being studied in human clinical trials [18,19].

The "centrosome", or MTOC, contains two cylinders in a peculiar perpendicular array, each composed of nine parallel, interconnected triplets of microtubules, also known as the "centriole" (Figure 2). The MTOC centriole has been hypothesized, and to some degree demonstrated experimentally, to act as the cell's cytoskeletal "brain", and "eye", capable of detecting electromagnetic signals in the visible and near-visible range [20]. Moreover, linked by microtubule spindles, the two cylinders replicate and separate to initiate cell division/mitosis and establish daughter cell shape. Single cylinders with nine parallel microtubule doublets and a central doublet also occur by themselves, providing the structure for cilia and flagella, which protrude, membrane-covered, from cell surfaces to serve sensing and cellular locomotion functions. Other cytoskeletal appendages include filopodia, composed mostly of actin, which are anchored, initiated, and sometimes occupied by microtubules. SARS-CoV-2 causes infected human cells to produce multiple filopodia laden with virus to extend outward toward neighboring cells, invade these cells, and spread the virus.

The MTOC appears to be the cytoskeletal command center, regulating cellular function and activities through microtubules and actin. Theoretical models [21–23] have suggested that microtubules process information, with individual tubulin subunit states representing information, and each tubulin dimer having two protruding, negatively charged C-termini "tails". In this paper we consider how the SARS-CoV-2 virus may co-opt the cytoskeleton by docking inside centriole barrels of the MTOC, forming interactions between the virus's positively charged spike ("S") proteins and the negatively charged C-termini of the microtubules making up the centriole. In this manner, we suggest, the virus can modulate and interrupt MTOC and microtubule function, behaving in much the same way as fingers on a keyboard.

1.3. Molecular Architecture of Centriole and Cartwheel

In human cells, the centriole is a cylindrical structure, typically 450 nm in length and with inner and outer diameters of ~130 nm and ~250 nm, respectively. Centrioles and procentrioles are characterized by a nine-fold radial symmetric arrangement of triplet microtubules that is polarized along its height, with the base referred to as the proximal end and the tip as the distal end. The proximal region of the centriole is defined by the presence of the cartwheel structure, which serves as an essential seed for centriole formation and is thought to participate in establishing the nine-fold symmetry of the entire organelle (Figure 5a) [24,25]. The cartwheel is built from a central hub (20~25 nm in diameter) from which nine spokes (40~45 long) emanate and connect through a pinhead (~20 nm) with A-microtubule of the peripheral microtubule triplets (Figure 5b). Each hub ring can accommodate nine homodimers of SAS-6, a protein that is essential for cartwheel assembly. The A-microtubule from one triplet is connected with the C-microtubule of the next triplet in a clockwise manner via a so-called A–C linker. The pinhead and the A–C linker, which are connected through the triplet base, extend more distal than the cartwheel and co-exist with the inner scaffold structure (the cartwheel is indirectly connected to the A–C linker through a flexible triplet-base structure extending from the pinhead). A close-up view of the A-microtubule and its associated key proteins on human centrioles and procentrioles, such as hSAS-6, CEP135, and CPAP, is shown in Figure 5c–g [26–28]. In this regard, a recent study applied cryo-electron tomography to four procentriole cartwheels of human cells (note that mature human centrioles lack cartwheels) in order to determine the cartwheel length and to better understand the relationship between the A–C linker and the cartwheel.

This study found that the cartwheel length extends 189 +/− 9 nm and spans 70% of the length of the A–C linker (270 +/− 26 nm of the proximal region) in humans, where their study considered a procentriole length of ~410 nm [29].

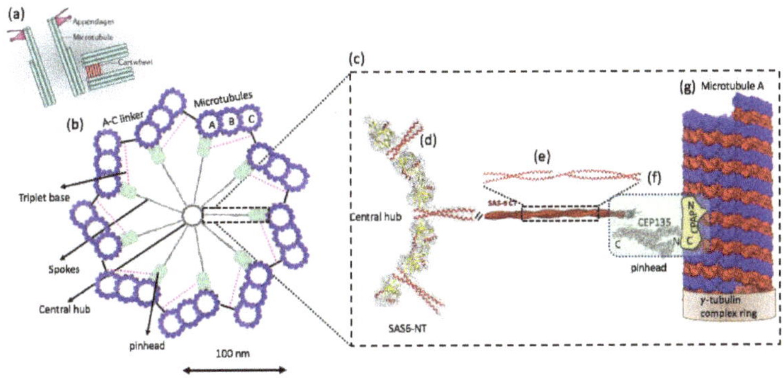

Figure 5. (**a**) Schematic representation of mother and daughter centriole pair in a human cell. (**b**) Schematic representation of the cartwheel and its associated proteins viewed from the proximal end (left). The cartwheel consists of a central hub (20~25 nm in diameter) from which nine spokes (40~45 long) emanate and connect through a pinhead (~20 nm) with A-microtubule of the peripheral microtubule triplets (blue). The A-microtubule from one triplet is connected with the C-microtubule of the next triplet located clockwise via a so-called A–C linker. The pinhead and the A–C linker are connected through the triplet base. The pink dashed line indicates the putative position of the triplet base. (**c**) A close-up view of the A-microtubule and its associated key proteins, such as hSAS-6, CEP135, and CPAP (right). A-microtubule is rotated 90 degrees from the view on the left. (**d**) The cartwheel's central hub comprises SAS-6 homo-oligomers [26,30]. The central hub is not a continuous tube but rather a structure exhibiting periodicities along its 100 nm height. (**e**) The spokes extending outward from the hub are homodimeric SAS-6 coiled-coils (SAS-6 NT). (**f**) SAS-6 NT extends [30] into a region known as the "pinhead", i.e., the celadon green color in the low magnification view (left) and in the rectangular box (right). CEP135 protein is located at the pinhead of the cartwheel and serves as a bridge between the spokes and the outer microtubules (possibly the A-tubule). It interacts with SAS-6 (via its C-terminal domain), microtubules s, and SAS-4 (via its N-terminus) [26]. CPAP carries both a tubulin dimer-binding domain [31] and a microtubule-binding domain [32] and is associated with the γ-tubulin complex [33]. (**g**) A-microtubule is a long, hollow cylinder made up of polymerized α- (blue) and β-tubulin (red) dimers. The outer diameter of a microtubule is approximately 25 nm, while the inner diameter is approximately 17 nm. The proximal end of the A-microtubule in a nascent human procentriole is capped by a structure similar to that of the γ-tubulin ring complex [34], a known microtubule nucleator in animal cells. ((**a**) [24] and part of (**e**) are adopted from [28] with author permission).

2. Materials and Methods

The VMD molecular graphics software package is used for both the execution of Adaptive Poisson–Boltzmann Solver (APBS) [35] and the visualization of the resulting electrostatic potentials. AMBER force field [36] parameters such as atomic charges and radii are assigned using PDB2PQR webserver (http://server.poissonboltzmann.org, accessed on 29 May 2021) [37].

3. Results

3.1. Possible SARS-CoV 2 Invasion in Procentriole and Interaction with Microtubules

Electron micrographs of negative-stained SARS-CoV-2 particles show a generally spherical shape with some pleomorphism. The diameter of SARS-CoV-2, meanwhile, ranges between 50 nm and 140 nm [38]. As mentioned in the previous section, in human cells the centriole is a cylindrical structure, typically 450 nm in length and with inner and outer diameters of ~130 nm and ~250 nm, respectively. By comparing the size of the two structures, we can conclude that SARS-CoV-2 can physically fit within the centriole. In this study, we propose that the spike protein's tip, which is seen to be relatively positively charged, docks onto the microtubule with the negatively charged C-terminus of the tubulin due to electrostatic attraction (Figure 6).

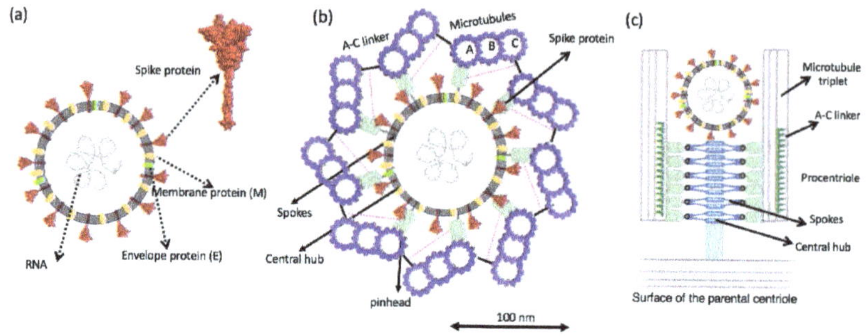

Figure 6. (**a**) Schematic of structural proteins of the SARS-CoV-2 virion. (**b**) Top view cartwheel complexed with the invading SARS-CoV-2 virion. Trimeric spike (S) protein on the surface of the SARS-CoV-2 is shown in red. (**c**) Side view of cartwheel-containing procentriole in complex with SARS-CoV 2 virion. (Part of (**c**) is adopted from [25] with author permission).

Regarding protein–protein interactions, many studies have examined the short-range interaction, which is dependent on the specific, complementary structure of each protein. However, the weak electrostatic binding interaction, which is probably effective over a long range, holds the same level of importance. A recent study showed that neutralizing the positively charged polybasic cleavage sites ($R_{682}RAR_{685}$ for each spike subunit), which are distributed approximately 10 nm away from the receptor-binding domain (RBD) of the spike protein, results in a weakening (by 34%) of the spike protein−ACE2 binding energy. It should be noted in this regard that ACE2 is highly negatively charged (−28 e), entailing that the extracellular membrane surface will be highly negatively charged. The dramatic drops observed in the RBD−ACE2 binding energy after neutralizing the polybasic sites with negatively charged mutations may be largely attributable to a change in (long-range) Coulomb interactions between ACE2 and the spike protein due to the RBD domain of the spike protein becoming more negative [39]. Microtubules are highly negatively charged due to the negative charge of the αβ-tubulin heterodimer itself (20–30 e−). However, a large portion of the tubulin charge, at least 40%, is concentrated in the non-structured C-terminal tails, which are rich in Glu and Asp amino acids.

Several studies have characterized the interactions of various microtubule-associated proteins and the role of the negatively charged C-terminal tubulin tails in binding positively charged domains of these proteins to the tubulin surface [40,41]. Electrostatic potential at the binding interface provides insights into the role of electrostatic in the binding. Detailed information about the spike protein's electrostatic surface can be obtained via the APBS calculation and the visualization of the resulting electrostatic potentials (see Figures 7 and 8).

3.2. Electrostatic Potential at the Binding Interface

We performed the APBS calculations and visualized the resulting electrostatic potentials for the SARS-CoV-2 Spike protein (Figure 7a–c) and tubulin protein (Figure 7d). The electrostatic potential map of the spike protein (red = negative charge, blue = positive charge) highlights a trefoil of positive charge (blue) around the central point of the bulbous head (i.e., the central area of the receptor-binding domain). The C-terminals (two tails) of α and β tubulin monomers, meanwhile, feature the negative charge. Figure 8 provides a schematic representation of the attraction between the negative C-terminals of tubulin profilament and the central point of the receptor-binding domain of the spike protein embedded in the membrane of the SARS-CoV-2 virion.

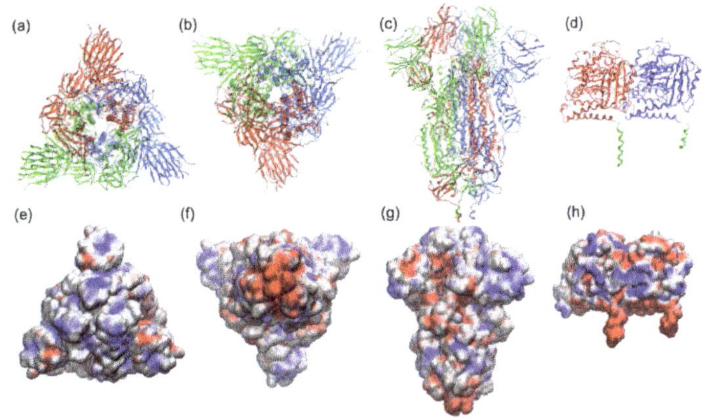

Figure 7. Structure and electrostatic map of the SARS-CoV-2 spike-S1 protein as viewed from (**a**,**e**) top (crown), (**b**,**f**) bottom (junction of S1 and HR2 linker; HR2 linker is not shown here), and, in (**c**,**g**), side, based on Protein Data Bank (PDB) entry 6VXX. The trimeric assembly of the spike protein (each monomer is colored in red, blue, and green) is presented as cartoon (first row) and as the electrostatic potential map (second row) highlighting a trefoil of positive charge (blue) around the central point. (**d**,**h**) Structure and electrostatic map of tubulin protein (PDB: 1jff) (α and β tubulin are in blue and red, respectively, and C-terminals are depicted in green). The VMD molecular graphics software package was used for both the execution of APBS (Adaptive Poisson–Boltzmann Solver) [35] and the visualization of the resulting electrostatic potentials. AMBER force field [36] parameters such as atomic charges and radii are assigned using PDB2PQR webserver (http://server.poissonboltzmann.org, accessed on 29 May 2021) [37].

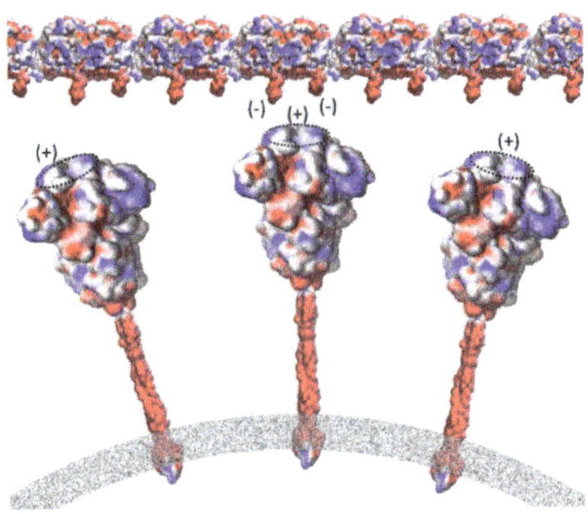

Figure 8. Interaction between the negative C-terminals of tubulin and the central point of the receptor-binding domain of the spike protein embedded in the membrane of the SARS-CoV-2 virion. Mapping of electrostatic potential (red = negative charge, blue = positive charge) was carried out using the VMD molecular graphics software package for both the execution of APBS [35] and the visualization of the resulting electrostatic potentials. Force field [36] parameters such as atomic charges and radii were assigned using the PDB2PQR webserver (http://server.poissonboltzmann.org, accessed on 29 May 2021) [37].

4. Discussion

4.1. Snatching and Hijacking Microtubules

While the virus may be difficult to kill or avoid, most infections are not serious if the following three factors are minimized: (1) excessive viral load due to host cell entry, intra-cellular transport, and replication; (2) cell-to-cell spread by filopodia; and (3) immune dysfunction, ranging from hyper-inflammatory cytokine storm to inadequate response [42]. These factors all depend on the microtubules and MTOC, underscoring the potential benefit of therapies aimed not at the virus, but at microtubules and MTOC of the host cell on which the virus depends. There are at least three types of microtubule-based activities hijacked by SARS-CoV-2, as briefly described in the following subsections.

4.1.1. Entry and Transport Inside the Host Cell

SARS-CoV-2 attacks a host cell through the coronavirus spike (S) protein, which mediates host cell attachment and viral entry [43]. The S protein attaches to membrane receptor proteins, primarily the ACE2 receptor, which in turn is associated with β-tubulin and anchored by intra-cellular microtubules. The virus then invaginates with surrounding membrane into the cell (becoming a "virion") and then attaches to a microtubule via a motor protein. The motor protein then moves along the microtubule to bring the virus to the MTOC near the nucleus at the cell center that redirects virions. From there, virions may enter the nucleus, where coronavirus RNA exits the virion for reverse transcription and replication to a DNA replicon. This is subsequently transcribed back into RNA, which exits the nucleus into the MTOC. From there it is transported by different motor proteins moving along microtubules to various cell locations, including back to the cell membrane for egress from the cell. Virions also travel to other locations inside host cells, transported to specific areas by motor proteins along multiple cytoskeletal structures in what can be described as an on-demand "ride-share" system for virion transport inside host cells. Viruses rearrange these cytoskeletal filaments so that they can either utilize them as tracks or move them aside when they represent barriers (Figure 1).

4.1.2. MTOC (Microtubule Organizing Center) and Cytokine Storm

The centrosome, or MTOC, organizes and directs intra-cellular movement and activities. For example, in mitosis/cell division, centrosomes replicate and separate to pull apart duplicate sets of chromosomes by means of microtubule mitotic spindles to establish daughter cell genome and architecture. In motile cells such as fibroblasts, which move into wounds to enable healing, the centrosome/MTOC is located in the front end of the cell, directing the "leading edge" of the actin to move in the proper direction. In the immune cells that move about to find infectious agents and antigen-containing cells, the centrosome/MTOC is behind the nucleus, still playing a key role in directional guidance and motility. However, when an immune cell, such as a killer T cell, reaches an antigen-containing cell, it forms an "immunological synapse" (IS), and the centrosome/MTOC moves to the front, near the IS. From there, the centrosome/MTOC directs release of cytolytic substances, including cytokines, tumor necrosing factor, and interleukins, in a phenomenon referred to as "cytokine storm" (Figure 3).

4.1.3. Invasive Filopodia

Cells hijacked by SARS-CoV-2 grow arm-like cytoskeletal extensions, or filopodia, which extend outward from the cell and carry the virus from cell to cell, enabling local spread. Filopodia are composed largely of actin but are initiated, steered, and occupied by direct interactions with microtubules. SARS-CoV-2 is thought to control microtubules by controlling or activating kinases—enzymes that regulate cell functions by a process of phosphorylation, depositing high-energy phosphate groups. However, SARS-CoV-2 may directly access microtubule information codes to regulate functions. Figure 1a depicts a filopodium from another cell carrying the virus to a host cell, while Figure 1b depicts a filopodium carrying a virus out of the cell.

4.2. Therapeutic Approaches

Since SARS-CoV-2 is highly dependent on microtubule activities, therapies aimed at microtubules may be efficacious in treating COVID-19, either alone or in combination with other interventions. These potential therapies include (1) anti-microtubule drugs (see Section 4.2.1), such as colchicine, taxol, and others (but with a much lower toxicity risk, e.g., noscapine), that impair the microtubule and MTOC function on which virus proliferation depends; (2) low-intensity ultrasound vagal stimulation at the neck surface to enlist anti-inflammatory vagal actions (see Section 4.2.2); and (3) low-intensity ultrasound to the spleen to modulate inflammatory actions of immune cells, possibly dampening the effects on MTOC-mediated cytokine storm (see Section 4.2.3), where the latter two approaches use low-intensity ultrasound to resonate with microtubules' characteristic mechanical vibrations.

The cholinergic anti-inflammatory pathway (CAIP) has been proposed as a principal mechanism by which the brain, through the vagus nerve, modulates the immune system in the spleen [44]. According to this mechanism, in response to infection or injury, the parasympathetic vagus nerve transmits signals from the brain to the adrenergic splenic nerve, which consequently interacts with splenic immune cells. Stimulation of the vagus and spleen nerves triggers this neural-immune reflex and dampens the inflammatory response to infection or tissue injury [45,46]. Vagus nerve stimulation (VNS), it should be noted, refers to any technique that stimulates the vagus nerve (e.g., electrical or ultrasound impulses).

It is worth mentioning that neurons (also called nerve cells) are the fundamental units of the brain and nervous system, and they rely on microtubules for their structure and functions. Microtubules form bundles that are particularly prominent in neurons and in the nervous system. Non-invasive, painless ultrasound to modulate microtubules in the vagus nerve may be a preferable option as a therapeutic approach for patients infected with SARS-CoV-2 [47–49].

4.2.1. Anti-Microtubule Drugs

Colchicine destabilizes microtubules by preventing free tubulin dimers from being incorporated into the microtubule structure and thus prevents the assembly/disassembly cycles which enable immune cell movement. Accordingly, colchicine is effective against gout, a painful inflammation of joint spaces due to the build-up of uric acid crystals. By disabling microtubules, colchicine blocks immune cells from entering and inflaming joints to fight the uric acid crystals, which are relatively harmless. In this case, the body's immune response causes severe pain and swelling that is more problematic than any effect of the uric acid crystals themselves. A recent open-label, randomized clinical study of colchicine in COVD-19 patients showed significant improvement [50]. While colchicine is somewhat toxic, with renal and other effects, new, less toxic colchicine-like drugs are currently under development. Moreover, there are a number of anti-cancer drugs that affect microtubule stability, impairing mitosis to fight malignancy, that could also be efficacious against COVID-19. One such example is noscapine [51]. We also would like to mention that our study is indirectly corroborated by the efficacy shown in clinical trials of anti-microtubule drugs. Recently, COLCORONA (Colchicine Coronavirus SARS-CoV2), which is a microtubule-targeting compound that disrupts mitotic spindle poles in human cells, has been selected for the phase 3 trial to treat COVID-19 (https://clinicaltrials.gov/ct2/show/NCT04322682, accessed on 29 May 2021) [52].

4.2.2. Vagal Modulation with Ultrasound

The vagus nerve, the longest nerve in the nervous system, innervates visceral organs and tissues throughout the body, mediating parasympathetic effects, boosting the immune system, and blocking inflammation. Electromagnetic stimulation of the vagus nerve at the neck reduces systemic inflammation and is being tried in COVID-19 patients [53].

Ultrasound consists of mechanical waves above audible range, i.e., from 20,000 Hz to hundreds of megahertz but primarily hundreds of thousands of kilohertz to the low megahertz range. This technology has been safely used for medical imaging for nearly

100 years. Moreover, high-intensity ultrasound has been used to cause lesions by heating, e.g., to ablate tumors, and mid-range intensities from the scalp have been used to open the blood–brain barrier to allow drugs to enter the brain. Low-intensity, sub-thermal intensities, meanwhile, have been used to cause healing of peripheral nerves, bone, and other tissues and, when delivered (focused or unfocused) to the brain, can cause mood elevation and physiological effects. Although the mechanism underlying the healing and mood enhancement effected by low-intensity ultrasound is uncertain, it may involve microtubules, known to have resonances in megahertz range. Moreover, the application of low-intensity ultrasound to chondrocyte and osteoblast cells has been shown to cause rearrangement of the cytoskeleton [54–56].

4.2.3. Splenic Ultrasound

The spleen produces and stores blood cells, including immune cells, and also has significant vagal innervation. In an animal model of systemic inflammatory arthritis, ultrasound to the spleen showed significant resolution of arthritis in all joints [46]. In that particular study, the mechanism was attributed to vagal stimulation, but it could well involve modulation, rearrangement, and repurposing of microtubules and MTOC in immune cells that then promulgate throughout the body. In the case of COVID-19, splenic ultrasound may change cytoskeletal plasticity in immune cells, inhibiting the ability of the virus to control the MTOC and/or promote anti-inflammatory actions [57].

5. Conclusions

The damage to host cells that leads to clinical symptoms in COVID-19 is governed largely by three factors: (1) excessive virus load due to host cell entry, intra-cellular transport, and replication; (2) cell-to-cell spread by filopodia; and (3) immune dysfunction, ranging from hyper-inflammatory cytokine storm to inadequate response [42]. These factors all depend on the SARS-CoV-2 virus co-opting microtubules and MTOCs (also consisting of microtubules) for their own purposes, and this suggests possible benefits of therapies aimed not at the virus but at the microtubules and MTOC of the host cell on which the virus depends. In this regard, clinical trials for colchicine and other anti-microtubule drugs as pharmacological agents aimed at preventing and limiting COVID-19 infections are underway [50]. As a complement to these efforts, we propose clinical studies for the application of low-intensity ultrasound to the vagus nerve at the neck and to the spleen (either through the abdomen or posteriorly through the ribcage) as a therapeutic for COVID-19 patients.

In fact, vagal stimulation at the neck with electromagnetic devices is already being tested in COVID-19 patients [53], and vagal ultrasound has been shown to reduce inflammation in animals [58]. In humans, anecdotally, vagal ultrasound is sedating and safe, resulting in a slight decrease in heart rate, and is easy to apply at the neck. Ultrasound to the spleen, meanwhile, has been shown to improve inflammatory arthritis in animals, an effect attributed to vagal stimulation in the spleen [46].

Ultrasound consists of mechanical oscillations from 20 kHz to hundreds of megahertz, ~100 million per second. The mechanism underlying the physiological effects of ultrasound is not yet known, but it is presumed to involve microtubules, as they have resonant oscillations in frequencies ranging from tens of kilohertz to megahertz, gigahertz, and even terahertz. Centrioles/MTOC are known to be sensitive to terahertz and gigahertz, abruptly changing direction of cell movement in response to infrared light [20]. In cell cultures, ultrasound has been shown to cause microtubule-dependent neurite sprouting and optimal functional rearrangement of microtubule configurations [59,60].

The main purpose and goal of this approach is for the ultrasound waves to resonate, modulate, and normalize microtubules and MTOC, which normally regulate immune and other cells but, in the case of viral infection, have been hijacked by the virus for its own purposes. Specific resonance frequencies and patterns of pulsation for microtubule have been identified that may be optimal [12–14], and nonspecific ultrasound frequencies (e.g., from 8 MHz to 500 kHz) transmitted to the brain through the scalp and skull have been shown to enhance mood in human volunteers [61,62]. As such, the application of nonspecific ultrasound frequencies to the vagus nerve and/or spleen is reasonable to consider as a therapeutic for COVID-19 patients.

Vagus nerve stimulation could be a promising adjunctive therapy for the treatment of COVID-19 patients. It can also be beneficial when the patient is not responsive to other kinds of therapies, such as anti-inflammatory drugs (including NSAIDs and glucocorticoids), or is suffering severe side effects as a result. The main advantages of low-intensity ultrasound are that it is safe, painless, non-invasive, relatively inexpensive, and readily available. Ultrasound machines are found in virtually every hospital emergency department and intensive care unit and are commonly used for imaging in sick patients. While we wait for effective antiviral drugs, and for those who do contract COVID-19, we propose a non-traditional approach that addresses and normalizes the hijacked microtubule and MTOC cytoskeleton.

Author Contributions: Conceptualization: S.H. and J.A.T.; visualization: M.A.; data curation: M.A.; formal analysis: M.A.; writing—original draft preparation: M.A.; writing—review and editing: S.H., J.A.T. and M.A.; supervision: S.H. and J.A.T. All authors have read and agreed to the published version of the manuscript.

Funding: J.A.T. gratefully acknowledges support for his research from NSERC (Canada).

Institutional Review Board Statement: Not applicable.

Informed Consent Statement: Not applicable.

Data Availability Statement: The datasets used and analyzed in the current study are available from the corresponding author upon request.

Acknowledgments: The assistance with artwork for this paper provided by Edgar Mendoza is greatly appreciated.

Conflicts of Interest: The authors declare no conflict of interest.

References

1. Klein, S.; Cortese, M.; Winter, S.L.; Wachsmuth-Melm, M.; Neufeldt, C.J.; Cerikan, B.; Stanifer, M.L.; Boulant, S.; Bartenschlager, R.; Chlanda, P. SARS-CoV-2 structure and replication characterized by in situ cryo-electron tomography. *Nat. Commun.* **2020**, *11*, 5885. [CrossRef] [PubMed]
2. Boson, B.; Legros, V.; Zhou, B.; Siret, E.; Mathieu, C.; Cosset, F.-L.; Lavillette, D.; Denolly, S. The SARS-CoV-2 envelope and membrane proteins modulate maturation and retention of the spike protein, allowing assembly of virus-like particles. *J. Biol. Chem.* **2021**, *296*, 100111. [CrossRef] [PubMed]
3. Bracquemond, D.; Muriaux, D. Betacoronavirus Assembly: Clues and Perspectives for Elucidating SARS-CoV-2 Particle Formation and Egress. *mBio* **2021**, *12*, e02371-21. [CrossRef]
4. Pizzato, M.; Baraldi, C.; Boscato Sopetto, G.; Finozzi, D.; Gentile, C.; Gentile, M.D.; Marconi, R.; Paladino, D.; Raoss, A.; Riedmiller, I.; et al. SARS-CoV-2 and the Host Cell: A Tale of Interactions. *Front. Virol.* **2022**, *1*, 220–225. [CrossRef]
5. Trougakos, I.P.; Stamatelopoulos, K.; Terpos, E.; Tsitsilonis, O.E.; Aivalioti, E.; Paraskevis, D.; Kastritis, E.; Pavlakis, G.N.; Dimopoulos, M.A. Insights to SARS-CoV-2 life cycle, pathophysiology, and rationalized treatments that target COVID-19 clinical complications. *J. Biomed. Sci.* **2021**, *28*, 9. [CrossRef]
6. Jackson, C.B.; Farzan, M.; Chen, B.; Choe, H. Mechanisms of SARS-CoV-2 entry into cells. *Nat. Rev. Mol. Cell Biol.* **2022**, *23*, 3–20. [CrossRef]
7. Hassan, Z.; Kumar, N.D.; Reggiori, F.; Khan, G. How Viruses Hijack and Modify the Secretory Transport Pathway. *Cells* **2021**, *10*, 2535. [CrossRef]
8. Ghosh, S.; Dellibovi-Ragheb, T.A.; Kerviel, A.; Pak, E.; Qiu, Q.; Fisher, M.; Takvorian, P.M.; Bleck, C.; Hsu, V.W.; Fehr, A.R.; et al. β-Coronaviruses Use Lysosomes for Egress Instead of the Biosynthetic Secretory Pathway. *Cell* **2020**, *183*, 1520–1535.e14. [CrossRef]
9. Wen, Z.; Zhang, Y.; Lin, Z.; Shi, K.; Jiu, Y. Cytoskeleton—A crucial key in host cell for coronavirus infection. *J. Mol. Cell Biol.* **2020**, *12*, 968–979. [CrossRef]
10. October 2020 Newsletter: Coronavirus and the Cytoskeleton. Available online: https://www.cytoskeleton.com/coronavirus-newsletter-cyto (accessed on 29 May 2021).
11. Simpson, C.; Yamauchi, Y. Microtubules in Influenza Virus Entry and Egress. *Viruses* **2020**, *12*, 117. [CrossRef]

12. Sahu, S.; Ghosh, S.; Ghosh, B.; Aswani, K.; Hirata, K.; Fujita, D.; Bandyopadhyay, A. Atomic water channel controlling remarkable properties of a single brain microtubule: Correlating single protein to its supramolecular assembly. *Biosens. Bioelectron.* **2013**, *47*, 141–148. [CrossRef] [PubMed]
13. Sahu, S.; Ghosh, S.; Hirata, K.; Fujita, D.; Bandyopadhyay, A. Multi-level memory-switching properties of a single brain microtubule. *Appl. Phys. Lett.* **2013**, *102*, 123701. [CrossRef]
14. Saxena, K.; Singh, P.; Sahoo, P.; Sahu, S.; Ghosh, S.; Ray, K.; Fujita, D.; Bandyopadhyay, A. Fractal, Scale Free Electromagnetic Resonance of a Single Brain Extracted Microtubule Nanowire, a Single Tubulin Protein and a Single Neuron. *Fractal Fract.* **2020**, *4*, 11. [CrossRef]
15. Craddock, T.J.A.; Kurian, P.; Preto, J.; Sahu, K.; Hameroff, S.R.; Klobukowski, M.; Tuszynski, J. Anesthetic Alterations of Collective Terahertz Oscillations in Tubulin Correlate with Clinical Potency: Implications for Anesthetic Action and Post-Operative Cognitive Dysfunction. *Sci. Rep.* **2017**, *7*, 9877. [CrossRef]
16. Hameroff, S.; Penrose, R. Consciousness in the universe: A Review of the "Orch OR" Theory. *Phys. Life Rev.* **2014**, *11*, 39–78. [CrossRef]
17. Perricone, C.; Triggianese, P.; Bartoloni, E.; Cafaro, G.; Bonifacio, A.F.; Bursi, R.; Perricone, R.; Gerli, R. The anti-viral facet of anti-rheumatic drugs: Lessons from COVID-19. *J. Autoimmun.* **2020**, *111*, 102468. [CrossRef]
18. Elshafei, M.N.; Khalil, A.; El-Bardissy, A.; Danjuma, M.; Ahmed, M.B.; Mohamed, M.F. The efficacy of colchicine in the management of coronavirus disease 2019: A Protocol for Systematic Review and Meta-Analysis. *Medicine* **2020**, *99*, e21911. [CrossRef]
19. Sapey, E.; Gallier, S.; Mainey, C.; Nightingale, P.; McNulty, D.; Crothers, H.; Evison, F.; Reeves, K.; Pagano, D.; Denniston, A.K.; et al. Ethnicity and risk of death in patients hospitalised for COVID-19 infection in the UK: An observational cohort study in an urban catchment area. *medRxiv* **2020**, *7*, e000644. [CrossRef]
20. Albrecht-Buehler, G. Rudimentary Form of Cellular "Vision". *Proc. Natl. Acad. Sci. USA* **1992**, *89*, 8288–8292. [CrossRef]
21. Hameroff, S.; Nip, A.; Porter, M.; Tuszynski, J. Conduction pathways in microtubules, biological quantum computation, and consciousness. *Biosystems* **2002**, *64*, 149–168. [CrossRef]
22. Craddock, T.J.A.; Tuszynski, J.A.; Hameroff, S. Cytoskeletal Signaling: Is Memory Encoded in Microtubule Lattices by CaMKII Phosphorylation? *PLoS Comput. Biol.* **2012**, *8*, e1002421. [CrossRef] [PubMed]
23. Tuszyński, J.A.; Brown, J.A.; Hawrylak, P. Dielectric polarization, electrical conduction, information processing and quantum computation in microtubules. Are they plausible? *Philos. Trans. R. Soc. London. Ser. A Math. Phys. Eng. Sci.* **1998**, *356*, 1897–1926. [CrossRef]
24. Gönczy, P. Towards a molecular architecture of centriole assembly. *Nat. Rev. Mol. Cell Biol.* **2012**, *13*, 425–435. [CrossRef] [PubMed]
25. Guichard, P.; Hamel, V.; Gönczy, P. The Rise of the Cartwheel: Seeding the Centriole Organelle. *BioEssays* **2018**, *40*, e1700241. [CrossRef]
26. Lin, Y.-C.; Chang, C.-W.; Hsu, W.-B.; Tang, C.-J.C.; Lin, Y.-N.; Chou, E.-J.; Wu, C.-T.; Tang, T. Human microcephaly protein CEP135 binds to hSAS-6 and CPAP, and is required for centriole assembly. *EMBO J.* **2013**, *32*, 1141–1154. [CrossRef]
27. Jana, S.C.; Marteil, G.; Bettencourt-Dias, M. Mapping molecules to structure: Unveiling secrets of centriole and cilia assembly with near-atomic resolution. *Curr. Opin. Cell Biol.* **2014**, *26*, 96–106. [CrossRef]
28. Cottee, M.A.; Muschalik, N.; Wong, Y.L.; Johnson, C.M.; Johnson, S.; Andreeva, A.; Oegema, K.; Lea, S.M.; Raff, J.W.; Van Breugel, M. Crystal structures of the CPAP/STIL complex reveal its role in centriole assembly and human microcephaly. *eLife* **2013**, 2, e01071. [CrossRef]
29. Klena, N.; Le Guennec, M.; Tassin, A.; Hoek, H.V.D.; Erdmann, P.S.; Schaffer, M.; Geimer, S.; Aeschlimann, G.; Kovacik, L.; Sadian, Y.; et al. Architecture of the centriole cartwheel-containing region revealed by cryo-electron tomography. *bioRxiv* **2020**, *39*, e106246. [CrossRef]
30. van Breugel, M.; Hirono, M.; Andreeva, A.; Yanagisawa, H.-A.; Yamaguchi, S.; Nakazawa, Y.; Morgner, N.; Petrovich, M.; Ebong, I.-O.; Robinson, C.V.; et al. Structures of SAS-6 Suggest Its Organization in Centrioles. *Science* **2011**, *331*, 1196–1199. [CrossRef]
31. Hung, L.-Y.; Chen, H.-L.; Chang, C.-W.; Li, B.-R.; Tang, T.K. Identification of a Novel Microtubule-destabilizing Motif in CPAP That Binds to Tubulin Heterodimers and Inhibits Microtubule Assembly. *Mol. Biol. Cell* **2004**, *15*, 2697–2706. [CrossRef] [PubMed]
32. Hsu, W.-B.; Hung, L.-Y.; Tang, C.-J.C.; Su, C.-L.; Chang, Y.; Tang, T.K. Functional characterization of the microtubule-binding and -destabilizing domains of CPAP and d-SAS-4. *Exp. Cell Res.* **2008**, *314*, 2591–2602. [CrossRef] [PubMed]
33. Hung, L.-Y.; Tang, C.-J.C.; Tang, T.K. Protein 4.1 R-135 Interacts with a Novel Centrosomal Protein (CPAP) Which Is Associated with the γ-Tubulin Complex. *Mol. Cell. Biol.* **2000**, *20*, 7813–7825. [CrossRef] [PubMed]
34. Guichard, P.; Chrétien, D.; Marco, S.; Tassin, A.-M. Procentriole assembly revealed by cryo-electron tomography. *EMBO J.* **2010**, *29*, 1565–1572. [CrossRef] [PubMed]
35. Baker, N.A.; Sept, D.; Joseph, S.; Holst, M.J.; McCammon, J.A. Electrostatics of nanosystems: Application to microtubules and the ribosome. *Proc. Natl. Acad. Sci. USA* **2001**, *98*, 10037–10041. [CrossRef] [PubMed]
36. Case, D.A.; Betz, R.M.; Botello-Smith, W.; Cerutti, D.S.; Cheatham, T.E., III; Darden, T.A.; Duke, R.E.; Giese, T.J.; Gohlke, H.; Goetz, A.W.; et al. *AMBER 2014*; University of California: San Francisco, CA, USA, 2014.
37. Dolinsky, T.J.; Nielsen, J.E.; McCammon, J.A.; Baker, N.A. PDB2PQR: An automated pipeline for the setup of Poisson-Boltzmann electrostatics calculations. *Nucleic Acids Res.* **2004**, *32*, W665–W667. [CrossRef]
38. Bar-On, Y.M.; Flamholz, A.; Phillips, R.; Milo, R. SARS-CoV-2 (COVID-19) by the numbers. *eLife* **2020**, *9*, e57309. [CrossRef]

39. Qiao, B.; de la Cruz, M.O. Enhanced Binding of SARS-CoV-2 Spike Protein to Receptor by Distal Polybasic Cleavage Sites. *ACS Nano* **2020**, *14*, 10616–10623. [CrossRef]
40. Skiniotis, G.; Cochran, J.C.; Müller, J.; Mandelkow, E.; Gilbert, S.P.; Hoenger, A. Modulation of kinesin binding by the C-termini of tubulin. *EMBO J.* **2004**, *23*, 989–999. [CrossRef]
41. Wang, Z.; Sheetz, M.P. The C-Terminus of Tubulin Increases Cytoplasmic Dynein and Kinesin Processivity. *Biophys. J.* **2000**, *78*, 1955–1964. [CrossRef]
42. Mathew, D.; Giles, J.R.; Baxter, A.E.; Oldridge, D.A.; Greenplate, A.R.; Wu, J.E.; Alanio, C.; Kuri-Cervantes, L.; Pampena, M.B.; D'Andrea, K.; et al. Deep immune profiling of COVID-19 patients reveals distinct immunotypes with therapeutic implications. *Science* **2020**, *369*, eabc8511. [CrossRef]
43. Shang, J.; Wan, Y.; Luo, C.; Ye, G.; Geng, Q.; Auerbach, A.; Li, F. Cell Entry Mechanisms of SARS-CoV-2. *Proc. Natl. Acad. Sci. USA* **2020**, *117*, 11727–11734. [CrossRef] [PubMed]
44. Borovikova, L.V.; Ivanova, S.; Zhang, M.; Yang, H.; Botchkina, G.I.; Watkins, L.R.; Wang, H.; Abumrad, N.; Eaton, J.W.; Tracey, K.J. Vagus nerve stimulation attenuates the systemic inflammatory response to endotoxin. *Nature* **2000**, *405*, 458–462. [CrossRef] [PubMed]
45. Andersson, U.; Tracey, K.J. Neural reflexes in inflammation and immunity. *J. Exp. Med.* **2012**, *209*, 1057–1068. [CrossRef] [PubMed]
46. Zachs, D.P.; Offutt, S.J.; Graham, R.S.; Kim, Y.; Mueller, J.; Auger, J.L.; Schuldt, N.; Kaiser, C.R.W.; Heiller, A.P.; Dutta, R.; et al. Noninvasive ultrasound stimulation of the spleen to treat inflammatory arthritis. *Nat. Commun.* **2019**, *10*, 951. [CrossRef] [PubMed]
47. Matamoros, A.J.; Baas, P.W. Microtubules in health and degenerative disease of the nervous system. *Brain Res. Bull.* **2016**, *126*, 217–225. [CrossRef]
48. Huang, L.; Peng, Y.; Tao, X.; Ding, X.; Li, R.; Jiang, Y.; Zuo, W. Microtubule Organization Is Essential for Maintaining Cellular Morphology and Function. *Oxidative Med. Cell. Longev.* **2022**, *2022*, 1623181. [CrossRef]
49. Almeida-Souza, L.; Timmerman, V.; Janssens, S. Microtubule dynamics in the peripheral nervous system: A Matter of Balance. *BioArchitecture* **2011**, *1*, 267–270. [CrossRef]
50. Deftereos, S.G.; Giannopoulos, G.; Vrachatis, D.A.; Siasos, G.D.; Giotaki, S.G.; Gargalianos, P.; Metallidis, S.; Sianos, G.; Baltagiannis, S.; Panagopoulos, P.; et al. Effect of Colchicine vs Standard Care on Cardiac and Inflammatory Biomarkers and Clinical Outcomes in Patients Hospitalized with Coronavirus Disease 2019: The GRECCO-19 Randomized Clinical Trial. *JAMA Netw. Open* **2020**, *3*, e2013136. [CrossRef]
51. Mahmoudian, M. The Anti-Cancer Activity of Noscapine: A Review. *Recent Patents Anti-Cancer Drug Discov.* **2009**, *4*, 92–97. [CrossRef]
52. Jaunky, D.B.; Larocque, K.; Husser, M.C.; Liu, J.T.; Forgione, P.; Piekny, A. Characterization of a recently synthesized microtubule-targeting compound that disrupts mitotic spindle poles in human cells. *Sci. Rep.* **2021**, *11*, 23665. [CrossRef]
53. Staats, P.; Giannakopoulos, G.; Blake, J.; Liebler, E.; Levy, R.M. The Use of Non-invasive Vagus Nerve Stimulation to Treat Respiratory Symptoms Associated With COVID-19: A Theoretical Hypothesis and Early Clinical Experience. *Neuromodul. Technol. Neural Interface* **2020**, *23*, 784–788. [CrossRef] [PubMed]
54. Xu, M.; Wang, L.; Wu, S.; Dong, Y.; Chen, X.; Wang, S.; Li, X.; Zou, C. Review on experimental study and clinical application of low- intensity pulsed ultrasound in inflammation. *Quant. Imaging Med. Surg.* **2021**, *11*, 443–462. [CrossRef] [PubMed]
55. Juan, E.J.; González, R.; Albors, G.; Ward, M.P.; Irazoqui, P. Vagus nerve modulation using focused pulsed ultrasound: Potential applications and preliminary observations in a rat. *Int. J. Imaging Syst. Technol.* **2014**, *24*, 67–71. [CrossRef] [PubMed]
56. Puleo, C.; Cotero, V. Noninvasive Neuromodulation of Peripheral Nerve Pathways Using Ultrasound and Its Current Therapeutic Implications. *Cold Spring Harb. Perspect. Med.* **2020**, *10*, a034215. [CrossRef] [PubMed]
57. Graham, R.S.; Zachs, D.P.; Cotero, V.; D'Agostino, C.; Ntiloudi, D.; Kaiser, C.R.W.; Graf, J.; Wallace, K.; Ramdeo, R.; Coleman, T.R.; et al. First-in-Human Demonstration of Splenic Ultrasound Stimulation for Non-Invasively Controlling Inflammation. *medRxiv* **2020**. [CrossRef]
58. Wasilczuk, K.M.; Bayer, K.C.; Somann, J.P.; Albors, G.O.; Sturgis, J.; Lyle, L.T.; Robinson, J.P.; Irazoqui, P.P. Modulating the Inflammatory Reflex in Rats Using Low-Intensity Focused Ultrasound Stimulation of the Vagus Nerve. *Ultrasound Med. Biol.* **2019**, *45*, 481–489. [CrossRef]
59. Hauser, J.; Hauser, M.; Muhr, G.; Esenwein, S. Ultrasound-induced modifications of cytoskeletal components in osteoblast-like SAOS-2 cells. *J. Orthop. Res.* **2009**, *27*, 286–294. [CrossRef]
60. Noriega, S.; Hasanova, G.; Subramanian, A. The Effect of Ultrasound Stimulation on the Cytoskeletal Organization of Chondrocytes Seeded in Three-Dimensional Matrices. *Cells Tissues Organs* **2013**, *197*, 14–26. [CrossRef]
61. Hameroff, S.; Trakas, M.; Duffield, C.; Annabi, E.; Gerace, M.B.; Boyle, P.; Lucas, A.; Amos, Q.; Buadu, A.; Badal, J.J. Transcranial Ultrasound (TUS) Effects on Mental States: A Pilot Study. *Brain Stimul.* **2013**, *6*, 409–415. [CrossRef]
62. Sanguinetti, J.L.; Hameroff, S.; Smith, E.; Sato, T.; Daft, C.M.W.; Tyler, W.J.; Allen, J.J.B. Transcranial Focused Ultrasound to the Right Prefrontal Cortex Improves Mood and Alters Functional Connectivity in Humans. *Front. Hum. Neurosci.* **2020**, *14*, 52. [CrossRef]

life

MDPI

Communication

Prevention and Treatment of Life-Threatening COVID-19 May Be Possible with Oxygen Treatment

Jukka Ylikoski [1,2,3], Jarmo Lehtimäki [2,3], Rauno Pääkkönen [1] and Antti Mäkitie [1,*]

1 Department of Otorhinolaryngology—Head and Neck Surgery, University of Helsinki and Helsinki University Hospital, 00029 Helsinki, Finland; jukka.ylikoski@fimnet.fi (J.Y.); rauno.paakkonen@ains.fi (R.P.)
2 Helsinki Ear Institute, 00420 Helsinki, Finland; jarmo.lehtimaki@gmail.com
3 Salustim Group Inc., 90440 Kempele, Finland
* Correspondence: antti.makitie@helsinki.fi; Tel.: +358-50-428-6847

Abstract: Most SARS CoV-2 infections probably occur unnoticed or cause only cause a mild common cold that does not require medical intervention. A significant proportion of more severe cases is characterized by early neurological symptoms such as headache, fatigue, and impaired consciousness, including respiratory distress. These symptoms suggest hypoxia, specifically affecting the brain. The condition is best explained by primary replication of the virus in the nasal respiratory and/or the olfactory epithelia, followed by an invasion of the virus into the central nervous system, including the respiratory centers, either along a transneural route, through disruption of the blood-brain barrier, or both. In patients, presenting with early dyspnea, the primary goal of therapy should be the reversal of brain hypoxia as efficiently as possible. The first approach should be intermittent treatment with 100% oxygen using a tight oronasal mask or a hood. If this does not help within a few hours, an enclosure is needed to increase the ambient pressure. This management approach is well established in the hypoxia-related diseases in diving and aerospace medicine and preserves the patient's spontaneous breathing. Preliminary research evidence indicates that even a small elevation of the ambient pressure might be lifesaving. Other neurological symptoms, presenting particularly in long COVID-19, suggest imbalance of the autonomous nervous system, i.e., dysautonomia. These patients could benefit from vagal nerve stimulation.

Keywords: SARS CoV-2; hyperbaric oxygen; autonomous nerve system; brain hypoxia; dysautonomia

Citation: Ylikoski, J.; Lehtimäki, J.; Pääkkönen, R.; Mäkitie, A. Prevention and Treatment of Life-Threatening COVID-19 May Be Possible with Oxygen Treatment. *Life* **2022**, *12*, 754. https://doi.org/10.3390/life12050754

Academic Editors: Silvia De Francia and Sarah Allegra

Received: 19 March 2022
Accepted: 12 May 2022
Published: 19 May 2022

Publisher's Note: MDPI stays neutral with regard to jurisdictional claims in published maps and institutional affiliations.

1. Introduction

Many countries are still struggling with COVID-19. The disease has caused many fatalities because no effective treatment has been available. At the beginning of the COVID-19 pandemic, the causative agent, SARS-CoV-2, was found to show great similarities with 2002/3 SARS-CoV and 2012 MERS-CoV. Consequently, severe COVID-19 was primarily regarded as a pulmonary one-organ disease, with pneumonia and bronchiolitis leading to dyspnea, and in the most severe cases to acute respiratory distress syndrome (ARDS) and septic shock. The SARS-CoV-2 virus was thought, in severe cases, to affect part of the innate immune response and to activate an inflammatory cascade stimulating the release of cytokines and chemokines, particularly within the lungs [1–3]. This would lead to a robust inflammatory response that, if it was not controlled, could result in a "cytokine storm" with detrimental systemic consequences [3].

However, even from the beginning of the pandemic, COVID-19 surprised by somewhat commonly causing a wide spectrum of clearly extrapulmonary symptoms that do not fit the aforementioned pulmonary concept. There were several early reports of patients with COVID-19 seeking medical attention who presented themselves with pure neurological manifestations at onset with nonneurological features first manifesting days later. It was proposed to term these cases "neuro-COVID syndrome" [4].

Among these early symptoms were impaired consciousness, fatigue, and headache, which all pointed to impaired oxygenation in the central nervous system (CNS). These usually appeared in combination with respiratory distress as the usual cause for hospitalization. We and others have stated that these symptoms are best explained by early neuroinvasion by SARS-CoV-2 of the CNS, including respiratory centers, leading to brain hypoxia [5–7]. In the first report consisting of 214 hospitalized severely affected patients, during January–February 2020 in Wuhan, 36% had neurological manifestations, and in some, these were the only symptoms. Impaired consciousness was seen in 15% of the cases, and it occurred early in the illness, i.e., during the first 1–2 days [5]. It is obvious that in such cases the primary target of therapy should be brain hypoxia and not a pulmonary disease. Therefore, the prevailing global principle in anesthesiology concerning oxygenation therapy as distinct from oxygen treatment, i.e., "as little oxygen as possible", might not be accurate for hypoxic COVID-19 patients. Based on our background experience in diving, aviation, and other potentially hypoxia-generating conditions, we have analyzed the reports of the clinical course, including signs and symptoms, of COVID-19. Our analyses have revealed that most of the early neurological symptoms of COVID-19 resemble those of mild brain hypoxia. Therefore, we propose that it is the oxygen deficiency in tissues, particularly in the brain, that should be the main target of therapies.

2. Potential Pathophysiological Mechanisms of COVID-19

2.1. Early Neurological (Extrapulmonary) Symptoms

Initially, COVID-19 was predominantly characterized as a respiratory illness targeting the upper airways, similar to other human coronaviruses. Clinical findings in people infected with SARS-CoV-2 range from an asymptomatic course to severe pneumonia requiring mechanical ventilation. The pathophysiology and severity of COVID-19 illness vary among patients and depend in part on underlying risk factors and chronic diseases. Its pathogenesis follows that of other respiratory viruses typically replicating in the epithelia of the nasal cavity or nasopharynx. Even from the early reports from Wuhan, it was clear that the great majority of COVID-19 cases show only symptoms of a mild, common cold-type upper respiratory infection [8] and recover within one or two weeks. However, these reports also described a significant proportion of hospitalized patients and even about 25% of severe cases presenting with neurological symptoms, including anosmia, impaired consciousness, and dyspnea, but normal pulmonary CT scans [8,9]. It was reported that most (up to 89%) of COVID-19 patients admitted to the intensive care unit could not breathe spontaneously [9]. There were also reports of COVID-19 patients presenting with asymptomatic (silent) hypoxia in whom their early respiratory dysfunction was described as a "cessation of spontaneous breathing." These features suggested a central apnea or failure in the feedback loop from the pulmonary receptors to the respiratory centers, mediated by the vagus nerve [10]. Note that hypoxia may also result from a combination of these two.

So far, several pieces of evidence have clearly shown that SARS-CoV-2 affects not only the respiratory tract but also the CNS, resulting in dyspnea and clearly neurological symptoms such as loss of smell and taste, headache, fatigue, impaired consciousness, nausea, and vomiting in more than one third of individuals with COVID-19 [11,12]. Therefore, it was hypothesized, first from Wuhan and later on in many other reports, that SARS-CoV-2 can be neurotropic, entering the body through the nose and spreading to the CNS, including the respiratory centers in the brain stem and medulla. Accordingly, the most dangerous symptom, the initial respiratory failure, would be neurogenic in origin [4–7,13,14]. Note that even in the early reports on COVID-19, impaired consciousness, in some patients the only symptom, was described as occurring in 15–50% of cases [5,6,15,16]. It is particularly noteworthy that most neurological manifestations could occur very early in the illness.

From accumulated human and experimental research data, we have constructed the putative, simplified pathogenetic pathway of SARS-CoV-2 (Figure 1). It seems clear that the virus enters the body by first attacking and replicating in the respiratory or olfactory epithelia of the nasal cavity and causing a common cold-like upper respiratory infection, with the highest viral replication occurring in the nose at day four [17]. Olfactory engagement seems apparent from the common occurrence of anosmia that has been described as occurring in more than 90% of cases seeking medical attention [9,18]. The olfactory epithelium contains, in addition to sustentacular and microvillar cells, olfactory sensory neurons. Their peripheral olfactory cilia are protected from the external air by only a thin mucous blanket. Therefore, the olfactory nerve has been described as a shortcut into the CNS for several viral diseases [19]. The SARS-CoV-2 spike (S)-protein that binds to its specific receptor ACE2, in concert with host proteases—principally TMPRSS2 (promotes cellular entry)—is co-expressed in epithelial type II pneumocytes in the lungs and in ciliated and goblet cells of the nasal epithelium [20]. Olfactory epithelia as the entry site of SARS-CoV-2 into the CNS were strongly supported by a recent autopsy material study of 33 COVID-19 victims. By using immunohistochemistry, in situ hybridization, and electron microscopy, it was possible to visualize the presence of intact CoV particles together with SARS-CoV-2 RNA in the olfactory mucosa [21]. That study also revealed viral particles and RNA in neuroanatomical areas receiving olfactory tract projections that may suggest SARS-CoV-2 neuroinvasion into deeper parts of the brain, including respiratory centers in the thalamus and brain stem, through axonal transport. This is not unexpected, as human coronaviruses have been shown to be potentially neurotropic and induce immune overactivation [22]. Another possible path for CNS spread is the hematogenous route, which involves early viral crossing of the blood-brain barrier (BBB). In general, the effect of COVID-19 on the brain may take several forms, some via direct infection and others via secondary mechanisms, e.g., immune response or respiratory failure-induced hypoxia. Direct presence of SARS-CoV-2 in the brain has been demonstrated through the detection of SARS-CoV-2 RNA in the cerebrospinal fluid of infected patients [23].

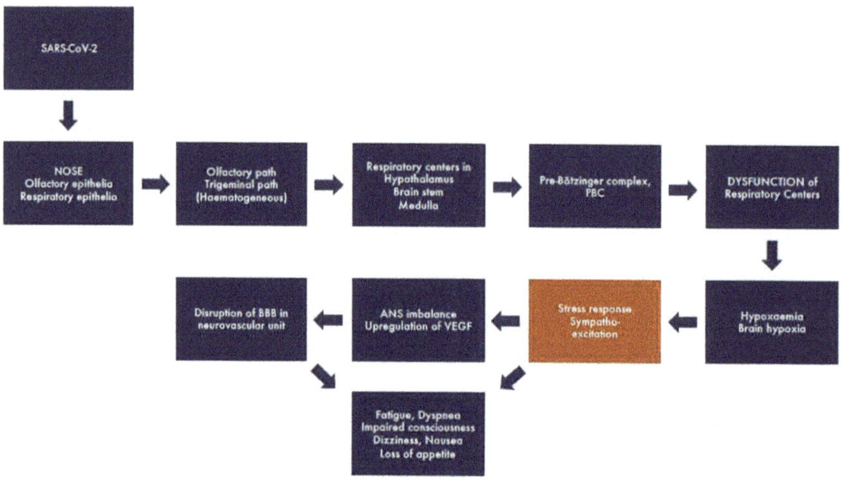

Figure 1. Schematic representation showing the putative brain pathway of SARS-CoV2 in COVID-19. The virus enters the respiratory or olfactory epithelia of the nasal cavity, spreads (by trans-synaptic migration) to the brain through the olfactory or trigeminal tracts, and infects deeper parts of the brain, including respiratory centers in the thalamus, brain stem, and medulla. This leads to dysfunction of the respiratory centers, including Pre-Bötzinger complex, causing respiratory distress and hypoxaemia/hypoxia and leading to a strong stress response associated with sympathoexcitation. Dysfunction of the ANS triggers upregulation of VEGF and disruption of the BBB, further leading to the worsening of hypoxia, resulting in both acute and long-standing neurological symptoms.

Regarding the extent of SARS-CoV-2 being able to affect the brainstem, it has been hypothesized that the respiratory breakdown in COVID-19 patients may be caused at least in part by SARS-CoV-2 infecting and destroying respiratory centers in the medulla oblongata and the pons [14]. Two sets of neuronal networks within the brainstem are crucial to the generation of respiratory rhythm, the Pre-Bötzinger complex (PBC), the pacemaker of the respiratory rhythm generator—also proposed as the kernel of respiration—and the retrotrapezoid nucleus/parafacial respiratory group [24]. When, the PBC was shut down in a mouse model of SARS CoV, it caused lethality due to respiratory failure [25]. It was hypothesized that SARS CoV-2 behaves similarly and that the destruction of the respiratory center (PBC) in the brainstem could be accountable for respiratory breakdown in COVID-19 patients [9,14].

It has also been suggested that the virus might enter the lowest region of the brain stem, the dorsal vagal complex (DVC), located in the medulla oblongata, which is involved in the control of several autonomic activities including breathing, food intake, nausea, and vomiting: these are all frequent extrapulmonary symptoms of COVID-19 [26]. In the DVC, the nucleus of tractus solitarius (NTS) is known, in addition to the hypothalamus, to be involved in food intake. The loss of appetite, sometimes a prominent symptom in COVID-19, means that the crosstalk between the hypothalamus and DVC has been broken as in pathological states such as under stress [27]. It is well established that another component of the DVC, the area postrema lying beneath the fourth cerebral ventricle, plays a crucial role in the elicitation of nausea and vomiting. This structure, together with DVC, NTS, and the dorsal motor nucleus of the vagus (DMNV), forms neurocircuits that have been classified as the "emetic chemoreceptor trigger zone" [28]. Interestingly, the eventual involvement of the subnucleus of NTS, the gelatinous nucleus, has been documented in the respiratory failure of sudden infant death syndrome (SIDS). [27]. Taken together, this neuroinvasive propensity leading to respiratory dysfunction has been demonstrated as a common feature of CoVs. Furthermore, hypoxemia and subsequent hypoxia are known to induce a stress reaction with the imbalance of the ANS and in the CNS, particularly of the central autonomic network (CAN) [29]. Hypoxia itself, or in combination with ANS/CAN imbalance, is a well-known cause of BBB disruption, leading to a multitude of local and systemic manifestations (Figure 1).

2.2. *Significance of Virus Genotypes for Virulence Symptoms and Neurotropism*

A critical initial step of infection is the interaction of the virus with receptors on host cells. The target tissues for viral infection, i.e., tissue tropism, is determined by the availability of virus receptors and entry cofactors on the surfaces of host cells. In the case of SARS-CoV-2 and other coronaviruses, the receptor binding occurs through the spike (S) protein on the virus surface. Both SARS-CoV-2 and the related SARS-CoV bind to ACE2 on human cells. ACE2, however, is expressed at low protein levels in respiratory and olfactory epithelial cells [30]. Although previous analyses have revealed that TMPRSS2, the primary protease important for SARS-CoV-2 entry, is highly expressed in different tissues, it was presumed that additional cofactors are required to facilitate virus-host cell interactions in cells with low ACE2 expression. It was shown that the membrane protein neuropilin-1 (NRP1) promotes SARS-CoV-2 entry by interacting with the SARS-CoV-2 S-protein and that NRP1 could thus represent an ACE2 potentiating factor by promoting the interaction of the virus with ACE2 [31,32].

It has been presumed that the pathogenic pathways and high transmission potentials of human coronaviruses are facilitated by an interplay between epigenetics and coronavirus infection. SARS-CoV-2 utilizes multiple ways for cellular entry (both nonendosomal and endosomal) and potentially uses various means of epigenetic control to inhibit the initiation of the host innate immune response. During the course of the pandemic, this virus efficiently has undergone genomic rearrangements, thereby developing important means for immunological escape. Such mutations have been especially effectively revealed by performing genome analyses of the SARS-CoV-2 sequences over the course of the COVID-19 pandemic in Costa Rica with a population of five million. Those analyses reveal the detection of mutations in line with other studies but also point out the local increases, particularly in the detection of Spike-T1117I variant. These constant genomic rearrangements may offer some explanation for the variations in transmission rates, symptoms, tropisms, and virulence of the SARS-CoV-2 [33–35].

2.3. The Role of the Blood-Brain Barrier

It is well-known that the BBB is a dynamic platform, collectively referred to as the neurovascular unit (NVU), responsible for the exchange of substances between the blood and the brain parenchyma and that it is an essential functional gatekeeper for the CNS. The propensity less commonly attributed to the BBB is its responsibility for the exchange of oxygen, which plays a critical role in the maintenance of brain homeostasis. Dysfunction of the BBB/NVU is a characteristic of several neurovascular pathologies. Moreover, physiological changes, environmental factors, nutritional habits, and psychological stress can modulate the tightness of the barrier. Mild inflammation is often associated with reduced BBB integrity, as observed for instance in obesity or psychosocial stress. Among extrinsic insults known to induce BBB breakdown, hypoxia is probably the most well characterized, but many knowledge gaps remain. Hypoxia can disrupt the BBB and result in increased permeability, vasogenic edema, and tissue damage [36]. Of the different types of hypoxia, hypobaric hypoxia (HH) is probably the best characterized because under hypobaric conditions, during ascent, it possible to monitor the gradual progression of HH. The proposed mechanisms for NVU damage by HH, and hypoxia in general, include induction of increased expression hypoxia-inducible factor-1 (Hif-1), enhanced endothelial transcytosis and oxidative stress [37]. It is well known that increased Hif-1 leads to increased expression of vascular endothelial growth factor (VEGF) in activated astrocytes, which further leads to NVU damage through changes in its tight junctions. One major mechanism associated with hypoxia-induced BBB opening is enhanced endothelial transcytosis that may be mediated by factors such as nitric oxide, calcium influx, or the release of inflammatory cytokines [38]. This may be a major mechanism because different cellular components of NVU show distinct differences in sensitivity to oxygen deprivation, so that endothelial cells (ECs) are markedly more sensitive than are pericytes or astrocytes. It is currently unclear whether the neurological symptoms in COVID-19 are a direct result of neural infection or secondary to endothelial cell infection, hypoxia, or circulating pro-inflammatory cytokines.

Several reports bring up the hypothesis that COVID-19, because it produces protean manifestations ranging from head to toe, represents an endothelial disease [39]. A similar pathophysiological mechanism has been proposed for long COVID-19. That hypothesis states that the initial pathology is due to the virus binding to the ACE-2 protein on ECs lining blood vessels and entering these cells in order to replicate, in turn releasing the immune response and thereby symptoms [40]. It even states that after initiating this immunologic cascade, the nascent virus spreads further into the nasopharyngeal tract. The early effect on the EC system is particularly attractive because it could explain, through dysregulation of the BBB and further dysautonomia, the relatively commonly occurring neurological symptoms of long COVID-19. The early EC effect by SARS-CoV-2, however, seems to be a rarity because virus RNA has usually not been found in the blood of patients with early COVID-19 [17,41–43]. However, several studies have found SARS-CoV-2 viremia and as such strong support for virus brain entry across the BBB at later stages of severe COVID-19 [44–47].

2.4. Role of Autonomic Dysfunction in Symptoms of COVID-19 and Long COVID-19

All our unconscious bodily functions are controlled by the ANS, and particularly by the CAN [29]. The most common cause for the dysfunction of the CAN is stress, the major cause of deteriorating health conditions and illnesses. It is well-known that infections, including viral ones, are associated with oxidative stress and subsequent reactive oxygen species. Recent research, by investigating small RNAs, powerful stress markers in the blood samples of patients with moderate or severe COVID-19, has revealed that the cells of COVID-19 patients undergo tremendous stress [47].

The CAN initially reacts to stressor effects with a sympathetic fight/flight response that is restored to normal by the parasympathetic nervous system's relax/digest response [48]. Many symptoms and illnesses result from the inability of the parasympathetic activity to restore the ANS balance (for review see [49]). These two circuits, the sympathetic and parasympathetic systems, are constantly interacting by heart rate variability (HRV) as a read-out of ANS balance; thus, HRV may consequently serve as a measure of stress [50].

The vagal system, with the vagus nerve terminating at the DVC in the brain stem, is known to be the key factor in most aspects of respiration. It both transmits the sensory information from pulmonary chemoreceptors to the CNS and is responsible for the activity that provides appropriate stimuli to the nuclei of the respiratory centers of the brain stem. This activity includes stimuli to NTS and nucleus ambiguus to give efferent commands to the respiratory muscles to function effectively. Interestingly, in some cases, the initial respiratory failure in COVID-19 can be a weakness or paresis of the diaphragm muscle, which is one of the two pumps necessary for life. Phrenic nerve paralysis has been described in a COVID-19 patient [51]. The paralysis of the diaphragm muscle was the cause of respiratory failure in connection with bulbar polio virus. Notably, the symptoms of "long polio", which can occur as long as 15 years after the acute stage, resemble those of long COVID-19 (fatigue, headache, musculoskeletal pains) [52].

Since the COVID-19 pandemic began, there has been a concern that survivors might be at an increased risk of neurological disorders. This concern, initially based on findings from infections with other coronaviruses, was rapidly followed by case series studies of the current pandemic. Multiple studies reported on the long-term outcomes of SARS survivors in Toronto in 2003. Most patients had persistent functional disabilities and were unable to return to their work. Their persisting debilitating physical symptoms included musculoskeletal pains, profound fatigability, shortness of breath, psychological distress, and major sleep problems [53,54]. These neurological long-term sequelae strongly suggest that they had been caused by an infection or inflammation in the CNS as a causative factor.

Following the initial surge of infections by SARS-CoV-2, focus has shifted to managing the longer-term sequelae of illness survivors. Post-acute COVID-19 syndrome (known colloquially as long COVID-19) is emerging as a prevalent syndrome. It encompasses a plethora of debilitating symptoms (including fatigue, breathlessness, pain, palpitations, sleep disturbance, and cognitive impairment ("brain fog"), which can last for weeks or months following the acute stage, even after a mild illness [55,56]. These symptoms thus greatly resemble those seen in follow-up studies of SARS2002/3 survivors [53,54]. As in the case of SARS-CoV, these neurological long-term sequelae of SARS-CoV-2 strongly suggest that their cause is an infection or inflammation in the CNS, not only in the lungs. Most patients recover from COVID-19 within a few weeks, but in surprisingly many, the (neurological) symptoms can continue for a long time [55,56]. These symptoms refer to imbalance in the CAN (with sympathetic dominance), "dysautonomia." This dysautonomia is supposed to be caused by cerebral hypoperfusion that leads to an overactive sympathetic system (fight or flight) with correspondingly reduced parasympathetic (relax) activity [57].

3. Therapies of COVID-19 and Long COVID-19

Most SARS CoV-2 infections probably occur unnoticed or cause only mild common cold symptoms that need no treatment. Pneumonia is probably the most common complication, but in the same way as ARDS occurs at later stages and thus cannot be responsible for early neurological symptoms such as impaired consciousness beginning during the first days of the disease, sometimes without any other symptoms.

It is well established that COVID-19 currently lacks curative therapy. However, the above-described mechanisms of the early dyspnea and hypoxia being due to a dysfunction of the central regulatory mechanisms of respiration shift the main focus of therapeutic efforts to the stage of early brain hypoxia. This pathophysiological concept suggests that therapeutically, the most important step would be to correct the hypoxemia and subsequent brain hypoxia by optimizing the efficacy of oxygen treatment in preserving the spontaneous breathing in affected individuals.

The role of the ANS, and particularly its CAN partition, in the disease process of COVID-19 appears to be crucial from the first symptoms to the final stage. Therefore, returning autonomic imbalance to normal might be important as a part of the therapeutic regimen, particularly in the patient population suffering from long COVID-19.

4. Oxygen Treatment

4.1. Principles of the Delivery of Oxygen into the Body and Its Administration

All of the body's tissues rely on a continuous oxygen supply at a rate that matches their changing metabolic demands. The amount of dissolved oxygen within the tissues and the cells depends crucially on the atmospheric partial pressure of oxygen and how effectively respiration is able to deliver oxygen from air to the blood plasma. Partial pressure of oxygen (PO_2) depends only on the atmosphere's barometric pressure (BP), and at normal BP conditions the content of inspired oxygen is 20.8%. The oxygen delivery chain begins in the nose, where the inspired air is warmed, humidified, and delivered by convection through the trachea to the lung alveoli and further to circulation, with the destination being the mitochondria (Table 1).

Table 1. Atmospheric/ambient pressures under different baric conditions and corresponding oxygen tensions in alveoli, arteries, extracellular fluids of tissues, and mitochondria. ATA = atmospheres absolute; AP = atmospheric pressure; PO_2 = partial pressure of oxygen; PAO_2 = partial pressure of oxygen in alveoli; PaO_2 = partial pressure of oxygen in arteries; PtO_2 = partial pressure of oxygen in tissues; PmO_2 = partial pressure of oxygen in mitochondria. * Estimations through extrapolations.

Atmospheric/Ambient	Oxygen Tension of Inspired Air, PO_2	Oxygen Tension in Alvoli, PAO_2	Oxygen Tension in Arteries, PaO_2	Oxygen Tension in Tissues, PtO_2	Oxygen Tension in Mitochondria, PmO_2
Pressure (AP) mmHg	mmHg	mmHg	mmHg	mmHg	mmHg
2.5 ATA, AP 1875					
15 m diving, 100% O_2 breathing	1875	1284 *	1274	250–500 [58,59]	80–125 *
	236	162 *	152	30–60 *	
5 m diving, 100% O_2 breathing	1125	771 *	761 (59)	150–304 *	50–76 *
1.3 ATA, AP 988 air breathing	207	158 *	148 [60,61]	30–60 *	10–15 *
3 m diving, 100% O_2 breathing	975	668 *	658	130–263 *	45–65 *
Sea level (1.0 ATA), AP 760 mmHg					
Air breathing [62–64] 20–50 y	160	102–110	97–99	20–40	7.5–11
>64 y	-"-	-"-	82–93	16–37 *	
100% O_2 breathing [63,64]	760	674	516	207 *	77 *
Dead Sea * −457 m AP 802 mmHg Air	167	114	104	42	15

At sea level in normal BP (760 mmHg, 1 ATA, atmospheric absolute), the PO_2 of inspired air is 160 mmHg. Along its route down to the alveoli, the PO_2 is reduced through various resistances, and final reduction takes place in alveoli due to dead space and the mixing of inspired and expired gases, resulting in a partial pressure of oxygen in alveoli (PAO_2) of about 110 mmHg (Table 1; Figure 2). From the alveoli, oxygen diffuses across the alveolar-capillary membrane to pulmonary circulation with only a small reduction in partial pressure (PaO_2 about 100 mmHg). From the pulmonary capillaries and arteries, the oxygen is carried to all parts of the body in two forms—a major fraction (up to 99%) that is bound to hemoglobin (Hb) and a small free fraction that dissolves in the plasma. Therefore, the number of red blood cells will dominantly affect the total capacity of oxygen delivery. However, at an elevated partial pressure of oxygen (such as during hyperbaric conditions), the dissolved amount can become significant. In all cases, the diffusion gradients are the oxygen's driving force from the plasma to the mitochondria. Thus, the free dissolved fraction only is transported to mitochondria. While it is transferred to various tissues, the PaO_2 is evenly reduced so that its level of extracellular tissue fluids (PtO_2) at sea level is 20–40 mmHg. From there, PtO_2 further reduces as oxygen diffuses to cells and mitochondria (partial pressure of oxygen 7.5–11 mmHg) (Table 1; Figure 2).

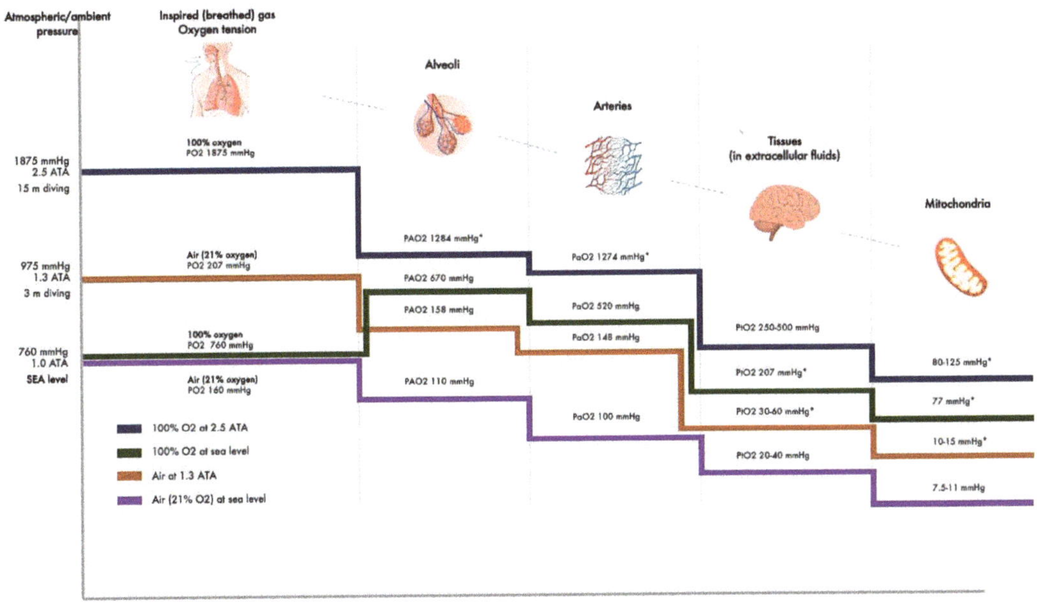

Figure 2. Oxygen tensions in the body. Reduction of oxygen tensions at different levels of airways, arteries, and tissues after breathing air at sea level and 1.3 ATA and breathing 100% oxygen at sea level and 2.5 ATA. (ATA = atmospheres absolute; PO_2 = partial pressure of oxygen; PAO_2 = partial pressure of oxygen in alveoli; PaO_2 = partial pressure of oxygen in arteries; PtO_2 = partial pressure of oxygen in tissues. PmO_2 = partial pressure of oxygen in mitochondria. * Estimations through extrapolations).

Note that the figures above apply to healthy young individuals in normal BP conditions. Factors that reduce PaO_2 include age, humidification, and barometric changes. It has been shown that alveolar diffusion capacity decreases with aging, about 0.24 mmHg per year. Consequently, PaO_2 in pulmonary capillaries in persons < 24 years is 99 mmHg, and 82–93 mmHg in those > 64 years (Table 1; Figure 2) [65]. Humidification in the nose will add of water vapor to inspired air, and its pressure is constant at 47 mmHg at normal body temperature (37 °C) [66]. This leads to a reduction of PAO_2 by about 10 mmHg- and a corresponding reduction of PaO_2.

In oxygen therapy planning, it remains important to recognize that the pressure (PO_2), and therefore the oxygen concentration, will determine the efficacy of oxygen treatment, i.e., how much oxygen will be transferred from the lung alveoli through the alveolar membrane to the arterial (capillary) blood. The quantity of dissolved oxygen in blood plasma, not the hemoglobin saturation, determines how much oxygen is diffused to the tissue. Further, according to Henry's law, the pressure of oxygen determines the quantity of dissolved oxygen in blood plasma.

Currently, most COVID-19 patients who are hospitalized because of respiratory distress are treated according to modifications of guidelines originally published by the British Thoracic Society (BTS) in 2008 [67] and of the newer modification of these BTS guidelines constituting of oxygen delivery through high-flow nasal oxygen (HFNO) with an O_2 flow of 40–60 L per minute [68]. This HFNO method, in addition to providing more O_2, also necessitates the use of some elevated pressure, and thereby, lung alveoli are opened more efficiently. These guidelines stress the importance of the prevention of tissue hypoxia. However, this oxygen treatment strategy is targeted at correcting oxygen deficiency primarily in one tissue, blood (hypoxemia). Notably, blood is one of the tissues that is extremely resistant to oxygen deficiency, at least compared with brain tissue.

Many patients undoubtedly benefit from the current treatment consisting of respiratory assistance and supplemental oxygen. However, if the tissue hypoxia is severe enough, this treatment may remain inadequate. This is because with employed extra-oxygen methods (including HFNO and loose masks), the content of inspired oxygen is only slightly increased. With a nasal cannula set at 2 L/min, oxygen tension of the inspired air ranges anywhere between 24% and 35%, leading only to a small increase in PaO_2 [69].

This might be sufficient to prevent tissue hypoxia, but it may not be enough to correct it. The only means to effectively correct tissue hypoxia is to increase the content of plasma dissolved oxygen. This is possible by providing the patient's alveoli with 100% oxygen by using a tight oronasal mask or hood, thereby by practicing conventional oxygen therapy, as for example with severe pulmonary damage [70]. Using this method, the plasma PaO_2 can be increased about fivefold (from about 100 to 500 mmHg) [71] (Table 1; Figure 2). An appropriate dose could be 40–60 min of 100% oxygen given twice per day. If this does not help within a few hours, an enclosure is needed to increase the ambient pressure. Hypoxemia with COVID-19, as reported, is usually accompanied by an increased alveolar-to-arterial oxygen gradient, signifying either ventilation–perfusion mismatch or intrapulmonary shunting [72,73]. Preliminary experience from Wuhan suggests that HBOT may solve this problem [74].

Oxygen was first used as a specific treatment by John S. Haldane, often called the father of oxygen therapy, in Ypres, Belgium, for victims of chlorine gas attack in 1915. In the first instance of chemical warfare in history, the German forces bombarded the British front lines with 6000 pressurized bottles containing poisonous chloride gas. Thousands of young men died, while those who survived suffered from severe pulmonary damage. Haldane designed a tightly fitting oronasal mask, connected an oxygen bottle to it, and rescued a large number of victims with oxygenation therapy using 100% oxygen. The method developed by Haldane began to be used in other medical emergencies such as in CO poisoning and also in diving, aviation, and by mountaineers.

4.2. *Oxygen Toxicity*

Already Haldane was already very well aware that long-term 100% oxygen treatment can be accompanied by pulmonary damage and other adverse effects. He stressed that all therapies must be given in appropriated dosages and that the negative effects of hyperoxia may be avoided if hypoxia is confirmed before oxygen therapy is initiated. It is indisputable that long-term 100% oxygen breathing can cause pulmonary damage [75]. The mechanism is still uncertain, but the strongest candidate is the so-called absorption collapse theory. The normal maximal lung capacity is 3.5–4.5 L, but at rest we only breathe about 0.5 L and thus use only a small part of our lungs. The peripheral ("resting") parts contain mainly nitrogen (78% of air) that keeps them open. Long-term oxygen breathing slowly leads to the replacement of nitrogen by oxygen in these pulmonary "resting areas". The ensuing diffusion of oxygen through alveolar membranes into blood capillaries will lead to decrease in the alveolar gas pressure, and further, to their gas pressure vacuum and collapse (atelectasis or absorption collapse). This causes local hypoxia and subsequent further cellular damage with free radicals that invite inflammatory cells to the scene, and an infection arises. The problem can be easily avoided by taking air breaks in oxygen breathing, thereby filling the alveoli with nitrogen. Air breaks have been used in hyperbaric oxygenation therapy (HBOT) for decades, and this has diminished the myth of oxygen itself being toxic.

When, in 1977, in addition to pulmonary damage theory, a highly recognized biochemist named Irwin Fridovich [76] reported that oxygen therapy created free radicals, a general perception of oxygen toxicity arose. Although Fridovich admitted that he was wrong in another article two years later (1979) [77] and that it was not the oxygen but, on the contrary, hypoxia that creates free radicals, the general perception of oxygen toxicity has remained the generally accepted "fact." That concept has continued to live a life of its own, particularly among anesthesiologists, to this day.

4.3. Oxygen Treatment under Increased Ambient Pressure

From the above, it can be concluded that the amount of blood PaO_2 can be increased and thereby tissue hypoxia corrected by using 100% oxygen with a tight oronasal mask. However, a much more rapid and efficient method would involve elevating the ambient pressure at which oxygen is administered. A large amount of reliable human experimental knowledge is available, particularly from aviation and mountain climbing, about the importance of pressure for brain oxygenation. It was as early as in 1862 when the British hydrogen ballooners James Glaisher and Henry Coxwell described their experience in *Lancet* [78]. They climbed to 9000 m, where the content of oxygen remains at about 21% but the partial pressure falls because the atmospheric pressure is only one third of that at sea level. Glaisher suddenly lost consciousness, and all of Coxwell's limbs became paralyzed. However, he was able to open the gas valve by drawing the valve rope with his teeth. Only after a descent of 50 m did their consciousness and bodily functions recover, and they survived. That was not the case for many other early balloonists who were supposed to have died of cold and brain hypoxia.

It seems clear that if brain hypoxia plays a role in the early extrapulmonary symptoms of COVID-19, it should be the primary target of treatment. Noninvasive brain oxygen state monitoring has been relatively reliably used with commercial near-infrared spectroscopy (NIRS) brain oximeters [for review see [79]]. Surprisingly, to the best of our knowledge, the use of NIRS on COVID-19 patients has not been reported so far.

The significance of pressure elevation is easy to understand from the fact that by pressurizing air (20.8% oxygen) to such a low overpressure as 1.3 ATA (corresponding to 3 m diving), the partial pressure of oxygen (PaO_2) in blood rises from 95 mmHg to 148 mmHg or 50% (Figure 2) [60]. This elevation might be enough for most COVID-19 patients for survival, particularly if the oxygen partial pressure of the inspired air is also elevated e.g., up to 30–40%. In conventional HBOT, (2.5 ATA, corresponding to a 15 m dive), the amount of blood plasma dissolved oxygen rises 20-fold (i.e., to 6 mL oxygen in 100 mL blood).

Modern aircrafts resemble pressure chambers as they are usually pressurized to about 2400 m above sea level (from about 0.33 ATA/250 mmHg at 10,000 m to 0.76 ATA/577 mmHg at 2400 m) and the cabin pressure is thus elevated by about 0.45 ATA. The proposal to use some of several hundreds of ground-bound aircrafts in British airports to treat critically ill COVID-19 patients in 1.3–1.5 ATA elevated pressure was made in June 2020 by Dr. Philip James [80]. This proposal was made after reports from China, where oxygen treatment using HBOT when necessary was adopted in Wuhan at the end of March 2020 with dramatically good results [74]. In total, 4400 COVID-19-related deaths were reported in China up to March 2020; there have only been about 250 more to date.

In addition to reports from China, there are already multiple case-control studies reporting the use of HBOT for patients with COVID-19. All these studies state that HBOT is a safe and promising alternative for the treatment of COVID-19 patients. Some studies specifically describe patients reporting the prompt resolution of labored breathing following a single HBOT treatment [81–85]. In one very recent study, aimed at becoming RCT-compatible, it was possible to correct severe hypoxemia with mild HBOT (0.45 ATA overpressure) in three days compared with nine days in the controls [86]. The elevation of the pressure by 0.45 ATA (corresponding to an altitude of 2400 m) is equal to that used in airplanes prophylactically to correct potentially fatal hypobaria. Another recent study used conventional HBOT, 10 sessions at 2.4 ATA, to treat long COVID-19 patients with disabling fatigue and found statistically significant beneficial effects [87]. In a recent case report, HBOT was successfully applied to treat long COVID-19 symptoms, leading to improvements in cognition and cardiopulmonary function. This effect was proposed to be due to the ability of HBOT to reverse hypoxia, reduce neuroinflammation, and induce neuroplasticity [88].

Additional evidence of the potential usefulness of HBOT in brain hypoxia or ischemia comes from a large number of clinical studies demonstrating beneficial effects of HBOT. Similar to COVID-19, especially long COVID-19, which is characterized by reduced memory and cognitive functions in addition to tiredness, brain hypoxia has been proposed to be the underlying mechanism in such neurological diseases as Alzheimer's disease, traumatic brain injury, cerebral palsy, and stroke (see Fisher and Barak 2020 for review) [89].

The most convincing evidence comes from management of stroke patients, where a considerable amount of preclinical research supports the post-stroke use of HBOT for damaged brain tissue. However, earlier controlled clinical trials of HBOT for stroke patients have yielded nonconclusive and somewhat contradicting results. This has changed along with a more recent breakthrough as prospective, randomized controlled studies have presented convincing evidence that HBOT can be the coveted neurotherapeutic method for brain repair [90–93].

The authors of COVID-19-HBOT studies conclude in common that well-defined RCT-compatible clinical trials are urgently needed [82–86]. However, designing such a trial may be difficult. Not only is it challenging to design a placebo treatment, but it might be even more difficult to randomize patients for the active vs. placebo treatment when the primary end point is death. HBOT has been accepted worldwide as the gold standard treatment for scuba-diving associated decompression (divers) disease, without any RCT evidence—because there also, the primary endpoint would be death. We propose the corresponding strategy—an elevated ambient pressure (1.3–1.5 ATA) for the treatment of COVID-19 patients with respiratory distress, using the higher pressures used in "conventional" HBOT (2.0–2.5 ATA) if necessary.

5. Reversing the ANS/CAN Imbalance in COVID-19 and Long COVID-19

Most of the early extrapulmonary symptoms of COVID-19 can be due to sympathetic dominance (and reduced parasympathetic function). Therefore, returning the ANS imbalance to normal by improving parasympathetic activity might be an important part of the therapeutic regimen. In addition, both the severe CNS and pulmonary disease are probably caused, and the pulmonary infection accentuated, through an inflammatory reflex mechanism due to inadequate immunological defense by the neuro-immune axis [94]. Excessive inflammation plays an important role in the pathogenesis of common and debilitating diseases, including septic shock [95].

Inflammation and VNS

Although the common pathways between stress exposure and pathophysiological processes leading to tissue damage are still debatable, several results indicate that stress can activate an inflammatory response in the brain and in the periphery [96]. Stress-induced pro-inflammatory factors play an important role in this damaging process. In common, over-activated immune systems, increased sympathetic nervous system activity, and reduced glucocorticoid (GC) responsiveness may work in tandem in the activation of inflammatory responses during stress. As the vagal system with the vagus nerve in front is responsible for parasympathetic activity, neuromodulation via vagal nerve stimulation (VNS) can serve as a targeted treatment in stressful conditions.

When the stimulation patterns and dynamics of functional networks during VNS were examined by fMRI, the vagus nerve was found to convey signals to the brain through the polysynaptic neuronal pathways by projecting to the brainstem nuclei (nucleus tractus solitarius, NTS, locus coeruleus), subcortical areas, and lastly, the cortex [97,98], thus covering the entire CAN. fMRI and a spatially independent component analysis were utilized in a recent experimental study [99]. That study demonstrated that VNS activated 15 out of 20 brain networks and that the activated regions covered >75% of the brain volume. There is strong preclinical scientific evidence for the beneficial role of VNS in the treatment of immunologic reflex-associated disorders, particularly rheumatoid arthritis (reviewed by Tracey, [100].

VNS has been conventionally performed for more than two decades to treat severe epilepsy and depression by applying an electrode surgically implanted into the cervical trunk of the vagus nerve. More recently, it has been shown by electrophysiological and neuroimaging studies that transcutaneous auricular VNS (taVNS) of the auricular branch of the vagus nerve (ABVN) activates central vagal pathways similar to the VNS with an implanted electrode [101,102].

In COVID-19/long COVID-19, taVNS may be especially effective, perhaps due to a dual action: it may attenuate the underlying neuroinflammation or inflammatory process in parallel or subsequent to stress response. As a method, taVNS is safer than VNS because the ABVN has no efferent neurons. In the pathogenesis of atrial fibrillation (AF), another common medical entity, accumulating evidence indicates that the inflammatory pathways play a significant role [103]. In a recent RCT-compatible clinical trial, chronic, intermittent taVNS resulted in significant reduction of inflammatory markers [104]. Based on its anti-inflammatory effects, taVNS targeted to the cervical part of the vagus nerve received emergency approval by the Federal Drug Administration in July 2020 for the treatment of asthmatic COVID-19 patients.

Earlier studies had failed to convincingly demonstrate that taVNS activates the crucial brainstem nuclei such as NTS. This has now changed: it was recently demonstrated using an ultrahigh-field (7T) fMRI that taVNS evokes activation in the ipsilateral NTS and upstream monoaminergic source nuclei of the brainstem [105]. Importantly, NTS activity is known to be modulated by respiration, both through the bottom-up afferent pathway from pulmonary stretch receptors and aortic baroreceptors and through the top-down effects from respiratory nuclei in the medulla [106]. Thus, taVNS treatment protocols should include instructions for slow breathing ("respiratory VNS") [106].

The symptoms of "long COVID-19" are best explained by imbalance in the ANS, dysautonomia. This could be reversed by increasing parasympathetic activity, which can be done using various behavioral methods such as relaxation and breathing exercises (e.g., meditations) but more efficiently by weak electrical stimulation of the vagal system/nerve [107,108].

6. Brain Hypoxia May Be the Major Cause for COVID-19 and Long COVID-19 Symptoms

There is strong evidence that the symptomatology of both acute COVID-19 and long COVID-19 can best be explained by a pathophysiological mechanism in which brain hypoxia is a crucial component. Lack of oxygen in the brain, including the brain stem and medullary respiration centers, causes symptoms of acute severe COVID-19, and hypoperfusion-induced brain injury also manifests itself as symptoms of long COVID-19. This mechanism becomes obvious if the symptomatology is compared with the two best-known, most common conditions causing brain hypoxia, the carbon monoxide (CO) poisoning and hypobaric hypoxia (HH) due to altitude elevation. Small concentrations of CO in the inspired air cause headache and fatigue, symptoms that are also the first signal of brain hypoxia in mountain climbing as well as in aviation and ballooning. In mountain climbing, the first symptom of hypobaric hypoxia is headache. If the ascension is continued, the next symptoms will be fatigue and high-altitude pulmonary edema (HAPE). The next step is unconsciousness. HAPE is induced by a hypoxic environment [109,110] and is characterized by interstitial edema and a large amount of exudation in the pulmonary alveoli [111]. The frequently described "ground-glass opacity" in chest CTs of hospitalized COVID-19 patients also, at least if it occurs at the early stage of COVID-19, suggests

HAPE-type pathogenesis and thereby may be a sign merely of hypoxia rather than of viral pneumonia. From HAPE and altitude medicine in general, it is known that even a small increase in pressure may be life-saving [78]. The symptoms of mild CO poisoning, HH, COVID-19, and long COVID-19 are listed in Table 2. It is obvious that the CNS-related symptoms of COVID-19 match perfectly with those of mild CO poisoning and HH.

Table 2. Lists of the most common CNS-related symptoms of mild CO poisoning, hypobaric hypoxia (HH), and COVID-19/long COVID-19. The symptoms of mild CO poisoning were listed in their order of prevalence by Haldane (1895). Order of prevalence was also intended by the authors in HH and COVID-19. Symptoms suggest hypoxia, primarily affecting the brain. In mountain climbing, the first symptom of HH is headache; the next are fatigue and high-altitude pulmonary edema (HAPE). The next step is unconsciousness. HAPE is induced by a hypoxic environment and is characterized by interstitial edema, shown as "ground-glass opacity" in chest CT.

Mild CO-Poisoning [112,113]	Aviation/Ballooning/Mountain Climbing	COVID-19/Long Covid
(in order of prevalence)	(Hypobaric hypoxia) [114–117]	[55,56,118–123]
Fatigue/lethargy	Visual disturbances	Anosmia
Headache	Headache	Fatigue
Numbness and tingling	Fatigue, lethargy	Headache
"Brain fog"	Dizziness, nausea	Dyspnea
Dizziness, nausea	Impaired fine touch & motor skills	"Brain fog"
Sleep disturbances	Personality & mood changes	Impaired consciousness
Palpitations	Sensory loss	Dizziness, nausea, tinnitus
Visual impairments	Confusion	Palpitations
Loss of consciousness	Loss of consciousness	Sleep disturbances
		Neuropsychological symptoms

Mild CO poisoning was also described by Haldane (1895) [112] as "the silent killer", which brings to mind recent reports of "silent hypoxia" (and thereby a potential silent killer) in COVID-19. It is now also known that CO intoxication can cause, in survivors, serious brain damage into the midbrain the same as that seen in traumatic brain injuries. The mechanism of this was proposed already decades ago to be injury to the blood-brain-barrier (BBB) [124]. The BBB plays a crucial role in maintaining homeostasis within the CNS, and BBB breakdown is clearly evident in many neurological disorders. It has been shown in many recent studies. One recent in vitro study used a six-cell-type neurovascular unit human 3D organoid model containing brain microvascular endothelial cells, pericytes, astrocytes, oligodendrocytes, microglia, and neurons that recapitulates characteristics of BBB dysfunction under hypoxic physiological conditions to show that exposure to hypoxia results in BBB breakdown and subsequent its dysfunction [125]. Because the verification of brain damage usually remains undetected in clinical evaluations, COVID-19 patients with long-standing CNS-symptoms are often stigmatized as psychosomatic exaggerators or even simulants.

Thus, we feel strongly that in pathophysiological mechanisms of both acute and long COVID-19, early neuroinvasion of the virus is involved and that this leads to brain hypoperfusion and hypoxia. It has now been reported that long COVID-19 (in particular) has taught us to better understand the putative pathogenic mechanisms of such (for school medicine) "contested diseases" as POTS (postural orthostatic tachycardia syndrome), CFS (chronic fatigue syndrome), neuroborreliosis, and fibromyalgia. All these conditions are best explained by functional disturbances to the CAN [57], dysautonomia, and specifically the associated sympathetic dominance. Symptoms comparable with those of the present long COVID-19 were reported in long term follow-up studies of patients of the SARS 2002/3 epidemic in Toronto. We propose that the crucial component in the pathophysiologic mechanism of these "contested diseases" is the injury of BBB [124]. If that is true, it might teach us more about the pathogenetics of other presently enigmatic neurodegenerative diseases.

7. Significance of Chest CT in COVID-19

Diagnosis of COVID-19 infection has largely been based on RT-PCR amplification of viral nucleic acid from the upper respiratory tract swabs. In addition, the main hallmark of COVID-19 pneumonia is the presence of ground-glass opacity (GGO) in lung computer tomography (CT). GGO, however, is not only a hallmark of pneumonia-type infection. It is merely a general sign of interstitial pulmonary edema occurring from diverse causes, of which the best characterized is the aforementioned HAPE from HH. Acute GGO has also been described as a serious complication in conditions such as during forceful swimming in triathlons and scuba diving and following electroconvulsive therapy (ECT), epileptic seizures, and CNS insults in general [126–131]. A characteristic feature of GGOs occurring in these connections is that they rapidly resolved with appropriate therapy, usually consisting of oxygen and diuretics. This may indicate that hypoxia is a component in such pathophysiologic GGO mechanisms because GGO in HAPE behaves similarly, resolving rapidly after only a relatively small descent.

There are various theories concerning the pathogenetic mechanisms of GGO, again best studied in HAPE, using animal models. Although the results of animal studies cannot be directly applied to human, it is interesting that, according to such studies, the brain's extremely complicated glymphatic systems are presented as playing a primary role, and in treatment aiming to resolve GGO, e.g., in rats, the positioning of animals, i.e., supine vs. sideways, plays an important role. Furthermore, there seem to be hitherto unknown mechanisms between the brain and the lungs manifesting in patients as GGOs after ECTs and various brain insults, named "neurogenic pulmonary edema" [131]. Abdennour et al. [132] even state that the brain and the lungs interact early and rapidly when hit by a disease process and furthermore that local brain inflammation spreads rapidly to the lung.

It is generally held that the diagnosis of COVID-9 is based primarily on a combination of symptoms and positive results of virus PCR. In addition, chest CT plays a major role in the diagnostic workup, even if it is not recommended routinely in mild disease and it not infrequently shows GGO, which finding is interpreted as indicating severe disease. However, several reports suggest that GGO is relatively common in asymptomatic carriers of SARS-CoV-2. One of the first such reports, based on a multicenter study in China, performed chest-CT imaging for 411 symptomatic and 100 asymptomatic individuals, all PCR positive [133]. The surprise finding was GGO in 60% of asymptomatic individuals. About 25% of these later developed symptoms, suggesting GGO as the presymptomatic finding. Most of asymptomatic carriers with GGOs came from high-altitude areas, which supports the idea that (hypobaric) hypoxia might play a role in the development of GGO. Two other studies, consisting of 60 and 64 asymptomatic PCR positive carriers on whom chest CT had been performed, described GGO findings in 47 and 60%, respectively [134,135]. This caused De Smet et al. [134] to propose that in a pandemic setting, such incidental CT-GGO findings should be reported as "compatible with COVID-19 pneumonia" rather than as "viral pneumonia" and that non-infectious lung diseases should be excluded from the diagnosis. From a therapeutic viewpoint, it would be crucial to know whether early GGO in chest CT is a sign of viral pneumonia or of (brain) hypoxia.

8. Conclusions and Future Prospects

The COVID-19 pandemic is beginning to end, and novel therapeutic methods will not help many patients. For the future, however, it would be extremely important to learn the pathophysiological mechanisms of the disease and thereby design better treatments. It seems evident that the earliest symptoms of potentially severe COVID-19, including respiratory distress, are best explained by primary replication of SARS-CoV-2 in the nasal and/or olfactory epithelia, followed by invasion of the virus into the CNS, including the brainstem, and that ARDS is a later complication.

It seems plausible that SARS-CoV2 is neurotropic, first attacking the mucous membranes of the nasal cavity in the same way as the other six coronaviruses. From the nose, there is a short path to the brain, including the respiratory centers where the virus causes a mild infection or perhaps only injures the BBB. These lead to imbalance of the ANS and malfunction of the CAN with reduced parasympathetic (vagal) tone. The dysfunction of the vagal system can lead to inefficient respiration with subsequent hypoxemia and ultimately brain hypoxia that further worsens the efficacy of respiration. If the early, centrally triggered brain hypoxia were recognized, that would automatically lead to a change in therapeutic strategy from providing only supplemental oxygen to real oxygen therapy. Preliminary evidence on the use of elevated ambient pressure for COVID-19-associated respiratory failure is very promising.

Several recent studies on long COVID-19 offer additional support for the idea that sympathovagal imbalance plays a crucial role. Therefore, VNS, or in practice taVNS, might offer a new, targeted therapeutic tool. Furthermore, taVNS is very patient-friendly and low-cost. However, as there are currently no appropriate online biomarkers available for taVNS, there is still a great need for additional research to find the optimal therapeutic regimen as well as better stimulating devices.

It will also be important that the medical community change its prevailing opinion and understand that oxygen itself is not toxic, that rather, with adequate dosing it has a crucial curative role. Many illnesses in which hypoxia may play an important pathogenetic role could be treated with pure oxygen and, if necessary, with elevated pressure. In the presence of a proven oxygen deficiency, it is not ethical to withhold additional oxygen, and at the correct dosage, which, as the Chinese studies have shown by measurements, may require the use of a pressure chamber. Although there is an enormous amount of reliable data on the importance of ambient pressure for the appropriate oxygenation of tissues, particularly for the most sensitive of them, the brain, it appears that the real significance of pressure in respiratory function is not fully understood even by most competent medical professionals. However, most of us may also not be aware that we have practically all been pressurized with 0.4–0.5 ATA and have breathed pressurized air in an airplane "hyperbaric chamber". Airplane cabin pressurization is necessary (for survival) because atmospheric pressure falls during ascent (one millibar per 10 m). Correspondingly, the pressure increases when one descends from the sea level. The lowest place on earth is the Dead Sea, situated 457 m below sea level, which means that the average atmospheric pressure there is 802 mmHg. When a group of hypoxemic (COPD) patients (from Jerusalem, +850 m) stayed there for three weeks, subjective well-being and all measured functional parameters improved significantly [136]. Similar beneficial effects were also reported in patients with cystic fibrosis [137]. It seems likely that the improvement in tissue oxygen content using even compressed air at 1.3–1.5 ATA would be valuable judged on the well-established benefit of descent in high altitude pulmonary and cerebral edema. This pressure increase might be life-saving for many COVID-19 and other critically ill patients. If this alternative were included in the therapeutic regimen of COVID-19, it might be that this disease could be reduced in severity and become just another "flu", i.e., influenza.

Author Contributions: Conceptualization, J.Y., J.L., R.P. and A.M.; methodology, J.Y. and J.L.; writing—original draft preparation, J.Y. and A.M.; writing—review and editing, J.Y., J.L., R.P. and A.M.; project administration, J.Y. and A.M. All authors have read and agreed to the published version of the manuscript.

Funding: This research was funded by the Research Foundation of the Finnish Otolaryngological Association.

Institutional Review Board Statement: The study was conducted in accordance with the Declaration of Helsinki.

Informed Consent Statement: Not applicable.

Data Availability Statement: Not applicable.

Conflicts of Interest: J.Y. and J.L. are board members of Helsinki Ear Institute Inc. and Salustim Group.

References

1. Perlman, S.; Netland, J. Coronaviruses post-SARS: Update on replication and pathogenesis. *Nat. Rev. Microbiol.* **2009**, *7*, 439–450. [CrossRef] [PubMed]
2. Channappanavar, R.; Perlman, S. Pathogenic human coronavirus infections: Causes and consequences of cytokine storm and immunopathology. *Semin. Immunopathol.* **2017**, *39*, 529–539. [CrossRef] [PubMed]
3. Guo, Y.-R.; Cao, Q.-D.; Hong, Z.-S.; Tan, Y.-Y.; Chen, S.-D.; Jin, H.-J.; Tan, K.-S.; Wang, D.-Y.; Yan, Y. The origin, transmission and clinical therapies on coronavirus disease 2019 (COVID-19) outbreak—An update on the status. *Mil. Med. Res.* **2020**, *7*, 11. [CrossRef]
4. Desai, I.; Manchanda, R.; Kumar, N.; Tiwari, A.; Kumar, M. Neurological manifestations of coronavirus disease 2019: Exploring past to understand present. *Neurol. Sci.* **2021**, *42*, 773–785. [CrossRef]
5. Mao, L.; Jin, H.; Wang, M.; Hu, Y.; Chen, S.; He, Q.; Chang, J.; Hong, C.; Zhou, Y.; Wang, D.; et al. Neurologic manifestations of hospitalized patients with coronavirus disease 2019 in Wuhan, China. *JAMA Neurol.* **2020**, *77*, 683–690. [CrossRef]
6. Roman, G.C.; Spencer, P.S.; Reis, J.; Buguet, A.; Faris, M.E.A.; Katrak, S.M.; Lainez, M.; Medina, M.T.; Meshram, C.; Mizusawa, H.; et al. WFN Environmental Neurology Specialty Group The neurology of COVID-19 revisited: A proposal from the Environmental Neurology Specialty Group of the World Federation of Neurology to implement international neurological registries. *J. Neurol. Sci.* **2020**, *414*, 116884. [CrossRef]
7. Ylikoski, J.; Markkanen, M.; Mäkitie, A. Pathophysiology of the COVID-19—Entry to the CNS through the nose. *Acta Oto-Laryngol.* **2020**, *140*, 886–889. [CrossRef]
8. Guan, W.J.; Ni, Z.Y.; Hu, Y.; Liang, W.H.; Qu, C.Q.; He, J.X.; Liu, L.; Shan, H.; Lei, C.L.; Hui, D.S.C.; et al. Clinical characteristics of coronavirus disease 2019 in China. *N. Engl. J. Med.* **2020**, *382*, 1708–1720. [CrossRef]
9. Li, Y.C.; Bai, W.Z.; Hashikawa, T. The neuroinvasive potential of SARS-CoV2 may play a role in the respiratory failure of COVID-19 patients. *J. Med. Virol.* **2020**, *92*, 552–555. [CrossRef]
10. Chang, R.B.; Strochlic, D.E.; Williams, E.K.; Umans, B.; Liberles, S.D. Vagal sensory neuron subtypes that differentially control breathing. *Cell* **2015**, *161*, 622–633. [CrossRef]
11. Huang, C.; Wang, Y.; Li, X.; Ren, L.; Zhao, J.; Hu, Y.; Zhang, L.; Fan, G.; Xu, J.; Gu, X.; et al. Clinical features of patients infected with 2019 novel coronavirus in Wuhan, China. *Lancet* **2020**, *395*, 497–506. [CrossRef]
12. Cardona, G.C.; Pájaro, L.D.; Marzola, I.D.; Villegas, Y.R.; Salazar, L.R. Neurotropism of SARS-CoV-2: Mechanisms and manifestations. *J. Neurol. Sci.* **2020**, *412*, 116824. [CrossRef] [PubMed]
13. Bougakov, D.; Podell, K.; Goldberg, E. Multiple Neuroinvasive Pathways in COVID-19. *Mol. Neurobiol.* **2020**, *58*, 564–575. [CrossRef] [PubMed]
14. Gandhi, S.; Srivastava, A.K.; Ray, U.; Tripathi, P.P. Is the collapse of the respiratory center in the brain responsible for respiratory breakdown in COVID-19 patients? *ACS Chem. Neurosci.* **2020**, *11*, 1379–1381. [CrossRef] [PubMed]
15. Roberts, C.M.; Levi, M.; McKee, M.; Schilling, R.; Lim, W.S.; Grocott, M.P. Covid-19: A complex multisystem clinical syndrome. *Br. J. Anaesth.* **2020**, *125*, 238. [CrossRef]
16. Helms, J.; Kremer, S.; Merdji, H.; Clere-Jehl, R.; Schenck, M.; Kummerlen, C.; Collange, O.; Boulay, C.; Fafi-Kremer, S.; Ohana, M.; et al. Neurologic features in severe SARS-CoV-2 infection. *N. Engl. J. Med.* **2020**, *382*, 2268–2270. [CrossRef]
17. Wölfel, R.; Corman, V.M.; Guggemos, W.; Seilmaier, M.; Zange, S.; Müller, M.A.; Niemeyer, D.; Jones, T.C.; Vollmar, P.; Rothe, C.; et al. Virological assessment of hospitalized patients with COVID-2019. *Nature* **2020**, *581*, 465–469. [CrossRef]
18. Lechien, J.R.; Chiesa-Estomba, C.M.; De Siati, D.R.; Horoi, M.; Le Bon, S.D.; Rodriguez, A.; Dequanter, D.; Blecic, S.; El Afia, F.; Distinguin, L.; et al. Olfactory and gustatory dysfunctions as a clinical presentation of mild-to-moderate forms of the coronavirus disease (COVID-19): A multi-center European study. *Eur. Arch. Oto-Rhino-Laryngol.* **2020**, *277*, 2251–2261. [CrossRef]
19. van Riel, D.; Verdijk, R.; Kuiken, T. The olfactory nerve: A shortcut for influenza and other viral diseases into the central nervous system. *J. Pathol.* **2015**, *235*, 277–287. [CrossRef]
20. Ziegler, C.G.K.; Allon, S.J.; Nyquist, S.K.; Mbano, I.M.; Miao, V.N.; Tzouanas, C.N.; Cao, Y.; Yousif, A.S.; Bals, J.; Hauser, B.M.; et al. SARS-CoV-2 Receptor ACE2 is an Interferon-Stimulated Gene in Human Airway Epithelial Cells and is detected in Specific Cell Subsets Across Tissues. *Cell* **2020**, *181*, 1016–1035. [CrossRef]
21. Meinhardt, J.; Radke, J.; Dittmayer, C.; Franz, J.; Thomas, C.; Mothes, R.; Laue, M.; Schneider, J.; Brünink, S.; Greuel, S.; et al. Olfactory transmucosal SARS-CoV-2 invasion as a port of central nervous system entry in individuals with COVID-19. *Nat. Neurosci.* **2020**, *24*, 168–175. [CrossRef] [PubMed]
22. Desforges, M.; Le Coupanec, A.; Brison, É.; Meessen-Pinard, M.; Talbot, P.J. Neuroinvasive and neurotropic human respiratory coronaviruses: Potential neurovirulent agents in humans. In *Infectious Diseases and Nanomedicine I*; Adhikari, R., Thapa, S., Eds.; Springer: New Delhi, India, 2014; Volume 807, pp. 75–96. [CrossRef]
23. Wang, H.Y.; Li, X.L.; Yan, Z.R.; Sun, X.P.; Han, J.; Zhang, B.W. Potential neurological symptoms of COVID-19. *Ther. Adv. Neurol. Disord.* **2020**, *13*, 1756286420917830. [CrossRef] [PubMed]
24. Smith, J.C.; Ellenberger, H.H.; Ballanyi, K.; Richter, D.W.; Feldman, J.L. PreBötzinger complex: A brainstem region that may generate respiratory rhythm in mammals. *Science* **1991**, *254*, 726–729. [CrossRef] [PubMed]
25. Burgold, T.; Voituron, N.; Caganova, M.; Tripathi, P.P.; Menuet, C.; Tusi, B.K.; Spreafico, F.; Bévengut, M.; Gestreau, C.; Buontempo, S.; et al. The H3K27 demethylase JMJD3 is required for maintenance of the embryonic respiratory neuronal network, neonatal breathing, and survival. *Cell Rep.* **2012**, *2*, 1244–1258. [CrossRef]

26. Chigr, F.; Merzouki, M.; Najimi, M. Comment on "The neuroinvasive potential of SARS-CoV-2 may play a role in the respiratory failure of COVID-19 patients". *J. Med. Virol.* **2020**, *92*, 703–704. [CrossRef]
27. Chigr, F.; Rachidi, F.; Tardivel, C.; Najimi, M.; Moyse, E. Modulation of orexigenic and anorexigenic peptides gene expression in the rat DVC and hypothalamus by acute immobilization stress. *Front. Cell. Neurosci.* **2014**, *8*, 198. [CrossRef]
28. Babic, T.; Browning, K.N. the role of vagal neurocircuits in the regulation of nausea and vomiting. *Eur. J. Pharmacol.* **2014**, *722*, 38–47. [CrossRef]
29. Benarroch, E.E. The central autonomic network: Functional organization, dysfunction, and perspective. *Mayo Clin. Proc.* **1993**, *68*, 988–1001. [CrossRef]
30. Benvenuto, D.; Giovanetti, M.; Ciccozzi, A.; Spoto, S.; Angeletti, S.; Ciccozzi, M. The 2019-new coronavirus epidemic: Evidence for virus evolution. *J. Med. Virol.* **2020**, *92*, 455–459. [CrossRef]
31. Daly, J.L.; Simonetti, B.; Klein, K.; Chen, K.-E.; Kavanagh Williamson, M.; Antón-Plágaro, C.; Shoemark, D.K.; Simón-Gracia, L.; Bauer, M.; Hollandi, R.; et al. Neuropilin-1 is a host factor for SARS-CoV-2 infection. *Science* **2020**, *370*, 861–865. [CrossRef]
32. Cantuti-Castelvetri, L.; Ojha, R.; Pedro, L.D.; Djannatian, M.; Franz, J.; Kuivanen, S.; Van Der Meer, F.; Kallio, K.; Kaya, T.; Anastasina, M.; et al. Neuropilin-1 facilitates SARS-CoV-2 cell entry and infectivity. *Science* **2020**, *370*, 856–860. [CrossRef] [PubMed]
33. Molina-Mora, J.A.; González, A.; Jiménez-Morgan, S.; Cordero-Laurent, E.; Brenes, H.; Soto-Garita, C.; Sequeira-Soto, J.; Duarte-Martínez, F. Clinical profiles at the time of diagnosis of COVID-19 in Costa Rica during the pre-vaccination period using a machine learning approach. *medRxiv* **2021**. [CrossRef]
34. Molina-Mora, J.A.; González, A.; Jiménez-Morgan, S.; Cordero-Laurent, E.; Brenes, H.; Soto-Garita, C.; Sequeira-Soto, J.; Duarte-Martínez, F. *Metagenomic Pipeline for Identifying Co-Infections among Distinct SARS-CoV-2 Variants of Concern: Study Cases from Alpha to Omicron*; Research Square: Durham, NC, USA, 2022. [CrossRef]
35. Molina-Mora, J.A. Insights into the mutation T1117I in the spike and the lineage B.1.1.389 of SARS-CoV-2 circulating in Costa Rica. *Gene Rep.* **2022**, *27*, 101554. [CrossRef] [PubMed]
36. Engelhardt, S.; Huang, S.F.; Patkar, S.; Gassmann, M.; Ogunshola, O.O. Differential responses of blood-brain barrier associated cells to hypoxia and ischemia: A comparative study. *Fluids Barriers CNS* **2015**, *12*, 4. [CrossRef] [PubMed]
37. Segarra, M.; Aburto, M.R.; Acker-Palmer, A. Blood–Brain Barrier Dynamics to Maintain Brain Homeostasis. *Trends Neurosci.* **2021**, *44*, 393–405. [CrossRef]
38. Kaur, C.; Ling, E.A. Blood brain barrier in hypoxic-ischemic conditions. *Curr. Neurovascular Res.* **2008**, *5*, 71–81. [CrossRef]
39. Lüscher, T.F. Understanding COVID-19: In the end it is the endothelium—What else? *Eur. Heart J.* **2020**, *41*, 3023–3027. [CrossRef]
40. Jarrott, B.; Head, R.; Pringle, K.G.; Lumbers, E.R.; Martin, J.H. "LONG COVID"—A hypothesis for understanding the biological basis and pharmacological treatment strategy. *Pharmacol. Res. Perspect.* **2022**, *10*, e00911. [CrossRef]
41. Yu, F.; Yan, L.; Wang, N.; Yang, S.; Wang, L.; Tang, Y.; Gao, G.; Wang, S.; Ma, C.; Xie, R.; et al. Quantitative Detection and Viral Load Analysis of SARS-CoV-2 in Infected Patients. *Clin. Infect. Dis.* **2020**, *71*, 793–798. [CrossRef]
42. Xie, C.; Jiang, L.; Huang, G.; Pu, H.; Gong, B.; Lin, H.; Ma, S.; Chen, X.; Long, B.; Si, G.; et al. Comparison of different samples for 2019 novel coronavirus detection by nucleic acid amplification tests. *Int. J. Infect. Dis.* **2020**, *93*, 264–267. [CrossRef]
43. Wu, J.; Liu, J.; Li, S.; Peng, Z.; Xiao, Z.; Wang, X.; Yan, R.; Luo, J. Detection and analysis of nucleic acid in various biological samples of COVID-19 patients. *Travel Med. Infect. Dis.* **2020**, *37*, 101673. [CrossRef] [PubMed]
44. Andersson, M.I.; Arancibia-Carcamo, C.V.; Auckland, K.; Baillie, J.K.; Barnes, E.; Beneke, T.; Bibi, S.; Brooks, T.; Carroll, M.; Crook, D.; et al. SARS-CoV-2 RNA detected in blood products from patients with COVID-19 is not associated with infectious virus. *Wellcome Open Res.* **2020**, *5*, 181. [CrossRef] [PubMed]
45. Song, E.; Zhang, C.; Israelow, B.; Lu-Culligan, A.; Prado, A.V.; Skriabine, S.; Lu, P.; Weizman, O.-E.; Liu, F.; Dai, Y.; et al. Neuroinvasion of SARS-CoV-2 in human and mouse brain. *J. Exp. Med.* **2021**, *218*, e20202135. [CrossRef] [PubMed]
46. Krasemann, S.; Haferkamp, U.; Pfefferle, S.; Woo, M.S.; Heinrich, F.; Schweizer, M.; Appelt-Menzel, A.; Cubukova, A.; Barenberg, J.; Leu, J.; et al. The blood-brain barrier is dysregulated in COVID-19 and serves as a CNS entry route for SARS-CoV-2. *Stem Cell Rep.* **2022**, *17*, 307–320. [CrossRef]
47. Liu, X.; Wen, Y.Z.; Huang, Z.L.; Shen, X.; Wang, J.H.; Luo, Y.H.; Chen, W.X.; Lun, Z.R.; Li, H.B.; Qu, L.H.; et al. SARS-CoV-2 Causes a Significant Stress Response Mediated by Small RNAs in the Blood of COVID-19 Patients. *Mol. Ther. Nucleic Acids* **2022**, *27*, 751–762. [CrossRef]
48. Selye, H. Stress and the General Adaptation Syndrome. *Br. Med. J.* **1950**, *1*, 1383–1392. [CrossRef]
49. McEwen, B.S.; Akil, H. Revisiting the Stress Concept: Implications for Affective Disorders. *J. Neurosci.* **2020**, *40*, 12–21. [CrossRef]
50. Akselrod, S.; Gordon, D.; Ubel, F.A.; Shannon, D.C.; Berger, A.C.; Cohen, R.J. Power spectrum analysis of heart rate fluctuation: A quantitative probe of beat-to-beat cardiovascular control. *Science* **1981**, *213*, 220–222. [CrossRef]
51. Maurier, F.; Godbert, B.; Perrin, J. Respiratory Distress in SARS-CoV-2 without Lung Damage: Phrenic Paralysis Should Be Considered in COVID-19 Infection. *Eur. J. Case Rep. Intern. Med.* **2020**, *21*, 7:001728. [CrossRef]
52. Dalakas, M.C.; Elder, G.; Hallett, M.; Ravits, J.; Baker, M.; Papadopoulos, N.; Albrecht, P.; Sever, J. A long-term follow-up study of patients with post-poliomyelitis neuromuscular symptoms. *N. Engl. J. Med.* **1986**, *314*, 959–963. [CrossRef]
53. Herridge, M.S.; Cheung, A.M.; Tansey, C.M.; Matte-Martyn, A.; Diaz-Granados, N.; Al-Saidi, F.; Cooper, A.B.; Guest, C.B.; Mazer, C.D.; Mehta, S.; et al. One-Year Outcomes In Survivors Of The Acute Respiratory Distress Syndrome. *N. Engl. J. Med.* **2003**, *348*, 683–693. [CrossRef] [PubMed]

54. Moldofsky, H.; Patcai, J. Chronic widespread musculoskeletal pain, fatigue, depression and disordered sleep in chronic post-SARS syndrome; a case-controlled study. *BMC Neurol.* **2011**, *11*, 37. [CrossRef] [PubMed]
55. Taquet, M.; Geddes, J.R.; Husain, M.; Luciano, S.; Harrison, P.J. 6-month neurological and psychiatric outcomes in 236,379 survivors of COVID-19: A retrospective cohort study using electronic health records. *Lancet Psychiatry* **2021**, *8*, 416–427. [CrossRef]
56. Graham, E.L.; Clark, J.R.; Orban, Z.S.; Lim, P.H.; Szymanski, A.L.; Taylor, C.; DiBiase, R.M.; Jia, D.T.; Balabanov, R.; Ho, S.U.; et al. Persistent Neurologic symptoms and cognitive dysfunction in non-hospitalized Covid-19 "long haulers". *Ann. Clin. Transl. Neurol.* **2021**, *8*, 1073–1085. [CrossRef]
57. Benarroch, E.E. Postural tachycardia syndrome: A heterogeneous and multifactorial disorder. *Mayo Clin. Proc.* **2012**, *87*, 1214–1225. [CrossRef] [PubMed]
58. Lamm, K.; Lamm, C.; Arnold, W. Effect of isobaric oxygen versus hyperbaric oxygen on the normal and noise-damaged hypoxic and ischemic guinea pig inner ear. *Hyperb. Oxyg. Ther. Otorhinolaryngol.* **1998**, *54*, 59–85.
59. Thom, S.R. Hyperbaric oxygen therapy. *J. Intensive Care Med.* **1989**, *4*, 58–74. [CrossRef]
60. James, P.B. *The Ultimate Oxygen Machine. Oxygen and the Brain: The Journey of Our Lifetime the Brain*; Best Publishing Company: North Palm Beach, FL, USA, 2014; pp. 23–29.
61. Montgomery, D.; Goldberg, J.; Amar, M.; Lacroix, V.; Lecomte, J.; Lambert, J.; Vanasse, M.; Marois, P. Effects of hyperbaric oxygen therapy on children with spastic diplegic cerebral palsy: A pilot project. *Undersea Hyperb. Med.* **1999**, *26*, 235–242.
62. Mellemgaard, K. The alveolar-arterial oxygen difference: Its size and components in normal man. *Acta Physiol. Scand.* **1966**, *67*, 10–20. [CrossRef]
63. Nunn, J.F.; Bergman, N.A.; Coleman, A.J. Factors influencing the arterial oxygen tension during anaesthesia with artificial ventilation. *Br. J. Anaesth.* **1965**, *37*, 898–914. [CrossRef]
64. Comroe, J.H., Jr.; Dripps, R.D., Jr. The oxygen tension of arterial blood and alveolar air in normal subjects. *Am. J. Physiol.* **1944**, *142*, 700–706. [CrossRef]
65. Crapo, R.O.; Jensen, R.L.; Hegewald, M.; Tashkin, D.P. Arterial bood gas reference values for sea level and an altitude of 1400 m. *Am. J. Respir. Crit. Care Med.* **1999**, *160*, 1525–1531. [CrossRef] [PubMed]
66. Ortiz-Prado, E.; Dunn, J.F.; Vasconez, J.; Castillo, D.; Viscor, G. Partial pressure of oxygen in the human body: A general review. *Am. J. Blood Res.* **2019**, *9*, 1–14. [PubMed]
67. O'Driscoll, B.R.; Howard, L.S.; Davison, A.G. British Thoracic Society Guideline for emergency oxygen use in adult patients. *Thorax* **2008**, *63*, 734–735.
68. Matthay, M.A. Saving Lives with High-Flow Nasal Oxygen. *N. Engl. J. Med.* **2015**, *372*, 2225–2226. [CrossRef] [PubMed]
69. Bazuaye, E.A.; Stone, T.N.; Corris, P.A.; Gibson, G.J. Variability of inspired oxygen concentration with nasal cannulas. *Thorax* **1992**, *47*, 609–611. [CrossRef]
70. Haldane, J.S. The therapeutic administration of oxygen. *Br. Med. J.* **1917**, *1*, 181–183. [CrossRef]
71. Nunn, J.F.; Williams, I.P.; Jones, J.G.; Hewlett, A.M.; Hulands, G.H.; Minty, B.D. Detection and reversal of pulmonary absorption collapse. *Br. J. Anaesth.* **1978**, *50*, 91–99. [CrossRef]
72. Tobin, M.J.; Laghi, F.; Jubran, A. Ventilatory failure, ventilator support, and ventilator weaning. *Compr. Physiol.* **2012**, *2*, 2871–2921.
73. Tobin, M.J. Basing respiratory management of COVID-19 on physiological principles. *Am. J. Respir. Crit. Care Med.* **2020**, *201*, 1319–1320. [CrossRef]
74. Harch, P.G. Hyperbaric oxygen treatment of novel coronavirus (COVID-19) respiratory failure. *Med. Gas Res.* **2020**, *10*, 61–62. [CrossRef] [PubMed]
75. Nunn, J.F.; Coleman, A.J.; Sachithanandan, T.; Bergman, N.A.; Laws, J.W. Hypoxaemia and atelectasis produced by forced inspiration. *Brit. J. Anaesth.* **1965**, *37*, 3–12. [CrossRef] [PubMed]
76. Fridovich, I. Oxygen is Toxic! *Bioscience* **1977**, *27*, 462–466. [CrossRef]
77. Fridovich, I. Hypoxia and oxygen toxicity. *Adv. Neurol.* **1979**, *26*, 255–259.
78. Glaisher, J. Notes of effects experienced during recent balloon ascents. *Lancet* **1862**, *80*, 559–560. [CrossRef]
79. Benni, P.B.; MacLeod, D.; Ikeda, K.; Lin, H.M. A validation method for near-infrared spectroscopy based tissue oximeters for cerebral and somatic tissue oxygen saturation measurements. *J. Clin. Monit. Comput.* **2018**, *32*, 269–284. [CrossRef]
80. James, P.B. Intermittent high dosage oxygen treats COVID-19 infection: The Chinese studies. *Med. Gas Res.* **2020**, *10*, 63. [CrossRef]
81. Thibodeaux, K.; Speyrer, M.; Raza, A.; Yaakov, R.; E Serena, T. Hyperbaric oxygen therapy in preventing mechanical ventilation in COVID-19 patients: A retrospective case series. *J. Wound Care* **2020**, *29*, S4–S8. [CrossRef]
82. Kjellberg, A.; De Maio, A.; Lindholm, P. Can hyperbaric oxygen safely serve as an anti-inflammatory treatment for COVID-19? *Med. Hypotheses* **2020**, *144*, 110224. [CrossRef]
83. Guo, D.; Pan, S.; Wang, M.; Guo, Y. Hyperbaric oxygen therapy may be effective to improve hypoxemia in patients with severe COVID-2019 pneumonia: Two case reports. *Undersea Hyperb. Med.* **2020**, *47*, 181–187. [CrossRef]
84. Gorenstein, S.A.; Castellano, M.L.; Slone, E.S.; Gillette, B.; Liu, H.; Alsamarraie, C.; Jacobson, A.M.; Wall, S.P.; Adhikari, S.; Swartz, J.L.; et al. Hyperbaric oxygen therapy for COVID-19 patients with respiratory distress: Treated cases versus propensity-matched controls. *Undersea Hyperb. Med.* **2020**, *47*, 405–413. [CrossRef] [PubMed]
85. Boet, S.; Etherington, C.; Djaiani, G.; Tricco, A.C.; Sikora, L.; Katznelson, R. Efficacy and safety of hyperbaric oxygen treatment in SARS-COV-2 (COVID-19) pneumonia: A systematic review. *Diving Hyperb. Med. J.* **2021**, *51*, 271–281. [CrossRef] [PubMed]

86. Cannellotto, M.; Duarte, M.; Keller, G.; Larrea, R.; Cunto, E.; Chediack, V.; Mansur, M.; Brito, D.M.; García, E.; Di Salvo, H.E.; et al. Hyperbaric oxygen as an adjuvant treatment for patients with COVID-19 severe hypoxaemia: A randomised controlled trial. *Emerg. Med. J.* **2022**, *39*, 88–93. [CrossRef] [PubMed]
87. Robbins, T.; Gonevski, M.; Clark, C.; Baitule, S.; Sharma, K.; Magar, A.; Patel, K.; Sankar, S.; Kyrou, I.; Ali, A.; et al. Hyperbaric oxygen therapy for the treatment of long COVID: Early evaluation of a highly promising intervention. *Clin. Med.* **2021**, *21*, e629–e632. [CrossRef] [PubMed]
88. Bhaiyat, A.M.; Sasson, E.; Wang, Z.; Khairy, S.; Ginzarly, M.; Qureshi, U.; Fikree, M.; Efrati, S. Hyperbaric oxygen treatment for long coronavirus disease-19: A case report. *J. Med. Case Rep.* **2022**, *16*, 80. [CrossRef] [PubMed]
89. Fischer, I.; Barak, B. Molecular and Therapeutic Aspects of Hyperbaric Oxygen Therapy in Neurological Conditions. *Biomolecules* **2020**, *10*, 1247. [CrossRef]
90. Efrati, S.; Fishlev, G.; Bechor, Y.; Volkov, O.; Bergan, J.; Kliakhandler, K.; Kamiager, I.; Gal, N.; Friedman, M.; Ben-Jacob, E.; et al. Hyperbaric oxygen induces late neuroplasticity in post stroke patients—randomized, prospective trial. *PLoS ONE* **2013**, *8*. [CrossRef]
91. Boussi-Gross, R.; Golan, H.; Fishlev, G.; Bechor, Y.; Volkov, O.; Bergan, J.; Friedman, M.; Hoofien, D.; Shlamkovitch, N.; Ben-Jacob, E.; et al. Hyperbaric oxygen therapy can improve post concussion syndrome years after mild traumatic brain injury—Randomized prospective trial. *PLoS ONE* **2013**, *8*, e79995. [CrossRef]
92. Efrati, S.; Ben-Jacob, E. Reflections on the neurotherapeutic effects of hyperbaric oxygen. *Expert Rev. Neurother.* **2014**, *14*, 233–236. [CrossRef]
93. Boussi-Gross, R.; Golan, H.; Volkov, O.; Bechor, Y.; Hoofien, D.; Beeri, M.S.; Ben-Jacob, E.; Efrati, S. Improvement of memory impairments in poststroke patients by hyperbaric oxygen therapy. *Neuropsychology* **2015**, *29*, 610–621. [CrossRef]
94. Pavlov, V.A.; Tracey, K.J. Neural circuitry and immunity. *Immunol. Res.* **2015**, *63*, 38–57. [CrossRef] [PubMed]
95. Nathan, C.; Ding, A. Nonresolving inflammation. *Cell* **2010**, *140*, 871–882. [CrossRef] [PubMed]
96. Calcia, M.A.; Bonsall, D.R.; Bloomfield, P.S.; Selvaraj, S.; Barichello, T.; Howes, O.D. Stress and neuroinflammation: A systematic review of the effects of stress on microglia and the implications for mental illness. *Psychopharmacology* **2016**, *233*, 1637–1650. [CrossRef] [PubMed]
97. Henry, T.R. Therapeutic mechanisms of vagus nerve stimulation. *Neurology* **2002**, *59*, S3–S14. [CrossRef]
98. Ressler, K.J.; Mayberg, H.S. Targeting abnormal neural circuits in mood and anxiety disorders: From the laboratory to the clinic. *Nat. Neurosci.* **2007**, *10*, 1116–1124. [CrossRef]
99. Cao, J.; Lu, K.-H.; Powley, T.L.; Liu, Z. Vagal nerve stimulation triggers widespread responses and alters large-scale functional connectivity in the rat brain. *PLoS ONE* **2017**, *12*, e0189518. [CrossRef]
100. Tracey, K.J. Neurons Are the Inflammatory Problem. *Cell* **2018**, *173*, 1066–1068. [CrossRef]
101. Kraus, T.; Hossl, K.; Kiess, O.; Schanze, A.; Kornhuber, J.; Forster, C. BOLD fMRI deactivation of limbic and temporal brain structures and mood enhancing effect by transcutaneous vagus nerve stimulation. *J. Neural. Transm.* **2007**, *114*, 1485–1493. [CrossRef]
102. Frangos, E.; Ellrich, J.; Komisaruk, B.R. Non-invasive access to the vagus nerve central projections via electrical stimulation of the external ear: fMRI evidence in humans. *Brain Stimul.* **2015**, *8*, 624–636. [CrossRef]
103. Hu, Y.F.; Chen, Y.J.; Lin, Y.J.; Chen, S.A. Inflammation and the pathogenesis of atrial fibrillation. *Nat. Rev. Cardiol.* **2015**, *12*, 230–243. [CrossRef]
104. Stavrakis, S.; Stoner, J.A.; Humphrey, M.B.; Morris, L.; Filiberti, A.; Reynolds, J.C.; Elkholey, K.; Javed, I.; Twidale, N.; Riha, P.; et al. TREAT AF (Transcutaneous Electrical Vagus Nerve Stimulation to Suppress Atrial Fibrillation): A Randomized Clinical Trial. *Clin. Electrophysiol.* **2020**, *6*, 282–291. [CrossRef] [PubMed]
105. Sclocco, R.; Garcia, R.G.; Kettner, N.W.; Isenburg, K.; Fisher, H.P.; Hubbard, C.S.; Ay, I.; Polimeni, J.R.; Goldstein, J.; Makris, N.; et al. The influence of respiration on brainstem and cardiovagal response to auricular vagus nerve stimulation: A multimodal ultrahigh-field (7T) fMRI study. *Brain Stimul.* **2019**, *12*, 911–921. [CrossRef] [PubMed]
106. Gerritsen, R.J.S.; Band, G.P.H. Breath of Life: The Respiratory Vagal Stimulation Model of Contemplative Activity. *Front. Hum. Neurosci.* **2018**, *12*, 397. [CrossRef] [PubMed]
107. Verbanck, P.; Clarinval, A.M.; Burton, F.; Corazza, F.; Nagant, C.; Cheron, G. Transcutaneous Auricular Vagus Nerve Stimulation (tVNS) can Reverse the Manifestations of the Long-COVID Syndrome: A Pilot Study. *Front. Neurology Neurosci. Res.* **2021**, *2*, 100011. [CrossRef]
108. Ylikoski, J.; Markkanen, M.; Pirvola, U.; Lehtimäki, J.A.; Ylikoski, M.; Jing, Z.; Sinkkonen, S.T.; Mäkitie, A. Stress and Tinnitus; Transcutaneous Auricular Vagal Nerve Stimulation Attenuates Tinnitus-Triggered Stress Reaction. *Front. Psychol.* **2020**, *11*, 570196. [CrossRef]
109. Basnyat, B. Acute high-altitude illnesses. *N. Engl. J. Med.* **2013**, *369*, 1664–1667.
110. Schoene, R. Illnesses at high altitude. *Chest* **2008**, *134*, 402–416. [CrossRef]
111. Bhagi, S.; Srivastava, S.; Singh, S.B. High-altitude pulmonary edema: Review. *J. Occup. Health* **2014**, *56*, 235–243. [CrossRef]
112. Haldane, J.S. The action of carbonic oxide on man. *J. Physiol.* **1895**, *18*, 430–462. [CrossRef]
113. *WHO Guidelines for Indoor Air Quality: Selected PollutantsCarbon Monoxide*; World Health Organization: Geneva, Switzerland, 2010.
114. Shaw, D.M.; Cabre, G.; Gant, N. Hypoxic Hypoxia and Brain Function in Military Aviation: Basic Physiology and Applied Perspectives. *Front. Physiol.* **2021**, *12*, 665821. [CrossRef]

115. West, J.B. Early history of high-altitude physiology. *Ann. N. Y. Acad. Sci.* **2016**, *1365*, 33–42. [CrossRef] [PubMed]
116. Virués-Ortega, J.; Buela-Casal, G.; Garrido, E.; Alcázar, B. Neuropsychological functioning associated with high-altitude exposure. *Neuropsychol. Rev.* **2004**, *14*, 197–224. [CrossRef] [PubMed]
117. Monge, C.C.; Whittembury, J. Chronic mountain sickness. *Johns Hopkins Med. J.* **1976**, *139*, 87–89. [CrossRef] [PubMed]
118. Arons, M.M.; Hatfield, K.M.; Reddy, S.C.; Kimball, A.; James, A.; Jacobs, J.R.; Taylor, J.; Spicer, K.; Bardossy, A.C.; Oakley, L.P.; et al. Presymptomatic SARS-CoV-2 infections and transmission in a nursing facility. *N. Engl. J. Med.* **2020**, *382*, 2081–2090. [CrossRef]
119. Whittaker, A.; Anson, M.; Harky, A. Neurological manifestations of COVID-19: A systematic review and current update. *Acta Neurol. Scand.* **2020**, *142*, 14–22. [CrossRef]
120. Chuang, D.T.; Aydemir, S.; Magda, P.; Thomas, C.; Zarnegar, R. Neurological manifestations as primary presentation of COVID-19 in hospitalized patients. *Acta Neurol. Scand.* **2021**, *143*, 569–574. [CrossRef]
121. WHO. Naming the Coronavirus Disease (COVID-19) and the Virus that Causes it. Available online: https://www.who.int/emergencies/diseases/novel-coronavirus-2019/technical-guidance/naming-the-coronavirus-disease-(COVID-2019)-and-the-virus-that-causes-it/ (accessed on 20 November 2020).
122. Romero-Sanchez, C.M.; Diaz-Maroto, I.; Fernandez-Diaz, E.; Sánchez-Larsen, Á.; Layos-Romero, A.; García-García, J. Neurologic manifestations in hospitalized patients with COVID-19: The ALBACOVID registry. *Neurology* **2020**, *95*, e1060–e1070. [CrossRef]
123. Richardson, S.; Hirsch, J.S.; Narasimhan, M.; Crawford, J.M.; McGinn, T.; Davidson, K.W.; Barnaby, D.P.; Becker, L.B.; Chelico, J.D.; Cohen, S.L.; et al. Presenting characteristics, comorbidities, and outcomes among 5700 patients hospitalized with COVID-19 in the New York City area. *JAMA* **2020**, *323*, 2052–2059. [CrossRef]
124. Olesen, S.-P. Rapid increase in blood-brain barrier permeability during severe hypoxia and metabolic inhibition. *Brain Res.* **1986**, *368*, 24–29. [CrossRef]
125. Nzou, G.; Wicks, R.T.; VanOstrand, N.R.; Mekky, G.A.; Seale, S.A.; El-Taibany, A.; Wicks, E.E.; Nechtman, C.M.; Marrotte, E.J.; Makani, V.S.; et al. Multicellular 3D Neurovascular Unit Model for Assessing Hypoxia and Neuroinflammation Induced Blood-Brain Barrier Dysfunction. *Sci. Rep.* **2020**, *10*, 9766. [CrossRef]
126. Shupak, A.; Weiler-Ravell, D.; Adir, Y.; Daskalovic, Y.I.; Ramon, Y.; Kerem, D. Pulmonary oedema induced by strenuous swimming: A field study. *Respir. Physiol.* **2000**, *121*, 25–31. [CrossRef]
127. Adir, Y.; Shupak, A.; Gil, A.; Peled, N.; Keynan, Y.; Domachevsky, L.; Weiler-Ravell, D. Swimming-induced pulmonary edema: Clinical presentation and serial lung function. *Chest* **2004**, *126*, 394–399. [CrossRef] [PubMed]
128. Slade, J.B., Jr.; Hattori, T.; Ray, C.S.; Bove, A.A.; Cianci, P. Pulmonary edema associated with scuba diving: Case reports and review. *Chest* **2001**, *120*, 1686–1694. [CrossRef] [PubMed]
129. Yamanashi, H.; Koyamatsu, J.; Nobuyoshi, M.; Murase, K.; Maeda, T. Exercise-Induced Pulmonary Edema in a Triathlon. *Case Rep. Med.* **2015**, *2015*, 968152. [CrossRef]
130. Manne, J.R.; Kasirye, Y.; Epperla, N.; Garcia-Montilla, R.J. Non-cardiogenic pulmonary edema complicating electroconvulsive therapy: Short review of the pathophysiology and diagnostic approach *Clin. Med. Res.* **2012**, *10*, 131–136. [CrossRef]
131. Arakaki, W.; Kinjo, T.; Nakamura, H.; Fujita, J. Seizure followed by lung edema: An intriguing link between the brain and the lung. *Clin. Case Rep.* **2020**, *8*, 2291–2292. [CrossRef]
132. Abdennour, L.; Zeghal, C.; Dème, M.; Puybasset, L. Interaction brain-lungs. *Ann. Fr. D'anesthesie Reanim.* **2012**, *31*, e101–e107. [CrossRef]
133. Kong, W.; Wang, Y.; Hu, J.; Chughtai, A.; Pu, H. Comparison of clinical and epidemiological characteristics of asymptomatic and symptomatic SARS-CoV-2 infection: A multi-center study in Sichuan Province, China. *Travel Med. Infect. Dis.* **2020**, *37*, 101754. [CrossRef]
134. De Smet, K.; De Smet, D.; Ryckaert, T.; Laridon, E.; Heremans, B.; Vandenbulcke, R.; Demedts, I.; Bouckaert, B.; Gryspeerdt, S.; Martens, G.A. Diagnostic performance of chest CT for SARS-CoV-2 infection in individuals with or without COVID-19 symptoms. *Radiology* **2021**, *298*, E30–E37. [CrossRef]
135. Uysal, E.; Kilinçer, A.; Cebeci, H.; Özer, H.; Demir, N.A.; Öztürk, M.; Koplay, M. Chest CT findings in RT-PCR positive asymptomatic COVID-19 patients. *Clin. Imaging* **2021**, *77*, 37–42. [CrossRef]
136. Kramer, M.R.; Springer, C.; Berkman, N.; Glazer, M.; Bublil, M.; Bar-Yishay, E.; Godfrey, S. Rehabilitation of hypoxemic patients with COPD at low altitude at the Dead Sea, the lowest place on earth. *Chest* **1998**, *113*, 571–575. [CrossRef] [PubMed]
137. Goldbart, A.D.; Cohen, A.D.; Weitzman, D.; Tal, A. Effects of rehabilitation winter camps at the Dead Sea on European cystic fibrosis patients. *Isr. Med. Assoc. J.* **2007**, *9*, 806–809. [PubMed]

life

Article

Can Lung Imaging Scores and Clinical Variables Predict Severe Course and Fatal Outcome in COVID-19 Pneumonia Patients? A Single-Center Observational Study

Ivan Skopljanac [1,2], Mirela Pavicic Ivelja [2,3,4,*], Danijela Budimir Mrsic [2,4,5], Ognjen Barcot [6], Irena Jelicic [3], Josipa Domjanovic [7] and Kresimir Dolic [2,4,5]

1 Department of Pulmology, University Hospital of Split, 21000 Split, Croatia; ivan.skopljanac@gmail.com
2 School of Medicine, University of Split, 21000 Split, Croatia; danijelabudimir@gmail.com (D.B.M.); kdolic79@gmail.com (K.D.)
3 Department of Infectious Diseases, University Hospital of Split, 21000 Split, Croatia; irenajelicic80@gmail.com
4 University Department of Health Studies, University of Split, 21000 Split, Croatia
5 Clinical Department of Diagnostic and Interventional Radiology, University Hospital of Split, 21000 Split, Croatia
6 Department of Surgery, University Hospital of Split, 21000 Split, Croatia; ognjen.barcot@gmail.com
7 Department of Nephrology, University Hospital of Split, 21000 Split, Croatia; josipa.domjanovic@gmail.com
* Correspondence: mpavivelj@kbsplit.hr

Abstract: COVID-19 prediction models mostly consist of combined clinical features, laboratory parameters, and, less often, chest X-ray (CXR) findings. Our main goal was to propose a prediction model involving imaging methods, specifically ultrasound. This was a single-center, retrospective cohort observational study of patients admitted to the University Hospital Split from November 2020 to May 2021. Imaging protocols were based on the assessment of 14 lung zones for both lung ultrasound (LUS) and computed tomography (CT), correlated to a CXR score assessing 6 lung zones. Prediction models for the necessity of mechanical ventilation (MV) or a lethal outcome were developed by combining imaging, biometric, and biochemical parameters. A total of 255 patients with COVID-19 pneumonia were included in the study. Four independent predictors were added to the regression model for the necessity of MV: LUS score, day of the illness, leukocyte count, and cardiovascular disease ($\chi2 = 29.16$, $p < 0.001$). The model accurately classified 89.9% of cases. For the lethal outcome, only two independent predictors contributed to the regression model: LUS score and patient's age ($\chi2 = 48.56$, $p < 0.001$, 93.2% correctly classified). The predictive model identified four key parameters at patient admission which could predict an adverse outcome.

Keywords: lung ultrasound; COVID-19; prognostic; pneumonia; CT; chest X-ray

Citation: Skopljanac, I.; Pavicic Ivelja, M.; Budimir Mrsic, D.; Barcot, O.; Jelicic, I.; Domjanovic, J.; Dolic, K. Can Lung Imaging Scores and Clinical Variables Predict Severe Course and Fatal Outcome in COVID-19 Pneumonia Patients? A Single-Center Observational Study. *Life* **2022**, *12*, 735. https://doi.org/10.3390/life12050735

Academic Editors: Silvia De Francia and Sarah Allegra

Received: 20 April 2022
Accepted: 11 May 2022
Published: 15 May 2022

1. Introduction

Since the end of 2019, the world has battled the growing coronavirus disease 2019 (COVID-19) epidemic caused by severe acute respiratory syndrome coronavirus 2 (SARS-CoV-2), with more than 290 million confirmed cases worldwide [1]. The virus is designated as severe acute respiratory syndrome coronavirus 2 (SARS-CoV-2), and the disease is coronavirus disease 2019 (COVID-19). Although the understanding of COVID-19 is evolving, there are still major issues concerning the immune response to SARS-CoV-2, COVID-19 diagnosis, management, prevention, and emerging new variants which make it obvious that SARS-CoV-2 is here to stay and potentially becoming endemic [2,3].

Thoracic imaging, including chest X-ray (CXR) or computed tomography (CT), is essential in the diagnosis of COVID-19 pneumonia. However, previous studies showed that CXR was less sensitive in the detection of COVID-19 lung disease compared to CT, with a reported baseline CXR sensitivity of 69% [4]. The sensitivity of CT is 83–100%, considering the results of reverse transcription-polymerase chain reaction (RT-PCR) tests as the gold standard for diagnosis of COVID-19 [5]. High-resolution CT (HRCT) is the "gold" standard imaging method to evaluate the severity of lung involvement in COVID-19 patients. Despite the above, CXR is still extensively being used in the diagnosis of COVID-19 pneumonia due to its wide availability and relative inexpensiveness. COVID-19 pneumonia changes on CXR are typically ill-defined bilateral alveolar opacities of peripheral distribution. Similarly, the most commonly reported HRCT findings of COVID-19 pneumonia include airspace opacities (ground-glass and/or consolidation), typically subpleural and multilobar involvement, sometimes associated with septal thickening [6].

In addition to the aforementioned imaging methods that use ionizing radiation, bedside lung ultrasonography (LUS) is a rapid, non-ionizing, repeatable, and reliable examination technique that is increasingly used by clinicians in the COVID-19 pandemic, especially for hospitalized and critically ill patients reducing the need for transportation. Growing evidence shows that LUS sensitivity is close to that of chest CT and is much higher than that of CXR; its usefulness for the management of patients with COVID-19 pneumonia, from diagnosis to monitoring, follow-up, and even outcome prediction, is also demonstrated [7–12].

The vast majority of patients with more severe symptoms of the disease have one or more comorbidities, such as obesity and cardiovascular disease, with high mortality among elderly patients [13]. Biochemical and hematological laboratory factors such as lymphopenia, elevated serum ferritin, d-dimer, troponin, C-reactive protein, lactate dehydrogenase, and IL-6 are associated with severe disease, poor prognosis, and increased mortality [14–16].

Identifying risk factors at hospital admission that can predict the clinical course of the disease would help physicians to provide appropriate and timely therapeutic interventions. The COVID-19 prediction models developed so far have mostly combined clinical features, laboratory parameters, and, less often, CXR findings [17–20]. To the best of our knowledge, none of the proposed combined prediction models involved lung ultrasound, and that was the main goal of our work.

2. Materials and Methods

2.1. Study Design

This was a single-center, retrospective cohort observational study.

2.2. Inclusion and Exclusion Criteria

The study included a consecutive cohort of patients admitted with the diagnosis of COVID-19 pneumonia in the University Hospital of Split, Croatia, from November 2020 to May 2021, before the COVID-19 vaccine was widely available. Inclusion criteria were WHO diagnostic criteria for pneumonia COVID-19, SARS-CoV-2 infection confirmed by PCR [1], and the existing LUS exam after admission. Exclusion criteria were pulmonary edema associated with heart failure, severe lung emphysema, chronic interstitial lung disease, severe hemodynamic instability and inability to change body position, severe chest deformity, extensive subcutaneous emphysema, any other pulmonary diseases impeding ultrasound image acquisition (i.e., significant pleural effusion, previous pneumonectomy), and an inability to undergo LUS examination (Figure 1).

2.3. Outcomes

The study's primary outcome was the definition of biometric (e.g., age), biochemical (e.g., spO2, LDH), and radiological predictive factors (LUS, CXR, and CT scores) for the necessity of mechanical ventilation (MV) in the treatment of pneumonia or a lethal outcome for the patients.

The secondary outcomes were the comparisons of the reliability and applicability of the three radiological scores assessed.

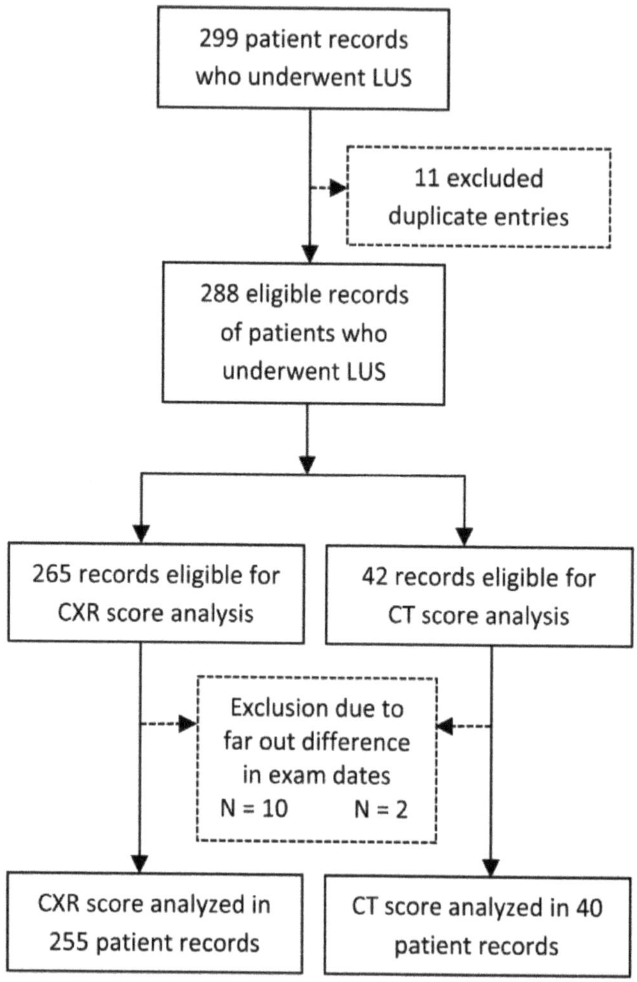

Figure 1. Flowchart of the records included or excluded from the analysis. Legend: CT score—Computerized tomography score; CXR score—chest X-ray score; LUS score—Lung ultrasound score.

2.4. Data Extraction

Patient demographics, comorbidities, symptoms, laboratory tests, imaging findings, treatment modalities, disease severity, and mortality data were extracted from electronic medical records.

2.5. Acquisition Protocol

Lung ultrasound examinations were performed by a trained sonographer (IS) on admission to the hospital using ultrasound equipment (Toshiba Nemio XG Istyle, Tokyo, Japan) with a 1–6 MHz convex transducer. The extent and severity of pulmonary infiltrations were described by a numerically repeatable LUS (Lung Ultrasound Score) coefficient proposed for COVID-19 pneumonia by Soldati et al. [21]. Fourteen areas (three posterior, two lateral, and two anterior for each lung) were examined completely intercostally to cover the widest possible area with a single scan. Changes were scored from 0 to 3. Zero (0) is a regular finding, the existence of a regular and not thickened pleural line, with a sliding sign and the presence of A-lines; 1: denotes an irregular pleural line with some B lines suggesting some loss of aeration; 2: suggests a severe loss of aeration by a broken pleural line and small-to-large consolidated areas with associated areas of white below the consolidated area; 3: is attributed if the scanned area shows large dense consolidations which signify a complete loss of aeration called "white lung" (Table 1). For each patient, the stated scores in all 14 zones were added up (ranging from 0 to 42) to obtain the total LUS score [21]. According to the part of the lung they are positioned in, the 14 areas were grouped in apical, middle, and basal for further statistical analysis.

Table 1. Grading of LUS score, CT score, and CXR score.

Score/ Grade	LUS Score	CT Score	CXR Score
0	regular finding: the existence of a regular and not thickened pleural line, with the sliding sign, and the presence of A-lines	no abnormalities	no abnormalities
1	some loss of aeration: irregular pleural line with some B lines	prevalent ground-glass opacities (GGOs)	interstitial infiltrates
2	severe loss of aeration: broken pleural line; small-to-large, consolidated areas with associated areas of white below the consolidated area	GGOs mixed with consolidations	interstitial and alveolar infiltrates (interstitial predominance)
3	complete loss of aeration: scanned area shows large dense consolidations; "white lung"	prevalent consolidations	interstitial and alveolar infiltrates (alveolar predominance)

Acronyms: CT—computerized tomography; CXR—Chest X-ray; LUS—lung ultrasound.

CT scans were acquired during hospitalization, as indicated by attending physicians, thus for a limited number of patients. Scanning was acquired on 128 slice multi-slice CT (Philips, Ingenuity Elite). Fourteen CT areas (three posterior, two lateral, and two anterior for each lung) on non-contrast native scans corresponded to fourteen LUS areas. CT findings in each area were classified as follows: 0—no abnormalities, 1—prevalent ground-glass opacities (GGOs), 2—GGOs mixed with consolidations, and 3—prevalent consolidations (Table 1) [22]. For each patient, these scores in all 14 areas were added up (ranging from 0 to 42) to obtain a total CT score.

Chest X-rays were also acquired upon hospital admission, but the CXR score was assessed retrospectively. For CXR scoring, we used the Brixia score [23]. Chest X rays were divided into 3 zones per lung (upper, middle, and lower) in a total of 6 zones: the upper zone extends above the inferior wall of the aortic arch; the mid-zone is the space below the inferior wall of the aortic arch and above the inferior wall of the right inferior pulmonary vein; the lower zone extends below the inferior wall of the right inferior pulmonary vein [23]. Given that, we agreed that anatomically upper CXR zones (zone 1—left, 2—right) approximately corresponded to areas 6 (left) and 3 (right) on the LUS and CT; middle CXR zones (3—left, 4—right) corresponded to areas 2, 10, 14 (left) and 5, 8, 12 (right) on the LUS and CT, and lower CXR zones (5—left and 6—right) corresponded to areas 4, 9, 13 (left) and 1, 7, 11 (right) on the LUS and CT. CXR changes within each zone were scored: 0—no abnormalities, 1—interstitial infiltrates, 2—interstitial and alveolar infiltrates (interstitial predominance), and 3—interstitial and alveolar infiltrates (alveolar predominance) see Table 1. For each patient, the stated scores in all 6 zones were added up (ranging from 0 to 18) to obtain the total CXR score.

2.6. Bias

All patient records with an LUS exam were consecutively analyzed without exclusions to minimize selection bias. The only reduction in sample numbers was due to the availability of the other radiological scores, duplicate entries, or based on statistical tests (far out values detection).

The radiological exams were performed by a single clinician (LUS score—IS; CXR score—DBM; CT score—DBM), thus not as a two-person consensus. However, both authors (IS, DBM) are highly experienced in their respective clinical fields. HRCT and LUS exams were blinded and performed without knowledge of laboratory parameters, current treatment, and further involvement in the treatment of the patient. However, the decision to mechanically ventilate the patient was up to the third party without knowledge (blinded) of the LUS score. The CXR and CT scores were assessed upon writing this manuscript with a clinician blinded to any of the observed predictors or clinical outcomes.

2.7. Study Size

This was a sample of consecutive patients collected during the 6-month pandemic peak in Croatia, and the expected enrollment was over 240 cases. We expected the area under the receiver operating curve to be above 0.7 with a ratio of positive outcome cases around 1:10. To accommodate statistical power of 80%, this required a minimum sample of 190 patients.

2.8. Statistical Analysis

Descriptive statistics were performed: categorical data were presented by absolute and relative frequencies; continuous data with normal distribution were presented as mean and standard deviation (SE) when highly variable by the median and interquartile range (IQR). The outliers and far-out values were detected by the Tukey method [24]. The normality of the distribution of continuous variables was tested by the Shapiro–Wilk test. The *t*-test was used to compare the means and the Mann–Whitney U test to compare the medians between two groups, while the one-way Chi-square test was used to compare proportions for dichotomous variables. Logistic regression analysis (univariate, multivariate—stepwise method) was used to analyze independent factors associated with the necessity for mechanical ventilation or lethal outcome. The continuous variables included in the models were added to the combined model with their respective slope coefficients as follows: combined score = $\beta 1 \times \text{var1} + \beta 2 \times \text{var2} + \beta 3 \times \text{var3} + \ldots + \beta n \times \text{var}n$. The receiver operating curve (ROC) was used to determine the optimal threshold, the area under the curve (AUC), specificity, and sensitivity of the tested parameters. Regression analysis was used to describe the relationship between radiological scores. The type I error (alpha) was set to 0.05, and the type II error (beta, statistical power) was set to 80%. The statistical software used for analysis was MedCalc® Statistical Software version 19.6 (MedCalc Software Ltd., Ostend, Belgium; https://www.medcalc.org, accessed on 15 July 2021).

2.9. Reporting

We reported the study in line with the STROBE reporting guideline for cohort studies; the STROBE checklist is available in Supplementary File S1.

3. Results

3.1. Patients and Characteristics

We examined 299 patient records hospitalized in the University Hospital of Split due to COVID-19 from November 2020 to May 2021. Eleven duplicate records or control LUS were excluded, and the rest of the 288 eligible records of patients who underwent LUS were analyzed (Figure 1). Out of 265 records of patients that had a chest X-ray, 10 were excluded due to the dates between LUS and CXR exams being too far apart. Two records out of 42 patients who underwent CT scans were excluded for the same reason. Finally, 255 cases with CXR and LUS and 40 with CT and LUS scores were analyzed.

Patient characteristics between groups not requiring and requiring MV differed only when cardiovascular disease, hemiplegia, or leukemia were present as a negative prognostic factor. Almost the same was noticed in the case of a death outcome (Table 2). Similarly, factors for both the necessity for MV and fatal outcome were the age of the patients and the day of the illness on admission (Table 2). Furthermore, a difference was found in biochemical parameters such as elevated LDH in the group requiring MV, lower spO2, and higher troponin among patients with a fatal outcome. Leukocyte counts were lower in both mechanically ventilated and deceased patients, and respective LUS scores were significantly higher (Table 2).

Table 2. Patient characteristics according to need for mechanical ventilation or death outcome.

	No MV n/N (%), Median (IQR)	MV Required n/N (%), Median (IQR)	p *	No Death n/N (%), Median (IQR)	Death n/N (%), Median (IQR)	p *
Comorbidity/habit						
Arterial hypertension	132/253 (52.2%)	14/28 (50.0%)	0.827	126/253 (49.8%)	20/28 (71.4%)	0.030
Cardiovascular disease	34/253 (13.4%)	8/28 (28.6%)	0.033	31/253 (12.3%)	11/28 (39.3%)	0.000
COPD	14/253 (5.5%)	1/28 (3.6%)	0.662	14/253 (5.5%)	1/28 (3.6%)	0.662
CVI or TIA	3/253 (1.2%)	0/28 (0.0%)	0.563	3/253 (1.2%)	0/28 (0.0%)	0.563
Dementia	3/253 (1.2%)	0/28 (0.0%)	0.563	1/253 (0.4%)	2/28 (7.1%)	0.001
Diabetes	50/253 (19.8%)	6/28 (21.4%)	0.835	50/253 (19.8%)	6/28 (21.4%)	0.835
Hemiplegia	0/253 (0.0%)	1/28 (3.6%)	0.003	0/253 (0.0%)	1/28 (3.6%)	0.003
Kidney failure	12/253 (4.7%)	1/28 (3.6%)	0.780	10/253 (4.0%)	3/28 (10.7%)	0.107
Leukemia	3/253 (1.2%)	2/28 (7.1%)	0.024	4/253 (1.6%)	1/28 (3.6%)	0.451
Liver failure	7/253 (2.8%)	0/28 (0.0%)	0.374	6/253 (2.4%)	1/28 (3.6%)	0.700
Lymphoma	7/253 (2.8%)	0/28 (0.0%)	0.374	6/253 (2.4%)	1/28 (3.6%)	0.700
Malignancy	25/253 (9.9%)	4/27 (14.8%)	0.425	22/253 (8.7%)	7/28 (25.0%)	0.008
Myocardial infarct	10/253 (4.0%)	3/28 (10.7%)	0.107	10/253 (4.0%)	3/28 (10.7%)	0.107
Peptic ulcer	2/253 (0.8%)	1/28 (3.6%)	0.175	2/253 (0.8%)	1/28 (3.6%)	0.175
Peripheral vascular disease	10/253 (4.0%)	1/28 (3.6%)	0.922	9/253 (3.6%)	2/28 (7.1%)	0.354
Rheumatological disease	8/253 (3.2%)	0/28 (0.0%)	0.341	8/253 (3.2%)	0/28 (0.0%)	0.341
Smoker	33/243 (13.6%)	4/27 (14.8%)	0.860	34/246 (13.8%)	3/24 (12.5%)	0.858
Biometrics						
Female gender	68/252 (27.0%)	10/28 (35.7%)	0.952	66/252 (26.2%)	12/28 (42.9%)	0.063
Age (years)	62 (55–70)	69 (62–75)	0.004	62 (54–68)	75 (69–82)	0.000
Weight (kg)	91 (84–103)	94 (82–103)	0.627	91 (84–103)	92 (80–97)	0.388
Height (cm)	180 (173–186)	175 (170–186)	0.356	179 (173–187)	176 (168–180)	0.181
BMI (kg/m^2)	28.3 (26.1–30.9)	29.4 (27.8–33.1)	0.088	28.4 (26.1–31.1)	28.7 (26.9–32.9)	0.803
Day of the illness	9 (6–12)	6 (5–8)	0.002	9 (6–12)	7 (4–10)	0.035
Biochemical parameters						
CRP (mg/L)	76.3 (47.8–136.8)	98 (67.8–146.6)	0.101	79.9 (49.3–136.8)	95.8 (61.4–146.6)	0.416
D-dimer (μg/L)	0.92 (0.61–1.49)	1.05 (0.61–2.20)	0.529	0.88 (0.60–1.48)	1.48 (0.72–2.66)	0.051
LDH (U/L)	351 (282–424)	434 (306–456)	0.043	355 (285–429)	366 (285–448)	0.545
Leukocyte count (10^9/L)	7.9 (5.7–10.8)	5.7 (4.5–8.7)	0.010	7.9 (5.7–10.6)	5.7 (4.4–10.5)	0.042
Lymphocytes (%)	13.0 (8.4–18.4)	16.3 (10.8–23.9)	0.103	13.0 (8.5–18.4)	16.3 (9.2–23.8)	0.272
Neutrophils (%)	81.2 (74.6–86.9)	75.8 (70.9–83.1)	0.117	81.2 (74.8–86.7)	75.8 (69.4–83.0)	0.123
pO2 (kPa)	7.30 (6.51–7.97)	7.27 (6.18–7.78)	0.599	7.30 (6.50–7.95)	7.20 (6.06–7.97)	0.506
spO2 (%)	91 (88–93)	89 (83–94)	0.151	91 (88–93)	88 (82–93)	0.022
hs-Troponin (ng/L)	9.45 (6.50–15.30)	10.9 (8.0–20.9)	0.216	9.25 (6.55–14.65)	14.6 (8.60–44.10)	0.012
Radiological scores						
LUS score	25 (19–31)	31 (26–37)	0.001	25 (18–31)	32 (26–36)	0.000
CXR score	6 (4–10)	8 (6–10)	0.125	6 (4–10)	8 (6–12)	0.016
CT score	22.0 ± 8.0	29.0 ± 2.0	0.146	21.6 ± 7.9	29.0 ± 3.8	0.050

* for dichotomous variables (n/N) one-way classification Chi-squared test was used; for continuous variables without normal distribution, Mann–Whitney U-test was used, and respective *p*-values stated. Acronyms: BMI—body mass index; COPD—chronic obstructive pulmonary disease; CXR score—chest X-ray score; CT score—chest computerized tomography score; IQR—interquartile range; LUS score—lung ultrasound score; MV = 1—mechanical ventilation required, MV = 2—MV not required, death = 0—no death, death = 1—death outcome; SD—standard deviation; TIA—transitory ischemic attack.

3.2. *Predictors of Necessity for MV or a Lethal Outcome*

When combined in stepwise multivariate analysis, both CXR and CT scores were excluded from models for either MV or lethal outcome. Only the LUS score was retained in both models as follows (univariate analysis details available in Supplementary File S2).

Four independent predictors gave a unique statistically significant contribution to the regression model for the necessity of MV, and these are LUS score, day of the illness at admission, leukocyte count, and presence of cardiovascular disease ($\chi 2 = 29.16$, $p < 0.001$). The model accurately classified 89.9% of cases (Table 3, Figure 2A).

Table 3. Stepwise multivariate logistic regression according to the necessity of MV or lethal outcome.

Variable	Coefficient	*p*	Odds Ratio	95% CI	Cutoff	Sensitivity	Specificity
Mechanical ventilation							
LUS score	0.101	<0.001	1.11	1.04–1.17	>27	72.4%	60.8%
Day of illness	−0.131	0.022	0.88	0.78–0.98	≤7	72.4%	64.1%
Leukocyte count	−0.160	0.021	0.85	0.74–0.98	≤6.3	64.3%	66.7%
Cardiovascular disease present	1.019	0.041	2.77	1.04–7.35	positive	28.6%	86.6%
	Correctly Classified		**AUC**	**95% CI**	**Cutoff**	**Sensitivity**	**Specificity**
Combined score	88.9%		0.807	0.755–0.851	>0.51	85.7%	66.4%
Death	**Coefficient**	*p*	**Odds Ratio**	**95% CI**	**Cutoff**	**Sensitivity**	**Specificity**
Age	0.153	<0.0001	1.17	1.10–1.24	>65	89.3%	65.4%
LUS score	0.088	0.005	1.09	1.03–1.16	>29	69.0%	72.2%
	Correctly Classified		**AUC**	**95% CI**	**Cutoff**	**Sensitivity**	**Specificity**
Combined score	93.2%		0.859	0.814–0.901	>9.6	96.5%	57.9%

Acronyms: AUC—area under receiver operating curve; CI—confidence interval; CT score—chest computerized tomography score; LUS score—lung ultrasound score; CXR score—chest X-ray score.

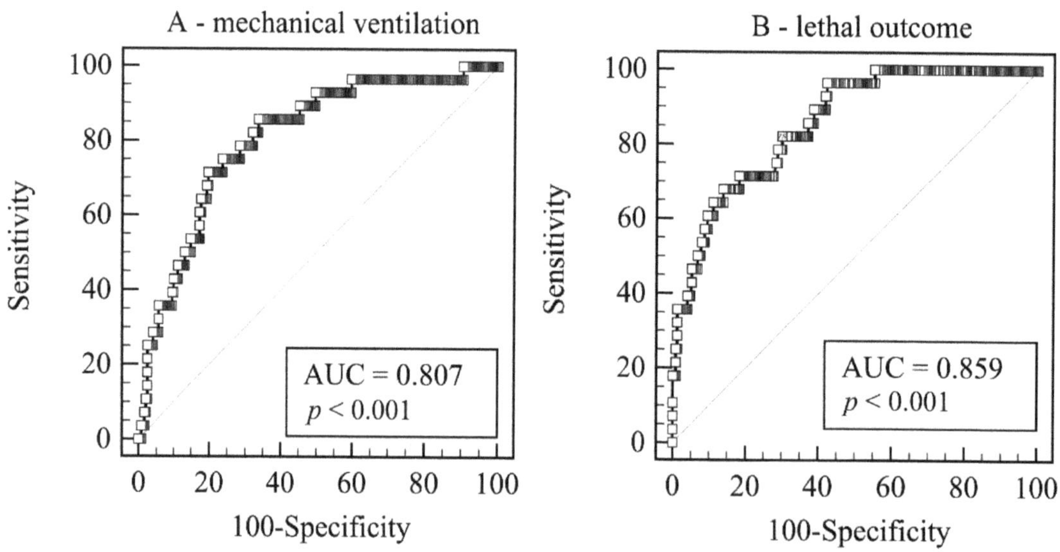

Figure 2. Receiver operating curves of the combined scores. (**A**) Predicting the need for mechanical ventilation. (**B**) Predicting the lethal outcome. Legend: AUC—Area under hierarchical receiving operator curve.

For the lethal outcome, only two independent predictors gave a unique statistically significant contribution to the regression model: LUS score and age of the patient ($\chi 2 = 48.56$, $p < 0.001$). The model accurately classified 93.2% of cases (Table 3, Figure 2B).

3.3. Relationship between LUS and CXR Scores

The regression model between the CXR score and LUS score demonstrated a strong trend (slope 0.160, 95%CI 0.109 to 0.212, $p < 0.001$); however, there was significant variability around the regression line ($R^2 = 0.128$; Supplementary Files S2 and S3).

Prediction models for MV based on LUS score (AUC = 0.693 ± 0.058) and CXR score (AUC = 0.586 ± 0.054) showed no significant difference of 0.106, $p = 0.136$ (Figure 3A).

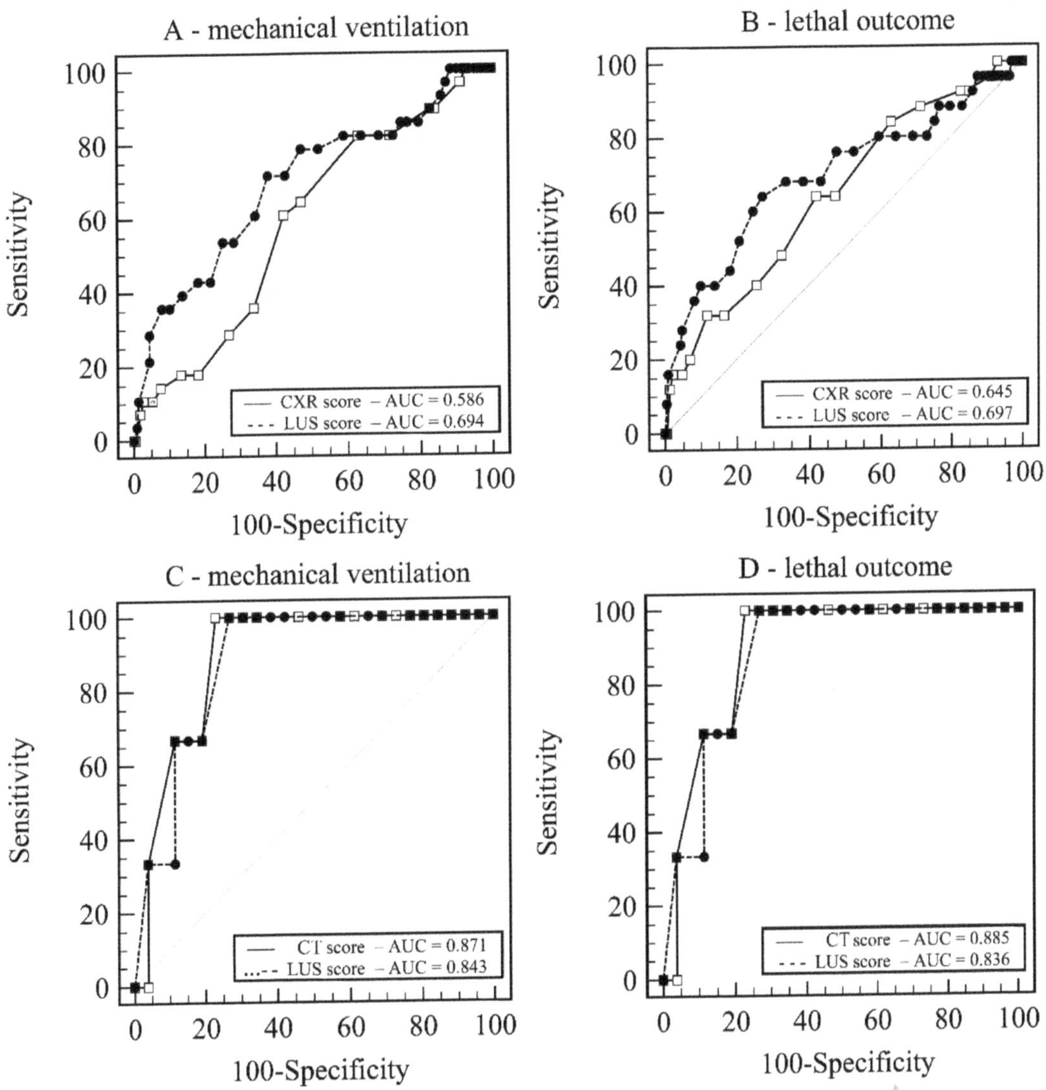

Figure 3. Receiver operating curves. (**A**) LUS score vs. CXR score—prediction of mechanical ventilation. (**B**) LUS score vs. CXR score—prediction of lethal outcome. (**C**) LUS score vs. CT score—prediction of mechanical ventilation. (**D**) LUS score vs. CT score—prediction of lethal outcome. Legend: AUC—Area under hierarchical receiving operator curve; CT score—Computerized tomography score; CXR score—chest X-ray score; LUS score—Lung ultrasound score.

Additionally, models for death based on LUS score (AUC = 0.697 ± 0.064) and CXR score (AUC = 0.645 ± 0.059) showed no significant difference of 0.052, $p = 0.449$ (Figure 3B).

3.4. Relationship between LUS and CT Scores

The regression model between the CT score and LUS score demonstrated a strong trend (slope 0.502, 95%CI 0.292 to 0.711, $p < 0.0001$); however, there was significant variability around the regression line (R2 = 0.396; Supplementary File S2).

Prediction models for MV based on LUS score (AUC = 0.871 ± 0.090) and CT score (AUC = 0.843 ± 0.077) showed no significant difference of 0.023, p = 0.819 (Figure 3C).

Additionally, models for death based on LUS score (AUC = 0.885 ± 0.057) and CT score (AUC = 0.836 ± 0.080) showed no significant difference of 0.049, p = 0.582 (Figure 3D).

4. Discussion

The results of our study draw us to conclusions defining four independent predictors regarding the regression model for the necessity of MV: LUS score, day of the illness at admission, leukocyte count, and presence of cardiovascular disease. For the lethal outcome, the two independent predictors were the LUS score and the age of the patient. Although LUS' predictive value was shown in several previous studies, as far as we know, it was not combined with demographics, clinical data, and laboratory parameters, creating prediction models for disease severity.

The average CXR score was not previously shown to be different between the group of patients who needed MV and the group of patients who did not. Although it is a possible prognostic factor for death in the univariate model, the CXR score was dropped from the multivariate model as insignificant in favor of, for example, the patient's age. Therefore, we can conclude that the CXR score is inferior to the LUS score in the prediction of the need for MV or death.

For both the CT score and the CXR score, we showed a significant correlation with the LUS score by linear regression, although in both cases, there was significant variability. We attributed this to subjectivity in the scoring system based on the decision of the clinician interpreting the finding, the lower sensitivity and specificity of the CXR, and the time interval between CT, LUS, and CXR acquisition, in comparison with other studies that used shorter time delays among them [25]. On the other hand, the narrower CXR score scale (0–18) allowed for a smaller distribution of the total score. It was expected that the relative cutoff value (out of maximal score) would be higher in CXR than in the LUS score, which did not prove to be exact. The cutoff CXR score for the prediction of MV was already at 28% of the maximum score, while for the fatal outcome, it was at 38% and was therefore clinically completely irrelevant for these critical outcomes for the remaining upper ~60% of the score scale. We believe that based on plain CXR, very little can be described, such as the gentle difference of changes in the lung parenchyma essential for the outcomes. The non-specificity of CXR is a possible reason for the relatively low cutoff values obtained.

On the other hand, a much better correlation between LUS score and CT score, with significantly less variability and equal width of the scoring scale, speaks in favor of LUS as an excellent method of choice in monitoring patients with pneumonia; this was confirmed in other studies as well [25,26]. Despite this, previously published studies of predictive models involving radiological imaging have exceptionally included ultrasound more commonly than CT or chest CXR. Huang J. et al. suggested a diagnostic model obtained to distinguish between moderate and severe/critical COVID-19 using CT as a part of the diagnostic procedure. Their model included CT imaging in all patients and even those with mild symptoms [27]. Although CT is the gold standard, of course, the widespread use of such a model, especially in times of intense epidemics, is not possible. The Dutch COVID-19 risk model presented by Schalekamp et al. used similar demographic and laboratory parameters combined with the chest CXR score, which, unlike ours, used four zones in determining the CXR score. As in our study, patients who developed critical illness had leukopenia and higher lactate dehydrogenase levels more often. Contrary to our results, no significant difference was found for the duration of symptoms, although the proportion of patients with a symptom duration of more than 7 days was slightly smaller among those with critical disease [27]. As previously reported, age was a strong predictor of a lethal outcome in our population as well [28].

Multiple studies reported a significant association between elevated leukocyte counts and decreased lymphocyte counts among patients with severe cases of COVID-19 compared with those with mild cases [15,29]. However, in our study population, a decreased leukocyte count was predictive of more severe disease.

Various biochemical markers were researched for their predictive usefulness, for example, LDH, CRP, and D-dimer [30–32]. In our study, patients who needed MV had significantly higher LDH values, although LDH did not appear to have predictive power in regression analysis. CRP and D-dimer were not significantly higher in the MV or deceased groups.

In our study, a lower day of illness at hospital admission showed to be a predictive factor for disease severity (MV), as it might suggest a more rapidly evolving and progressive disease in these patients. Similar results concerning onset to hospitalization time were found as a part of the risk nomogram established to predict the incidence of severe or critical COVID-19 in elderly patients in the study by Zeng et al. [20].

The percentage of smokers in the hospitalized population was significantly lower than in the overall population of Croatia, regarding which there is still controversy in the available literature, but most studies still link smoking to a higher risk of severe COVID-19 [32–35].

The proportion of patients who ended up on MV or died was similar to our previous study [11]. Only the therapeutic options in the patients changed, so we believe that the lower predictability of the LUS score in this study could be related to a modified corticosteroid therapy that the patients received (methylprednisolone 1 mg/kg vs. dexamethasone 6 mg used previously), which might have affected the clinical course and mortality of this disease.

4.1. Strengths

Our study was conducted on a large sample of hospitalized patients, and data were analyzed without exclusion by clinical characteristics using only statistically relevant and transparent figures.

4.2. Limitations

As mentioned in the bias section of the methods, we did not do a double, independent check of the radiological scores, and this could be a limiting factor. Although we demonstrated a strong association between the LUS score and the CT score, the sample for the CT score was relatively small and did not reach the test strength of 80%.

5. Conclusions

The predictive model obtained in our study identified four key parameters at the patient admission to the hospital, LUS score, day of the illness, leukocyte count, and presence of cardiovascular disease, that can predict an adverse COVID outcome. Hopefully, it will help physicians provide appropriate and timely therapeutic interventions.

Supplementary Materials: The following supporting information can be downloaded at: https://www.mdpi.com/xxx/s1, Supplementary File S1—STROBE checklist, Supplementary File S2—Univariate regression of predictors for mechanical ventilation or death, Supplementary File S3—Scatter diagram and regression line between LUS score and CXR score.

Author Contributions: Conceptualization, I.S., M.P.I., K.D. and D.B.M.; methodology, I.S., I.J., J.D. and D.B.M.; software, O.B.; validation, I.S. and O.B.; formal analysis, I.S., I.J., J.D. and D.B.M.; investigation, I.S., I.J., J.D. and D.B.M.; resources, I.S., I.J., J.D. and D.B.M.; data curation, I.S., I.J., J.D., D.B.M. and O.B.; writing—original draft preparation, I.S., M.P.I., O.B. and D.B.M.; writing—review and editing, I.S., I.J., J.D., M.P.I., D.B.M., O.B. and K.D.; visualization I.S., I.J., J.D. and D.B.M.; supervision, K.D. and M.P.I.; project administration, I.S., M.P.I. and D.B.M.; funding acquisition, K.D. All authors have read and agreed to the published version of the manuscript.

Funding: This research received no external funding.

Institutional Review Board Statement: The study was approved by the Institutional Ethics Board of University Hospital Centre Split, Croatia (Klasa: 500-03/21-01/01, Ur.br. 2181-147-01/06/M.S.-20-02, 29 January 2021).

Informed Consent Statement: Written informed consent was not required for participants with emerging infectious diseases, LUS is a standard of care in our center, and this study had no interventions.

Data Availability Statement: The datasets used and/or analyzed during the current study are available from the corresponding author on reasonable request.

Conflicts of Interest: The authors declare no conflict of interest.

References

1. WHO. Diagnostic Testing for SARS-CoV-2. Available online: https://www.who.int/publications/i/item/diagnostic-testing-for-sars-cov-2 (accessed on 18 April 2021).
2. Cascella, M.; Rajnik, M.; Cuomo, A.; Dulebohn, S.C.; Di Napoli, R.S. *Features, Evaluation, and Treatment of Coronavirus*; StatPearls: Treasure Island, FL, USA, 2020.
3. Pascarella, G.; Strumia, A.; Piliego, C.; Bruno, F.; del Buono, R.; Costa, F.; Scarlata, S.; Agrò, F.E. COVID-19 diagnosis and management: A comprehensive review. *J. Intern. Med.* **2020**, *288*, 192–206. [CrossRef] [PubMed]
4. Borakati, A.; Perera, A.; Johnson, J.; Sood, T. Diagnostic accuracy of X-ray versus CT in COVID-19: A propensity-matched database study. *BMJ Open* **2020**, *10*, e042946. [CrossRef] [PubMed]
5. Kovács, A.; Palásti, P.; Veréb, D.; Bozsik, B.; Palkó, A.; Kincses, Z.T. The sensitivity and specificity of chest CT in the diagnosis of COVID-19. *Eur. Radiol.* **2021**, *31*, 2819–2824. [CrossRef] [PubMed]
6. Kaufman, A.E.; Naidu, S.; Ramachandran, S.; Kaufman, D.S.; Fayad, Z.A.; Mani, V. Review of radiographic findings in COVID-19. *World J. Radiol.* **2020**, *12*, 142–155. [CrossRef]
7. Jackson, K.; Butler, R.; Aujayeb, A. Lung ultrasound in the COVID-19 pandemic. *Postgrad. Med. J.* **2021**, *97*, 34–39. [CrossRef]
8. Lieveld, A.W.; Kok, B.; Schuit, F.H.; Azijli, K.; Heijmans, J.; van Laarhoven, A.; Assman, N.L.; Kootte, R.S.; Olgers, T.J.; Nanayakkara, P.W.; et al. Diagnosing COVID-19 pneumonia in a pandemic setting: Lung Ultrasound versus CT (LUVCT)—A multicentre, prospective, observational study. *ERJ Open Res.* **2020**, *6*, 00539-2020. [CrossRef]
9. Ottaviani, S.; Franc, M.; Ebstein, E.; Demaria, L.; Lheure, C.; Debray, M.; Khalil, A.; Crestani, B.; Borie, R.; Dieudé, P. Lung ultrasonography in patients with COVID-19: Comparison with CT. *Clin. Radiol.* **2020**, *75*, 877.e1–877.e6. [CrossRef]
10. Nouvenne, A.; Zani, M.D.; Milanese, G.; Parise, A.; Baciarello, M.; Bignami, E.G.; Odone, A.; Sverzellati, N.; Meschi, T.; Ticinesi, A. Lung Ultrasound in COVID-19 Pneumonia: Correlations with Chest CT on Hospital admission. *Respir. Int. Rev. Thorac. Dis.* **2020**, *99*, 617–624. [CrossRef]
11. Skopljanac, I.; Ivelja, M.; Barcot, O.; Brdar, I.; Dolic, K.; Polasek, O.; Radic, M. Role of Lung Ultrasound in Predicting Clinical Severity and Fatality in COVID-19 Pneumonia. *J. Pers. Med.* **2021**, *11*, 757. [CrossRef]
12. Gutsche, H.; Lesser, T.G.; Wolfram, F.; Doenst, T. Significance of Lung Ultrasound in Patients with Suspected COVID-19 Infection at Hospital Admission. *Diagnostics* **2021**, *11*, 921. [CrossRef]
13. Mohammad, S.; Aziz, R.; Al Mahri, S.; Malik, S.S.; Haji, E.; Khan, A.H.; Khatlani, T.S.; Bouchama, A. Obesity and COVID-19: What makes obese host so vulnerable? *Immun. Ageing* **2021**, *18*, 1. [CrossRef] [PubMed]
14. Ji, D.; Zhang, D.; Xu, J.; Chen, Z.; Yang, T.; Zhao, P.; Chen, G.; Cheng, G.; Wang, Y.; Bi, J.; et al. Prediction for Progression Risk in Patients With COVID-19 Pneumonia: The CALL Score. *Clin. Infect. Dis. Off. Publ. Infect. Dis. Soc. Am.* **2020**, *71*, 1393–1399. [CrossRef] [PubMed]
15. Huang, G.; Kovalic, A.J.; Graber, C.J. Prognostic Value of Leukocytosis and Lymphopenia for Coronavirus Disease Severity. *Emerg. Infect. Dis.* **2020**, *26*, 1839–1841. [CrossRef]
16. Luo, X.; Zhou, W.; Yan, X.; Guo, T.; Wang, B.; Xia, H.; Ye, L.; Xiong, J.; Jiang, Z.; Liu, Y.; et al. Prognostic Value of C-Reactive Protein in Patients with Coronavirus 2019. *Clin. Infect. Dis. Off. Publ. Infect. Dis. Soc. Am.* **2020**, *71*, 2174–2179. [CrossRef]
17. Fink, D.L.; Khan, P.Y.; Goldman, N.; Cai, J.; Hone, L.; Mooney, C.; El-Shakankery, K.H.; Sismey, G.; Whitford, V.; Marks, M.; et al. Development and internal validation of a diagnostic prediction model for COVID-19 at time of admission to hospital. *QJM Mon. J. Assoc. Physicians* **2021**, *114*, 699–705. [CrossRef]
18. Guan, X.; Zhang, B.; Fu, M.; Li, M.; Yuan, X.; Zhu, Y.; Peng, J.; Guo, H.; Lu, Y. Clinical and inflammatory features based machine learning model for fatal risk prediction of hospitalized COVID-19 patients: Results from a retrospective cohort study. *Ann. Med.* **2021**, *53*, 257–266. [CrossRef] [PubMed]
19. Qi, J.; He, D.; Yang, D.; Wang, M.; Ma, W.; Cui, H.; Ye, F.; Wang, F.; Xu, J.; Li, Z.; et al. Severity-associated markers and assessment model for predicting the severity of COVID-19: A retrospective study in Hangzhou, China. *BMC Infect. Dis.* **2021**, *21*, 774. [CrossRef] [PubMed]
20. Zeng, F.; Deng, G.; Cui, Y.; Zhang, Y.; Dai, M.; Chen, L.; Han, D.; Li, W.; Guo, K.; Chen, X.; et al. A predictive model for the severity of COVID-19 in elderly patients. *Aging* **2020**, *12*, 20982–20996. [CrossRef]

21. Soldati, G.; Smargiassi, A.; Inchingolo, R.; Buonsenso, D.; Perrone, T.; Briganti, D.F.; Perlini, S.; Torri, E.; Mariani, A.; Mossolani, E.E.; et al. Proposal for International Standardization of the Use of Lung Ultrasound for Patients With COVID-19: A Simple, Quantitative, Reproducible Method. *J. Ultrasound Med. Off. J. Am. Inst. Ultrasound Med.* **2020**, *39*, 1413–1419. [CrossRef]
22. Smargiassi, A.; Soldati, G.; Torri, E.; Mento, F.; Milardi, D.; del Giacomo, P.; de Matteis, G.; Burzo, M.L.; Larici, A.R.; Pompili, M.; et al. Lung Ultrasound for COVID-19 Patchy Pneumonia: Extended or Limited Evaluations. *J. Ultrasound Med. Off. J. Am. Inst. Ultrasound Med.* **2021**, *40*, 521–528. [CrossRef]
23. Borghesi, A.; Maroldi, R. COVID-19 outbreak in Italy: Experimental chest X-ray scoring system for quantifying and monitoring disease progression. *La Radiol. Med.* **2020**, *125*, 509–513. [CrossRef] [PubMed]
24. Tukey, J.W. *Exploratory Data Analysis*; THETA: Reading, MA, USA, 1977; Volume 2.
25. Gibbons, R.C.; Magee, M.; Goett, H.; Murrett, J.; Genninger, J.; Mendez, K.; Tripod, M.; Tyner, N.; Costantino, T.G. Lung Ultrasound vs. Chest X-Ray Study for the Radiographic Diagnosis of COVID-19 Pneumonia in a High-Prevalence Population. *J. Emerg. Med.* **2021**, *60*, 615–625. [CrossRef] [PubMed]
26. Rizzetto, F.; Perillo, N.; Artioli, D.; Travaglini, F.; Cuccia, A.; Zannoni, S.; Tombini, V.; di Domenico, S.L.; Albertini, V.; Bergamaschi, M.; et al. Correlation between lung ultrasound and chest CT patterns with estimation of pulmonary burden in COVID-19 patients. *Eur. J. Radiol.* **2021**, *138*, 109650. [CrossRef] [PubMed]
27. Schalekamp, S.; Huisman, M.; van Dijk, R.A.; Boomsma, M.F.; Jorge, P.J.F.; de Boer, W.S.; Herder, G.J.M.; Bonarius, M.; Groot, O.A.; Jong, E.; et al. Model-based Prediction of Critical Illness in Hospitalized Patients with COVID-19. *Radiology* **2021**, *298*, E46–E54. [CrossRef] [PubMed]
28. Mi, J.; Zhong, W.; Huang, C.; Zhang, W.; Tan, L.; Ding, L. Gender, age and comorbidities as the main prognostic factors in patients with COVID-19 pneumonia. *Am. J. Transl. Res.* **2020**, *12*, 6537–6548. [PubMed]
29. Tong, X.; Cheng, A.; Yuan, X.; Zhong, X.; Wang, H.; Zhou, W.; Xu, X.; Li, Y. Characteristics of peripheral white blood cells in COVID-19 patients revealed by a retrospective cohort study. *BMC Infect. Dis.* **2021**, *21*, 1236. [CrossRef]
30. Chen, R.; Liang, W.; Jiang, M.; Guan, W.; Zhan, C.; Wang, T.; Tang, C.; Sang, L.; Liu, J.; Ni, Z.; et al. Risk Factors of Fatal Outcome in Hospitalized Subjects with Coronavirus Disease 2019 From a Nationwide Analysis in China. *Chest* **2020**, *158*, 97–105. [CrossRef]
31. Liu, F.; Li, L.; Xu, M.; Wu, J.; Luo, D.; Zhu, Y.; Li, B.; Song, X.; Zhou, X. Prognostic value of interleukin-6, C-reactive protein, and procalcitonin in patients with COVID-19. *J. Clin. Virol. Off. Publ. Pan Am. Soc. Clin. Virol.* **2020**, *127*, 104370. [CrossRef]
32. Martha, J.W.; Wibowo, A.; Pranata, R. Prognostic value of elevated lactate dehydrogenase in patients with COVID-19: A systematic review and meta-analysis. *Postgrad. Med. J.* **2021**. [CrossRef]
33. Ismail, N.; Hassan, N.; Abd Hamid, M.H.N.; Yusoff, U.N.; Khamal, N.R.; Omar, M.A.; Wong, X.C.; Pathmanathan, M.D.; Mohd Zin, S.; Muhammad Zin, F.; et al. Association of smoking and severity of COVID-19 infection among 5889 patients in malaysia: A multi-center observational study. *Int. J. Infect. Dis.* **2022**, *116*, 189–196. [CrossRef]
34. Zhao, Q.; Meng, M.; Kumar, R.; Wu, Y.; Huang, J.; Lian, N.; Deng, Y.; Lin, S. The impact of COPD and smoking history on the severity of COVID-19: A systemic review and meta-analysis. *J. Med. Virol.* **2020**, *92*, 1915–1921. [CrossRef] [PubMed]
35. European Commission; Directorate-General for Communication; Directorate-General for Health and Food Safety. Attitudes of Europeans towards tobacco and electronic cigarettes: Report, European Commission. 2021. Available online: https://op.europa.eu/en/publication-detail/-/publication/c070c04c-6788-11eb-aeb5-01aa75ed71a1 (accessed on 19 April 2022). [CrossRef]

life

MDPI

Article

The Influence of Sex, Gender, and Age on COVID-19 Data in the Piedmont Region (Northwest Italy): The Virus Prefers Men

Silvia De Francia [1,*], Alessandro Ferretti [2], Francesco Chiara [1], Sarah Allegra [1], Daniele Mancardi [1], Tiziano Giacomo Allice [3], Maria Grazia Milia [3], Gabriella Gregori [3], Elisa Burdino [3], Claudio Avanzini [3], Valeria Ghisetti [3] and Alessandra Durio [4]

1 Department of Biological and Clinical Sciences, University of Turin, S. Luigi Gonzaga Hospital, 10043 Orbassano, Italy; 336124@edu.unito.it (F.C.); sarah.allegra@unito.it (S.A.); daniele.mancardi@unito.it (D.M.)
2 Department of Physics, University of Turin, 10125 Turin, Italy; alessandro.ferretti@unito.it
3 Laboratory of Microbiology and Virology ASL Turin, 10149 Turin, Italy; tiziano.allice@gmail.com (T.G.A.); mariagrazia.milia@aslcittaditorino.it (M.G.M.); gabriella.gregori@aslcittaditorino.it (G.G.); elisa.burdino@aslcittaditorino.it (E.B.); claudio.avanzini@aslcittaditorino.it (C.A.); valeria.ghisetti@aslcittaditorino.it (V.G.)
4 Department of Economics and Statistics "Cognetti de Martiis", University of Turin, 10153 Turin, Italy; alessandra.durio@unito.it
* Correspondence: silvia.defrancia@unito.it

Abstract: Several important sex and gender differences in the clinical manifestation of diseases have been known for a long time but are still underestimated. The infectious Coronavirus 2019 disease pandemic has provided evidence of the importance of a sex and gender-based approach; it mainly affected men with worse symptomatology due to a different immune system, which is stronger in women, and to the Angiotensin-converting enzyme 2 and Transmembrane protease serine 2 roles which are differently expressed among the sexes. Additionally, women are more inclined to maintain social distance and smoke less. Analysis of data on the infectious Coronavirus 2019 disease testing from people admitted to the Amedeo di Savoia Hospital, a regional referral center for infectious diseases, has been applied to the whole of 2020 data (254,640 records). A high percentage of data in the dataset was not suitable due to a lack of information or entering errors. Among the suitable samples, records have been analyzed for positive/negative outcomes, matching records for unique subjects (N = 123,542), to evaluate individual recurrence of testing. Data are presented in age and sex-disaggregated ways. Analyses of the suitable sample also concerned the relation between testing and hospital admission motivation and symptoms. Our analysis indicated that a sex and gender-based approach is mandatory for patients and the National Health System's sustainability.

Keywords: sex; gender; Coronavirus infectious disease 2019; differences; tailored approach

Citation: De Francia, S.; Ferretti, A.; Chiara, F.; Allegra, S.; Mancardi, D.; Allice, T.G.; Milia, M.G.; Gregori, G.; Burdino, E.; Avanzini, C.; et al. The Influence of Sex, Gender, and Age on COVID-19 Data in the Piedmont Region (Northwest Italy): The Virus Prefers Men. *Life* **2022**, *12*, 643. https://doi.org/10.3390/life12050643

Academic Editor: Alfredo De Giorgi

Received: 7 April 2022
Accepted: 24 April 2022
Published: 26 April 2022

1. Introduction

Important sex and gender differences are observed in the frequency, symptoms, and severity of several diseases, in addition to the response to treatments and adverse drug reactions. A sex and gender-based approach to clinical practice can significantly contribute to health promotion by improving the appropriateness of care and, therefore, providing benefits for patients and the National Health System's sustainability. This is true also in the context of the infectious Coronavirus 2019 disease (COVID-19). What was discovered as a cluster of patients with a mysterious respiratory illness in Wuhan, China, in December 2019, was later identified as COVID-19. The pathogen of severe acute respiratory syndrome coronavirus 2 (SARS-CoV-2), a novel Beta coronavirus, was subsequently isolated as the causative disease agent [1–7] and on 11 March 2020, the World Health Organization (WHO) declared COVID-19 a pandemic. A growing body of evidence reveals that the male sex is a risk factor for more severe disease: globally, approximately 60% of deaths from COVID-19 are reported in men [8]. Women seem to be more protected for different reasons. Past studies have shown that sex has a considerable effect on the outcome of infection and has been associated with underlying differences in immune responses leading to physiological and anatomical differences that may influence exposure, receptor recognition, clearance, and even transmission of microorganisms. The X-linked nature of immune response proteins deeply marks the difference: women mount a stronger immune response to infections and vaccinations and outlive men [9–17]. In addition to the different immune responses, Angiotensin-converting enzyme 2 (ACE2), a protein involved in blood pressure regulation and the cleavage of substrates acting in different physiological processes, plays a central role in COVID-19 sex-related progression. SARS-CoV-2 utilizes the ACE2 receptor as its main entry portal and, possibly, as a route to secondary "metastatic" end-organ disease [18–24]. The binding of COVID-19 spike protein to ACE2 induces the ACE2 down-regulation that leads to a decrease of angiotensin (1–7) production in the lung, igniting acute respiratory failure. Estrogen, in particular 17β-estradiol (E2), the main female sex hormone, upregulates the expression of ACE2 that, accordingly, is higher in females than in males. Therefore, E2 by ACE2 overexpression in the female sex could, at least partially, account for the better outcome and the lower death rate in female COVID-19 patients. SARS-CoV-2 interfaces furthermore with the renin-angiotensin-aldosterone system (RAAS) through ACE2 and there are concerns that RAAS inhibitors may change ACE2 expression and thus COVID-19 virulence. In addition, estrogens are believed to inhibit the activity or expression of different components of the RAAS system [25]. In the end, Transmembrane protease serine 2 (TMPRSS2) leads to continuous virus entry into cells, but its expression is upregulated by androgens [26,27]. Different immune responses, the RAAS system, ACE2, and TMPRSS2 role, and hormonal status are biological sex-related differences that count markedly towards the different COVID-19 progression among the sexes [28–32]. Gender, profoundly understudied, counts significantly in this context [33]. It may reflect behaviors that influence exposure to microorganisms, access to healthcare, or health-seeking behaviors that can affect the infection course. Women smoke less and show more compliance to basic rules of social distancing. Women use facial masks accurately and are more skilled with personal hygiene. Understanding these factors will not only help to gain a better knowledge of COVID-19 pathogenesis but will also guide the design of effective strategies for sex and gender-based personalized medicine. The supranational organization Global Health 50:50 [34] requested participating nations to report the sex and gender-disaggregated clinical data related to COVID-19 incidence and its mortality. However, to date, most clinical specialists continue to analyze data without any categorization. Our analysis highlights this point, recognizing a central role in categorizing data according to sex and gender differences.

The aim of this work was to analyze data on COVID-19 testing in the Piedmont region, northwest Italy, for people admitted to the Amedeo di Savoia Hospital, a regional referral center for infectious diseases. Data are referred to for the whole of 2020.

2. Materials and Methods

2.1. COVID-19 Testing

Analysis was performed on the COVID-19 testing dataset obtained from the database used by the Microbiology and Virology Laboratory of the Amedeo di Savoia Hospital, a regional reference center for infectious diseases. The original dataset constituted 254,640 records and 10 analytical variables referred for the whole of 2020 (from 1 January to 31 December). The data analyzed were not attributable to identity data (name and surname); each record was immediately encoded with a specific identification code. The statistical software used for analysis was R (R Core Team 2017) and its text-mining (TM) packages [35–38]. The use of these packages to perform the statistical analysis is based on the evidence that health care professionals produce abundant textual information in their daily clinical practice, stored in many different sources. The extraction of insights from all the gathered information, mainly unstructured and lacking normalization, is one of the major challenges in computational medicine. In this respect, TM assembles different techniques to derive valuable insights from unstructured textual data, so it becomes especially relevant in medical analyses. The work of cleaning and data editing on the COVID-19 testing dataset was carried out with handmade checks scrolling through every single record. This allowed us to (1) define new variables by cross-checking some information contained in the date of birth, tax code, and date of the test, and (2) allowed us to enhance some missing values to create a unique subject identifier. The dataset resulting from all editing operations was therefore made up of 251,657 records and 19 variables. We identified the frequency of regional distribution for patients tested for COVID-19 origin and analyzed some information about the test performed: the type of test done; the execution date, considered on four different year periods (I: February-May, II: June-August, III: September-October, and IV: November-December) decided retrospectively from evaluation of pandemic trend; and the positivity and negativity percentage rate related to the different defined periods.

2.2. Unique Subjects

To better understand the characteristics of the subjects who have undergone at least one COVID-19 test during 2020, unique subjects have been labeled, identifying them through the code-specific tax key. Data have been presented in a sex and age-disaggregated way, considering different age classes. The distribution of tests for unique subjects during 2020 has been analyzed and matched with the test result, selecting for each unique subject the first positive test obtained, if present, and for subjects never testing positive, their first result is listed in the dataset.

2.3. Epidemiological Criterion

Among the variables of the original dataset, there was that relating to the epidemiological criterion, an open and non-mandatory field. The operator could fill it out by writing text with the contents he/she considered most significant. This variable potentially contained lots of information about subjects tested for COVID-19, but this information was not directly analyzable with common statistical tools. Using TM, it was possible to maximize the information obtainable from this field. The nature of this field entailed a number of obstacles to the use of recorded information; first of all, not being a mandatory field, many records did not contain any text. Furthermore, being a free field, the contained information was dependent on the operator who compiled the record corresponding to the COVID-19 test done. What appeared in the text of the criterion field could be very detailed, including information on health, such as symptoms or concomitant pathologies, of the subject tested and/or the reasons for having done the test, but it could also be used only as a field for notes, containing telephone numbers, email contacts, or personal names of the attending physician or reference contact. Being a text field, it was edited by the operator and therefore subjected to typos and grammatical errors, the use of acronyms, abbreviations, and synonyms, generating significant confusion. To manage the information contained in the epidemiological criterion present in the dataset, we used TM techniques which allowed us to transform texts into structured data. At first, we had to clean the dataset, removing empty records for epidemiological criteria. The resulting dataset was therefore made up of 196,970 records. The corpus of documents that we analyzed consisted of 196,970 texts (the epidemiological criteria). We then proceeded to standardize the texts of each document through different operations: the conversion of all characters to uppercase or lowercase; the elimination of special characters such as punctuation, multiple spacing, symbols, or numbers; the elimination of all stop-words, the common language words that did not add meaning to the sentence content, such as articles, pronouns, adverbs; and the transformation of words into lemmas, that is the transformation of a word into its canonical form. We turned each document into a word vector (unigram) and created the bag of words (BofW) matrix of 196,970 rows, one for each criterion, and 23,022 columns, one for each different word contained throughout the corpus. In the cells of the BofW matrix, there was the number of times that the column word was present in the row criterion. The 23,022 words identified in this first phase form the dictionary of the document's corpus corresponding to the criteria. For the problems described above, these 23,022 words did not identify as many different concepts, many were synonymous, others were misspelled, and still, others were abbreviations. For example, in the resulting dictionary, we found 55 different ways in which the term "COVID" was spelled. There is no statistical method that can automatically identify these 55 words and reduce them all to the "COVID" unigram, it was, therefore, necessary to correct the dictionary manually by reading the 23,022 words and indicating their possible correspondence to its correction. Subsequently, we proceeded to correct the data with a string replacement algorithm. This correction operation has been repeated twice in trying to unify in a single term the plural, singular, male, and female forms. For example, different Italian words meaning "sanitary" (sanitari, sanitario, and sanitaria) have all been reduced to one term "sanitariao". In the end, we reduced the dictionary to 9683 unigrams.

2.4. Definition of Subject and Symptoms Categories

First, looking at the entire dictionary, we defined three subsets of terms corresponding to a separate given category of subjects. These three subsets are defined by the terms written with the same color in Figure 1a. The subset of the terms in blue identifies the category of healthcare workers, the terms in purple identify the category of those who had contact with a positive subject, and the subset in red is related to the category of assistance home guests. Afterward, if the criterion contained at least one of the terms of a subset, the dataset record was classified as a test mate to a subject of the corresponding category. In this way, we added to the original dataset three new dichotomous variables, one for each category, the result of structuring the information contained in the textual description of the criterion. Going into more detail, we identified unigrams related to the definition of three different subject categories: healthcare workers; assistance home guests; and contacts with a positive subject. We then proceeded to subject categories analysis, studying them separately.

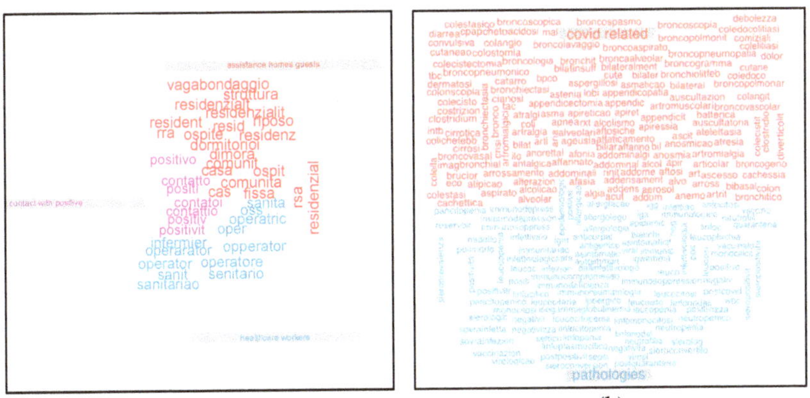

(**a**) (**b**)

Figure 1. (**a**) The clusters of terms defining different subjects' categories that have undergone the COVID-19 test. (**b**) The clusters of terms defining different symptom categories of subjects that have undergone the COVID-19 test.

A comparative study was carried out between categories and the presence of symptoms. Coming back to the entire dictionary, we defined two subsets of terms, each corresponding to a given macro category of symptoms. These two subsets were defined by the terms written with the same color in Figure 1b. The subset of the 597 terms in blue identifies the category of other different pathologies (P), while the 773 terms in red identify the category of COVID-19 related symptoms (S). Afterward, if the criterion contained at least one of the terms of a subset, the dataset record was classified as a test mate to a subject of the corresponding category. In this way, we added to the original dataset two new dichotomous variables, one for each symptom category. General analysis of symptoms, furthermore, was referred in a sex, age, and test outcome-disaggregated way.

3. Results

We showed the main results of the analyses performed. What is not included in the following figures and tables can be found in the Supplementary Paper Material. We did not report the *p*-values of statistical analyses performed because, with such high numbers, the difference between subpopulations considered was always statistically significant.

3.1. COVID-19 Testing

COVID-19 tests were made on patients coming from 18 different local health districts: 80% from the Turin city district and 20% attributable to patients coming from Turin province local health districts. Different types of tests were performed: for 18% of tests there was no information, 69.5% were nasal and nasopharyngeal tests, 9.3% were nasopharyngeal, 1.8% were nasal tests, and the residual percentage referred to a pharyngeal, bronchoaspirate, or bronchoalveolar washing or salivary tests. The date of test execution was analyzed in terms of frequency referred to the whole of 2020; the bar graph in Figure 2 shows the monthly time series of tests. In the same figure, the red line shows the monthly time series of tests with a negative outcome and the red line of tests with a positive outcome. There were two peaks in the incidence of infection coinciding, respectively, with the spring (April–May) and autumn months (October–November).

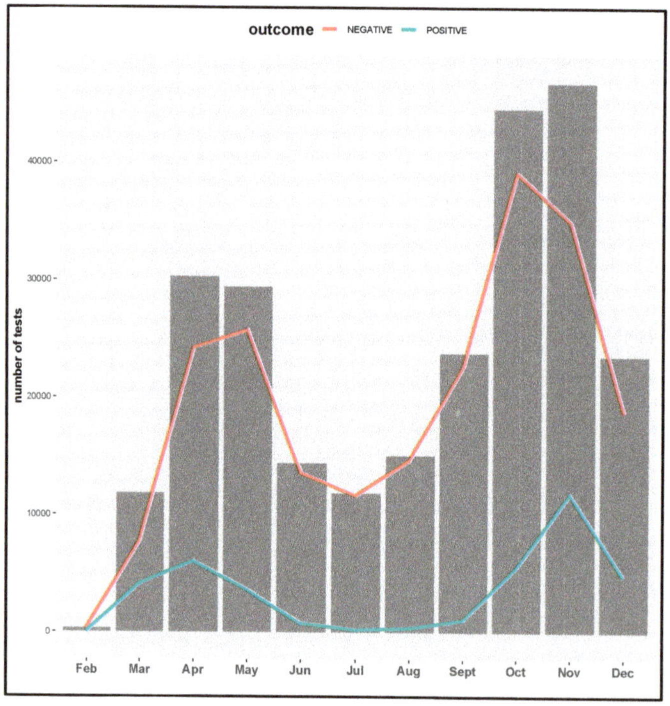

Figure 2. The monthly time series of COVID-19 tests.

With respect to the defined four periods of the year, we stated that in the first period, (I: February–May) 28.5% of tests were performed; in the second period (II: June–August), 16.4% of tests were performed; in the third period (III: September–October), 27.2% of tests were performed; and in the fourth period (IV: November–December), 27.9% of tests were performed. Positive and negative percentage rates related to different periods were analyzed: in the first period, 71,765 COVID-19 tests were done and 13,812 were positive (19.2%); in the second period, 41,222 COVID-19 tests were done and 1322 were positive (3.2%); in the third period, 68,355 COVID-19 tests were done and 6524 were positive (9.5%); in the fourth period, 70,315 COVID-19 tests were done and 16,582 were positive (23.6%).

3.2. Unique Subjects

A total of 123,542 unique subjects have been identified: 54.7% were female subjects and 45.3% were male. As distribution in the four periods, we observed that 31.9% of unique subjects were tested in the first period, 19.0% in the second, 26.6% in the third, and 22.4% in the fourth. A total of 80.6% of unique subjects (N = 99,595) were tested for COVID-19 detection in only one of the four periods, while 2% (N = 2501) were tested in each period, and 6% (N = 7413) were tested in three out of the four periods. Sex and age-disaggregated data analysis has been done. In Table 1, we provide the main age distribution parameters (minimum, maximum, mean, median, and quartiles) by sex and period. On average, a subject repeated the test during the year two times, furthermore, 90.5% of the subjects repeated the test no more than four times. However, 11.7% of unique female subjects were tested for COVID-19, more than 4 times that of men, 6.8%.

Table 1. Age distribution parameter by period and sex for unique subjects.

		Age						
	Period	**N°**	**Min.**	**1st Qu.**	**Median**	**Mean**	**3rd Qu.**	**Max.**
All Subjects (123,542)	I-FMAM	39,517	0	43	55	56.86	74	108
	II-JJA	23,490	0	30	49	49.53	69	106
	III-SO	32,840	0	22	40	41.07	58	104
	IV-ND	27,695	0	32	50	49.81	66	109
Male (55,974)	I-FMAM	15,510	0	42	56	56.25	73	108
	II-JJA	11,387	0	30	49	48.99	68	106
	III-SO	16,110	0	20	39	40.05	58	102
	IV-ND	12,967	0	31	49	48.74	65	101
Female (67,568)	I-FMAM	24,007	0	43	55	57.26	75	107
	II-JJA	12,103	0	31	48	50.03	71	103
	III-SO	16,730	0	24	41	42.06	58	104
	IV-ND	14,728	0	33	50	50.75	67	109

The number of women was higher than the number of men in each period-related percentage. The median age was, in general, a similar value among sexes (49 for males and 50 for females), but considering the age distribution in the four periods, it was shown that for both males and females, the average and median age was inversely correlated to the period, decreasing over time during 2020. Analysis of unique subjects for test results has also been done, selecting the first positive test obtained, if present. We observed 97,839 subjects who were never positive: 70,788 with a single test and 27,051 with multiple negative tests. We chose the first result listed in the dataset. Furthermore, we observed 25,703 subjects with at least one positive test: 9933 with only one positive test and 15,770 with multiple tests, positive and negative. For them, we chose their first positive test listed in the dataset. Then, we analyzed tests results in a sex and age-disaggregated way for each period. Regarding the whole year, 79.19% of the subjects had a negative result and 20.81% had a positive result. Looking at the distribution of positive tests in the periods, 34.9% were in period I, 1.5% in II, 19.6% in III, and 43.9% in IV. Considering the male percentage with positive tests, we observed growth from 33.6% (I) to 44.2% (IV), with an increase of 10.60%; for the positive female percentage, the increase was 7.7% (from 36.1% in period I to 43.8% in period IV). The average and median age for both positive and negative subjects, men and women, inversely correlated to period, decreasing over time during 2020. Going into more detail, considering the age distributions by period, sex, and test type outcome in Figure 3, we observed that in the I period, ages for positive tests (see red histograms) were higher for both females (left column) and males (right column), in the II, they are lower, starting from the age class of 18–25 years; in the III and IV periods, they are in the middle ages. Furthermore, focusing on the cumulative percentage data for positive subjects aged less than 25 years, we observed that only 4% of them were located in the I period, in the second they are 26.1%, in the third 21.1%, and in the fourth 11.1%.

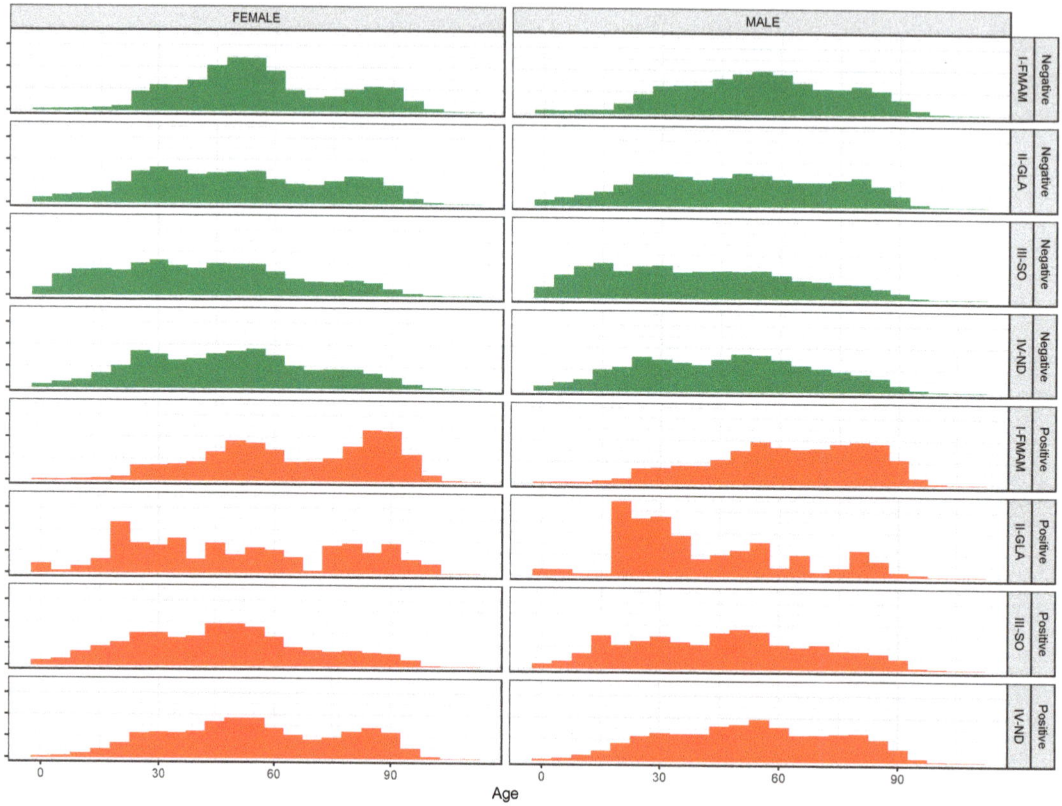

Figure 3. Age distributions by sex (left column = Female, right column = Male), outcome test type (green = Negative, red = Positive), and periods (two pairs of four rows).

3.3. Epidemiological Criterion

Using TM, we maximized the information obtainable from the epidemiological criterion field; we transformed texts into structured data through TM techniques. After dataset cleaning from empty spaces, the resulting dataset was made up of 196,970 records (21.7% of records were eliminated). After dataset standardization of texts in epidemiological criterion space and turning each criterion into a word vector (unigram), we created a BofW matrix composed of 196,970 rows and 23,022 columns. The 23,022 words identified have been corrected manually, indicating their possible correspondence to its correction. After being replaced with a string algorithm, we obtained 9683 unigrams. The first 100 most recurring in the criteria correspond to the words in the word cloud of Figure 4, where the font size representing the word depends on how often this word was used in the criteria, i.e., words written in large are present in many criteria.

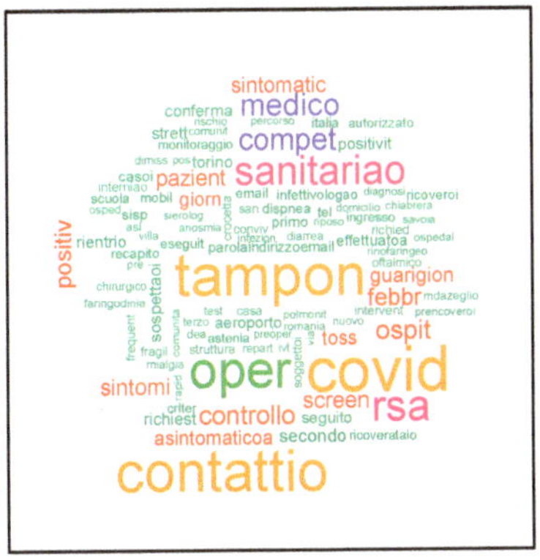

Figure 4. The most common words in the corpus of epidemiological criteria.

3.4. Analysis of Subjects' Categories

From the analysis of different subject categories in the 9683 unigrams dictionary identified from the TM analysis of epidemiological criterion, we obtained distinctive results. At first, a comparative study was carried out between different categories and test results obtained during 2020 and the four different year periods. For the three different subject categories identified (healthcare workers, assistance home guests, and contact with a positive subject) we obtained the following results.

Healthcare workers: 43,020 records throughout the period, 93.6% negative and 6.4% positive. The healthcare workers' proportion of positive tests in the different periods swung, high in the first (31.6%) and the fourth (23.5%) and lowest in the second period (8.4%). Comparing the healthcare workers with the other subject categories (see the top lines of Table 2), we observed that the percentage of positive tests for others was always higher (17.0% positive) than the percentage of positive healthcare workers (6.4%) and this happened for each period (I: 23.6% for others vs. 8.6% for healthcare workers; II: 3.6% for others vs. 1.2% for healthcare workers; III: 10.9% for others vs. 3.0% for healthcare workers; and IV: 32.2% for others vs. 13.3% for healthcare workers).

In order to study the category of the assistance home guests, we removed from the dataset all the records relating to healthcare workers. The following results refer, therefore, to 153,950 records related to subjects who are not healthcare workers. Assistance home guests had a total of 26,715 records throughout the period, 83.0% negative and 17.0% positive. The proportion of assistance home guests tested for COVID-19 in the different periods was decreasing, high in the first (23.3%), going down more and more until the last period (11.9%). Looking at the positive percentage compared with others (see central lines of Table 2), we observed that in the first and second periods, the proportion of positives among the assistance home guests was higher, while in the last two periods, the relationship was reversed (I: 22.7% for others vs. 26.7% for assistance home guests; II: 3.3% for others vs. 5.1% for assistance home guests; III: 11.9% for others vs. 5.1% for assistance home guests; and IV: 33.1% for others vs. 25.6% for assistance home guests).

Contact with a positive subject had 33,603 records throughout 2020. These subjects were tested for COVID-19 because they had contact with a positive subject. Positive tests resulting from this category were 23.7%, and 76.3% obtained negative results. The contact with a positive subject proportion tested for COVID-19 in the four different periods was very high in the last period (41.9%), very low in the second period (7.4%), and started to rise in the third (16.9%). Looking at the positive percentage compared with others who are not healthcare workers (see the bottom lines of Table 2), we observed that in the first, second, and third periods, the proportion of positives among contact with a positive subject category was higher, while in the last period the relationship was reversed (I: 22.6% for others vs. 26.6% for contacts with a positive subject; II: 3.5% for others vs. 5.2% for contacts with a positive subject; III: 10.2% for others vs. 14.7% for contacts with a positive subject; and IV: 33.2% for others vs. 30.8% for contacts with a positive subject).

Table 2. Test outcome by subjects' category in each year.

Category	Period	Category Belonging	% Negative	% Positive	Total Records
Healthcare Workers	I-FMAM	Yes	91.3	8.73	22,216
		No	76.4	23.64	48,170
	II-JJA	Yes	98.8	1.20	4568
		No	96.4	3.63	34,019
	III-SO	Yes	96.9	3.04	13,744
		No	89.0	10.96	44,735
	IV-ND	Yes	86.7	13.32	2492
		No	67.8	32.19	27,026
Assistance Home Guests	I-FMAM	Yes	73.3	26.70	11,239
		No	77.3	22.71	36,931
	II-JJA	Yes	94.9	5.09	6188
		No	96.7	3.31	27,831
	III-SO	Yes	94.9	5.11	6067
		No	88.1	11.88	38,668
	IV-ND	Yes	74.4	25.64	3221
		No	66.9	33.08	23,805
Contact with a Positive Subject	I-FMAM	Yes	73.4	26.60	12,207
		No	77.4	22.64	35,963
	II-JJA	Yes	94.8	5.19	2525
		No	96.5	3.51	31,494
	III-SO	Yes	85.3	14.74	7549
		No	89.8	10.19	37,186
	IV-ND	Yes	69.2	30.83	11,322
		No	66.8	33.18	15,704

3.5. *Analysis of Symptoms*

Considering the analysis of the symptoms in general, we observed that for the records having one of the words classified as symptoms S in the criterion were 14.6% of the total, while the P category percentage was lower (6.2%). The percentage of positives for those who had symptoms S was greater than that of positives who did not have symptoms (23.0% vs. 13.3%), while for the type P symptomatology, related to other pathologies, the situation was reversed. Among those who did not have symptoms, the positives were 15.1% while among those who had symptoms the positives were 9.3%.

We decided to analyze in detail the category of S symptoms. Looking at the data in different 2020 periods, we noticed that in the fourth period, the percentage of positives was greater among those who did not have symptoms (31.3% vs. 25.4%), which may be due to missing filled in data in the epidemiological criterion field, due to the huge stress at the end of 2020.

This trend was also observed in the second period (3.6% vs. 1.1%) but not in the first and in the third periods where the percentage of positives was greater among those who had symptoms (I: 30.3% vs. 15.9%; III: 20.9% vs. 7.8%). In the fourth period, many tests were for positive contact so there were many positive subjects who did not yet develop COVID-19 symptoms. For these reasons, we decided to focus our analysis on symptoms not in relation to different 2020 periods. General analysis of COVID-19 related symptoms was referred to in a sex, age, and outcome test-disaggregated way. For sex, we observed that of 196,970 tests, 41.1% were done on males and 58.9% on females. The positives among males were 16.2% and 13.6% among females. We noted that among males, the proportion of those who had symptoms was 17.3%, while among females it was 12.8%. For both sexes, we observed that the proportion of positives among those who had symptoms was higher than that of the positives who had no symptoms. The difference for males was 10.9% (25.2–14.4%) while for females it was 8.3% (20.9–12.6%). Comparing males and females with symptoms, 25.2% of males were positive while 20.9% of females were positive (see Figure 5).

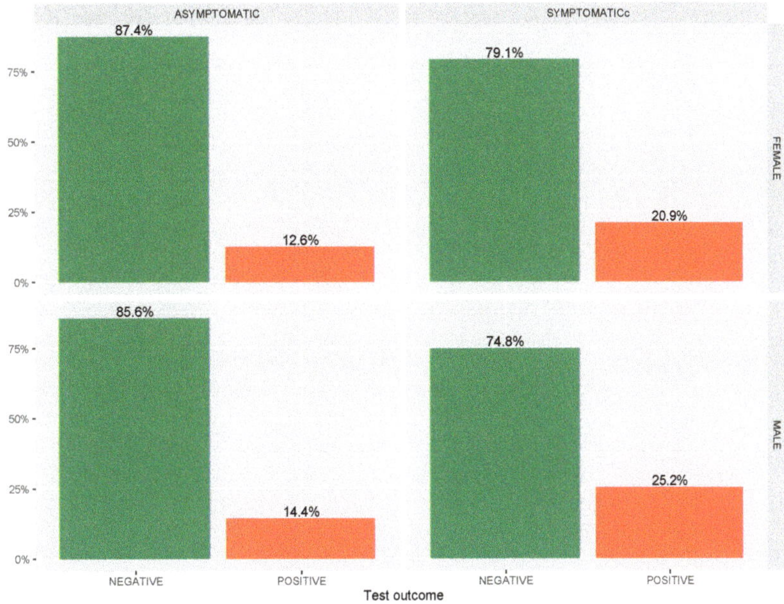

Figure 5. Test outcomes for COVID-19 related symptoms in a sex-disaggregated way.

Regarding age, we observed that it was a risk factor for more severe disease; positive asymptomatic were younger than those positive with COVID-19 related symptoms (42.5% of symptomatic positives were aged less than or equal to 55 years while 53.1% of asymptomatic positives were aged less than or equal to 55 years). Looking at the data disaggregated by sex in Table 3, we observed that for males, 40.9% of symptomatic positives were aged less than or equal to 55 years while 56.2% of asymptomatic positives were aged less than or equal to 55 years, and for females, 44.3% of symptomatic positives were aged less than or equal to 55 years while 50.7% of asymptomatic positives were aged less than or equal to 55 years. Additionally, the male sex was a risk factor for more severe disease in terms of symptoms.

Table 3. Test outcome by sex and age for asymptomatic and symptomatic COVID-19 related subjects.

	Sex	Age	Negative	Positive
Symptomatic	Male	From 0 to 55	5209 (49.8%)	1442 (40.9%)
		Greater than 55	5242 (50.2%)	2085 (59.1%)
	Female	From 0 to 55	6068 (51.8%)	1371 (55.7%)
		Greater than 55	5642 (48.2%)	1724 (55.7%)
Asymptomatic	Male	From 0 to 55	34,791 (60.6%)	5406 (56.2%)
		Greater than 55	22,609 (39.4%)	4214 (43.8%)
	Female	From 0 to 55	53,249 (60.2%)	6453 (70.7%)
		Greater than 55	35,189 (39.8%)	6276 (49.3%)

Regarding the presence of comorbidities, 5700 records were found to be related to this factor among subjects affected by COVID-19-related symptoms. Analyzing data in a sex-disaggregated way, we observed no differences; fewer positive men and women were affected by concomitant pathologies compared to positive subjects not affected by other pathologies.

A comparative study was carried out between categories and the presence of symptoms. For symptoms, we identified 1370 words divided into two macro-categories (S: COVID-19 related; P: other different pathologies).

A total of 3.0% of healthcare workers had COVID-19 related symptoms considering each period in all of 2020 (vs. 17.8% for others). The percentage of others who had symptoms was always higher than the percentage of healthcare workers with symptoms (I: 28.7% for others vs. 4.1% for healthcare workers; II: 13.7% for others vs. 2.7% for healthcare workers; III: 12.3% for others vs. 1.9% for healthcare workers; and IV: 12.8% for others vs. 0.9% for healthcare workers). This data may be explained by the need for healthcare workers to be tested for COVID-19 for monitoring purposes.

The 3.7% of assistance home guests who had COVID-19 related symptoms, considering all the 2020 and looking only at those we observed, was higher compared to others' percentage (40% vs. 22.4%).

The 18.2% of subjects who had contact with a positive subject had COVID-19 related symptoms, considering all of 2020 and looking only at those we observed, was higher compared to others' percentage (29.3% vs. 21.3%). Finally, the percentage of others who had symptoms was always higher than the percentage of those who had contact with a positive subject with symptoms, except in the last period (I: 29.8% for others vs. 26.6% for contact with a positive subject; II: 14.5% for others vs. 3.3% for contact with a positive subject; III: 13.6% for others vs. 5.0% for contact with a positive subject; and IV: 6.9% for others vs. 21.0% for contact with a positive subject).

4. Discussion

From the first 2019 reports from China, a sex imbalance with regard to detected cases and case fatality rate of COVID-19 was observed. As the disease spread across multiple continents, the *Global Health 50/50 research initiative* presented an impressive overview of sex-disaggregated data from countries worldwide, clearly demonstrating similar numbers of cases in women and men, but an increased case-fatality in men [34]. The sex disparity of COVID-19–related morbidity and mortality is likely explained by a combination of biological sex differences, such as hormonal and genetic (immune response, RAAS system, and the ACE2 and TMPRSS2 role), and gender-specific factors, such as differential behaviors and activities by social and cultural or traditional roles. Men are more likely to engage in poor health behaviors (smoking and alcohol consumption) and have higher age-adjusted rates of pre-existing co-morbidities associated with poor COVID-19 prognosis, including hypertension, cardiovascular disease, and chronic obstructive pulmonary disease [39].

Nevertheless, sex-disaggregated data are still not provided by all countries, and neither the interaction of sex and age is usually visible in the public databases.

We analyzed data on COVID-19 testing in the Piedmont region, northwest Italy, for people admitted to the Amedeo di Savoia Hospital, a regional referral center for infectious diseases, during 2020. Our undertaking was conducted to better understand the characteristics of the subjects who have undergone at least one COVID-19 test during 2020. During the time of tailored medicine, our intent was to study the variables of the whole population registered in the database used by the Microbiology and Virology Laboratory of the Amedeo di Savoia Hospital for COVID-19 testing.

From the analysis of the execution date of the test in four different year periods (I: February–May, II: June–August, III: September–October, and IV: November–December) decided retrospectively to evaluate the pandemic trend, we observed that in periods I, III, and IV the number of tests performed was almost the same, while in II (summer) it was lower. For the whole of 2020, we observed that the percentage of positive COVID-19 tests was 15.2%. In the IV period, we observed a higher positivity rate (23.6%) than the rate observed in the I period (19.2%). Questions still needed to be addressed are: who was tested in different periods? Certainly, in the first period, the number of COVID-19 tests made was very poor. Analysis of unique subjects showed the presence of a higher female percentage. This greater value may be due to the fact that throughout 2020, tests were made almost exclusively for health workers, mostly composed of female subjects, and in nursing homes, mainly inhabited by women, due to the greater presence of women in the older population segment. Analysis of test distribution for unique subjects during 2020 also showed that 8% of them had been monitored for COVID-19 detection in three or four different periods (N = 9914); this could be related to the need to control infection in affected patients or to the work-related monitoring for health personnel. Analyzing data in a sex and age-disaggregated way, we observed major tests done on the female unique subject. The number of women was higher than the number of men in each period-related percentage which is realistic if we think about women's major involvement in the healthcare workload. Looking at the age distribution in the four periods, we observed a decrease over time during 2020 for both sexes, verifying a potential and progressive expansion of the tests to younger subjects. This may be due to the involvement of different categories of people not specifically requested to be tested for COVID-19 for professional reasons. The major female connection in COVID-19 tests, due to their major involvement in the healthcare workload, is also confirmed by the analysis of test repetition: in all periods, 11.7% of unique female subjects were tested for COVID-19 detection more than four times, vs. 6.8% of men. Analysis of test results further described the population through the positive and negative rates of infection of the population involved. In spite of the greater number of tests conducted on women, the percentage of positives among sexes was similar and we observed growth over time during 2020 from 33.6% (I) to 44.2% (IV) for men, with an increase of 10.6%, while an increase of 7.7% (from 36.1% in period I to 43.8% in period IV) was observed for women in terms of positivity rate. The observation about age distribution was confirmed also considering positive and negative subjects: the age distribution decreases, for both men and women, over time during 2020, possibly due to the admission to tests, as already mentioned, for younger subjects, i.e., students. Going into more detail, focusing on cumulative percentage data for positive subjects aged less than 25 years, we observed that fewer percentages were located in the I and IV periods, corresponding to the distance learning periods for students during the year. Analysis of epidemiological criteria was difficult because it was characterized by little standardized, very variable, and error-rich data, due to the open and non-mandatory nature of this field. The operator could fill it out by writing text with the contents he/she considered most significant, but this variable was potentially full of information that is not directly analyzable with common statistical tools. Through TM techniques, it was possible to transform texts into structured data. This underlined the importance of specific training for operators in entering additional information on data collection platforms. A uniform and codified system would have allowed us to perform an easier analysis, as well as allow more results to be obtained. However, TM techniques are not within everyone's reach. The analysis of the epidemiological criterion through TM was conducted with different purposes. In the reduced 9683 unigrams dictionary, analysis of different subject categories identified and matched with test results and the presence of symptoms, gave us some information about the trend of the vaccination campaign, which started at the end of 2020, indicating that the need for testing decreased. Comparing, furthermore, the percentage of symptoms present, which were always higher for others than for healthcare workers, gave us the information that healthcare workers had done a large amount of testing for monitoring. The data referring to assistance home guests brought us back to the terrible period we all lived in the management of the COVID-19 pandemic. Isolation of older patients and incorrect choices led to high positivity numbers. Assistance homes were, for improvised management, especially at the beginning of pandemic, outbreaks; if an old man had a fever, he had twice the probability of being positive as one who did not live there. Separately studying the category of those who had contact with a positive subject allowed us to trace a trend for the whole of the 2020 track, confirming the usefulness of lockdowns. Additionally, the relation of testing to symptoms could furthermore allow us to better underline the potential screening role of COVID-19 tests. As the first limit of this study, we should admit the nature of the field "epidemiological criterion", which we imagine is little or badly filled out by operators in times of great stress. In the end, and as a central point of the huge work conducted, interesting information on the disease outcome that we obtained from a general analysis of COVID-19 symptoms referred in a sex, age, and outcome test-disaggregated way, showed that male sex and older age were risk factors for more severe disease.

A further limitation of this study is the absence of correlation with mortality data, but this variable was not present in the original database since the census of the deceased was not among the objectives of the regional referral center for infectious diseases through the platform. Looking at the data collected for each Italian region by the National Istituto Superiore di Sanità referring to 2020, we observed a greater lethality of the male sex which is consistent with our observations [40].

5. Conclusions

Gender medicine does not exist [41]. What should definitively exist is a medical approach tailored to the variables and characteristics of each subject needing clinical assistance. We believe that observations obtainable from studies like ours could have a central role in a good preventive health policy. Statistical models like ours could be applied in general for human diseases, giving the opportunity to better understand the mechanisms underlying pathologies in the interest of the whole community. Regarding COVID-19, our study shows important implications for public health policy, illustrating the right direction for government policies for future pandemics and emphasizing intervention for those who need it most.

Supplementary Materials: The following supporting information can be downloaded at: https://www.mdpi.com/article/10.3390/life12050643/s1.

Author Contributions: Conceptualization, S.D.F. and A.D.; methodology, S.D.F., A.D., A.F. and D.M.; software, A.D.; formal analysis, A.D.; investigation, S.D.F., A.D., A.F. and D.M.; resources, F.C., S.A., T.G.A., M.G.M., G.G., V.G., C.A. and E.B.; data curation, S.D.F., A.D., A.F. and D.M.; writing—original draft preparation, S.D.F. and A.D.; writing—review and editing, S.D.F., A.D., D.M., A.F., F.C., S.A., T.G.A., M.G.M., G.G., V.G., C.A. and E.B.; supervision, S.D.F. and A.D. All authors have read and agreed to the published version of the manuscript.

Funding: This research received no external funding.

Institutional Review Board Statement: Not applicable.

Informed Consent Statement: Not applicable.

Data Availability Statement: All codes and routines used for data analysis are available to readers upon request to alessandra.durio@unito.it. The original database is not public and is property of the Piedmont Region.

Conflicts of Interest: The authors declare no conflict of interest.

References

1. Zhu, N.; Zhang, D.; Wang, W.; Li, X.; Yang, B.; Song, J.; Zhao, X.; Huang, B.; Shi, W.; Lu, R.; et al. A Novel Coronavirus from Patients with Pneumonia in China, 2019. *N. Engl. J. Med.* **2020**, *382*, 727–733. [CrossRef] [PubMed]
2. Perlman, S.; Netland, J. Coronaviruses post-SARS: Update on replication and pathogenesis. *Nat. Rev. Microbiol.* **2009**, *7*, 439–450. [CrossRef] [PubMed]
3. Channappanavar, R.; Perlman, S. Pathogenic human coronavirus infections: Causes and consequences of cytokine storm and immunopathology. *Semin. Immunopathol.* **2017**, *39*, 529–539. [CrossRef]
4. Guo, Y.-R.; Cao, Q.-D.; Hong, Z.-S.; Tan, Y.-Y.; Chen, S.-D.; Jin, H.-J.; Tan, K.-S.; Wang, D.-Y.; Yan, Y. The origin, transmission and clinical therapies on coronavirus disease 2019 (COVID-19) outbreak—An update on the status. *Mil. Med. Res.* **2020**, *7*, 11. [CrossRef]
5. Guan, W.J.; Ni, Z.Y.; Hu, Y.; Liang, W.H.; Qu, C.Q.; He, J.X.; Liu, L.; Shan, H.; Lei, C.L.; Hui, D.S.C.; et al. Clinical Characteristics of coronavirus disease in China. *N. Engl. J. Med.* **2020**, *382*, 1708–1720. [CrossRef]
6. Roberts, C.M.; Levi, M.; McKee, M.; Schilling, R.; Lim, W.S.; Grocott, M.P. COVID-19: A complex multisystem disorder. *Br. J. Anaesth.* **2020**, *125*, 238–242. [CrossRef]

7. Kong, W.; Wang, Y.; Hu, J.; Chughtai, A.; Pu, H. Comparison of clinical and epidemiological characteristics of asymptomatic and symptomatic SARS-CoV-2 infection: A multi-center study in Sichuan Province, China. *Travel Med. Infect. Dis.* **2020**, *37*, 101754. [CrossRef]
8. Gebhard, C.; Regitz-Zagrosek, V.; Neuhauser, H.K.; Morgan, R.; Klein, S.L. Impact of sex and gender on COVID-19 outcomes in Europe. *Biol. Sex Differ.* **2020**, *11*, 29. [CrossRef]
9. Klein, S.L.; Flanagan, K.L. Sex differences in immune responses. *Nat. Rev. Immunol.* **2016**, *16*, 626–638. [CrossRef] [PubMed]
10. Fischer, J.; Jung, N.; Robinson, N.; Lehmann, C. Sex differences in immune responses to infectious diseases. *Infection* **2015**, *43*, 399–403. [CrossRef]
11. Gadi, N.; Wu, S.C.; Spihlman, A.P.; Moulton, V.R. What's Sex Got to Do With COVID-19? Gender-Based Differences in the Host Immune Response to Coronaviruses. *Front. Immunol.* **2020**, *11*, 2147. [CrossRef] [PubMed]
12. Guerra-Silveira, F.; Abad-Franch, F. Sex Bias in Infectious Disease Epidemiology: Patterns and Processes. *PLoS ONE* **2013**, *8*, e62390. [CrossRef] [PubMed]
13. Moore, A.L.; Kirk, O.; Johnson, A.M.; Katlama, C.; Blaxhult, A.; Dietrich, M.; Colebunders, R.; Chiesi, A.; Lungren, J.D.; Phillips, A.N. Virologic, Immunologic, and Clinical Response to Highly Active Antiretroviral Therapy: The Gender Issue Revisited. *JAIDS J. Acquir. Immune Defic. Syndr.* **2003**, *32*, 452–461. [CrossRef] [PubMed]
14. Collazos, J.; Asensi, V.; Cartón, J.A. Sex differences in the clinical, immunological and virological parameters of HIV-infected patients treated with HAART. *AIDS* **2007**, *21*, 835–843. [CrossRef]
15. Fink, A.L.; Engle, K.; Ursin, R.L.; Tang, W.-Y.; Klein, S.L. Biological sex affects vaccine efficacy and protection against influenza in mice. *Proc. Natl. Acad. Sci. USA* **2018**, *115*, 12477–12482. [CrossRef]
16. David, R. Are men and women different? *Nat. Rev. Immunol.* **2009**, *9*, 605. [CrossRef]
17. Stawicki, S.P.; Jeanmonod, R.; Miller, A.C.; Paladino, L.; Gaieski, D.F.; Yaffee, A.Q.; De Wulf, A.; Grover, J.; Papadimos, T.J.; Bloem, C.; et al. The 2019–2020 Novel Coro-navirus (Severe Acute Respiratory Syndrome Coronavirus 2) Pandemic: A Joint American College of Academic International Medicine-World Academic Council of Emergency Medicine Multidisciplinary COVID-19 Working Group Consensus Paper. *J. Glob. Infect. Dis.* **2020**, *12*, 47–93. [CrossRef]
18. Oz, M.; Lorke, D.E.; Kabbani, N. A comprehensive guide to the pharmacologic regulation of angiotensin converting enzyme 2 (ACE2), the SARS-CoV-2 entry receptor. *Pharmacol. Ther.* **2021**, *221*, 107750. [CrossRef]
19. Zhang, H.; Penninger, J.M.; Li, Y.; Zhong, N.; Slutsky, A.S. Angiotensin-converting enzyme 2 (ACE2) as a SARS-CoV-2 receptor: Molecular mechanisms and potential therapeutic target. *Intensive Care Med.* **2020**, *46*, 586–590. [CrossRef]
20. Verdecchia, P.; Cavallini, C.; Spanevello, A.; Angeli, F. The pivotal link between ACE2 deficiency and SARS-CoV-2 infection. *Eur. J. Intern. Med.* **2020**, *76*, 14–20. [CrossRef]
21. Hanff, T.C.; Harhay, M.O.; Brown, T.S.; Cohen, J.B.; Mohareb, A.M. Is There an Association Between COVID-19 Mortality and the Renin-Angiotensin System? A Call for Epidemiologic Investigations. *Clin. Infect. Dis.* **2020**, *71*, 870–874. [CrossRef] [PubMed]
22. Gurwitz, D. Angiotensin receptor blockers as tentative SARS-CoV-2 therapeutics. *Drug Dev. Res.* **2020**, *81*, 537–540. [CrossRef] [PubMed]
23. Bukowska, A.; Spiller, L.; Wolke, C.; Lendeckel, U.; Weinert, S.; Hoffman, J.; Bornfleth, P.; Kutschka, I.; Gardemann, A.; Isermann, B.; et al. Protective regulation of theACE2/ACE gene expression by estrogen in human atrial tissue from elderly men. *Exp. Biol. Med.* **2017**, *242*, 1412–1423. [CrossRef] [PubMed]
24. Gagliardi, M.C.; Tieri, P.; Ortona, E.; Ruggieri, A. ACE2 expression and sex disparity in COVID-19. *Cell Death Discov.* **2020**, *6*, 37. [CrossRef]
25. Beyerstedt, S.; Casaro, E.B.; Rangel, É.B. COVID-19: Angiotensin-converting enzyme 2 (ACE2) expression and tissue susceptibility to SARS-CoV-2 infection. *Eur. J. Clin. Microbiol. Infect. Dis.* **2021**, *40*, 905–919. [CrossRef]
26. Jackson, C.B.; Farzan, M.; Chen, B.; Choe, H. Mechanisms of SARS-CoV-2 entry into cells. *Nat. Rev. Mol. Cell Biol.* **2022**, *23*, 3–20. [CrossRef]
27. Dong, M.; Zhang, J.; Ma, X.; Tan, J.; Chen, L.; Liu, S.; Xin, Y.; Zhuang, L. ACE2, TMPRSS2 distribution and extrapulmonary organ injury in patients with COVID-19. *Biomed. Pharmacother.* **2020**, *131*, 110678. [CrossRef]
28. Brodin, P. Immune determinants of COVID-19 disease presentation and severity. *Nat. Med.* **2021**, *27*, 28–33. [CrossRef]
29. Costagliola, G.; Spada, E.; Consolini, R. Age-related differences in the immune response could contribute to determine the spectrum of severity of COVID-19. *Immun. Inflamm. Dis.* **2021**, *9*, 331–339. [CrossRef]
30. Sette, A.; Crotty, S. Adaptive immunity to SARS-CoV-2 and COVID-19. *Cell* **2021**, *184*, 861–880. [CrossRef]
31. Mangge, H.; Kneihsl, M.; Schnedl, W.; Sendlhofer, G.; Curcio, F.; Domenis, R. Immune Responses against SARS-CoV-2—Questions and Experiences. *Biomedicines* **2021**, *9*, 1342. [CrossRef] [PubMed]
32. Lotter, H.; Altfeld, M. Sex differences in immunity. *Semin. Immunopathol.* **2019**, *41*, 133–135. [CrossRef] [PubMed]
33. Flak, M.B.; Neves, J.F.; Blumberg, R.S. Welcome to the Microgenderome. *Science* **2013**, *339*, 1044–1045. [CrossRef] [PubMed]
34. Global Health 50/50 (Globalhealth5050.org). Available online: http://globalhealth5050.org (accessed on 1 March 2021).
35. Feldman, R.; Sanger, J. *The Text Mining Handbook: Advanced Approaches in Analyzing Unstructured Data*, 1st ed.; Cambridge University Press: New York, NY, USA, 2007; pp. 1–410.
36. Feinerer, I.; Hornik, K.; Meyer, D. Text Mining Infrastructure in R. *J. Stat. Softw.* **2008**, *25*, 1–54. [CrossRef]
37. Luque, C.; Luna, J.M.; Luque, M.; Ventura, S. An advanced review on text mining in medicine. *Wiley Interdiscip. Rev. Data Min. Knowl. Discov.* **2019**, *9*, e1302. [CrossRef]

38. Silge, J.; Robinson, D. *Text Mining with R: A Tidy Approach*, 1st ed.; O'Reilly Media: Sebastopol, CA, USA, 2017; pp. 1–174.
39. Yang, J.; Zheng, Y.; Gou, X.; Pu, K.; Chen, Z.; Guo, Q.; Ji, R.; Wang, H.; Wang, Y.; Zhou, Y. Prevalence of comorbidities and its effects in patients infected with SARS-CoV-2: A systematic review and meta-analysis. *Int. J. Infect. Dis.* **2020**, *94*, 91–95. [CrossRef]
40. Sorveglianza Integrata COVID-19: I Principali Dati Nazionali (iss.it). Available online: https://epicentro.iss.it/coronavirus/sars-cov-2-sorveglianza-dati (accessed on 9 September 2021).
41. Baggio, G. Gender medicine: A interdisciplinary approach to medicine. *G. Ital. Di Med. Del Lav. Ed Ergon.* **2017**, *39*, 196–198.

life

MDPI

Article

Cellular Immune Response in Patients Immunized with Three Vaccine Doses of Different Vaccination Schemes Authorized by the Chilean Ministry of Health in January 2022

Paz Beatriz Tabilo Valenzuela [1,*], Gabriela Flores Balter [1], Gustavo Saint-Pierre Contreras [1], Daniel Conei Valencia [2,3], Catalina Moreno Calderón [1], Constanza Bohle Venegas [1], Marcia Guajardo Rivera [1], Francisco Silva Ojeda [1] and Maria Jesus Vial Covarrubias [1,4]

1 Servicio de Laboratorio Clínico, Hospital Clínico Universidad de Chile, Santiago 8380000, Chile; gabriela.flores@ug.uchile.cl (G.F.B.); gsaintpierre@hcuch.cl (G.S.-P.C.); cmorenoc@hcuch.cl (C.M.C.); cbohle@hcuch.cl (C.B.V.); mguajardo@hcuch.cl (M.G.R.); fsilva@hcuch.cl (F.S.O.); mjesus@hcuch.cl (M.J.V.C.)
2 Programa de Doctorado en Ciencias Morfológicas, Universidad de La Frontera, Temuco 4811230, Chile; danielconei@ug.uchile.cl
3 Departamento de Ciencias de la Salud, Universidad de Aysén, Coyhaique 5951537, Chile
4 Jefe Departamento Laboratorio Clínico, Hospital Clínico Universidad de Chile, Santiago 8380000, Chile
* Correspondence: paztabilo@gmail.com or paztabilo@hcuch.cl; Tel.: +56-22-2978-8070

Citation: Tabilo Valenzuela, P.B.; Flores Balter, G.; Saint-Pierre Contreras, G.; Conei Valencia, D.; Moreno Calderón, C.; Bohle Venegas, C.; Guajardo Rivera, M.; Silva Ojeda, F.; Vial Covarrubias, M.J. Cellular Immune Response in Patients Immunized with Three Vaccine Doses of Different Vaccination Schemes Authorized by the Chilean Ministry of Health in January 2022. *Life* **2022**, *12*, 534. https://doi.org/10.3390/life12040534

Academic Editors: Silvia De Francia, Sarah Allegra and Daniele Focosi

Received: 22 February 2022
Accepted: 29 March 2022
Published: 5 April 2022

Abstract: In December 2019, a case of atypical pneumonia was reported in Wuhan, China. It was named COVID-19 and caused by SARS-CoV-2. In a few months, scientific groups around the world developed vaccines to reduce the disease's severity. The objective was to evaluate the humoral and cellular immune response post immunization with three different vaccination schedules administered in Chile until January 2022. Sixty volunteers were recruited with a three-dose schedule, who had no history of infection nor close contact with a positive patient. IgG against the spike antigenic domain was detected, and the neutralization capacity against two groups of variants, Original/Alpha and Beta/Gamma, was also measured. Finally, the cellular response with interferon release was measured through IGRA. Results showed that there were significant differences in the neutralizing antibodies for the original and alpha variant when comparing three Comirnaty doses with Coronavac and Vaxzevria. A high number of reactive subjects against the different SARS-CoV-2 variants, alpha, gamma, and delta, were observed, with no significant differences between any of the three schemes, confirming the existence of a cellular immune response against SARS-CoV-2. In conclusion, the three vaccine schemes generated a cellular immune response in these volunteers.

Keywords: SARS-CoV-2; COVID-19 vaccines; cell-mediated immunity; interferon-gamma release assays

1. Introduction

In December 2019, the scientific community was informed of an unidentified case of pneumonia in Wuhan, People's Republic of China. This disease had similar characteristics to those that occurred in Guangdong Province, China in 2003 [1]. It is called coronavirus disease (COVID-19), which is caused by SARS-CoV-2, and during these two years of pandemic, it has killed more than five million people in the world. To prevent the transmission of SARS-CoV-2, restrictions on mobility and social interactions have been imposed on populations and during 2021, emphasis has been placed on mass vaccination of the world's population [2,3].

SARS-CoV-2 is a positive single-stranded RNA virus with mantle and is covered by proteins with antigenic capacity such as spike (or protein S), which interact with its RBD domain with the cellular receptor ACE-2 to allow adsorption and subsequent viral replication in human cells [4–6]. Spike, in turn, is the molecule recognized by the immune system to prevent adsorption, that is, non-binding with the ACE-2 cell receptor, and therefore, it is a target for the design of COVID-19 vaccines, but also to adapt current platforms to mutations in this protein, particularly in the RBD domain [6,7]. T cells may provide protective immunity and limit serious disease in settings where antibody responses may be diminished [8]. In addition, T cells are capable of recognizing mutant variants of SARS-CoV-2 [9,10].

Two years after the beginning of the pandemic, scientists have learned how to recognize the virus's morphology and pathophysiology. It was initially supposed to be a severe respiratory disease, like SARS-CoV-1. The initial respiratory manifestations, including severe pneumonia, were predominant, but extrapulmonary signs and symptoms were described during the next months like anosmia, ageusia, cough, and odynophagia. Clinical experience demonstrated that there are acute and subacute symptoms that may even prolong in time, many of them associated with the inflammatory response driven by cytokines, including cytokine storm [11].

In the COVID-19 acute phase, pain can be presented in diverse forms. One of the first reports showed that muscular pain affected up to 44% of the patients. There have been different types of pain described: myalgias/arthralgias, odynophagia, abdominal and thoracic pain, and headache [12].

Some authors have described a relation between IL-10 systemic levels and COVID-19, particularly pain related, showing a strong negative correlation between IL-10 systemic levels and pain intensity. IL-10 has been described through the years as an anti-inflammatory and analgesic cytokine.

Not only has pain been related to COVID-19. A study conducted by the "Sex@COVID online survey" described the endothelial damage that was later associated with acute and long-term erectile dysfunction [13]. The authors suggest that this is a potential reason for mask use and vaccination [14].

Vaccination with a complete vaccination schedule is the most effective measure to prevent SARS-CoV-2 infection and to diminish the population's viral load. As of 31 December 2021, the WHO has approved 10 vaccines for clinical use [15]. Vaccine use, and especially the complete scheme (two doses), has been associated with a decrease in COVID-19 severe cases, including hospitalization and deaths [16–18]. Vaccines not only prevent mortality but also decrease symptoms that may mean a repercussion in daily life worldwide. It has been demonstrated that they also help in reducing long-term COVID-19 effects [19], inferior respiratory infections, and erectile dysfunction associated with COVID-19 [13].

Chile has been one of the countries with the highest number of vaccinated per capita. Vaccination of the fourth dose has now begun and up to 24 January 2022, 16.9 million people have completed a vaccination schedule (two doses) [20], that is, 88.2% of the target population, (considered between 3 years of age and older) [21]. This country has had a new impetus in its national vaccination program in the wake of the pandemic, adjusting public policies to initiate Phase 2 and 3 studies for different vaccine models circulating in the world, including ChAdOx1 nCoV-19 now known as Vaxzevria [22,23], Coronavac [24], and Cansino [25]. This allowed the country to position itself as a safe and reliable place to carry out scientific studies on SARS-CoV-2, which included studies associated with the success of the third dose in the Chilean population [26], particularly in the increase of the humoral and cellular immune system response when using an inactivated vaccine as a triple dose compared to two doses of the same model [26–28].

Currently, scientists have planned new vaccine models in order to find vaccines that generate a greater cellular immune response and a longer duration of the immune response against SARS-CoV-2. In other words, in the new models, it has been proposed that the measurement of not only the humoral immune response, but also the cellular immune response, would allow a better assessment of a possible prolonged immunological memory [28].

Due to this background and in order to complement the information obtained in the literature, we have proposed the evaluation of the humoral and cellular immune response of T lymphocytes post-immunization with SARS-CoV-2 vaccines, according to the three vaccination schemes administered in Chile. This measurement was carried out after the inoculation of the third dose, according to current ministerial regulations, in a voluntary population assigned to the Hospital Clínico Universidad de Chile during January 2022.

2. Materials and Methods

Prospective observational study carried out at the Hospital Clínico Universidad de Chile in Santiago de Chile during January 2022.

We recruited 60 volunteer participants with complete primary vaccination schemes and boosters (three vaccine doses). They had not been infected with SARS-CoV-2 nor had close contact with a confirmed case. This was certified by the Chilean ministry database called EPIVIGILA which has the record of every test performed in the country in public and private health centers as well as all the demographic and clinical information of each person who suffered COVID-19 or was in close contact with a confirmed person. Additionally, they should have not consumed immunosuppressive or immunomodulatory medication. These patients were divided by the three different vaccination schemes authorized in the country. These were Sinovac pharmaceuticals known as Coronavac, Pfizer with the BNT162b2 vaccine, currently Comirnaty, and finally Oxford-Astrazeneca with Vaxzevria.

From the initial sample, 4 subjects inoculated with the Comirnaty-Comirnaty-Comirnaty scheme had cellular immune response triggered by nucleocapsid protein, protein which is not present in this type of vaccine, so they probably had controlled exposure to the virus due to the hospital service where they worked (Intensive Care Unit). Therefore, in order to ensure comparison between the groups, 4 subjects were excluded from the other vaccination schemes, which were selected according to the data's dispersion based on confounding variables analysis.

These participants, after signing an informed consent, were studied for both humoral and cellular immunity against the SARS-CoV-2 virus.

The participants were analyzed in 3 groups, those with primary and booster vaccination schedules.

- Coronavac-Coronavac-Vaxzevria (n = 16)
- Comirnaty-Comirnaty-Comirnaty (n = 16)
- Coronavac-Coronavac-Comirnaty (n = 16)

2.1. Evaluation of Humoral Response

Humoral response was measured in all patients using two techniques: (1) patient antibody titer measurements in the Vitros XT7600 equipment from Ortho Clinical diagnostics, and (2) an IgG antibody neutralization reaction measured by immunofluorescence in an SD Biosensor cassette.

The IgG Detection Immunoassay against the antigenic domain corresponding to the spike protein (S1) is a quantitative assay. The result is considered reactive and provides a numerical value in antibody binding units (BAU) starting at 17.8 BAU/mL, to its upper limit of measurement up to 200 BAU/mL (VITROS Immunodiagnostic Products Anti-SARS-CoV-2 IgG Quantitative Reagent Pack, Ortho-Clinical Diagnostics, Illkirch-Graffenstaden, France).

According to the manufacturer, this test has 94% (95% CI, 90.1–96.7%) sensitivity and 99.5% (95% CI, 99.3–99.99%) specificity [29].

Additionally, the detection of neutralizing antibodies was performed by a cassette neutralization reaction, which was measured by immunofluorescence, SARS-CoV-2 nAb FIA (Standard F Sd Biosensor, Suwon, Korea). This kit is intended for use as an aid in identifying individuals with an adaptive immune response to SARS-CoV-2. A subset of secreted antibodies that have been demonstrated in the laboratory to prevent SARS-CoV-2 viral entry to human cells are termed neutralizing antibodies. Neutralizing antibody tests for SARS-CoV-2 use purified proteins based on the key viral recognition, docking, and infection through the interaction of the SARS-CoV-2 spike protein binding domain (RBD) and ACE2 receptor. This kit was compared with WHO Reference panel materials and other kits in the market such as GenScript kit (virus neutralization test sVNT, Piscataway, NJ, USA).

In this test, the result was interpreted as reactive over a 20 percent neutralization. According to the manufacturer, this test has 100% sensitivity and 99% specificity. This technique allows evaluating the neutralization capacity against two groups of variants, group 1 (FIA V1): Wuhan/Alpha (UK) and group 2 (FIA V2): Beta/Gamma Brazil/South Africa/Japan [30].

2.2. *Assessment of the Specific Cellular Response of SARS-CoV-2*

Cell-mediated immunity was assessed by measuring interferon gamma secreted by T cells in response to different antigens of the SARS-CoV-2 virus, using an IGRA kit specific for SARS-CoV-2 (interferon gamma release assay) detected by immunofluorescence (Covi-FERON FIA, SD Biosensor, Suwon, Korea).

IGRAs have been well studied and widely used in the field of tuberculosis diagnosis. In virology, this is their first time used. The production of interferon-gamma is a critical step in immunological defense mechanisms against SARS-CoV-2.

Whole blood tubes were used as samples. In each kit's tube, 1 mL of blood was injected. A total of 6 tubes per participant were used (nil tube, original protein S tube, variant protein S tube, Delta variant protein S tube, nucleocapsid protein tube, and mitogen tube), as specified by the manufacturer [31].

The kit includes the following tubes:

- Nil: patient's baseline gamma interferon measurement
- Original SP: Tube with spike antigen of ancestral virus (Wuhan/Hu-1/2019) and UK/alpha lineage B.1.1.7 variant.
- Variant SP: Tube with spike antigen South African variant/Beta lineage B.1.351 and Brazilian variant/gamma lineage P.1.
- Variant S Delta: Tube with spike antigen variant Delta lineage B.1.617.2.
- NP antigen: tube with antigen corresponding to the nucleocapsid protein.
- Mitogen: positive control tube to rule out lymphocytic anergy.

After collecting the sample, it was incubated from 16 to 24 h at 37 °C and then centrifuged for 15 min at 2200–2300× *g*.

IFN-gamma was detected by the F2400 equipment by means of a fluorescent signal, delivering results in concentration (IU/mL) through the integrated software of the equipment (SD Biosensor).

For the subsequent interpretation of the results, the "cut-off value charts for STANDARD F Covi-FERON FIA" provided by the manufacturer were used.

According to the instructions from the provider, the interferon gamma value obtained by the patient's tube was used and the result obtained in the Nil tube (patient's baseline) was subtracted.

2.3. *Statistical Analysis*

For the statistical analysis, a descriptive analysis of the variables was carried out, expressing the quantitative variables as mean ± standard deviation and the qualitative variables as total sample size and total percentage (%). An analysis of confounding variables was performed, which included age, comorbidities (hypertension, type 2 diabetes mellitus, and obesity), and habits (smoking). Subsequently, antibody levels (Wuhan, Brazil, U.K. and South African in a quantitative way were analyzed), the reactivity of cellular immunity in a qualitative way, levels of interferon (IFN), percentage of neutralization of IgG anti-spike V1 FIA and V2 FIA, and titers of total and diluted CVT2QN antibodies in a quantitative way. Finally, a qualitative analysis of antibody titers according to reactivity was made.

For data processing, the GraphPad Prism 9.3.1 pro× (GraphPad Software Inc., San Diego, CA, USA, 2021) was used. Shapiro–Wilk was used as a data normality test. Depending on the normality of the data, the one-way ANOVA test with Tukey's post-hoc test or the Kruskal–Wallis test with Dunn's multiple comparisons test was used for quantitative variables. For qualitative variables, Chi square was used. A simple linear regression analysis was performed to calculate 95% confidence intervals in interferon levels according to the Wuhan variant, compared to the other variants; and in the percentage of neutralization of IgG anti-spike V1 FIA and V2 FIA. For the correlation analysis, the Pearson or Spearman correlation test was used, depending on the normality of the data, determining an *r* value. A *p* value < or = 0.05 was considered statistically significant (Table S1).

2.4. Ethical Considerations and Disclosures

All patients gave their written informed consent to participate in this study, which was approved by the ethics committee of the Hospital Clínico Universidad de Chile (Act of approval No. 66, Exempt Resolution No. 1014, Certificate OAIC 1237/22) this study was conducted following the guidelines of the Declaration of Helsinki and following the standards of good clinical practice of the World Health Organization (WHO).

3. Results

There were no significant differences in the population in its general variables, except for age in those vaccinated with Vaxzevria booster dose (Table 1). In the comparison of the cellular immune response, there were statistically significant differences in the nucleocapsid protein tube variable, being lower in the group with the triple Comirnaty scheme (Table 2). When interpreted as a percentage, this also showed differences in the nucleocapsid for the same triple Comirnaty group (Table 3). The above is related to the number of subjects with this scheme, which were non-reactive from a qualitative point of view. Meanwhile, the mixed schemes had more reactive subjects (eight with Coronavac (two doses) + Vaxzeria scheme and seven with Coronavac (two doses) + Comirnaty scheme) (Figure 1).

Table 1. General background of the study population.

	Coronavac-Coronavac-Comirnaty	**Coronavac-Coronavac-Vaxzevria**	**Comirnaty-Comirnaty-Comirnaty**	***p* Value**
n	16	16	16	-
Age ($\overline{x} \pm SD$)	35.31 ± 7.382	59.13 ± 3.828	35.94 ± 8.290	<0.0001 †
Arterial Hypertension (%)	0 (0)	2 (12.5)	0 (0)	0.1241‡
Type 2 Diabetes Mellitus (%)	0 (0)	0 (0)	0 (0)	-
Obesity	0 (0)	2 (12.5)	0 (0)	0.1241 ‡
Smoking	0 (0)	2 (12.5)	0 (0)	0.1241 ‡

† One-way ANOVA t and Tukey's multiple comparisons. ‡ Chi-square test.

Table 2. Comparison of cellular immune response through IGRA between the three vaccination schedules in subjects studied at HCUCH in January 2022.

$\overline{x} \pm SD$	**Coronavac-Coronavac-Comirnaty**	**Coronavac-Coronavac-Vaxzevria**	**Comirnaty-Comirnaty-Comirnaty**	***p* Value**
Nil Tube	0.1628 ± 0.0712	0.1794 ± 0.0940	0.1450 ± 0.0000	>0.9999 §
Original SP Tube	3.236 ± 3.234	2.058 ± 1.876	3.206 ± 3.617	>0.9999 §
Variant SP Tube	3.057 ± 3.320	3.003 ± 2.972	2.811 ± 3.611	>0.9999 §
Spike Delta	0.9856 ± 1.132	0.9603 ± 1.280	0.7991 ± 1.167	>0.9999 §
NP Tube	0.8503 ± 1.879	1.172 ± 1.783	0.1588 ± 0.0536	0.0078 §
Mitogen Tube	10.000 ± 0.000	10.000 ± 0.000	10.000 ± 0.000	-

§ Kruskal–Wallis test and Dunn's multiple comparisons.

Table 3. Interpretation of cellular immunity response when comparing IGRA for the three vaccination schedules in subjects studied at HCUCH in January 2022.

$n \pm \%$	Coronavac-Coronavac-Comirnaty	Coronavac-Coronavac-Vaxzevria	Comirnaty-Comirnaty-Comirnaty	p Value
Alpha/Beta	14 (87.5)	11 (68.75)	12 (75)	0.4380 ‡
Gamma	15 (93.75)	13 (81.25)	12 (75)	0.3499 ‡
Delta	8 (50)	8 (50)	7 (43.75)	0.9199 ‡
Nucleocapsid	7 (43.75)	8 (50)	0 (0)	0.0040 ‡

‡ Chi-squared test.

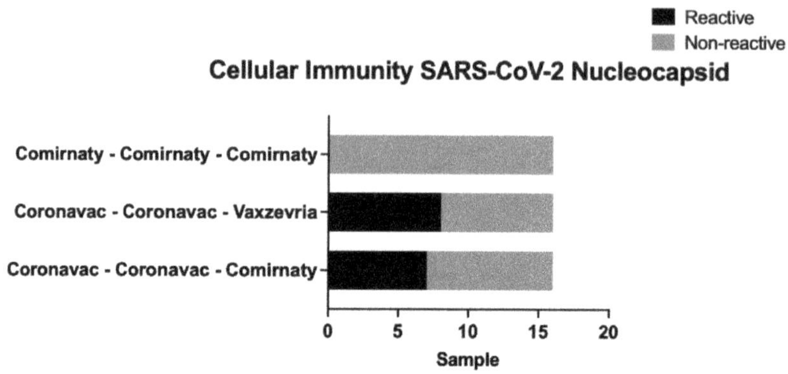

Figure 1. Comparison of cellular immunity against nucleocapsid (against NP) in subjects studied with the three vaccine dosage models at the HCUCH, January 2022.

Regarding interferon levels (IFN), there were significant differences in the nucleocapsid's IFN level measure, between the groups with triple Cominarty scheme and mixed schemes (Coronavac complete scheme with Vaxzevria or Comirnaty booster doses) (Figure 2). Comparing the neutralization's percentage, the group that presented the lowest values was the one immunized with Vaxzevria booster, with significant differences with the other vaccination schemes (Figure 3).

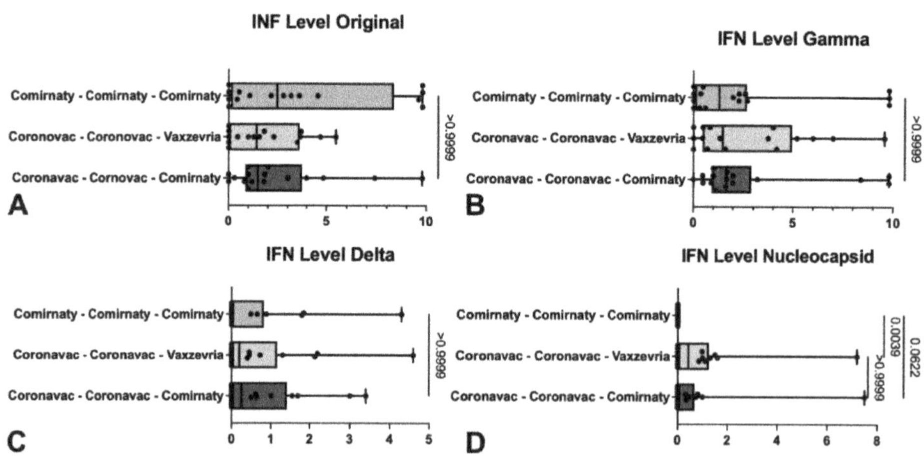

Figure 2. Comparison of cellular immune response through IGRA between the three vaccination schemes in subjects studied at HCUCH in January 2022.

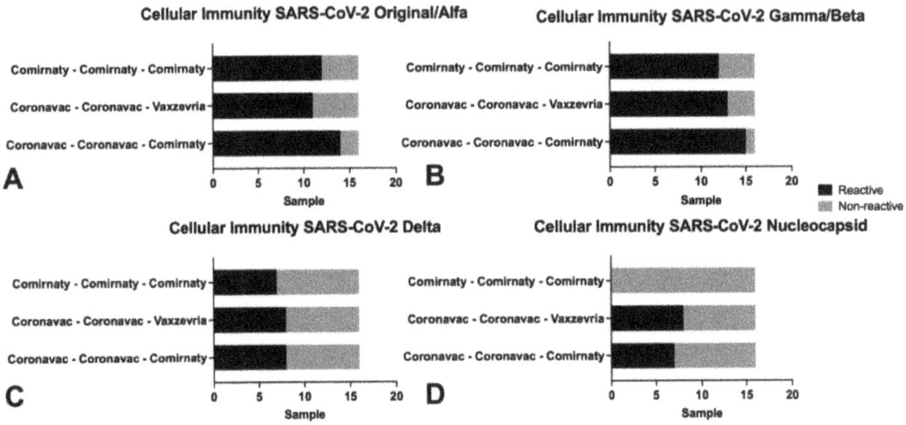

Figure 3. Interpretation of cellular immunity response when comparing IGRA for the three vaccination schemes in study subjects at HCUCH, January 2022.

Regarding the CVT2QN titles, quantitatively, this group also presented the lowest average, with significant differences from the other two schemes (Figure 4).

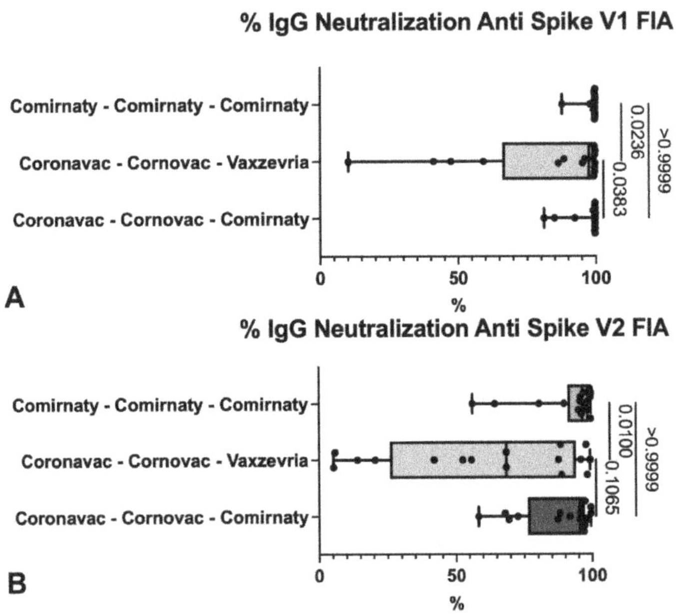

Figure 4. Comparison of the percentage of neutralization for two variants of spike protein according to the vaccine protocol under study. Subjects studied HCUCH, January 2022.

In the linear regression analysis, when comparing the data with the IFN levels in the original variant, the lowest value was presented by the Delta variant in the triple Comirnaty scheme with ($p = 0.0060$; $r = 0.6535$; CI 95% [0.2334–0.8900]) (Table 4). When comparing the neutralization percentage of V1 FIA with V2 FIA, both mixed Coronavac schemes presented similar values (Table 5).

Table 4. Linear correlation of IFN levels measured by immunization schedule in original variant (Wuhan) in study subjects, HCUCH January 2022.

	Coronavac-Coronavac-Comirnaty			Coronavac-Coronavac-Vaxzevria			Comirnaty-Comirnaty-Comirnaty		
	Brazil	**Delta**	**Nucleocapsid**	**Brazil**	**Delta**	**Nucleocapsid**	**Brazil**	**Delta**	**Nucleocapsid**
p value	0.0055	0.0060	0.0023	0.0082	0.1097	0.2210	0.9265	0.8794	>0.9999
r	0.6591	0.6535	0.7056	0.6352	0.4153	0.3239	−0.02510	−0.04127	-
IC 95%	0.2427–0.8704	0.2334–0.8680	0.3229–0.8900	0.2037–0.8601	−0.1013–0.7555	−0.2046–0.7062	−0.5144–0.4765	−0.5262–0.4639	-

Table 5. Linear correlation of the neutralization percentage by immunization schedule for V1 FIA vs. V2 FIA, in study subjects at HCUCH, January 2022.

	Coronavac-Coronavac-Comirnaty	Coronavac-Coronavac-Vaxzevria	Comirnaty-Comirnaty-Comirnaty
p value	<0.0001	<0.0001	0.0161
r	0.8238	0.8618	0.5902
IC 95%	0.5545–0.9369	0.6392–0.9512	0.1336–0.8401

4. Discussion

In Chile, different vaccination schemes were used according to existing agreements with the pharmaceutical industry, largely due to phase 2 and 3 studies carried out at the beginning of vaccine production. Due to this, the entire Chilean population has had access to vaccines since December 2020. According to the authorizations issued by the Chilean public health institute (ISP) and European community [32–34], certain restrictions were considered for its application such as the use of the Astrazeneca laboratory vaccine (Vaxzevria) for people older than 55 years. Because of these restrictions, there was a significant difference observed in the age of the subjects recruited for the study of cellular and humoral immunity that we carried out at the HCUCH. The rest of the pathologies evaluated when recruiting the subjects are comparable between groups, such as tobacco, type 2 diabetes, hypertension, and obesity.

We found four subjects in triple Comirnaty scheme who had a cellular immune response to nucleocapsid protein, which is a finding that made us modify the study subgroups, because of the unfeasible result due to the vaccination model administered (mRNA vaccines, which only synthesize spike protein).

We therefore decided to eliminate these four subjects from the sample who were probably asymptomatic in their hospital functions prior to the study in question.

This work has the advantage of being the first in South America to compare cellular immunity in three different vaccine schemes in a population that has used three doses of SARS-CoV-2 vaccines. Therefore, it was interesting to observe the response to each of the schemes.

In Table 1 and Figures 2 and 3, it is observed that only those vaccinated with Corona vac have a cellular immune response triggered by nucleocapsid antigenic protein. This is because of the vaccine's type administered, where Coronavac vaccine is a viral inactivated vaccine of the SARS-CoV-2 virus, unlike the other platforms that use only the spike protein fragment to generate an immune response. There are studies in which the cellular immune response has been evaluated in some specific populations like patients with comorbidities, such as solid and hematological cancer and transplants, without considering multiple vaccination schemes in these studies [35–37].

A high number of reactive subjects can be observed against the different SARS-CoV-2 variants, both ancestral, alpha, beta, gamma, and delta, with no significant differences between any of the three schemes.

When performing the analysis of the humoral response, already known by the previous studies carried out in Chile and in the world [38–40], this particular study has the advantage of identifying neutralizing antibodies against some SARS-CoV-2 variants. It was observed that, with a complete vaccination schedule with a booster dose, there are high titers of antibodies and within these, neutralizing type antibodies. This has not been widely studied in world literature, let alone in South America [40–42].

In our investigation, it was observed that there were significant differences for the neutralizing antibodies for the original and alpha variant when comparing triple Comirnaty with Coronavac (two doses) + Vaxzevria and the same when comparing the latter with Coronavac (two doses) + Comirnaty. A lower percentage of neutralization was observed in the Coronavac (two doses) + Vaxzevria user group (Table 6).

Table 6. Neutralization percentage of antibodies through IgG anti-spike measurement for variant V1 and V2 in the study's population at HCUCH, January 2022.

$\bar{x}\pm SD$	Coronavac-Coronavac-Comirnaty	Coronavac-Coronavac-Vaxzevria	Comirnaty-Comirnaty-Comirnaty	*p* Value
V1FIA (ancestral/alfa)	97.08 ± 5.847	82.51 ± 27.60	98.78 ± 2.953	0.0236 [§]
V2FIA (beta/gamma)	88.26 ± 13.38	61.80 ± 34.84	91.39 ± 13.13	0.0100 [§]

[§] Kruskal–Wallis test and Dunn's multiple comparisons.

According to the demographic data obtained, the only difference between the populations was their age, which could partially explain this difference in neutralizing antibodies, requiring further study to investigate other causes attributable to this difference. Similar results were obtained when looking for neutralizing antibodies for beta and gamma variants (FIA V2).

Regarding the measurement of antibody titers (Figure 5), this was done with an automated chemiluminescent immunometric technique with linear measurement up to 200 BAU/mL. Given the high titers of anti-spike protein antibody results, the team of specialists in the design of the technique was consulted, who recommended a dilution of 1/20 to achieve measurement linearity up to 4000 BAU/mL.

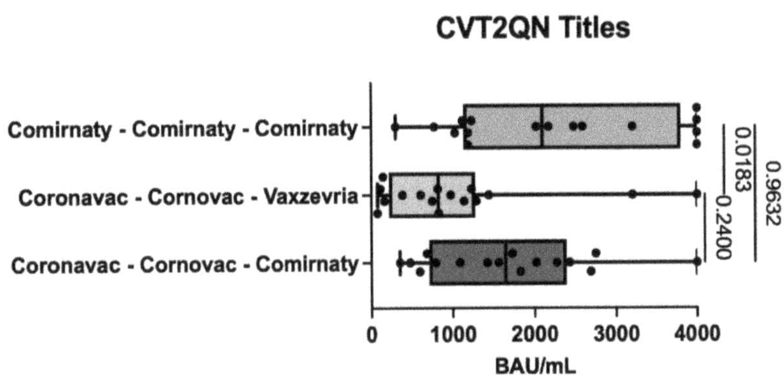

Figure 5. Measurement of antibody titers for the three vaccination schedules in study subjects at HCUCH, January 2022.

Unlike other articles, which compared the humoral immune response between different vaccines and found differences between them, as described in the investigation of Dashdorj et al. [43,44], in our study, we did not find statistically significant differences between the three vaccine schemes. All three schedules demonstrated excellent humoral response 2 months after the administration of the third dose (booster) (Figure 2). However, significant differences were observed in the measurement of neutralizing IgG antibodies against SARS-CoV-2 for the original and alpha variant (FIA V1) when compared to other variants, beta and gamma (FIA V2).

According to what was analyzed for the study of antibodies with the dilutions of the samples, it can be observed that all the vaccination schemes obtained high antibody titers, without great differences between groups. However, when performing the correlation analysis between the groups of vaccines with original variant and the other variants, such as beta and gamma, it was observed that there is a clear correlation for the Coronavac x2 + Comirnaty scheme and Coronavac x2 + Vaxzevria scheme, in which both have an r greater than 0, unlike triple Comirnaty, where there is no clear correlation, as shown in Table 5. This means, antibodies would only have a neutralizing effect for the original variant and not with delta in our group of subjects ($r = 0.5902$).

Within the limitations of our study, we highlight the low number of patients recruited per group with only 16 cases in each one, due to the limited access to IGRA and the recruitment time of the volunteers. However, we consider that this is pioneering work in South America that sets the start point on new studies in the cellular immune response against the COVID-19 pandemic.

To conclude, as a laboratory, we believe that the three schemes implemented by the Chilean government were useful in the prevention of SARS-CoV-2 from the perspective of the mechanism and functioning of RNA viruses from viral adsorption, neutralizing spike, preventing binding to ACE 2, defending through cellular immunity, and interferon effect. We believe that the next step will be the analysis of different vaccination scheme groups, evaluating the effect on the possibility of contagion and seeing its effect on the reduction of symptoms and the severity of the disease.

5. Conclusions

The vaccination schemes evaluated in this study showed a good humoral and cellular response, since there were no significant differences in antibody titers, neutralization percentage, or cellular immunity presence between the groups.

It is necessary to increase research on the cellular immune response as a complement to the humoral immunity obtained by SARS-CoV-2 vaccines. The new vaccines in development seek to enhance cellular immunity and studies like this show that this particular IGRA technique is a simple way to study this type of immunity in low and medium complexity laboratories from developing countries, as it is a friendly kit and easy to perform in any laboratory.

Supplementary Materials: The Database used in this investigation is available at https://www.mdpi.com/article/10.3390/life12040534/s1, Table S1: Cellular and Humoral immune response to different Chilean vaccines schemes Database.

Author Contributions: Conceptualization, P.B.T.V. and G.F.B.; methodology, P.B.T.V.; software, D.C.V.; validation, P.B.T.V., D.C.V. and G.F.B.; formal analysis, D.C.V.; investigation, P.B.T.V. and G.S.-P.C.; resources, P.B.T.V. and M.J.V.C.; data curation, D.C.V. and G.F.B.; writing—original draft preparation, P.B.T.V., G.S.-P.C. and G.F.B.; writing—review and editing, P.B.T.V., G.S.-P.C., G.F.B. and F.S.O.; visualization, D.C.V., P.B.T.V., G.S.-P.C. and C.B.V.; supervision, P.B.T.V., M.J.V.C., F.S.O. and M.G.R.; project administration, P.B.T.V., C.M.C. and G.F.B.; funding acquisition, P.B.T.V. and M.J.V.C. All authors have read and agreed to the published version of the manuscript.

Funding: This research received no external funding.

Institutional Review Board Statement: The study was conducted according to the guidelines of the Declaration of Helsinki, and approved by the Ethics Committee of the Hospital Clínico Universidad de Chile (Act of approval No. 66, Exempt Resolution No. 1014, Certificate OAIC 1237/22).

Informed Consent Statement: Informed consent was obtained from all subjects involved in the study.

Data Availability Statement: Publicly available datasets were analyzed in this study.

Acknowledgments: We thank each of the participants in this study, as well as all the staff in our laboratory who allowed us to carry out this research. We thank SD Biosensor DiagnoChile and Ortho Clinical Diagnostics for the technical and logistical assistance with the diagnostic kits to obtain these results.

Conflicts of Interest: The authors declare that the research was carried out in the absence of commercial or financial relationships that could be interpreted as a possible conflict of interest.

References

1. Wang, L.; Wang, Y.; Ye, D.; Liu, Q. Review of the 2019 novel coronavirus (SARS-CoV-2) based on current evidence. *Int. J. Antimicrob. Agents* **2020**, *55*, 105948. [CrossRef] [PubMed]
2. Chung, Y.-S.; Lee, N.-J.; Woo, S.H.; Kim, J.-M.; Kim, H.M.; Jo, H.J.; Park, Y.E.; Han, M.-G. Validation of real-time RT-PCR for detection of SARS-CoV-2 in the early stages of the COVID-19 outbreak in the Republic of Korea. *Sci. Rep.* **2021**, *11*, 14817. [CrossRef] [PubMed]
3. Nature. Available online: https://www.nature.com/articles/d41586-022-00104-8 (accessed on 18 February 2022).
4. Hu, B.; Guo, H.; Zhou, P.; Shi, Z.-L. Characteristics of SARS-CoV-2 and COVID-19. *Nat. Rev. Microbiol.* **2021**, *19*, 141–154. [CrossRef] [PubMed]
5. Park, T.; Lee, S.; Kim, S.; Jeong Kim, M.; Kim, H.; Jun, S.; Kim, S.; Kim, B.; Changkyun Park, E.; Park, D. Spike protein binding prediction with neutralizing antibodies of SARS-CoV-2. *bioRxiv* **2020**. [CrossRef]
6. Al-Zyoud, W.; Haddad, H. Dynamics prediction of emerging notable spike protein mutations in SARS-CoV-2 implies a need for updated vaccines. *Biochimie* **2021**, *191*, 91–103. [CrossRef]
7. Hastie, K.M.; Li, H.; Bedinger, D.; Schendel, S.L.; Dennison, S.M.; Li, K.; Rayaprolu, V.; Yu, X.; Mann, C.; Zandonatti, M.; et al. Defining variant-resistant epitopes targeted by SARS-CoV-2 antibodies: A global consortium study. *Science* **2021**, *374*, 472–478. [CrossRef]
8. Manisty, C.; Otter, A.D.; Treibel, T.A.; McKnight, Á.; Altmann, D.M.; Brooks, T.; Noursadeghi, M.; Boyton, R.J.; Semper, A.; Moon, J.C. Antibody response to first BNT162b2 dose in previously SARS-CoV-2-infected individuals. *Lancet* **2021**, *397*, 1057–1058. [CrossRef]
9. Tarke, A.; Sidney, J.; Methot, N.; Yu, E.D.; Zhang, Y.; Dan, J.M.; Goodwin, B.; Rubiro, P.; Sutherland, A.; Wang, E.; et al. Impact of SARS-CoV-2 variants on the total CD4+ and CD8+ T cell reactivity in infected or vaccinated individuals. *Cell Rep. Med.* **2021**, *2*, 100355. [CrossRef] [PubMed]
10. Geers, D.; Shamier, M.C.; Bogers, S.; den Hartog, G.; Gommers, L.; Nieuwkoop, N.N.; Schmitz, K.S.; Rijsbergen, L.C.; van Osch, J.A.; Dijkhuizen, E.; et al. SARS-CoV-2 variants of concern partially escape humoral but not T-cell responses in COVID-19 convalescent donors and vaccinees. *Sci. Immunol.* **2021**, *6*, eabj1750. [CrossRef]
11. Richter, D.; Guasti, L.; Koehler, F.; Squizzato, A.; Nistri, S.; Christodorescu, R.; Dievart, F.; Gaudio, G.; Asteggiano, R.; Ferrini, M. Late phase of COVID-19 pandemic in General Cardiology. A position paper of the ESC Council for Cardiology Practice. *ESC Hearth Fail.* **2021**, *8*, 3483–3494. [CrossRef]
12. Cascella, M.; Del Gaudio, A.; Vittori, A.; Bimonte, S.; Del Prete, P.; Forte, C.A.; Cuomo, A.; De Blasio, E. COVID-Pain: Acute and Late-Onset Painful Clinical Manifestations in COVID-19–Molecular Mechanisms and Research Perspectives. *J. Pain Res.* **2021**, *14*, 2403–2412. [CrossRef]
13. Bussmann, A.J.C.; Ferraz, C.R.; Lima, A.V.A.; Castro, J.G.S.; Ritter, P.D.; Zaninelli, T.H.; Saraiva-Santos, T.; Verri, W.A.; Borghi, S.M. Association between IL-10 systemic low level and pain during symptomatic SARS-CoV-2 infection. *Pain Pract.* **2022**. [CrossRef] [PubMed]
14. Sansone, A.; Mollaioli, D.; Ciocca, G.; Colonnello, E.; Limoncin, E.; Balercia, G.; Jannini, E.A. "Mask up to keep it up": Preliminary evidence of the association between erectile dysfunction and COVID-19. *Andrology* **2021**, *9*, 1053–1059. [CrossRef] [PubMed]
15. Parra-Lucares, A.; Segura, P.; Rojas, V.; Pumarino, C.; Saint-Pierre, G.; Toro, L. Emergence of SARS-CoV-2 Variants in the World: How Could This Happen? *Life* **2022**, *12*, 194. [CrossRef] [PubMed]
16. Cavanaugh, A.M.; Spicer, K.B.; Thoroughman, D.; Glick, C.; Winter, K. Reduced Risk of Reinfection with SARS-CoV-2 After COVID-19 Vaccination—Kentucky, May–June 2021. *MMWR. Morb. Mortal. Wkly. Rep.* **2021**, *70*, 1081–1083. [CrossRef] [PubMed]
17. Xing, K.; Tu, X.-Y.; Liu, M.; Liang, Z.-W.; Chen, J.-N.; Li, J.-J.; Jiang, L.-G.; Xing, F.-Q.; Jiang, Y. Efficacy and safety of COVID-19 vaccines: A systematic review. *Chin. J. Contemp. Pediatr.* **2021**, *23*, 221–228.
18. Haas, E.J.; Angulo, F.J.; McLaughlin, J.M.; Anis, E.; Singer, S.R.; Khan, F.; Brooks, N.; Smaja, M.; Mircus, G.; Pan, K.; et al. Impact and effectiveness of mRNA BNT162b2 vaccine against SARS-CoV-2 infections and COVID-19 cases, hospitalisations, and deaths following a nationwide vaccination campaign in Israel: An observational study using national surveillance data. *Lancet* **2021**, *397*, 1819–1829. [CrossRef]
19. Ledford, H. Do vaccines protect against long COVID? What the data say. *Nature* **2021**, *599*, 546–548. [CrossRef]
20. Our World in Data. Coronavirus Pandemic (COVID-19). 2022. Available online: https://ourworldindata.org/coronavirus (accessed on 18 February 2022).
21. Vacunas MINSAL. Available online: https://www.minsal.cl/wp-content/uploads/2021/12/Vacunaci%C3%B3n-contra-SARS-CoV-2-en-pre-escolares-3-a-5-a%C3%B1os.pdf (accessed on 18 February 2022).
22. Guelfenbein, C.I.; Torretti, J.P.T.; de Pablo, M.E.S. Vacunas SARS CoV-2, estudios en fase III. *Rev. Chile. Infectol.* **2021**, *38*, 88–98. [CrossRef]
23. Knoll, M.D.; Wonodi, C. Oxford-AstraZeneca COVID-19 vaccine efficacy. *Lancet* **2021**, *397*, 72–74. [CrossRef]
24. Jara, A.; Undurraga, E.A.; González, C.; Paredes, F.; Fontecilla, T.; Jara, G.; Pizarro, A.; Acevedo, J.; Leo, K.; Leon, F.; et al. Effectiveness of an Inactivated SARS-CoV-2 Vaccine in Chile. *N. Engl. J. Med.* **2021**, *385*, 875–884. [CrossRef] [PubMed]

25. Halperin, A.S.; Ye, L.; MacKinnon-Cameron, D.; Smith, B.; Cahn, P.E.; Ruiz-Palacios, G.M.; Ikram, A.; Lanas, F.; Guerrero, M.L.; Navarro, S.R.M.; et al. Final efficacy analysis, interim safety analysis, and immunogenicity of a single dose of recombinant novel coronavirus vaccine (adenovirus type 5 vector) in adults 18 years and older: An international, multicentre, randomised, double-blinded, placebo-controlled phase 3 trial. *Lancet* **2021**, *399*, 237–248. [CrossRef] [PubMed]
26. Schultz, B.; Melo-González, F.; Duarte, L.; Gálvez, N.; Pacheco, G.; Soto, G.; Berríos-Rojas, R.; González, L.; Moreno-Tapia, D.; Rivera-Pérez, D.; et al. A booster dose of an inactivated vaccine increases neutralizing antibodies and T cell responses against SARS-CoV-2. *medRxiv* **2021**. [CrossRef]
27. Schiffner, J.; Backhaus, I.; Rimmele, J.; Schulz, S.; Möhlenkamp, T.; Klemens, J.M.; Zapf, D.; Solbach, W.; Mischnik, A. Long-Term Course of Humoral and Cellular Immune Responses in Outpatients After SARS-CoV-2 Infection. *Front. Public Health* **2021**, *9*, 732787. [CrossRef]
28. Havervall, S.; Ng, H.; Falk, A.J.; Greilert-Norin, N.; Månberg, A.; Marking, U.; Laurén, I.; Gabrielsson, L.; Salomonsson, A.; Aguilera, K.; et al. Robust humoral and cellular immune responses and low risk for reinfection at least 8 months following asymptomatic to mild COVID-19. *J. Intern. Med.* **2021**, *291*, 72–80. [CrossRef] [PubMed]
29. VITROS Immunodiagnostic Products Anti-SARS-CoV-2 IgG Quantitative Reagent Pack. Technical Document. Available online: https://techdocs.orthoclinicaldiagnostics.com/TechDocs/TechDocSearch.aspx?tID=0&culture=en-us (accessed on 24 January 2022).
30. STANDARD F SARS-CoV-2 nAb FIA. Technical Document. Available online: https://www.sdbiosensor.com/product/product_view?product_no=175# (accessed on 24 January 2022).
31. Covi-FERON FIA (IFN-Gamma) Technical Document. Available online: https://www.sdbiosensor.com/product/product_view?product_no=1241 (accessed on 24 January 2022).
32. Torjesen, I. Covid-19: AstraZeneca vaccine is approved in EU with no upper age limit. *BMJ* **2021**, *372*, n295. [CrossRef]
33. Dyer, O. Covid-19: Canada relaxes age limits on AstraZeneca vaccine as Ontario cases surge. *BMJ* **2021**, *373*, n1053. [CrossRef]
34. ISP Chile. Available online: https://www.ispch.cl/wp-content/uploads/2021/12/UCD289.pdf (accessed on 18 February 2022).
35. Herishanu, Y.; Avivi, I.; Aharon, A.; Shefer, G.; Levi, S.; Bronstein, Y.; Morales, M.; Ziv-Baran, T.; Arbel, Y.S.; Scarfò, L.; et al. Efficacy of the BNT162b2 mRNA COVID-19 vaccine in patients with chronic lymphocytic leukemia. *Blood* **2021**, *137*, 3165–3173. [CrossRef]
36. Re, D.; Barrière, J.; Chamorey, E.; Delforge, M.; Gastaud, L.; Petit, E.; Chaminade, A.; Verrière, B.; Peyrade, F. Low rate of seroconversion after mRNA anti-SARS-CoV-2 vaccination in patients with hematological malignancies. *Leuk. Lymphoma* **2021**, *62*, 3308–3310. [CrossRef]
37. Kamar, N.; Abravanel, F.; Marion, O.; Couat, C.; Izopet, J.; Del Bello, A. Three Doses of an mRNA Covid-19 Vaccine in Solid-Organ Transplant Recipients. *N. Engl. J. Med.* **2021**, *385*, 661–662. [CrossRef]
38. Bueno, S.M.; Abarca, K.; González, P.A.; Gálvez, N.M.S.; Soto, J.A.; Duarte, L.F.; Schultz, B.M.; Pacheco, G.A.; González, L.A.; Vázquez, Y.; et al. Safety and Immunogenicity of an Inactivated Severe Acute Respiratory Syndrome Coronavirus 2 Vaccine in a Subgroup of Healthy Adults in Chile. *Clin. Infect. Dis.* **2021**. [CrossRef] [PubMed]
39. Levin, E.G.; Lustig, Y.; Cohen, C.; Fluss, R.; Indenbaum, V.; Amit, S.; Doolman, R.; Asraf, K.; Mendelson, E.; Ziv, A.; et al. Waning Immune Humoral Response to BNT162b2 Covid-19 Vaccine over 6 Months. *N. Engl. J. Med.* **2021**, *385*, e84. [CrossRef] [PubMed]
40. Lustig, Y.; Sapir, E.; Regev-Yochay, G.; Cohen, C.; Fluss, R.; Olmer, L.; Indenbaum, V.; Mandelboim, M.; Doolman, R.; Amit, S.; et al. BNT162b2 COVID-19 vaccine and correlates of humoral immune responses and dynamics: A prospective, single-centre, longitudinal cohort study in health-care workers. *Lancet Respir. Med.* **2021**, *9*, 999–1009. [CrossRef]
41. Khoury, D.S.; Cromer, D.; Reynaldi, A.; Schlub, T.E.; Wheatley, A.K.; Juno, J.A.; Subbarao, K.; Kent, S.J.; Triccas, J.A.; Davenport, M.P. Neutralizing antibody levels are highly predictive of immune protection from symptomatic SARS-CoV-2 infection. *Nat. Med.* **2021**, *27*, 1205–1211. [CrossRef]
42. Souza, W.M.; Amorim, M.R.; Sesti-Costa, R.; Coimbra, L.D.; Brunetti, N.S.; Toledo-Teixeira, D.A.; de Souza, G.F.; Muraro, S.P.; Parise, P.L.; Barbosa, P.P.; et al. Neutralisation of SARS-CoV-2 lineage P.1 by antibodies elicited through natural SARS-CoV-2 infection or vaccination with an inactivated SARS-CoV-2 vaccine: An immunological study. *Lancet Microbe* **2021**, *2*, e527–e535. [CrossRef]
43. Dashdorj, N.J.; Wirz, O.F.; Röltgen, K.; Haraguchi, E.; Buzzanco, A.S.; Sibai, M.; Wang, H.; Miller, J.A.; Solis, D.; Sahoo, M.K.; et al. Direct comparison of antibody responses to four SARS-CoV-2 vaccines in Mongolia. *Cell Host Microbe* **2021**, *29*, 1738–1743.e4. [CrossRef]
44. Munro, A.P.S.; Janani, L.; Cornelius, V.; Aley, P.K.; Babbage, G.; Baxter, D.; Bula, M.; Cathie, K.; Chatterjee, K.; Dodd, K.; et al. Safety and immunogenicity of seven COVID-19 vaccines as a third dose (booster) following two doses of ChAdOx1 nCov-19 or BNT162b2 in the UK (COV-BOOST): A blinded, multicentre, randomised, controlled, phase 2 trial. *Lancet* **2021**, *398*, 2258–2276. [CrossRef]

life

MDPI

Article

Convalescent Plasma for Hospitalized COVID-19 Patients: A Single-Center Experience

Massimo Franchini [1,*], Claudia Glingani [1], Giuseppe De Donno [2,†], Giuseppe Lucchini [3], Massimiliano Beccaria [2], Massimo Amato [4], Gian Paolo Castelli [5], Leonardo Bianciardi [6], Mauro Pagani [7], Marco Ghirardini [8], Giuseppe Puma [9], Barbara Presciuttini [7], Maria Teresa Costantino [10], Marilena Frigato [10], Verena Crosato [11], Giorgio Tiecco [11], Alice Mulè [11], Dorothea Angela Papalia [11], Francesco Inglese [2], Fabio Spreafico [2], Martina Garuti [2], Antonietta Pecoriello [2], Giulia Cervi [2], Graziana Greco [2], Vanni Galavotti [2], Tiziana Santini [8], Angela Berselli [5], Carlo Montalto [5], Riccardo Bertoletti [12], Simona Aurelia Bellometti [12], Enrico Capuzzo [1], Dario Benazzi [4], Gianpaolo Grisolia [13], Fabio Pajola [12], Raffaello Stradoni [14], Matteo Zani [1], Adriano Verzola [15], Vito Codeluppi [16], Silvia Vesentini [16], Elisa Bellocchio [16], Marco Candini [16], Giorgina Ambrosi [16], Francesca Carandina [16], Cleante Scarduelli [17], Albino Reggiani [18], Salvatore Casari [9] and on behalf of Convalescent Plasma Study Group [‡]

1 Department of Transfusion Medicine and Hematology, Carlo Poma Hospital, Azienda Socio Sanitaria Territoriale of Mantova, 46100 Mantova, Italy; claudia.glingani@asst-mantova.it (C.G.); enrico.capuzzo@asst-mantova.it (E.C.); matteo.zani@asst-mantova.it (M.Z.)
2 Intensive Care Respiratory Unit, Carlo Poma Hospital, Azienda Socio Sanitaria Territoriale of Mantova, 46100 Mantova, Italy; giuseppe.dedonno@asst-mantova.it (G.D.D.); massimiliano.beccaria@asst-mantova.it (M.B.); francesco.inglese@asst-mantova.it (F.I.); fabio.spreafico@asst-mantova.it (F.S.); martina.garuti@asst-mantova.it (M.G.); antonietta.pecoriello@asst-mantova.it (A.P.); giulia.cervi@asst-mantova.it (G.C.); graziana.greco@asst-mantova.it (G.G.); vanni.galavotti@asst-mantova.it (V.G.)
3 Biostatistic Unit, Carlo Poma Hospital, Azienda Socio Sanitaria Territoriale of Mantova, 46100 Mantova, Italy; giuseppe.lucchini@asst-mantova.it
4 Emergency Department, Carlo Poma Hospital, Azienda Socio Sanitaria Territoriale of Mantova, 46100 Mantova, Italy; massimo.amato@asst-mantova.it (M.A.); dario.benazzi@asst-mantova.it (D.B.)
5 Department of Anesthesiology and Intensive Care, Carlo Poma Hospital, Azienda Socio Sanitaria Territoriale of Mantova, 46100 Mantova, Italy; gianpaolo.castelli@asst-mantova.it (G.P.C.); angela.berselli@asst-mantova.it (A.B.); carlo.montalto@asst-mantova.it (C.M.)
6 Anesthesiology and Intensive Care Unit, SS Trinità and San Marcellino Muravera (Cagliari) Hospital, ASL 8 Cagliari, 09043 Cagliari, Italy; leonardo.bianciardi@atssardegna.it
7 Internal Medicine Unit, Department of Medicine, Carlo Poma Hospital, Azienda Socio Sanitaria Territoriale of Mantova, 46100 Mantova, Italy; mauro.pagani@asst-mantova.it (M.P.); barbara.presciuttini@asst-mantova.it (B.P.)
8 Department of Medicine, Hospital of Asola, Azienda Socio Sanitaria Territoriale of Mantova, 46100 Mantova, Italy; marco.ghirardini@asst-mantova.it (M.G.); tiziana.santini@asst-mantova.it (T.S.)
9 Unit of Infectious Diseases, Carlo Poma Hospital, Azienda Socio Sanitaria Territoriale of Mantova, 46100 Mantova, Italy; giuseppe.puma@asst-mantova.it (G.P.); salvatore.casari@asst-mantova.it (S.C.)
10 Allergology and Clinical Immunology Unit, Department of Medicine, Carlo Poma Hospital, Azienda Socio Sanitaria Territoriale of Mantova, 46100 Mantova, Italy; mariateresa.costantino@asst-mantova.it (M.T.C.); marilena.frigato@asst-mantova.it (M.F.)
11 Infectious and Tropical Diseases Clinic, University of Brescia and Azienda Socio Sanitaria Territoriale Spedali Civili, 25123 Brescia, Italy; v.crosato@unibs.it (V.C.); g.tiecco@unibs.it (G.T.); a.mule@unibs.it (A.M.); d.papalia@unibs.it (D.A.P.)
12 Medical Direction, Azienda Socio Sanitaria Territoriale of Mantova, 46100 Mantova, Italy; riccardo.bertoletti@asst-mantova.it (R.B.); simona.bellometti@asst-mantova.it (S.A.B.); fabio.pajola@asst-mantova.it (F.P.)
13 Department of Obstetrics and Gynecology, Carlo Poma Hospital, Azienda Socio Sanitaria Territoriale of Mantova, 46100 Mantova, Italy; gianpaolo.grisolia@asst-mantova.it
14 General Direction, Azienda Socio Sanitaria Territoriale of Valcamonica, 25043 Breno, Italy; raffaello.stradoni@ats-montagna.it
15 Management Planning and Control Service, Carlo Poma Hospital, Azienda Socio Sanitaria Territoriale of Mantova, 46100 Mantova, Italy; adriano.verzola@asst-mantova.it
16 Department of Anesthesiology and Intensive Care, Destra Secchia Hospital, Azienda Socio Sanitaria Territoriale of Mantova, 46100 Mantova, Italy; vito.codeluppi@asst-mantova.it (V.C.);

Citation: Franchini, M.; Glingani, C.; Donno, G.D.; Lucchini, G.; Beccaria, M.; Amato, M.; Castelli, G.P.; Bianciardi, L.; Pagani, M.; Ghirardini, M.; et al. Convalescent Plasma for Hospitalized COVID-19 Patients: A Single-Center Experience. *Life* **2022**, *12*, 420. https://doi.org/10.3390/life12030420

Academic Editors: Silvia De Francia and Sarah Allegra

Received: 20 February 2022
Accepted: 11 March 2022
Published: 14 March 2022

silvia.vesentini@asst-mantova.it (S.V.); elisa.bellocchio@asst-mantova.it (E.B.);
marco.candini@asst-mantova.it (M.C.); giorgina.ambrosi@asst-mantova.it (G.A.);
francesca.carandina@asst-mantova.it (F.C.)

[17] Intensive Cardiopulmonary Rehabilitation Unit, Carlo Poma Hospital, Azienda Socio Sanitaria Territoriale of Mantova, 46100 Mantova, Italy; cleante.scarduelli@asst-mantova.it

[18] Cardiology Unit, Destra Secchia Hospital, Azienda Socio Sanitaria Territoriale of Mantova, 46100 Mantova, Italy; albino.reggiani@asst-mantova.it

* Correspondence: massimo.franchini@asst-mantova.it; Tel.: +39-0376-201234; Fax: +39-0376-220144

† In memory of dr. Giuseppe De Donno, a pioneer of COVID-19 convalescent plasma therapy in Italy.

‡ Consuelo Basili, Angela Bellani, Alberto Biondo, Enrico Burato, Irene Breoni, Beatrice Caruso, Maria Beatrice Casarini, Maria Giovanna Cavazzini, Margherita Chiamenti, Nicola Cicorella, Paolo Costa, Giovanni Carlo Fibbia, Elisa Firriolo, Sabrina Frattima, Rino Frizzelli, Angela Furini, Silvia Garilli, Gianni Gattuso, Susy Longhi, Francesca Losa, Claudio Martini, Fabia Mascaro, Linda Maulucci, Barbara Melara, Elena Miglioli, Alessandro Occhionero, Roberto Olivetti, Matteo Ortolani, Solidea Palin, Loredana Palvarini, Giorgio Perboni, Viviana Ravagnani, Francesca Saggiani, Cinzia Sissa, Piero Superbi, Marcello Vincenzi, Baoran Yang, Fabio Zacchi, Paolo Zampriolo (see Supplemantary Materials).

Abstract: In Winter 2020, Italy, and in particular the Lombardy region, was the first country in the Western hemisphere to be hit by the COVID-19 pandemic. Plasma from individuals recovered from COVID-19 (COVID-19 convalescent plasma, CCP) was the first therapeutic tool adopted to counteract the severe acute respiratory syndrome coronavirus-2 (SARS-CoV-2). In this retrospective cohort study, we report the experience of the city hospital of Mantua, Lombardy region, on the compassionate use of CCP in patients hospitalized for severe COVID-19. Between April 2020 and April 2021, 405 consecutive COVID-19 patients received 657 CCP units with a median anti-SARS-CoV-2 neutralizing antibody (nAb) titer of 160 (interquartile range (IQR), 80–320). Their median age was 68 years (IQR, 56–78 years), and 62% were males. At enrollment, 55% of patients had an increased body mass index (BMI), and 25.6% had at least three comorbidities. The 28-day crude mortality rate was 12.6% (51/405). Young age (<68 years), mild disease (admission to low-intensity departments) and early treatment (<7 days from symptoms onset) with high nAb titer (≥320) CCP were found as independently associated with a favorable response to CCP treatment. No safety concerns were recorded, with a rate of CCP-related adverse reactions (all of mild intensity) of 1.3%. In our real-life experience, the first in the western world, early administration of high-titer CCP was a safe and effective treatment for hospitalized COVID-19 patients.

Keywords: COVID-19; convalescent plasma; mortality; safety

1. Introduction

Severe acute respiratory syndrome coronavirus 2 (SARS-CoV-2) has caused the coronavirus disease 2019 (COVID-19) pandemic, which was first reported from Wuhan at the end of 2019 and spread all over the world within a few months [1,2]. At the time of writing, more than 6 million people have died from coronavirus disease 2019 (COVID-19), and more than 400 million people have been infected [3]. Clinicians and researchers have struggled to develop an effective therapeutic protocol to treat and contain the spread of this infectious disease, and more than 300 drugs have been or are being investigated under clinical trials in different parts of the world [4,5]. Among the various therapeutic and prophylactic strategies developed to contain the COVID-19 epidemic, passive immunization by transfusion of COVID-19 convalescent plasma (CCP) has been utilized, particularly during the first two pandemic waves, before the introduction into the market of monoclonal antibody (mAb)-based therapies and small-chemical antivirals [6].

The results from randomized and non-randomized controlled trials (RCTs) on CCP have produced mixed results, probably due to the wide heterogeneity in their study designs and patient populations enrolled [7]. Furthermore, these trials were heterogeneous with respect to the characteristics of the CCP used (e.g., in terms of antibody content and stratification of recipients according to their serological status or disease severity). Despite these limitations, there are currently positive signals from the literature on a beneficial effect of CCP when administered early (within 72 h from symptom onset) and with a high-titer (>1:160) of anti-SARS-CoV-2 neutralizing antibodies (nAb) [8,9].

In this study, we reported the results of the one-year experience (April 2020–April 2021) on the compassionate use of CCP in patients hospitalized for severe COVID-19 at the city hospital of Mantua, Italy.

2. Patients and Methods

2.1. Patient Selection

In this retrospective study, we reported all the consecutive patients admitted to the city hospital of Mantua for COVID-19 and transfused with CCP between 1 April 2020 and 30 April 2021. This period was divided into two phases: the first pandemic wave from April to July 2020 and the second–third pandemic wave from October 2020 to April 2021. The transfusion of CCP was performed within a compassionate use program authorized by both the local ethical committee and the hospital health management. The study was registered at clinicaltrials.gov as NCT05157165.

We enrolled in the CCP compassionate use program adult inpatients with a confirmed diagnosis of SARS-CoV-2 infection (i.e., with a nasopharyngeal swab positive for SARS-CoV-2 by polymerase chain reaction) and at least one of the following criteria indicative of severe COVID-19: (1) radiologically confirmed pneumonia; (2) oxygen saturation (SpO2) $\leq$ 93% at rest and in-room air; (3) partial pressure of oxygen (PaO2)/fraction of inspired oxygen (FiO2) $\leq$ 300 mmHg. Written informed consent was obtained before enrollment.

Exclusion criteria were: (1) patients under the age of 18 years; (2) patients with proven hypersensitivity or allergic reaction to plasma, blood product or immunoglobulins.

In addition to patients' demographic and physical data (age, sex and body mass index (BMI)) and associated drug therapy, patients' ABO blood group and comorbidities (obesity, arterial hypertension, cardiovascular diseases, chronic kidney disease, dyslipidemia, diabetes mellitus, chronic lung disease and cancer) were also considered. In addition to these parameters, we recorded the date of hospital admission and of discharge (or death), the degree of intensity of the hospital department of admission (low intensity: infectious disease and internal medicine departments, where patients have undergone medical therapy, including oxygen therapy at high flows, but not mechanical ventilation; intermediate-high intensity: respiratory intensive care unit, emergency medicine and intensive care unit where patients have undergone mechanical invasive and non-invasive mechanical ventilation), the date of CCP transfusion, the days between symptom onset and transfusion of the first CCP unit, the anti-SARS-CoV-2 nAb titer and the PaO2/FiO2 ratio before CCP transfusion.

2.2. Convalescent Plasma

CCP units were obtained from previously infected subjects who had recovered and cleared the virus. Eligible donors were either men or nulliparous women aged 18 to 65 years, weighing more than 50 kg, with a laboratory-confirmed diagnosis of SARS-CoV-2 infection that had completely resolved at least 14 days before donation, as confirmed by two consecutive negative SARS-CoV-2 PCR test results from nasopharyngeal swabs collected 24 h apart. CCP donors were enrolled on a voluntary basis and had to meet all standard plasma collection requirements as provided by the current Italian laws. All routine screening tests for blood donors, including ABO blood group typing; Rh phenotyping; complete blood count; and screening for human immunodeficiency virus; hepatitis B, C, A and E viruses; parvovirus B19; and syphilis were conducted according to Italian regulations and the indications of the Italian National Blood Center [10].

CCP was collected through a plasmapheresis procedure using the AURORA cell separator (Fresenius Kabi, Italy), processed, and stored in agreement with national CCP regulations [10,11]. A plasma volume of about 600 mL was collected during each procedure and immediately divided into two bags, each corresponding to a therapeutic CCP unit of 300 mL. Collected CCP had an anti-SARS-CoV-2 nAb titer of 1:80 or higher. The live authentic SARS-CoV-2 neutralization test for the titration of nAbs was performed at the Molecular Virology Unit of the University Hospital of Pavia and was based on the determination of cytopathic effect, as previously described [12,13]. Viral inactivation, by using the UVA photoinactivation method following the addition of amotosalen hydrochloride (Intercept-CERUS System, distributed in Italy by Kedrion SpA, Castelvecchio Pascoli, Lucca, Italy) was applied to all CCP units, in accordance with Italian regulation at that time, before freezing 300 mL aliquots.

The CCP transfusions were performed by medical and nursing staff. CCP recipients were transfused with one to three units of ABO type-compatible CCP, according to the clinical response, over a period of 3–5 days. All the procedures were performed in agreement with the routine procedures of the Transfusion Service of the city Hospital of Mantua. All donors provided their written consent after being thoroughly informed.

2.3. Outcomes

The primary outcome in our cohort of CCP-treated patients was the overall mortality at 28 days after hospitalization. A subgroup analysis was also performed to identify patient- and treatment-related factors possibly related to a worse prognosis (i.e., sex, age, ABO blood group, BMI, comorbidities, COVID-19 severity (measured as PaO2/FiO2 and intensity of admission department), days between symptom onset CCP transfusion, number and nAb titer of CCP units transfused).

A secondary outcome was the safety of CCP treatment, measured as the rate of adverse reactions to CCP transfusion. The type, degree and outcome of adverse events occurring during or after (within 72 h) CCP transfusion were recorded, according to the Guideline on Good Pharmacovigilance Practices of the European Medicines Agency (https://www.ema.europa.eu/en/documents/scientific-guideline/guideline-good-pharmacovigilance-practices-annex-i-definitions-rev-4_en.pdf, accessed on 19 February 2022). Serious transfusion reactions were defined as a transfusion-associated circulatory overload (TACO), transfusion-related acute lung injury (TRALI), severe allergic reaction, hypotensive reaction or death. By definition, all serious transfusion reactions occurred within 6 h of CCP transfusion. All transfusion reactions were collected using electronic case report forms recorded in the local management software system for transfusion services EmoNet (GPI Group, Trento, Italy). Adverse reactions to CCP transfusion were also recorded in a computer database using the national hemovigilance system of the transfusion network organized by the Italian National Blood Center (SISTRA).

2.4. Statistical Analysis

Continuous variables were reported as mean (±standard deviation) or median and interquartile range (IQR) as appropriate according to distribution, while categorical data are reported as numbers and percentages. Comparisons between groups were carried out with an independent t-test or Mann–Whitney U test for continuous variables and chi-squared test or Fisher's exact test for categorical variables, as appropriate. All statistical tests were two-sided, and associations were considered statistically significant when the values were below a nominal level of 0.05 ($p < 0.05$).

The multivariate analysis was conducted with the binary logistic regression model, with death as a dependent variable and using the following explanatory dichotomous variables: age (<68 years versus ≥68 years, the intensity of hospital department (low versus intermediate–high), days between symptoms onset and CCP transfusion (<7 versus ≥7) and CCP neutralizing titer (<320 versus ≥320). Calculations were performed with IBM SPSS Statistics software version 24.

3. Results

The baseline demographic and clinical characteristics of the 405 COVID-19 patients receiving CCP during the 12-month period of the study are reported in Table 1. All patients were of Caucasian ethnicity. The median age was 68 years (IQR, 56–78 years), with an excess of men over women (male/female ratio: 1.6). The patients' median body mass index (BMI) at enrollment was above the normal range (25.7; IQR 23.4–31.0), and more than half of them (153/278, 55.0%) were overweight or obese. Approximately a quarter of patients (83/324, 25.6%) had three or more comorbidities, classified as follows in order of frequency: hypertension (56.8%), dyslipidemia (33.0%), cardiovascular disease (29.6%), diabetes (21.6%), chronic lung disease (11.7%), cancer (10.2%) and chronic kidney disease (9.6%). Regarding the degree of COVID-19 severity, 30.6% (124/405) of patients were admitted to intermediate/high-intensity departments, a proportion similar to that of the more severe forms of COVID-19 (PaO2/FiO2 $\leq$ 150: 34.8% (141/405)). Thus, the intensity of the hospital department appeared to be a reliable surrogate of a patient's disease severity.

Table 1. Demographic and clinical characteristics of the 405 patients enrolled in the study.

Parameters	Results
Median age, years (IQR)	68 (56–78)
Males/females	251/154
Male/female ratio	1.6
Median BMI (kg/m^2) (IQR)	25.7 (23.4–31.0)
BMI (kg/m^2), n (%) [1]	
- Normal (18.5–24.9)	125 (45.0)
- Overweight (25.0–29.9)	71 (25.5)
- Grade 1 obesity (30.0–34.9)	51 (18.3)
- Grade 2 obesity (35–39.9)	23 (8.3)
- Grade 3 obesity (>40)	8 (2.9)
Associated comorbidities, n (%) [2]	
- Hypertension	184 (56.8)
- Dyslipidemia	107 (33.0)
- Cardiovascular disease	96 (29.6)
- Diabetes	70 (21.6)
- Chronic lung disease	38 (11.7)
- Cancer	33 (10.2)
- Chronic kidney disease	31 (9.6)
- <3 comorbidities	241 (74.4)
- >3 comorbidities	83 (25.6)
COVID-19 severity, n (%)	
- PaO2/FiO2 [3]	
> 200–300	76 (18.8)
100–200	281 (69.4)
< 100	48 (11.8)
- Hospital department	
Low intensity	281 (69.4)
Intermediate/high intensity	124 (30.6)
Concomitant therapies, n (%)	
- Antiviral agents [4]	84 (20.7)
- Antibiotics	264 (65.2)
- Hydroxychloroquine	52 (12.8)
- Steroids	332 (82.0)
- Anticoagulants [5]	388 (95.8)
Median interval between symptoms onset and CCP therapy, days (IQR)	7.5 (5–12)
ABO blood type, n (%)	
- O	158 (39.0)
- A	188 (46.4)
- B	44 (10.9)
- AB	15 (3.7)

Abbreviations: IQR, interquartile range; BMI, body mass index; CCP, COVID-19 convalescent plasma. [1] Data available on 278 patients. [2] Data available on 324 patients. [3] Measured before convalescent plasma transfusion. [4] Protease inhibitors and remdesivir. [5] Low molecular weight heparin.

Almost all patients (388/405, 95.8%) were under heparin-based anticoagulation at the time of CCP infusion. A substantial proportion of them was also receiving corticosteroids (332/405, 82.0%) and antibiotics (264/405, 65.2%). Thirty-nine percent of patients (158/405) belonged to O blood type, a percentage (39%) that is slightly lower than that observed in the population of healthy blood donors from the same geographical area (43.6%) [14]. These results are in accordance with previous evidence of a protective effect of O blood type against COVID-19 [14].

Regarding the CCP-related data, the median time between symptom onset and CCP therapy was 7.5 days (IQR 5–12 days), and the mean number of CCP units infused per patient was 1.6 (±0.6) units. Overall, 657 CCP units, with a median nAb titer of 160 (IQR 80–320), were transfused to the 405 COVID-19 patients. Of them, 181 patients (44.7%) received 1 unit; 195 patients (48.1%), 2 units; 28 patients (6.9%), 3 units; and 1 patient (0.3%), 4 units. The median CCP neutralizing titer was 160 (IQR 80–320). Seventy-nine patients (19.5%) were treated during the first pandemic wave (driven by multiple D614G strains), while most of them (326/405, 80.5%) received CCP during the second/third pandemic wave (driven in Italy by the SARS-CoV-2 Alpha variant of concern).

The 28-day crude mortality rate among the 405 CCP-treated patients was 12.6% (51/405) (Figure 1). A slightly higher number of deaths was recorded during the first pandemic wave as compared with the second-third ones (27/51 (52.9%) versus 24/51 (47.1%), P = NS). A statistically significant difference in the overall mortality rate was observed between CCP-treated patients and the entire cohort of patients admitted to the city hospital of Mantova for COVID-19 during the same period (51/405 (12.6%) versus 510/2738 (18.6%), P 0.003) [15].

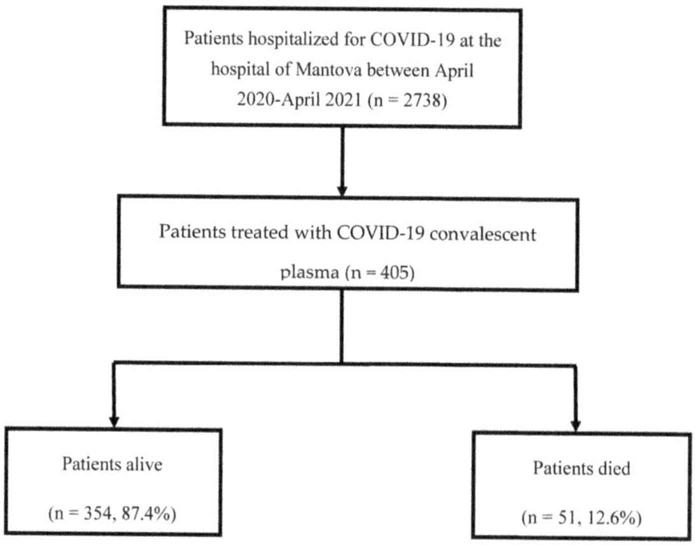

Figure 1. Flowchart of patients' enrollment.

Regarding the safety of CCP, 9 (1.3%) adverse reactions were recorded out of a total of 657 CCP units transfused. All cases were mild allergic reactions characterized by pruritus or rash, which rapidly faded with slowing of the CCP transfusion and after treatment with intravenous administration of antihistamine agents. In no case was it necessary to discontinue CCP transfusion. No cases of TRALI or other serious adverse events were recorded.

Table 2 reports a subgroup analysis that compared various variables in CCP-treated patients who survived (n = 354) or died (n = 51). Among deceased patients, 74.5% (38/51) were aged 68 years or older. At univariate analysis (died versus survived patients), death was significantly associated with an older median age (77 years versus 66 years, $p < 0.001$), a higher median BMI (31.1 Kg/m^2 versus 24.7 Kg/m^2, $p < 0.001$), a greater number of associated comorbidities (≥3 comorbidities: 74.2% versus 20.5%, $p < 0.001$), a more advanced disease (measured as median PaO2/FiO2 (92.0 versus 169.5, $p < 0.001$) and a higher intensity of hospital department (intermediate–high intensity: 49% versus 28%, $p = 0.004$)), in accordance with previous literature. In addition, deceased CCP-treated patients received CCP units later (16 days versus 7 days, $p < 0.001$) and with a less amount of nAb (mean nAb titer: 179.6 versus 227.2, $p = 0.04$) than alive patients. Interestingly, 82.3% (42/51) of deceased patients were transfused with CCP units with a nAb less than 320, while 90.2% (46/51) of them received CCP 7 days or more from the onset of symptoms. No deceased patient received CCP within 72 h from symptom onset. By contrast, no statistically significant difference between these two groups (alive and died) was observed regarding the sex and ABO blood group distribution and the mean number of CCP units transfused per patient.

Table 2. Subgroup analysis between CCP-treated patients alive and died.

Parameters	Alive n = 354 (87.4%)	Died n = 51 (12.6%)	*p*
Median age, years (IQR)	66 (55.2–76.3)	77 (67.8–81.8)	<0.001
Sex (males/females), number (ratio)	218/136 (1.6)	33/18 (1.8)	NS
Median BMI (kg/m^2) (IQR)	24.7 (22.9–28.6)	31.1 (27.5–35–5)	<0.001
Comorbidities, n (%)			
- <3	233/293 (79.5)	8/31 (25.8)	<0.001
- ≥3	60/293 (20.5)	23/31 (74.2)	
PaO2/FiO2, median (IQR)	169.5 (139.2–231.0)	92.0 (67.0–138.0)	< 0.001
PaO2/FiO2, n (%) [1]			
- <150	122/354 (34.5)	38/51 (74.5)	<0.001
- ≥150	232/354 (65.5)	13/51 (25.5)	
Hospital department, n (%)			
- Low intensity	255/354 (72.0)	26/51 (51.0)	0.004
- Intermediate/high intensity	99/354 (28.0)	25/51 (49.0)	
ABO blood type, n (%)			
- O	137/354 (38.7)	21/51 (41.2)	NS
- Non-O	217/354 (61.3)	30/51 (58.8)	
Days between symptoms onset and CCP therapy, median (IQR)	7 (4–10)	16 (11–29.5)	<0.001
Days between symptoms onset and CCP therapy			
- <7	126/354 (35.6)	5/51 (9.8)	<0.001
- ≥7	228/354 (64.4)	46/51 (90.2)	
Days between symptoms onset and CCP therapy			
- <3	39/354 (11.0)	0/51 (0)	0.01
- ≥3	315/354 (89.0)	51/51 (100)	
CCP units transfused, mean (±SD)	1.6 (±0.6)	1.7 (±0.8)	NS
CCP neutralizing titer, mean (±SD)	227.2 (±184.7)	179.6 (±170.3)	0.04
CCP neutralizing titer			
<160	116/354 (32.8)	23/51 (45.1)	NS
≥160	238/354 (67.2)	28/51 (54.9)	
CCP neutralizing titer			
<320	239/354 (67.5)	42/51 (82.3)	0.03
≥320	115/354 (32.5)	9/51 (17.7)	

Abbreviations: NS, not significant; CCP, COVID-19 convalescent plasma; BMI, body mass index; SD, standard deviation. [1] Measured before convalescent plasma transfusion.

The logistic regression model was statistically significant (Chi-squared test (5) = 43.102, $p < 0.001$). The model (Nagelkerke R^2) explained 19.0% of the variance in mortality and correctly classified 87.4% of cases. Of the five predictor variables analyzed (age ≥ or < 68 years, sex, ≥ or < 7 days from symptom onset and CCP transfusion, low or intermediate/high intensity of hospital department, nAb titer ≥ or < 320), four were statistically significant: age, days between symptoms onset and CCP transfusion, intensity of hospital department and CCP nAb titer (as shown in Table 3). Multivariate analysis was not possible for the variables "BMI" and "≥ or <3 comorbidities" because of the lack of data recorded. Patients with age over 68 years had 3.45 times higher odds of dying. Increasing days between symptoms onset and CCP transfusion (≥7 days) were associated with an increased likelihood of dying. Similarly, higher intensity of hospital department was associated with mortality, but increasing CCP nAb titer (≥320) was associated with a reduction in the likelihood of dying.

Table 3. Logistic regression predicting likelihood of death based on age, days between symptoms onset and CCP transfusion, intensity of hospital department and CCP neutralizing titer.

	B	**SE**	**Wald**	**df**	***p***	**OR**	**95% CI**	
Sex	−0.052	0.338	0.023	1	0.879	0.950	0.490	1.841
Age (≥68 years)	1.239	0.355	12.215	1	0.000	3.452	1.723	6.916
Days between symptoms onset and CCP transfusion (≥7 days)	1.589	0.496	10.260	1	0.001	4.897	1.853	12.945
Intensity of hospital department (high)	1.115	0.336	11.046	1	0.001	3.051	1.580	5.890
CCP neutralizing titer (≥320)	−0.829	0.412	4.052	1	0.044	0.437	0.195	.978
Constant	−4.141	0.590	49.246	1	0.000	0.016		

Abbreviations: OR, odds ratio; df, degrees of freedom; SE, standard error; CI, confidence interval.

4. Discussion

The first case of COVID-19 in Italy was diagnosed in the Lombardy region, in Codogno, on 20 February 2020. In the context of the region, there has subsequently been a diffusion of cases in a centrifugal direction starting from the Lodi area and from West to East. In this way, the first indigenous cases of the province of Mantua, at extreme East Lombardy, appeared a few days later (26 February). Thanks to this short delay, the health facilities of Mantua managed to develop an initial organizational adaptation to cope with the emergency by transforming the great majority of internal and surgery departments in COVID-19 areas [15]. In addition, to face the absolute lack at that time of effective treatments against this new serious infection, early efforts were directed towards the implementation of the collection of plasma from recovered individuals based on the first positive experiences from China and from previous pandemics [16–18]. We were the first center, along with the Hospital of Pavia, to use CCP in Italy and in western countries. Between the 12-month period (April 2020–April 2021) of the first three pandemic waves, more than 500 hundred CCP units were collected at the Transfusion center of Mantua hospital [19,20], allowing the treatment of 405 consecutive patients hospitalized for severe COVID-19 on the basis of a CCP compassionate use program.

The results of our cohort study are encouraging: the 12.6% mortality observed compares favorably with the overall mortality rate (18.6%) reported in COVID-19 patients hospitalized during the same period (April 2020–April 2021) in the same hospital [15]. Our data are in line with those recently published by the Registry of the Italian Veneto region [21], which confines with Lombardy region and Mantua province. During the one-year period (April 2020–April 2021), the investigators reported data analysis and clinical results in 1517 COVID-19 inpatients treated with high-titer CCP and observed a significantly reduced 30-day mortality in CCP-treated versus non-randomized, contemporary CCP-untreated patients (209/1517 (14%) versus 7700/31,071 (25%), $p < 0.001$). The mortality rate (13%) observed in the early experience on CCP EAP use in the USA [22], which led to the Emergency Use Authorization (EUA) by Food and Drug Administration (FDA), was very close to that reported in our study and the Veneto study.

Several factors, in our opinion, can be advocated to explain the discrepancy between mortality rates observed in real-life studies, as ours, and in RCTs, most of which failed to show a reduction in mortality and some discontinued for futility. Such reasons include the fact that CCP is not pharmaceutical but rather an artisanal product (it is produced at transfusion centers), and its nAb titer and absolute content in the cumulative volume varies widely [23]. Besides the CCP variability, differences in study design, patients' characteristics and disease severity could have played a role. Nevertheless, we would like to highlight that most of the published RCT showed signals of CCP efficacy, including reductions in mortality, when subgroup analyses were restricted to the early use of high-titer CCP [8].

Notably, nearly 80% of patients in our cohort received CCP during the second–third wave, and this fact could have contributed, along with the general improvement of the management of COVID-19 patients (i.e., widespread use of steroids and remdesivir, optimization of oxygen therapy, etc.) to the reduction of mortality for COVID-19 observed during the different waves in our hospital in low-intensity departments (from 21.9% in the first wave to 12.7% in the third wave) in a recently published analysis [15]. A similar finding, i.e., a strong inverse correlation between CCP use and mortality per hospital admission, was also observed in a recent publication reporting the US experience on EAP use of CCP in approximately 500,000 patients [24].

Multivariate analysis in our study showed that younger patients (<68 years) with less severe COVID-19 (i.e., admitted in low-intensity departments), where those who benefited the most from hyperimmune plasma-based therapy, particularly when it was administered early (<7 days from symptom onset) and at very high titer (>320). The latter finding is quite interesting as it confirms the most recent evidence from the literature showing a significant benefit of CCP among COVID-19 patients who received a larger amount of nAb [8,25].

Another noteworthy issue of our study is the safety of CCP transfusion. Adverse reactions, all of which were mild, were observed only in 1.3% of CCP transfused. This rate is similar to that reported in our previous meta-analysis on CCP-related adverse reactions [26]. All concerns about CCP safety have, however, been swept away by the recent results of the US Expanded Access Program (EAP), which reported an incidence of serious adverse events of less than 1% in over 100,000 CCP-treated patients, definitely proving the overall high safety profile of this passive immunotherapy [27].

In our study, we found a reduced incidence of O blood type among COVID-19 patients as compared with the control healthy population. By contrast, ABO blood type was not found to be associated with a favorable outcome in CCP-treated patients. These results are in line with the published literature data showing that O blood type protects against SARS-CoV-2 infection (the receptor-binding domain of SARS-CoV-2 preferentially binds the blood group A expressed on respiratory epithelial cells) [28] but not against the degree of COVID-19 severity [29].

In conclusion, based on our experience, the use of CCP in hospitalized COVID-19 patients was characterized by a high safety and efficacy profile, particularly when administered early and at high nAb titer. In this regard, as recently reported [30,31], the ideal clinical setting for high-titer CCP use is probably the treatment of early outpatients to minimize the risk of hospitalization. In this regard, the US FDA has updated its emergency use authorization on 27 December 2021, expanding use to outpatients (available at: https://www.fda.gov/media/141477/download, accessed on 28 December 2021). Our experience, in conjunction with that of other centers, could help to plan a proper CCP emergency use in the case of future SARS-CoV-2 variants of concern or pandemics from different pathogens.

Supplementary Materials: The following are available online at https://www.mdpi.com/article/10.3390/life12030420/s1.

Author Contributions: Conceptualization, M.F. (Massimo Franchini) and S.C.; data curation, C.G.; software and formal analysis, G.L.; writing—original draft preparation, M.F. (Massimo Franchini); writing—review and editing: G.D.D., M.B., M.A., G.P.C., L.B., M.P., M.G. (Marco Ghirardini), G.P., B.P., M.T.C., M.F. (Marilena Frigato), V.C. (Verena Crosato), G.T., A.M., D.A.P., F.I., F.S., M.G. (Martina Garuti), A.P., G.C., G.G. (Gianpaolo Grisolia), V.G., T.S., A.B., C.M., R.B., S.A.B., E.C., D.B., G.G. (Graziana Greco), F.P., R.S., M.Z., A.V., V.C. (Vito Codeluppi), S.V., E.B., M.C., G.A., F.C., C.S., A.R. and on behalf of Convalescent Plasma Study Group. All authors have read and agreed to the published version of the manuscript.

Funding: This research received no external funding.

Institutional Review Board Statement: The study was conducted according to the guidelines of the Declaration of Helsinki and approved by the Institutional Review Board of Azienda Socio Sanitaria Territoriale di Mantova (Protocol code 715 of 06/10/2020).

Informed Consent Statement: Informed consent was obtained from all subjects involved in the study.

Data Availability Statement: The data presented in this study are available on request from the corresponding author.

Acknowledgments: The authors acknowledge Fausto Baldanti and Cesare Perotti from the IRCCS Policlinico S. Matteo Pavia for their helpful collaboration. The authors acknowledge A.M.I.C.A. Association for its support.

Conflicts of Interest: The authors declare no conflict of interest.

References

1. Wang, D.; Hu, B.; Hu, C.; Zhu, F.; Liu, X.; Zhang, J.; Wang, B.; Xiang, H.; Cheng, Z.; Xiong, Y.; et al. Clinical Characteristics of 138 hospitalized patients with 2019 novel coronavirus-infected pneumonia in Wuhan, China. *JAMA* **2020**, *323*, 1061–1069. [CrossRef] [PubMed]
2. Wu, Z.; McGoogan, J.M. Characteristics of and important lessons from the coronavirus disease 2019 (COVID-19) outbreak in China: Summary of a report of 72 314 cases from the Chinese center for disease control and prevention. *JAMA* **2020**, *323*, 1239–1242. [CrossRef] [PubMed]
3. WHO. Coronavirus Disease (COVID-19). Available online: https://www.who.int/emergencies/diseases/novel-coronavirus-2019 (accessed on 31 August 2021).
4. Tobaiqy, M.; Qashqary, M.; Al-Dahery, S.; Mujallad, A.; Hershan, A.A.; Kamal, M.A.; Helmi, N. Therapeutic management of patients with COVID-19: A systematic review. *Infect Prev Pract.* **2020**, *2*, 100061. [CrossRef]
5. Heustess, A.M.; Allard, M.A.; Thompson, D.K.; Fasinu, P.S. Clinical management of COVID-19: A Review of pharmacological treatment options. *Pharmaceuticals* **2021**, *14*, 520. [CrossRef] [PubMed]
6. Franchini, M.; Liumbruno, G.M.; Piacentini, G.; Glingani, C.; Zaffanello, M. The three pillars of COVID-19 convalescent plasma therapy. *Life* **2021**, *11*, 354. [CrossRef] [PubMed]
7. Franchini, M.; Corsini, F.; Focosi, D.; Cruciani, M. Safety and efficacy of convalescent plasma in COVID-19: An overview of systematic reviews. *Diagnostics* **2021**, *11*, 1663. [CrossRef] [PubMed]
8. Focosi, D.; Franchini, M.; Pirofski, L.; Burnouf, T.; Paneth, N.; Joyner, M.J.; Casadevall, A. COVID-19 Convalescent Plasma and Clinical Trials: Understanding Conflicting Outcomes. *Clin. Microbiol. Rev.* **2022**, e0020021. [CrossRef] [PubMed]
9. Focosi, D.; Franchini, M. COVID-19 convalescent plasma therapy: Hit fast, hit hard! *Vox Sang* **2021**, *116*, 935–942. [CrossRef] [PubMed]
10. Franchini, M.; Marano, G.; Velati, C.; Pati, I.; Pupella, S.; Maria Liumbruno, G. Operational protocol for donation of anti-COVID-19 convalescent plasma in Italy. *Vox Sang.* **2021**, *116*, 136–137. [CrossRef] [PubMed]
11. Accorsi, P.; Berti, P.; de Angelis, V.; De Silvestro, G.; Mascaretti, L.; Ostuni, A. Position paper on the preparation of immune plasma to be used in the treatment of patients with COVID-19. *Transfus. Apher. Sci.* **2020**, *59*, 102817. [CrossRef]
12. Perotti, C.; Del Fante, C.; Baldanti, F.; Franchini, M.; Percivalle, E.; Nepita, E.V.; Seminari, E.; De Silvestri, A.; Bruno, R.; Klersy, C. Plasma from donors recovered from the new Coronavirus 2019 as therapy for critical patients with COVID-19 (COVID-19 plasma study): A multicentre study protocol. *Intern. Emerg. Med.* **2020**, *15*, 819–824. [CrossRef]
13. Perotti, C.; Baldanti, F.; Bruno, R.; Del Fante, C.; Seminari, E.; Casari, S.; Percivale, E.; Glingani, C.; Musella, V.; Belliato, V.; et al. Mortality reduction in 46 patients with severe COVID-19 treated with hyperimmune plasma. A proof-of-concept, single-arm, multicenter trial. *Haematologica* **2020**, *105*, 2834. [CrossRef]
14. Franchini, M.; Glingani, C.; Del Fante, C.; Capuzzo, M.; Di Stasi, V.; Rastrelli, G.; Vignozzi, L.; De Donno, G.; Perotti, C. The protective effect of O blood type against SARS-CoV-2 infection. *Vox Sang.* **2021**, *116*, 249–250. [CrossRef]

15. Casari, S.; Lucchini, G.; Burato, E.; Bertoletti, R.; Boscaini, R.; Bellani, A.; Lombardi, L.; Benedini, N.; Firriolo, E.; Vallicella, F.; et al. The COVID-19 Epidemic: The experience in the social and health care center of the Mantua District (Lombardy Region, Northern Italy). *Epidemiol. Prev.* **2021**, *54*, 559–567.
16. Ripoll, J.G.; van Helmond, N.; Senefeld, J.W.; Wiggins, C.C.; Klassen, S.A.; Baker, S.E.; Larson, K.F.; Murphy, B.M.; Andersen, K.J.; Ford, S.K.; et al. Convalescent plasma for infectious diseases: Historical framework and use in COVID-19. *Clin. Microbiol. Newsl.* **2021**, *15*, 23–32. [CrossRef] [PubMed]
17. Shen, C.; Wang, Z.; Zhao, F.; Yang, Y.; Li, J.; Yuan, J.; Wang, F.; Li, D.; Yang, M.; Xing, L.; et al. Treatment of 5 critically ill patients with COVID-19 with convalescent plasma. *JAMA* **2020**, *323*, 1582–1589. [CrossRef] [PubMed]
18. Mair-Jenkins, J.; Saavedra-Campos, M.; Baillie, J.K.; Cleary, P.; Khaw, F.-M.; Lim, W.S.; Makki, S.; Rooney, K.D.; Nguyen-Van-Tam, J.S.; Beck, C.R.; et al. The effectiveness of convalescent plasma and hyperimmune immunoglobulin for the treatment of severe acute respiratory infections of viral etiology: A systematic review and exploratory meta-analysis. *J. Infect. Dis.* **2015**, *211*, 80–90. [CrossRef]
19. Franchini, M.; Glingani, C. Safety of convalescent plasma transfusion. *Mayo Clin. Proc.* **2021**, *96*, 1368–1369. [CrossRef]
20. Del Fante, C.; Franchini, M.; Baldanti, F.; Percivalle, E.; Glingani, C.; Marano, G.; Mengoli, C.; Mortellaro, C.; Viarengo, G.; Perotti, C.; et al. A retrospective study assessing the characteristics of COVID-19 convalescent plasma donors and donations. *Transfusion* **2021**, *61*, 830–838. [CrossRef] [PubMed]
21. De Silvestro, G.; Marson, P.; La Raja, M.; Cattelan, A.M.; Guarnieri, G.; Monticelli, J.; Tiberio, I.; Vianello, A.; Gandini, G.; Gessoni, G.; et al. Veneto hospitals. outcome of SARS CoV-2 inpatients treated with convalescent plasma: One-year of data from the Veneto region (Italy) Registry. *Eur. J. Intern. Med.* **2021**, *97*, 42–49. [CrossRef]
22. Joyner, M.J.; Bruno, K.A.; Klassen, S.A.; Kunze, K.L.; Johnson, P.W.; Lesser, E.R.; Wiggins, C.C.; Senefeld, J.W.; Klompas, A.M.; Hodge, D.O.; et al. Safety update: COVID-19 convalescent plasma in 20,000 hospitalized patients. *Mayo Clin. Proc.* **2020**, *95*, 1888–1897. [CrossRef] [PubMed]
23. Franchini, M.; Mengoli, C.; Caruso, B.; Petilino, R.; Ballotari, A.; Glingani, C. Measuring accuracy of the neutralizing activity of COVID-19 convalescent plasma. *Clin. Chem. Lab. Med.* **2021**, *60*, e4–e6. [CrossRef] [PubMed]
24. Casadevall, A.; Dragotakes, Q.; Johnson, P.W.; Senefeld, J.W.; Klassen, S.A.; Wright, R.S.; Joyner, M.J.; Paneth, N.; Carter, R.E. Convalescent plasma use in the USA was inversely correlated with COVID-19 mortality. *eLife* **2021**, *10*, e69866. [CrossRef] [PubMed]
25. Körper, S.; Weiss, M.; Zickler, D.; Wiesmann, T.; Zacharowski, K.; Corman, V.M.; Grüner, B.; Ernst, L.; Spieth, P.; Lepper, P.M.; et al. CAPSID clinical trial group. Results of the CAPSID randomized trial for high-dose convalescent plasma in patients with severe COVID-19. *J. Clin. Investig.* **2021**, *131*, e152264. [CrossRef] [PubMed]
26. Franchini, M.; Cruciani, M. How safe is COVID-19 convalescent plasma? *Mayo Clin. Proc.* **2021**, *96*, 2279–2281. [CrossRef] [PubMed]
27. Senefeld, J.W.; Johnson, P.W.; Kunze, K.L.; Bloch, E.M.; van Helmond, N.; Golafshar, M.A.; Klassen, S.A.; Klompas, A.M.; Sexton, M.A.; Diaz Soto, J.C.; et al. Access to and safety of COVID-19 convalescent plasma in the United States expanded access program: A national registry study. *PLoS Med.* **2021**, *18*, e1003872. [CrossRef]
28. Wu, S.C.; Arthur, C.M.; Wang, J.; Verkerke, H.; Josephson, C.D.; Kalman, D.; Roback, J.D.; Cummings, R.D.; Stowell, S.R. The SARS-CoV-2 receptor-binding domain preferentially recognizes blood group A. *Blood Adv.* **2021**, *5*, 1305–1309. [CrossRef] [PubMed]
29. Franchini, M.; Cruciani, M.; Mengoli, C.; Marano, G.; Candura, F.; Lopez, N.; Pati, I.; Pupella, S.; De Angelis, V. ABO blood group and COVID-19: An updated systematic literature review and meta-analysis. *Blood Transfus.* **2021**, *19*, 317–326. [PubMed]
30. Franchini, M.; Glingani, C.; Morandi, M.; Corghi, G.; Cerzosimo, S.; Beduzzi, G.; Storti, A.; Di Stasi, V.; Rastrelli, G.; Vignozzi, L.; et al. Safety and efficacy of convalescent plasma in elderly COVID-19 patients: The RESCUE trial. *Mayo Clin. Proc. Innov. Qual. Outcomes* **2021**, *5*, 403–412. [CrossRef] [PubMed]
31. Sullivan, D.J.; Gebo, K.A.; Shoham, S.; Bloch, E.M.; Lau, B.; Shenoy, A.G.; Mosnaim, G.S.; Gniadek, T.J.; Fukuta, Y.; Patel, B.; et al. Randomized controlled trial of early outpatient COVID-19 treatment with high-titer convalescent plasma. *medRxiv*. 2021. Available online: https://www.medrxiv.org/content/10.1101/2021.12.10.21267485v1 (accessed on 19 February 2022).

life

MDPI

Article

SARS-CoV-2 Inactivation Simulation Using 14 MeV Neutron Irradiation

Fang Liu [1], Zhengtong Zhong [1], Bin Liu [1,*], Tianze Jiang [1], Hongchi Zhou [1], Guanda Li [1], Xin Yuan [1], Peiguang Yan [2], Fenglei Niu [1] and Xiaoping Ouyang [3,4]

1 Beijing Key Laboratory of Passive Safety Technology for Nuclear Energy, School of Nuclear Science and Engineering, North China Electric Power University, Beijing 102206, China; liuf@ncepu.edu.cn (F.L.); zhongzt@ncepu.edu.cn (Z.Z.); Jiangtz@ncepu.edu.cn (T.J.); zhouhongchi@ncepu.edu.cn (H.Z.); liguanda@ncepu.edu.cn (G.L.); yuanxin@ncepu.edu.cn (X.Y.); niufenglei@ncepu.edu.cn (F.N.)

2 Shenzhen Key Laboratory of Laser Engineering, College of Physics and Optoelectronic Engineering, Shenzhen University, Shenzhen 518060, China; yanpg@szu.edu.cn

3 Radiation Detection Research Center, Northwest Institute of Nuclear Technology, Xi'an 710024, China; oyxp2003@aliyun.com

4 Department of Engineering Physics, Tsinghua University, Beijing 100084, China

* Correspondence: liu_bin@ncepu.edu.cn; Tel.: +86-10-6177-1686

Abstract: The SARS-CoV-2 virus is deadly, contagious, can cause COVID-19 disease, and endangers public health and safety. The development of SARS-CoV-2 inactivation technology is crucial and imminent in current pandemic period. Neutron radiation is usually used to sterilize viruses because neutron radiation is 10 times more effective than gamma-rays in inactivating viruses. In this work we established a closed SARS-CoV-2 inactivation container model by the Monte Carlo method and simulated the inactivation performance by using several different neutrons sources. To study the effects of inactivation container factors, including the reflector thickness, the type of the reflector material, the SARS-CoV-2 layer area and the distance from the radiation source on the energy deposition of a single neutron particle in SARS-CoV-2 sample, we simulated the neutron energy deposition on a SARS-CoV-2 sample. The simulation results indicate that the saturated thicknesses of reflector materials for graphite, water and paraffin are approximately 30 cm, 15 cm, and 10 cm, respectively, and the energy deposition (radiation dose) becomes larger when the SARS-CoV-2 layer area is smaller and the SARS-CoV-2 layer is placed closer to the neutron source. The calculated single-neutron energy deposition on 10×10 cm^2 SARS-CoV-2 layer is about 3.0059×10^{-4} MeV/g with graphite as the reflection layer, when the 14 MeV neutron source intensity is 10^{12} n/s and the SARS-CoV-2 layer is 5 cm away from the neutron source. If the lethal dose of SARS-CoV-2 is assumed as the IAEA recommended reference dose, 25 kGy, the SARS-CoV-2 could be decontaminated in about 87 min, and the sterilization time could be less than 52 s if the 14 MeV neutron intensity is increased to 10^{14} n/s.

Keywords: SARS-CoV-2 inactivation; COVID-19; Monte Carlo simulation; neutron

Citation: Liu, F.; Zhong, Z.; Liu, B.; Jiang, T.; Zhou, H.; Li, G.; Yuan, X.; Yan, P.; Niu, F.; Ouyang, X. SARS-CoV-2 Inactivation Simulation Using 14 MeV Neutron Irradiation. *Life* **2021**, *11*, 1372. https://doi.org/10.3390/life11121372

Academic Editors: Silvia De Francia and Sarah Allegra

Received: 6 November 2021
Accepted: 7 December 2021
Published: 9 December 2021

1. Introduction

The severe acute respiratory syndrome coronavirus 2 (SARS-CoV-2) [1] is currently sweeping round the world and has caused the coronavirus disease 2019 (COVID-19) pneumonia (shown in Figure 1 [1]). This is a new strain of coronavirus, which is contagious, can spread rapidly and widely through carriers, and has threatened human lives globally.

Although the coronavirus was discovered in the 1930s [2], coronaviruses gained worldwide attention when the severe acute respiratory syndrome outbreak shook the world in 2003. Biologists' interest in this family of viruses grew in the aftermath of the epidemic of 2003, leading to the identification of many new coronavirus family members. The coronavirus break out in 2003 also gave a hint of the ability of coronaviruses to jump across species. Before gaining worldwide attention for threatening public health in 2003, the diseases associated with coronaviruses were mainly of veterinary interest. Coronaviruses can infect a wide variety of creatures, such as mammals and birds [3], causing respiratory and enteric diseases and, in some rarer cases, hepatitis and neurologic disease. Infection can be acute or persistent.

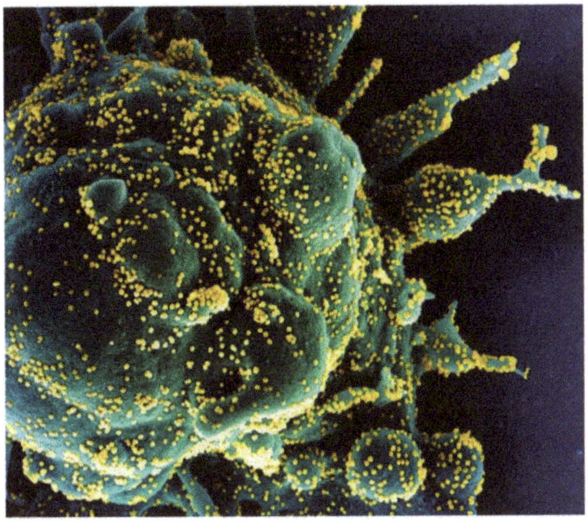

Figure 1. Scanning electron micrograph of a cell (green) heavily infected with SARS-CoV-2 virus particles (yellow) [1].

Coronaviruses are enveloped, spherical or pleiomorphic viruses, with a radius of approximately 60 nm. The most distinctive feature of coronaviruses is the club-shaped spike projections emanating from the surface of the virion. These spikes are a defining feature of the virion and give them the appearance of a corona, giving the name, coronaviruses. Within the shell of the virion is the nucleocapsid. Coronaviruses have helically symmetrical nucleocapsids, which is uncommon among positive-sense RNA viruses, but far more common in negative-sense RNA viruses. The coronavirus in the current pandemic can spread from an infected individual's mouth or nose in small liquid particles when they cough, sneeze, speak, sing or breathe in closed or open areas. These particles range from larger respiratory droplets to smaller aerosols. It is important to practice respiratory etiquette, for example by coughing into a flexed elbow, and if an infected person feels unwell, to stay home and self-isolate until they recover [4,5].

Most people infected with the coronavirus will experience mild to moderate respiratory illness and recover without requiring special medical treatment. However, some will become seriously ill and require medical attention. Older people and those with underlying medical conditions like cardiovascular disease, diabetes, chronic respiratory disease, or cancer are more likely to develop serious illness. However, any person, at any age, can get sick with coronavirus and become seriously ill or die. The coronavirus pandemic around world has caused millions of deaths as well as lasting health problems in some individuals who have survived the illness. Although COVID-19 vaccines have been authorized for emergency use by the most countries and vaccination programs are in progress across the many parts of the world, the development of SARS-CoV-2 inactivation technology is crucial and imminent in current pandemic period. Effective vaccines against SARS-CoV-2 in humans are still in research and development and some of the promising vaccines have been pre-approved and used in some countries [6,7]. When bulk goods, such as luggage, medical instrumentations, tools, etc., are contaminated by SARS-CoV-2 in public areas, the traditional medical procedures, including pasteurization, alcohol and ultraviolet irradiation, which are generally used to kill common pathogenic bacteria [8], just do not work. Using alcohol to sterilize SARS-CoV-2 contamination is only feasible in some cases of surface contamination; it does not work effectively against SARS-CoV-2 hidden inside sealed containers, luggage, food, and so on.

Compared with traditional sterilization methods, radiation sterilization technology [9] has the advantages of short time, high efficiency, no damage, low energy consumption, and suitability for large-scale sterilization. There are chemical and physical techniques for virus inactivation and radiation disinfection or sterilization is a physical method, acting mainly through the inactivation of viruses by gamma rays and electron beam irradiation [10]. There are many studies using irradiation technologies such as X-rays, neutron irradiation, etc., to inactivate different viruses [11–13].

The γ-ray irradiation technique has been used to kill bacteria and anthrax hidden in sealed metal equipment or large luggage because the γ-ray irradiation is more penetrative than electron beams [10]. However, theoretical study showed that the neutron irradiation yielded higher sterilization efficiency for anthrax spores than γ-ray irradiation. Neutrons can penetrate sealed equipment to kill both anthrax spores, not only on surfaces, but also those hidden inside bulk goods or luggage, because neutrons have no charge and can penetrate. Simulation study has shown that 2.5 MeV neutron irradiation from a D-D neutron generator can sterilize all anthrax spores in a sample within approximately 1 min [14].

The radiation particle type and energy affect the irradiation inactivation efficiency significantly because they have quite different weighting factors. The radiation weighting factor represents the relative radiation damage to the tissue or organ resulting from the unit deposition energy. The radiation damage depends not only on the deposition energy, but also on the radiation type and energy. The radiation weighting factor of photon and electron particles is 1, which is independent of the energy of the radiation. However, for neutron radiation, the weighting factor is energy-dependent, and its value may be from 5 to 20. The weighting factor of 14 MeV neutrons is 10, compared with electrons and γ-rays, the weighting factor of which is 1, meaning that the neutrons can cause 10 times the damage to organisms that γ-rays can for the same energy deposition or radiation dose. The virus inactivation is mainly caused by damage from the irradiation to the viral nucleic acid, including RNA and DNA. A single-strand break (for single-stranded viruses) or a double-strand break led by irradiation damage is enough to kill the viruses [15].

Several different neutron sources can create enough neutron irradiation to sterilize all kind of viruses. The common neutron sources used to sterilize the virus contamination are radioisotope neutron sources and monoenergetic neutron generators. There are many radioisotope neutron sources, the typical radioisotope neutrons sources used in nuclear technology are ^{252}Cf (Californium-252 neutron source) and ^{241}Am-Be (Americium beryllium neutron source). The half-life of ^{252}Cf neutron source is about 2.6 years, while the half-life of ^{241}Am-Be is 432.2 years. The neutron energy spectrum emitted from ^{252}Cf neutron source is similar to the fission neutron spectrum, with 2 MeV average neutron energy. ^{241}Am-Be neutron source emits both neutrons and α particles, with 4.5 MeV average neutron energy.

Neutron generators are electronic devices that contain compact linear particle accelerators and produce neutrons by fusing isotopes of hydrogen atoms together. The fusion reactions take place in these electronic devices by accelerating either deuterium, tritium, or a mixture of these two isotopes into a metal hydride target which also contains deuterium, tritium, or a mixture of these isotopes. There are two different types of monoenergetic neutron generators, D-D neutron generators and T–D neutron generators [16]. Fusion of deuterium atoms (D + D) results in the formation of a He-3 ion and a neutron with a kinetic energy of approximately 2.5 MeV. Fusion of a deuterium and a tritium atom (D + T) results in the formation of a He-4 ion and a neutron with a kinetic energy of approximately 14.1 MeV [17]. Neutron generators have wide applications in physics, nuclear technology, materials analysis, and medicine.

The use of neutron irradiation in the treatment of food and food packaging is regulated by the U.S. Food & Drug Administration (FDA), which controls the energy of neutrons used for inspection of food and packaged food ranging from 1 MeV to 14 MeV [18]. In the coronavirus inactivation simulation calculations, we choose a 14 MeV neutron generator to sterilize SARS-CoV-2, because it has high neutron yield and monochromatic energy spectrum and can generate pulsed neutrons. The 14 MeV neutron generator can be turned off to stop neutron generation, so it is easy to shield, store and transport. We set up a SARS-CoV-2 inactivation model and placed the neutron source above the SARS-CoV-2 layer at a height of 5 cm. In MCNP simulation calculations, we consider the effects of the simulation set-up components, including the reflector material, the reflector thickness, the SARS-CoV-2 layer area, and the distance between the samples and the neutron source [16].

In this study, we set up a simulation model by MCNP code [19] and calculated the single-neutron energy deposition, the effect of the reflector materials and reflector thickness, and the incident neutron energy that is needed to sterilize SARS-CoV-2 contamination. The simulation results present the effect of reflector materials, reflector thickness on the neutron energy deposition in SARS-CoV-2 layer and illustrate the available estimated sterilization time using 14 MeV neutron generator.

2. Materials and Methods

The SARS-CoV-2 virion is composed of 29 proteins in total including 16 non-structural proteins and nine accessory proteins [20], which are hydrogen (proton)-rich materials. Neutrons and hydrogen atoms have high nuclear interaction cross sections, which indicates the possibility of interaction between neutrons and hydrogen atoms is high and neutrons can cause damage to SARS-CoV-2 DNA strands. SARS-CoV-2 belongs to the coronavirus family that includes the other well-known viruses SARS-CoV and MERS-CoV [21], and is an enveloped, single-stranded, positive-sense RNA virus of similar size to other coronaviruses [20].

RNA has a single chain structure that is easy to break and recombine. When neutrons hit the SARS-CoV-2 virus, they will interact with the nuclei of the hydrogen, oxygen and carbon atoms within the proteins of the SARS-CoV-2 virion. The neutrons tend to knock protons out of hydrogen nuclei meanwhile and transfer their energy to the protons. The scattered protons are ionizing radiation and interact with biological molecules as they go through the SARS-CoV-2 layer. In this way, the protons deposit their energy to biological molecules along their tracks by ionizing or exciting surrounding viral molecules, and the deposited energy induces the biological damage. The damage severity is directly related to the local rate of energy deposition along the protons penetration track [17]. The biological damage can cause the single chain of RNA to break. The viruses lack enzymes and are therefore unable to repair any damage in their RNA and DNA [22]. This is the reason why we decide to use penetrating neutron radiation to sterilize SARS-CoV-2 contamination.

We set up a virus inactivation model shown in Figure 2 to simulate inactivation of SARS-CoV-2 using 14 MeV neutrons by MCNP code. This model consists of a stainless-steel container, a surrounding reflector layer, a SARS-CoV-2 layer, and 14 MeV neutron sources. We used water, paraffin, and graphite as the reflector materials because these three materials are hydrogen-rich and good for reflecting and "slowing down" neutrons.

Figure 2 shows the simulation model of neutron irradiation inactivation of SARS-CoV-2. In this model, the SARS-CoV-2 samples are supported by a stainless-steel frame at the bottom of the container; The SARS-CoV-2 layer area is 10×10 cm^2 and the thickness of the SARS-CoV-2 layer is 1 mm. We used a 14 MeV neutron source in the simulation model.

For the simulation calculations, we set up a simulation model in MCNP code to calculate the average energy deposition in the SARS-CoV-2 layer. For the simulation calculations, we wrote the input file in MCNP code to calculate the single-neutron energy deposition in the SARS-CoV-2 sample. The neutrons interact with the SARS-CoV-2 sample and deposit their energy within the SARS-CoV-2 sample mainly through neutron–proton collisions. We calculated the single-neutron energy deposition in the SARS-CoV-2 layer with different reflector thicknesses, various reflector materials, and different distances between the SARS-CoV-2 sample and the neutron source. In the simulation, we ignored the energy deposition of γ-ray irradiation which is created by the neutrons' interaction with the reflector materials.

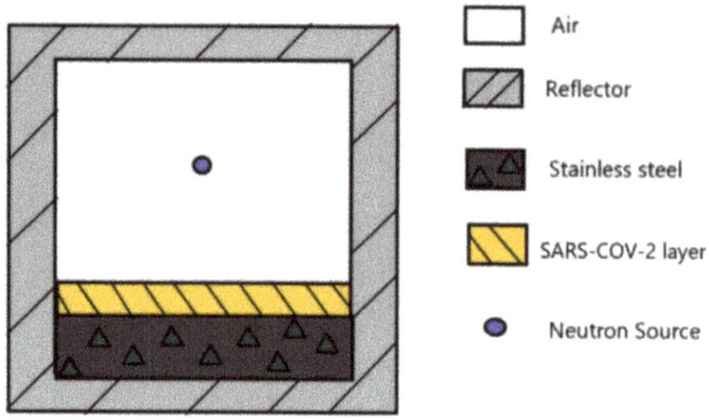

Figure 2. The SARS-CoV-2 inactivation simulation model.

3. Results and Discussions

We calculated the average energy deposition per particle in the SARS-CoV-2 sample using 14 MeV and 2.5 MeV neutrons, and 1.33 MeV gamma rays with the source 5 cm away from the SARS-CoV-2 sample, as shown in Figure 3. The calculation results show that the 14 MeV neutrons deposit more energy than the other two irradiation sources, therefore we choose 14 MeV neutrons to irradiate SARS-CoV-2 sample in the simulation model.

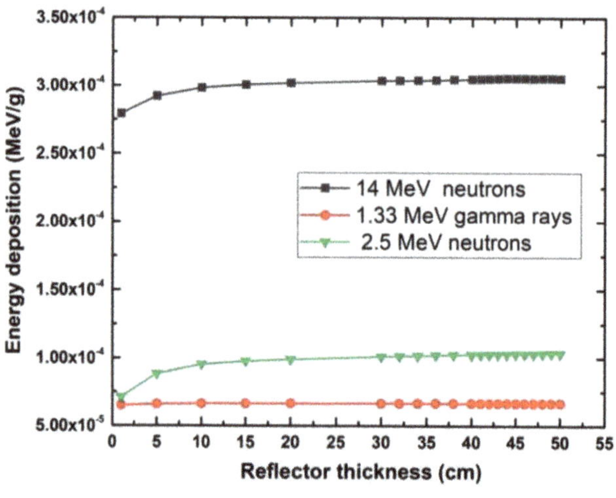

Figure 3. Single-energy deposition in the SARS-CoV-2 layer using three irradiation sources with graphite as the reflector.

Figure 4 shows that the reflector materials had an effect on the neutron energy deposition. Figure 5a–c show the effect of the distance between the neutron source and the SARS-CoV-2 sample on the neutron energy deposition with water, graphite, and paraffin as reflectors, respectively.

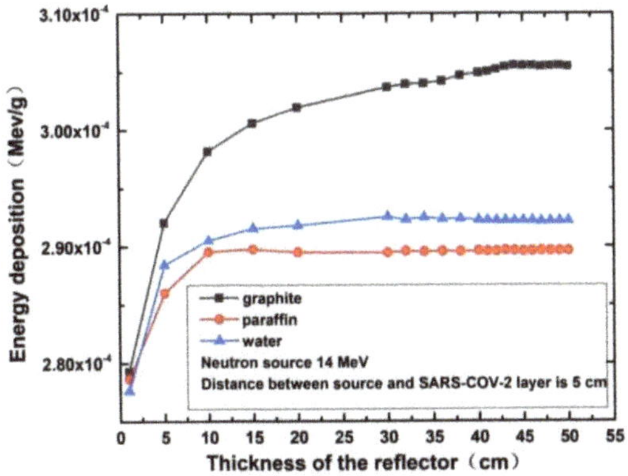

Figure 4. The single-energy deposition on SARS-CoV-2 layer for three reflecting materials.

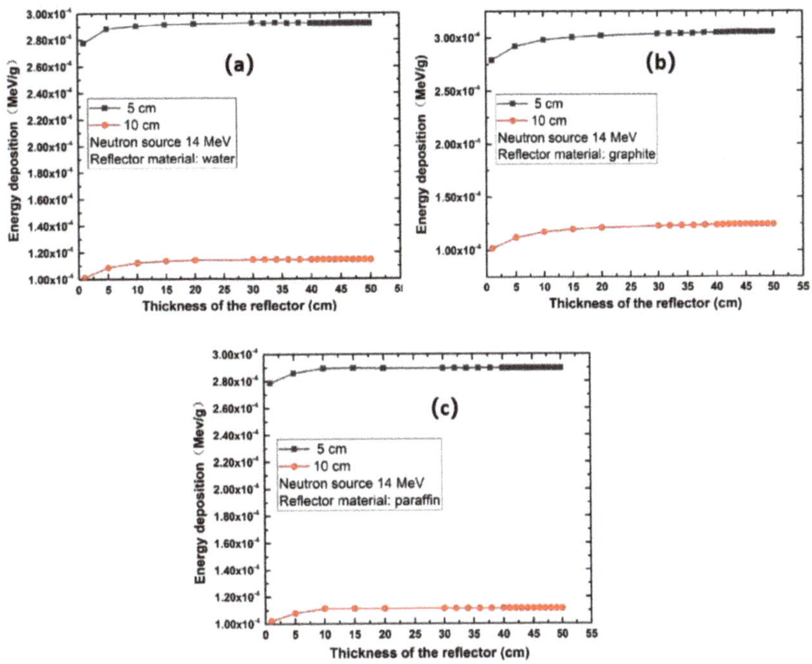

Figure 5. The single-energy deposition on SARS-CoV-2 layer using (**a**) water, (**b**) graphite, and (**c**) paraffin reflector.

3.1. The Effect of the Reflector Materials and Reflector Thickness on the Neutron Energy Deposition in the SARS-CoV-2 Layer

The simulation results of energy deposition using 14 MeV neutrons for three reflector materials are shown in Figure 4. We can see that the single-neutron energy deposition increases rapidly when the reflector thickness is less than 10 cm for the three materials used (water, paraffin or graphite). However, the single-neutron energy deposition shows almost no further increase when the thickness of reflector materials reaches a certain value; we call this reflector thickness the saturated thickness. Therefore, using reflector materials thicker than the saturated thickness to increase the single-neutron energy deposition is ineffective.

The results in Figure 4 show that the saturated thickness varies with different reflector materials. The saturated thicknesses of graphite, water, and paraffin are approximately 30 cm, 15 cm, and 10 cm, respectively. In the anthrax sterilization research using 2.5 MeV neutrons radiation, the saturated thickness was about 15 cm [17]. The saturated thickness is not closely related to the distance between the neutron source and the SARS-CoV-2 sample. From Figure 5a–c, we can see that the saturated thicknesses for graphite, water, and paraffin reflecting materials at 5 cm and 10 cm distances are still about 30 cm, 15 cm, and 10 cm, respectively, which shows that the saturated thickness is independent of the distance between the neutron source and the SARS-CoV-2 sample. This independence phenomenon also appeared in the anthrax sterilization work.

3.2. The Effect of the Distance between the Neutron Source and SARS-CoV-2 Sample on the Neutron Energy Deposition

We can see from the results shown in Figure 5a–c that the single-neutron energy deposition in the SARS-CoV-2 sample becomes greater when the SARS-CoV-2 layer is set much closer to the neutron source. At the same time, we find that these distance-related results are somewhat off the inverse-square law. However, the result is still reasonable because the SARS-CoV-2 sample was in a square shape, meaning each protein molecule had different distance to the neutron source, and we also used the reflector in the model. Both these factors can cause the neutron energy deposition to be off the inverse-square law somewhat. In real neutron-based inactivation of SARS-CoV-2 experimental set-up, it would be necessary to place the neutron source as close as possible to the SARS-CoV-2 area and to use a reflector to increase the energy deposition in the SARS-CoV-2 layer.

3.3. The Effect of the SARS-CoV-2 Layer Area on the Neutron Energy Deposition

Figure 6 illustrates the energy deposition on SARS-CoV-2 samples with different layer areas using 14 MeV neutrons and the graphite reflector. It is clear that the energy deposition in the SARS-CoV-2 sample with an area of 20×20 cm^2 is less than that of the 10×10 cm^2 SARS-CoV-2 layer. This is because the average distance between the protein molecules in the SARS-CoV-2 layer and the neutron source becomes larger with the increase in the SARS-CoV-2 layer area, and the lower average neutron penetration into the SARS-CoV-2 layer leads to lower energy deposition. Therefore, we suggest that medical products contaminated by SARS-CoV-2 should be placed around the neutron source in a circle to improve the sterilization efficiency.

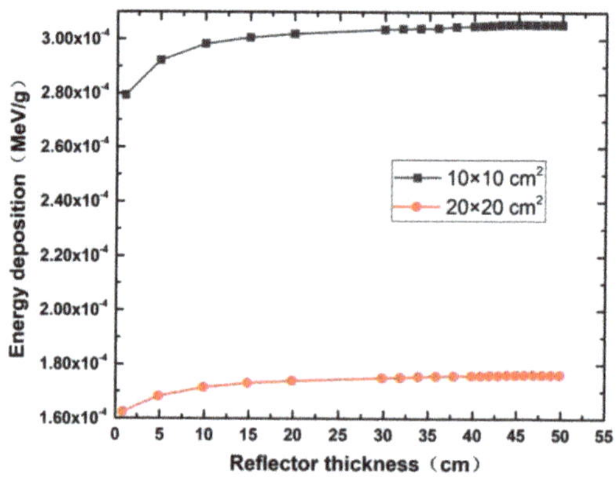

Figure 6. The single-energy deposition on SARS-CoV-2 samples with different layer areas using the graphite reflector.

3.4. *The Sterilization Time Estimation of SARS-CoV-2 Using Neutron Irradiation*

The virus damage induced by ionizing radiation is due to chemical alteration of the biological molecules, by the ionization or excitation caused by their interaction with the radiation. In the ionization or excitation process, the ionizing radiation particles deposit their energy onto the biological molecules. When a neutron collides with a proton in SARS-CoV-2, it transfers all its energy to the proton because a neutron has the same mass as a hydrogen nucleus (proton), thus it deposits all its energy in the SARS-CoV-2 virion. Therefore, compared with other forms of radiation, neutrons can create greater biological damage to SARS-CoV-2. Neutrons also collide with the oxygen and carbon nuclei in SARS-CoV-2. However, the oxygen and carbon nuclei are more than 10 times heavier than a neutron; therefore, during collision only a small part of the neutron's energy transfers to the oxygen and carbon nuclei. The energy transfer in neutron–proton collisions dominates the energy transfers to SARS-CoV-2 when irradiated by neutrons, so the neutron–proton collision is the dominant biological damage mechanism within SARS-CoV-2.

The inactivation cross-section of viruses relates to the irradiation dose. The typical survival curves of viruses [23] show that the survival rates of C16 bacteriophages [13] and the foot-and-mouth-disease picornavirus (FMDV) [24] reach the in order of 1.0×10^{-2} when the irradiation dose is 2 kGy and 10 kGy, respectively. The study of γ-ray irradiation used to disinfect anthrax suggested that a dose of 2.0×10^{6} rad was recommended to kill the most anthrax. The SI equivalent of 2.0×10^{6} rad is 20 kGy [9].

The International Atomic Energy Agency (IAEA) has suggested that when the pollution level and types of contaminated microorganisms cannot be confirmed, the standard irradiation dose can be set to 25 kGy [25]. The choice of 25 kGy for sterilization using gamma radiation was first suggested in 1959 by Artandli. The dose was proposed based on minimum killing dose of medical products for about 150 microbial species. 25 kGy was selected as the dose for sterilization because it is 40% above the minimum dose required to kill the resistant microorganisms [26]. Accordingly, 25 kGy is the minimum irradiation dose established for sterilization. Radiation sterilization at a dose of 25 kGy provides such a high safety factor that testing for sterility is generally considered superfluous. Therefore, it is reasonable to use the reference dose 25 kGy as the lethal dose for SARS-COV-2. When reference dose irradiated by γ rays is 25 kGy, the lethal dose for SARS-COV-2 irradiated by neutrons becomes 2.5 kGy, because the weighting factor of neutrons is 10 and γ rays' is 1. Thus, the necessary dose using neutrons irradiation to kill SARS-COV-2 virus is $25/10 = 2.5$ kGy and the no survival reference dose level D for SARS CoV-2 using neutron irradiation is:

$$D = 2500 \text{ Gy} \tag{1}$$

It can be seen from the simulation data shown in Figures 2–6 that the highest damage to the SARS-COV-2 caused by neutrons is with the 14 MeV neutron generator as the radiation source, graphite as the reflection layer (saturated thickness), the SARS-COV-2 layer area of 10×10 cm^2, the distance between neutron source and SARS-CoV-2 layer of 5 cm. Under these conditions, the energy deposition of a single neutron in SARS-COV-2 is about 3.0059×10^{-4} MeV/g (see Figure 6 for detail). Using this, we calculated the time needed to kill the SARS-CoV-2.

The estimated sterilization time of the SARS-COV-2 by using 14 MeV neutron irradiation can be calculated as [27]:

$$t = \frac{D(\text{Gy})}{d(\text{MeV/g}) \times 10^{6} \times 1.6 \times 10^{-19} \times 10^{3} \times Q \times A(\text{n/s}) \times 60} (\text{Minutes}) \tag{2}$$

where, D is the assumed lethal dose for SARS-COV-2, which is about 2500 Gy (2500 J/kg) provied in Equation (1); Q is the neutron weighting factor, which for 14.0 MeV energy is 10; d is the single-neutron energy deposition in the SARS-COV-2; and A is the intensity of radioactive neutrons.

The intensity of the neutron source A in this study is 10^{12} n/s, and the 14 MeV neutron energy deposition d is about 3.0059×10^{-4} MeV/g in Figure 6. We inserted these A and d parameters into Equation (2), and the estimated sterilization time is 87 min. The calculated sterilization time shrinks to about 52 s if the neutron intensity increases to 10^{14} n/s, which is feasible because a D-D neutron generator with 10^{14} n/s neutron intensity has been successfully developed [28].

4. Conclusions

In this paper, the effect of reflector materials, reflector thickness and the SARS-CoV-2 layer area on the neutron energy deposition in a SARS-CoV-2 layer are studied. In order to focus on the neutron energy deposition in the SARS-CoV-2 layer when the neutron irradiation is used to sterilize the SARS-CoV-2 and minimize the time of the simulation calculation, the energy deposition in the SARS-CoV-2 by other means, such as gamma rays from the surrounding reflector material was ignored when we used MCNP code to do the simulation calculation.

As can be seen from the simulation data, the greater the thickness of the reflector layer is, the more neutron energy is deposited. The neutron energy deposition in the SARS-CoV-2 sample increases with increasing thickness of the reflector, when the thickness of the reflector reaches the saturated thickness, further increases in the thickness of the reflector have little impact on the neutron energy deposition.

The different reflector materials also have different effects on the neutron energy deposition in the SARS-CoV-2 sample. Among all the three reflector materials, graphite gives the maximum neutron energy deposition in the SARS-CoV-2 sample. The saturated thickness of paraffin is only 33% that of graphite, and water is only 50% of graphite. Considering all the factors that affect the neutron energy deposition in the SARS-CoV-2 sample, graphite has the best reflecting effect to sterilize the SARS-CoV-2 viruses, but it has a high density, and is therefore hard to transport. Although water's neutron-reflecting effect is not as good as that of the graphite, water is readily available anywhere and anytime, thus water may be the ideal reflector material for sterilization of SARS-CoV-2 contamination.

Our calculation results also show that smaller the area of the SARS-CoV-2 layer is, the greater the average energy deposition. This is because the average distance between the protein molecules in SARS-CoV-2 layer and neutron source becomes greater as the SARS-CoV-2 layer area increases, and fewer neutrons penetrate the SARS-CoV-2 layer, leading to lower average neutron energy deposition in the SARS-CoV-2 sample.

In our simulation calculation, a 14 MeV neutron generator was used as the radiation source, the sample area of the SARS-CoV-2 layer was 10×10 cm^2, and the distance between the neutron source and the SARS-CoV-2 sample was set to 5 cm. The single-neutron energy deposition in the SARS-CoV-2 sample was about 3.0059×10^{-4} MeV/g. The lethal dose of the SARS-CoV-2 is 25 kGy. Our calculation results show that SARS-CoV-2 can be completely sterilized by 14 MeV neutron irradiation within about 87 min, and the estimated sterilization time shrinks to about 52 s if the neutron intensity is increased to 10^{14} n/s.

Author Contributions: Conceptualization, F.L. and B.L.; methodology, B.L. and Z.Z.; software, Z.Z. and T.J.; validation, Z.Z., H.Z., G.L., and X.Y.; formal analysis, F.L. and B.L.; investigation, F.L. and Z.Z.; resources, F.N.; data curation, Z.Z.; writing—original draft preparation, F.L. and Z.Z.; writing—review and editing, B.L. and P.Y.; visualization, B.L.; supervision, F.L.; project administration, X.O.; funding acquisition, F.L. All authors have read and agreed to the published version of the manuscript.

Funding: This work was supported by the Project of National Natural Science Foundation of China (Grant No. 11405055).

Institutional Review Board Statement: Not applicable.

Informed Consent Statement: Not applicable.

Data Availability Statement: Available on request.

Conflicts of Interest: The authors declare no conflict of interest.

References

1. Kramer, D. Beating back the coronavirus requires a bigger arsenal. *Phys. Today* **2021**, *74*, 20. [CrossRef]
2. Hudson, C.B.; Beaudette, F.R. Infection of the cloaca with the virus of infectious bronchitis. *Science* **1932**, *76*, 34. [CrossRef] [PubMed]
3. Casais, R.; Dove, B.; Cavanagh, D.; Britton, P. Recombinant avian infectious bronchitis virus expressing a heterologous spike gene demonstrates that the spike protein is a determinant of cell tropism. *J. Virol.* **2003**, *77*, 9084–9089. [CrossRef] [PubMed]
4. Perlman, S.; Netland, J. Coronaviruses post-SARS: Update on replication and pathogenesis. *Nat. Rev. Microbiol.* **2009**, *7*, 439–450. [CrossRef]
5. Enjuanes, L.; Almazan, F.; Sola, I.; Zuniga, S. Biochemical aspects of coronavirus replication and virus-host interaction. *Annu. Rev. Microbiol.* **2006**, *60*, 211–230. [CrossRef]
6. Gao, Q.; Bao, L.; Mao, H.; Wang, L.; Xu, K.; Yang, M.; Li, Y.; Zhu, L.; Wang, N.; Gao, H.; et al. Development of an inactivated vaccine candidate for SARS-CoV-2. *Science* **2020**, *369*, 77–81. [CrossRef] [PubMed]
7. Wang, H.; Zhang, Y.; Huang, B.; Deng, W.; Quan, Y.; Wang, W.; Xu, W.; Zhao, Y.; Li, N.; Zhang, J.; et al. Development of an inactivated vaccine candidate, BBIBP-CorV, with potent protection against SARS-CoV-2. *Cell* **2020**, *182*, 713–721. [CrossRef] [PubMed]
8. Sharp, D. The lethal action of short ultraviolet rays on several common pathogenic bacteria. *J. Bacteriol.* **1939**, *37*, 447–460. [CrossRef]
9. Horne, T.; Turner, G.; Willis, A. Inactivation of spores of Bacillus anthracis by G-radiation. *Nature* **1959**, *4659*, 475–476. [CrossRef] [PubMed]
10. Knoll, G. *Radiation and Detection and Measurement*, 3rd ed.; Wiley & Sons: New York, NY, USA, 1999.
11. Bachofer, C.S. Direct Effect of X-rays on bacterial viruses, modified by physical state, in relation to the target theory. *Science* **1953**, *117*, 280–282. [CrossRef] [PubMed]
12. Pollard, E. The action of ionizing radiation on viruses. *Adv. Virus Res.* **1954**, *2*, 109–151. [CrossRef]
13. Wollman, E.; Holweck, F.; Luria, S. Effect of radiations on bacteriophage C16. *Nature* **1940**, *145*, 935–936. [CrossRef]
14. Liu, B.; Wang, Q. Neutron Based Sterilization of Anthrax Contamination. *Health Phys.* **2006**, *90*, 80–84. [CrossRef] [PubMed]
15. Gialanella, G.; Grossi, G.F.; Macchiato, M.F.; Napolitano, M.; Speranza, P.R. Contributions of various types of damage to inactivation of T4 bacteriophage by protons. *Radiat. Res.* **1983**, *96*, 462–475. [CrossRef]
16. Reijonen, J.; Leung, K.N.; Firestone, R.B.; English, J.A.; Perry, D.L.; Smith, A.; Gicquel, F.; Sun, M.; Koivunoro, H.; Lou, T.-P.; et al. First PGAA and NAA experimental results from a compact high intensity D-D neutron generator. *Nucl. Inst. Methods* **2004**, *A511*, 598–602. [CrossRef]
17. Liu, B.; Xu, J.; Liu, T.; Ouyang, X. Monte Carlo N-particle simulation of neutron-based sterilization of anthrax contamination. *Natl. Libr. Med.* **2012**, *85*, e925–e932.
18. Department of Health and Human Services. FDA Title 21—Food and Drugs, Chapter I—Food and Drugs Administration Subchanpter B Food for Human Consumption, part 179 Irradiation in the Production, Processing and Handling of Food. Available online: https://www.accessdata.fda.gov/scripts/cdrh/cfdocs/cfcfr/CFRSearch.cfm?CFRPart=179&showFR---1 (accessed on 13 October 2021).
19. Cousins, C.; Bernardi, G.; Rehani, M.; International Commission on Radiological Protection. *1990 Recommendations of the International Commission on Radiological Protection*; ICRP Publication 60; Pergamon Press: Oxford, UK, 1991.
20. Michel, C.J.; Mayer, C.; Poch, O.; Thompson, J.D. Characterization of accessory genes in coronavirus genomes. *Virol. J.* **2020**, *17*, 131. [CrossRef] [PubMed]
21. Klein, S.; Cortese, M.; Winter, S.L.; Wachsmuth-Melm, M.; Neufeldt, C.J.; Cerikan, B.; Stanifer, M.L.; Boulant, S.; Bartenschlager, R.; Chlanda, P. SARS-CoV-2 structure and replication characterized by in situ cryo-electron tomography. *Nat. Commun.* **2020**, *11*, 5885. [CrossRef]
22. Lea, D.E. *Action of Radiations on Living Cells*; Cambridge University Press: New York, NY, USA, 1947.
23. Durante, M.; Schulze, K.; Incerti, S.; Francis, Z.; Zein, S.; Guzmán, C.A. Virus Irradiation and COVID-19 Disease. *Front. Phys.* **2020**, *8*, 565861. [CrossRef]
24. Smolko, E.E.; Lombardo, J.H. Virus inactivation studies using ion beams, electron and gamma irradiation. *Nucl. Instrum. Methods Phys. Res., B.* **2005**, *236*, 249–253. [CrossRef]
25. Tallentire, A.; IAEA. *Guidelines for Industrial Radiation Sterilization of Disposable Medical Products*; IAEA-TECDOC-539; International Atomic Energy Agency: Vienna, Austria, 1990.
26. Tallentire, A. The Spectrum of Microbial Radiation Sensitivity. *Radiat. Phys. Chem.* **1980**, *15*, 83–89. [CrossRef]
27. Xu, J.; Liu, B.; Wang, K.; Hu, W. MCNP Simulation of Neutron Generator Sterilization of Anthrax Contamination. *Chin. J. Radiol. Health* **2012**, *21*.
28. Leung, K.N.; Mok, P. A capacitor-free CMOS low-dropout regulator with damping-factor-control frequency compensation. *IEEE J. Solid-State Circuits* **2003**, *38*, 1691–1702. [CrossRef]

life

MDPI

Review

Safety of COVID-19 Vaccines: Spotlight on Neurological Complications

Giacomo Tondo [1], Eleonora Virgilio [2], Andrea Naldi [3,4], Angelo Bianchi [2] and Cristoforo Comi [1,*]

1 Neurology Unit, Department of Translational Medicine, S. Andrea Hospital, University of Piemonte Orientale, 13100 Vercelli, Italy
2 Neurology Unit, Department of Translational Medicine, Maggiore della Carità Hospital, University of Piemonte Orientale, 28100 Novara, Italy
3 Neurology Unit, San Giovanni Bosco Hospital, 10154 Turin, Italy
4 Department of Neuroscience "Rita Levi Montalcini", University of Turin, 10126 Turin, Italy
* Correspondence: cristoforo.comi@med.uniupo.it

Abstract: The COVID-19 pandemic has led to unprecedented demand on the global healthcare system. Remarkably, at the end of 2021, COVID-19 vaccines received approvals for human use in several countries worldwide. Since then, a solid base for response in the fight against the virus has been placed. COVID-19 vaccines have been shown to be safe and effective drugs. Nevertheless, all kinds of vaccines may be associated with the possible appearance of neurological complications, and COVID-19 vaccines are not free from neurological side effects. Neurological complications of COVID-19 vaccination are usually mild, short-duration, and self-limiting. However, severe and unexpected post-vaccination complications are rare but possible events. They include the Guillain-Barré syndrome, facial palsy, other neuropathies, encephalitis, meningitis, myelitis, autoimmune disorders, and cerebrovascular events. The fear of severe or fatal neurological complications fed the "vaccine hesitancy" phenomenon, posing a vital communication challenge between the scientific community and public opinion. This review aims to collect and discuss the frequency, management, and outcome of reported neurological complications of COVID-19 vaccines after eighteen months of the World Health Organization's approval of COVID-19 vaccination, providing an overview of safety and concerns related to the most potent weapon against the SARS-CoV-2.

Keywords: neurological adverse effects; SARS-CoV-2; vaccination; Guillain-Barré syndrome; Bell's palsy; Vaccine-Induced Thrombotic Thrombocytopenia; myelitis; multiple sclerosis

Citation: Tondo, G.; Virgilio, E.; Naldi, A.; Bianchi, A.; Comi, C. Safety of COVID-19 Vaccines: Spotlight on Neurological Complications. *Life* **2022**, *12*, 1338. https://doi.org/10.3390/life12091338

Academic Editors: Silvia De Francia and Sarah Allegra

Received: 17 July 2022
Accepted: 23 August 2022
Published: 29 August 2022

1. Introduction

The novel coronavirus SARS-CoV-2 causes a pulmonary-systemic coronavirus disease (COVID-19) which was firstly described in Wuhan, China, in December 2019. Fastly, COVID-19 became a worldwide health emergency, and, in March 2020, COVID-19 was declared a pandemic [1]. At the end of 2021, COVID-19 vaccines have been released for widespread vaccination, receiving approvals in several countries for human use and providing a solid hope in the battle against the virus [2]. Starting from the 31st December 2020, four major vaccine types have been used with an extensive distribution worldwide: the BNT162b2 (Pfizer–BioNTech), the mRNA1273 (Moderna), the ChAdOx1 nCov-19 (Oxford–AstraZeneca), and the Ad26.COV2.S (Janssen) vaccine [3]. The first two represent COVID-19 mRNA-based vaccines encoding the spike protein antigen of SARS-CoV-2; the other two vaccines are recombinant adenoviral vectors. The 2021 trials, including these four vaccines, reported high effectiveness ranging from 66.6 to 95% at preventing COVID-19 infections, while injection-site injury, pain and reactions, axillary lymphadenopathy, fatigue, myalgia, arthralgia, and headache were reported as the most frequent adverse effects [3]. Even if in the clinical trials the safety records were satisfying, since the beginning of the global vaccination program, vaccine hesitancy, mainly due to safety concerns, represented a challenge. Worries raised due to reported serious or even fatal events, such as anaphylaxis, myocarditis, and thrombotic events, primarily associated with the adenovirus-vectored vaccines [4]. However, the subsequent revisions of the UK Medicines and Healthcare Products Regulatory Agency and the European Medicines Agency concluded that the rate of fatal thrombotic events in the vaccinated population was non-significant compared with the general population, confirming a favorable risk/benefit for the vaccination [5]. Despite the reported COVID-19 vaccine acceptance rate of over 60% of the population in most countries, skepticism and concerns remain, and neurological adverse effects represent a particularly feared outcome [6]. Neurological complications of COVID-19 vaccines include central and peripheral nervous system manifestations, ranging from minimal and tolerable adverse effects to life-threatening toxicity.

Neurological side effects of SARS-CoV-2 vaccines are frequent but primarily non-serious [7]. They can be classified as mild or severe neurological events [8]. Mild or non-serious neurological events include weakness, muscle and joint pain, transient sensory symptoms, and headache [7]. These adverse reactions are the most frequently described adverse effects in the majority of reports. They usually occur acutely and are expected to be transient [7,9,10]. Minor adverse effects are generally more severe after the second dose than the first dose and seem to hit women more than men [11,12]. In the nationwide descriptive study of García-Grimshaw and colleagues, the overall incidence of non-serious neurological effects of COVID-19 vaccination with the Pfizer–BioNTech vaccine was 600.7 cases per 100,000 administered doses, and a headache was the most frequent disturbance [7].

Neurological severe adverse effects after COVID-19 vaccination are infrequent events [7]. Still, they are the main reason to create vaccination hesitancy. They include the Guillain-Barré syndrome (GBS), seizures and syncope, encephalitis, meningitis, myelitis, demyelinating disorders, myasthenic disorders, thrombocytopenia, cerebrovascular events, facial nerve palsy and other cranial nerve neuropathies [8,13]. The increased risk of cerebral venous sinus thrombosis (CVST), which initially was associated with the Oxford–AstraZeneca vaccine, is an example of a rare and severe adverse neurological effect. Another fearful neurological complication already reported in the Oxford–AstraZeneca clinical trial is transverse myelitis, diagnosed in two cases in the treatment arm [13]. GBS, which can result in life-threatening sequelae, has received particular attention, arousing concerns by the US Food and Drug Administration (FDA) due to the suspected link with the adenoviral vector vaccines [8]. Overall, vaccinations with vector-based vaccines seem burdened by more frequent or severe side effects than the mRNA vaccines [14,15]. While viral vector vaccine usually induces a strong immune response, thus potentially associated with a higher rate of adverse effects, and especially systemic reactions, mRNA vaccine is a novel vaccine platform with good immunogenicity and safety, showing a lower rate of complications, usually local side effects [16,17]. Nevertheless, the COVID-19 vaccine is generally well-tolerated, adverse effects are usually self-limiting, and there is no clear evidence of a higher rate of neurological disorders associated with the COVID-19 vaccination [9]. Conversely, it is not the first time in human history that vaccinations have been associated with developing neurological disorders. In people vaccinated against the pandemic influenza A (H1N1), a relatively increased risk for Bell's palsy and paresthesia was reported; however, the risk for other neurological complications such as GBS was found to be similar to that of unvaccinated people [18]. In the recent past, severe neurologic events have been speculated with the measles-mumps-rubella vaccinations. However, the association with several postulated complications, including encephalitis, aseptic meningitis, and autism, has been rebutted [19]. This narrative review aimed to collect and discuss literature data regarding neurological complications of SARS-CoV-2 vaccines after eighteen months of the World Health Organization's approval of COVID-19 vaccination. Here we provide an overview of the frequency, distribution, management, and outcome of neurological complications of COVID-19 vaccines. The last literature search was done on 30 June 2022.

2. Headache

Headache is one of the most frequent adverse effects of the COVID-19 vaccine, reported in approximately half of the vaccine recipients [20,21]. It usually occurs within 72 h after the vaccination. It resolves within hours or a few days later, manifesting as a single episode of moderate-to-severe intensity in most cases [22]. Pain characteristics are variable, but the headache is often bilateral, involving frontal and temporal regions, and is associated with fatigue and myalgia. The cumulative prevalence of all headache episodes, from mild to severe, is slightly higher after the second dose than in the first [22,23].

Headache occurs with any of the FDA and European Medicines Agency-approved vaccines; an Italian report describes the higher risk of developing a headache after the Oxford–AstraZeneca vaccine, followed by the Pfizer–BioNTech [24]. A history of headache, both migraine, and non-migraine types, is associated with a higher risk of suffering from headache after COVID-19 vaccination than controls without pre-existing headache [23]. A delayed presentation of the headache, occurring around one week after immunization, should be carefully evaluated. It has been associated with CVST in patients administered an adenovirus vector-based COVID-19 vaccine [25].

3. Cerebrovascular Events

It is unclear whether several vaccines may trigger cerebrovascular events [26]. As a fact, stroke occurrence has also been reported after the COVID-19 vaccination. Neurovascular thrombosis may involve both the arterial and venous systems, and different pathogenic mechanisms have been proposed. Ischemic and hemorrhagic strokes are reported, but—although rare—the most described event is CVST, which may occur as isolated or in association with the so call Vaccine-Induced Thrombotic Thrombocytopenia (VITT).

3.1. Vaccine-Induced Thrombotic Thrombocytopenia and Cerebral Venous Sinus Thrombosis

VITT is probably the most severe neurological complication of COVID-19 vaccines and is mostly—but not solely—related to adenoviral ones, particularly Oxford–AstraZeneca [27,28]. Cases associated with the Moderna vaccine are anecdotal, and no definite conclusions can be drawn [29–31]. VITT consists of immune-mediated thrombocytopenia with thrombosis with onset usually within 2–3 weeks after the first dose of vaccination [32]. The term VITT was coined because of its similarity with heparin-induced thrombocytopenia (HIT), which is caused by the emergence of antibodies activating platelets after heparin administration [33,34]. In this syndrome, heparin becomes immunogenic by binding the platelet factor 4 (PF4), thus inducing antibody formation [35]. As a result, platelets are activated, and thrombocytopenia develops, leading to potential thrombosis, predominantly in the deep veins system or in pulmonary embolism [36]. Unlike HIT, patients with VITT do not have previous exposure to heparin. However, similarly to HIT, high levels of antibodies directed to PF4-polyanion complexes are detectable. The mechanisms by which vaccines promote the development of antibodies are still unclear. Several pathogenic hypotheses have been proposed, and both vaccines and host factors seem to be involved. The PF4 binding to adenovirus vectors and pro-inflammatory and immunogenic signals are the most reported [37–39].

These IgG autoantibodies cause platelet activation via the FcγIIa receptor, stimulating the immune response (activation of monocytes, neutrophils, and endothelial cells) and platelet consumptions, leading to increased risk of thrombosis [40].

The effects of these interactions may be detected by laboratory tests, and high D-dimer levels, positive antibodies against PF4 (detected by enzyme-linked immunosorbent assay, ELISA), and low platelet count are considered the hallmark for serological diagnosis [4,41]. Compared with HIT, thrombosis in patients with VITT may occur at unusual sites, including CVST, splanchnic, portal, or hepatic veins [42,43]. Reasons for this specificity are still unknown, and research is ongoing to clarify the question [44]. Thrombosis in cerebral veins may be massive, thus explaining the potential catastrophic evolution of VITT [43,45]. Brain ischemia, intracerebral hemorrhage (ICH), and subarachnoid hemorrhage may complicate

CVST, requiring a prompt intervention [46,47]. Neurological symptoms recall those of CVST and depend on the territories involved in the thrombosis. The clinical onset is often insidious with malaise and progressive worsening headache, not responding to analgesic treatments. Focal neurological symptoms, seizures, vomiting, blurred vision, and consciousness impairment may suddenly develop. Jointly with other systemic symptoms, abdominal, back, and chest pain may coexist, reflecting thrombosis in other sites [48].

Due to the severity of the condition, prompt recognition of the syndrome is crucial to timely and correctly contrast thrombocytopenia and thrombosis. Whether the fatality rate was initially very high, mortality has significantly decreased over time, explaining the improvement in the early identification and intervention [49]. Several—constantly evolving—recommendations released from expert consensus and international societies are available for VITT management [50–52]. The hallmark of therapies includes the administration of high-dose intravenous immunoglobulin and non-heparin anticoagulation, depending on platelet count, clinical status, and residual organ function of the patient [41,52–54]. Some authors reported successful mechanical thrombectomy for massive CVST [45,55]. Even if knowledge is limited regarding the indications and timing of endovascular procedure in CVST, it may be considered an option in this life-threatening condition. Overall, management of VITT requires a multidisciplinary approach at centers with neuro-interventional and surgical experience [56]. VITT is a rare adverse event after COVID-19 adenoviral vaccines. The incidence seems to range from 1/125,000 to 1 in 1 million vaccinated cases [30,41]. A recent metanalysis estimated a higher incidence of VITT, reaching 28/100,000 incidence, and most frequently presented with CVT following deep vein thrombosis/pulmonary thromboembolism and splanchnic vein thrombosis, and about one-third of patients had a fatal outcome [43]. Risk factors are still unknown, although female and younger age have been potentially identified in the first descriptions [57].

CVTS may also occur outside the context of VITT [28,58–61]. Non-VITT CVST patients seem to be older than VITT-associated, with a lower number of veins thrombosed at first diagnosis and a lower rate of superimposed ICH. As a consequence, outcomes seemed to be more favorable [62]. Without any hematologic disorder associated with CVST vaccine-related, management does not differ from standard cases.

Several studies tried to identify a distinctive pattern of CVST-related VITT compared to non-VITT CVST. They found higher mortality or dependency after discharge and a shorter time interval between vaccination and clinical onset in the CVST-VITT group. In addition, CVST-VITT was more often complicated by ICH, explaining the worst outcome for these patients [63]. Interestingly, CVST with thrombocytopenia was almost exclusively described after vector-based vaccination, reported in 57% of cases of CVST following the Oxford-AstraZeneca vaccine [64].

3.2. *Ischemic and Hemorrhagic Strokes*

Ischemic strokes have been reported after the COVID-19 vaccination, and several descriptions are available [5,65,66]. The association with adenoviral vaccines seems more frequent, with a clinical onset within three weeks since the inoculation [67]. Both lacunar infarction and large vessel occlusion are reported, mainly involving the middle cerebral artery and its vascular territories [5,67]. Women below 60 years old seem more affected [67]. Management of acute ischemic stroke does not differ from routine clinical practice, and reperfusion therapies are indicated within international guidelines and recommendations. Both intravenous thrombolysis (IVT) and mechanical thrombectomy (MT) have been performed in ischemic stroke patients after Oxford–AstraZeneca and Pfizer–BioNTech vaccinations, with variable outcomes [5,65]. Cerebral arterial thrombosis has also been reported jointly with CVST or in the context of VITT [68–72]. In these cases, treatments for acute ischemic strokes have to be weighted considering the neuro-imaging findings and laboratory test results, particularly for low platelet counts that may contraindicate IVT [73,74]. Overall, arterial thrombosis causing stroke seems much less frequent than venous events [75].

Hemorrhagic events include ICH and SAH. Isolated ICH outside any known coagulation disorder is rarely reported [76,77]. In these cases, uncertainties regarding the causal relationship between vaccination and cerebral bleeding remain. Instead, to date, there is no description of isolated SAH after COVID-19 vaccination. Conversely, ICH and SAH are frequently described as a complication of CVST, as previously mentioned. Therefore, they are mostly reported after adenoviral vaccines, particularly Oxford–AstraZeneca, in people below 60 years old and within two weeks after the vaccination [67]. Management of ICH and SAH follows the general standard of care indications. The use of tranexamic acid is reported, as well as neurosurgical procedures in case of severe ICH, including hematoma drainage, external ventricular drain, and decompressive craniectomy [67,76]. However, aggregate data indicate that cerebrovascular events are not increased after COVID-19 mRNA vaccines, as reported in several series [78–80]. The mechanisms of ischemic and hemorrhagic stroke related to COVID-19 vaccines are still unclear, and, at least in part, they may overlap with those of VITT. Hypercoagulable states related to the vaccine's inflammatory process may promote clots formation and cause strokes [81]. Beyond the factors previously described for VITT, alterations in protein S levels, thrombocytopenia, hypofibrinogenemia, folate deficiency with elevated homocysteine, factor XIII deficiency, and antiphospholipid antibodies have been detected in the context of strokes after COVID-19 vaccination [39]. In addition, both vaccine components and host factors may produce an autoimmune response, leading to thrombosis.

4. Multiple Sclerosis

Multiple sclerosis is a chronic autoimmune disease of CNS, recognizing a genetic background on which several environmental factors, mainly infections, can act as a disease trigger [82,83]. In the past, vaccines (such as hepatitis B, yellow fever vaccination, and human papillomavirus) were awarded as a possible determinant for MS onset. However, several studies have failed to show an association [84]. Currently, only live attenuated vaccines are generally contraindicated in MS patients, particularly those receiving immunosuppressants. Live attenuated vaccines have the potential to cause an infection if the vaccine is administered during treatment with an immunosuppressant, the ability of the immune response of the MS patient to clear the infection could be impaired, possibly resulting in a worsening of MS symptoms. Nowadays, most international and national consensus from MS experts recommend COVID-19 vaccination, and mRNA vaccines are considered safe [84,85].

We found several studies reporting the new onset of MS after SARS-Cov2 immunization [86–93]. All cases were relapsing-remitting MS. Demographic and clinical characteristics are comparable to the general MS population ranging from optic neuritis (ON) to transverse myelitis (TM) or brainstem syndrome. The neurological manifestation presented from one day after the first vaccine dose to several weeks after the second immunization showing a temporal variability [89,91]. The case described by Halva et al. displayed a familiarity with MS [87]. The majority showed intrathecal immunoglobulin synthesis. The systematic review by Ismail et al. considered twelve MS-like presentations (both new-onset and relapses) correlated to the COVID19 vaccine reported up to September 2021. Only one patient among the new-onset was not vaccinated with mRNA [88]. The majority of the described cases of MS after the COVID-19 vaccine already presented high dissemination in space (DIS) and time (DIT) at diagnosis (i.e., previous history of neurological symptoms, concomitant enhancing and non-enhancing lesions, black holes lesions in T1 MRI); therefore, we could assume that most probably they already had the autoimmune disease long before the vaccine administration. Indeed, MS can sometimes be diagnosed in preclinical phases (the so-called radiologically isolated syndrome—RIS), demonstrating that the disease is pathologically and radiologically present long before the clinical manifestations. Even though only one case of the presented had a previous history of the clinically isolated syndrome (ON with unremarkable previous brain MRI) [91], several of the reported patients already showed a high DIS at diagnosis [86,87,91]. Furthermore, real-world data do not corroborate the hypothesis of a higher risk of new MS onset or MS relapse associated with the COVID-19 vaccination [91,94–96].

5. Neuromyelitis Optica Spectrum Disorder and MOG Antibody Disease

Neuromyelitis Optica Spectrum disorder (NMOSD) is a demyelinating disease that frequently presents with extensive longitudinal TM (LETM) and monolater or bilateral ON. The illness pathogenesis recognizes the presence of antibodies against aquaporin-4 (Abs-AQP4) in serum and CSF, and the disease is primarily idiopathic [97]. NMOSD is often in differential diagnosis with MOG antibody disease (MOGAD), displaying very different histopathology and prognosis [98]. NMOSD and MOGAD are, therefore, antibodies-mediated autoimmune syndromes. Seven cases of AQP4-NMOSD were reported, one with inactivated vaccine, one with a viral vector, and four with mRNA vaccines [99–102]. Latency ranged from one day after the first dose to eighteen days after. One patient presented with area postrema and hypothalamic syndrome. Five patients showed signs of TM (one with short TM, one with short TM and brainstem syndrome, and three with LETM). Two Abs-AQP4 monolateral ON were also reported after mRNA vaccine [103,104]. One case of Abs-AQP4 LETM NMOSD associated with Sjogren's disease after 18 days from Pfizer–BioNTech vaccination was described in a 64-year-old male with no medical history [88,89].

Less commonly, MOG antibodies positive syndromes were described [105,106]. Two LETM after Oxford–AstraZeneca vaccination [105,106], one with incomplete recovery after steroid and plasma exchange were reported [106].

For NMOSD and MOGAD post-vaccination, the outcome was favorable for the majority of the patients. Little information about long-term prognosis or treatment is at the moment available.

6. Myelitis

Post-vaccination acute TM cases have been reported following vaccination against hepatitis B, diphtheria, tetanus, and influenza. However, the pathogenesis remains unclear. It is assumed that recombinant or live-attenuated viruses can induce autoimmunity through molecular mimicry or epitopes spreading mechanisms. Several cases have been described of TM after COVID-19 immunization with different types of vaccine. A review by Garg et al. reports that adenoviral vector-based COVID-19 vaccines are more frequently associated with the spreading of TM [8].

Three TM cases were reported in the four RCT Oxford–AstraZeneca. They occurred on days 10, 14, and 64 after vaccination, respectively [107]. However, only one patient was considered possibly related to the vaccine, with an independent neurological committee considering the most likely diagnosis to be an idiopathic, short segment TM. Conversely, two additional cases were later determined unrelated to immunization based on the previous history of undiagnosed MS or extreme latency of the event [107,108]. In addition, national vigilance boards registered 45 cases of myelitis in the United Kingdom and nine patients in Germany after the Oxford–AstraZeneca vaccination [106,109]. More recently, other case reports all over the world suggested cases of TM after the Oxford–AstraZeneca vaccine [109–115]: some authors described cervical TM, other dorsal TM, mostly short TM but also LETM ranging from day fourafter the first dose to day fourteen. Veggezzi et al. suggested a correlation, although specifying that the definition of acute vaccine-induced myelitis according to the WHO criteria remained unclear [116]. To note, the majority of patients had an unremarkable medical history, and usually, a favorable outcome occurred with treatments except for the case of Notghi et al.: a 58-year-old man with a history of pulmonary sarcoidosis developed seven days after vaccination, a T2-T10 LETM and deteriorated even after steroids showing an extension up to C1. Neurosarcoidosis was ruled out with thorax CT and CT PET. The patient finally improved with plasma exchange sessions [114]. We may speculate that the severity of the presentation might be related to an autoimmune predisposition of the subject and the extension of the neurological event (LETM rather than TM).

On the other hand, in stage III clinical trial for Pfizer–BioNTech, 18,860 patients were vaccinated, and four (<0.01%) developed serious adverse events related to the vaccine: one patient reported transient leg paresthesia) [117]. He was not diagnosed with TM, and it is unclear if any MRI or CSF analysis was undertaken. However, real-world data on MS, MOGAD, NMOSD, and TM populations (963 patients), exposed to the Pfizer–BioNTech vaccine has been reassuring. Vaccine-associated new or worsening neurological symptoms occurred in less than ~15% of patients, commonly early in the post-vaccination period (within the first week) and mostly self-resolving within two weeks [95,96]. Other single cases of TM after Pfizer–BioNTech vaccine and after Moderna could be found in the literature. Cases are from all over the world, both occurring in females and males, mostly of young age, but TM in 69-, 75-, 81-, and 85-year old-patients were also described [118–124]. Symptoms onset ranged from three days from the first dose to four days after the second immunization, and even though long-term data are missing outcome was usually good except for very old patients or patients with LETM. Two cases after the Moderna vaccine were also described by Fitzsimmons et al. and Gao et al.: a 63-year-old male who developed symptoms of TM 17 h after the second dose of an mRNA vaccine (Moderna) and a 76-year-old female with a cervical gd+ LETM developed six days after Moderna vaccine [125]. However, he presented low vitamin B12, which could have influenced the presentation. A case series of CNS demyelinating disease after an mRNA vaccine was recently published. It included one case of a Caucasian man diagnosed with TM [89].

Tahir et al. reported a case of a 44-year-old woman presenting with C3-T1 LETM after receiving the Janssen immunization ten days before presentation. Interestingly while undergoing plasma exchange, she developed Bell's palsy. A second MRI after plasma exchange showed an improvement [126].

7. Optic Neuritis

ON is an isolated inflammation of the optic nerve in its course (anterior, retrobulbar, chiasmatic, i.e.,). It could represent a manifestation of MS, NMOSD, or recognize metabolic or infectious etiology, but ON may be monophasic and isolated in fewer cases. We found eleven idiopathic ON after COVID-19 immunization, the majority being females, with a good outcome.

The first case of left-isolated ON one week after a single dose of the Janssen vaccine and a prompt steroid response was reported. MOG Abs were not tested, and isoelectrofocusing showed intrathecal synthesis indicating a chronic inflammatory process. Therefore, strict follow-up of the patients is mandatory to intercept a possible evolution in clinically defined MS or MOGAD [127].

Barone et al. reported that two patients aged 48 and 31 years presented with acute optic neuritis a few days (twelve and nine) after the first dose of mRNA vaccination. The first patient had a typical presentation of unilateral ON and was treated with steroids with partial recovery. MRI was unremarkable. The second patient initially had a transient loss of vision after exposure to high temperature (defined by the Authors as Uhthoff's phenomenon), followed by persistent monocular dyschromatopsia and central scotoma diagnosed with ON. No information regarding CSF of both patients, MRI, or the outcome of the second patient is reported [128]. Four ON in Germany following vaccination with Pfizer–BioNTech were reported in the study of Kaulen et al. [122]. In Koh hospital-based study, covering a four-month period during which 1,398,074 people received at least one dose of COVID-19 mRNA vaccine, 457 patients with a spectrum of neurological disorders were recorded. Two female patients (0.4%) developed ON (48 and 62 years). One was idiopathic, occurring thirty-three days after the second dose. However, the second one developed an ON one day after the first vaccine dose and was later diagnosed with NMOSD AQP4+ Abs [103]. The authors categorized the likelihood of vaccination-associated or a concurrent and coincidental illness using the WHO Adverse Events Following Immunization (AEFI) causality assessment, Brighton criteria framework (Certain, Probable, Possible,

Unlikely, Unrelated) and concluded the two ON to be likely coincidental and classified as "unlikely" [103].

Arnao et al. reported the first case of a bilateral retrobulbar ON in a healthy woman after exposition to the first dose of Oxford–AstraZeneca vaccine [129]. Antibodies tested negative. After steroids, she fully recovered.

Finally, a retrospective study by Assiri et al. described 18 patients referred to or presented to the Saudi German Hospitals in Jeddah, Saudi Arabia, with neurological complications from March to September 2021 after receiving a COVID-19 vaccine. Ten patients received the Oxford–AstraZeneca vaccine, and eight the Pfizer–BioNTech vaccine [5]. Of the 18 patients, 2 presented with ON (one bilateral and one monolater) fourteen and nineteen days after vaccination. Both recovered completely.

8. Myasthenia Gravis

Few cases of new-onset myasthenia gravis (MG) post-immunization have been described since March 22. All of these reports concern patients with late-onset MG AchR positive.

First, an 82-year-old presented with bulbar symptoms. He was admitted four weeks after receiving the first dose and two days after receiving the second dose of the Pfizer–BioNTech vaccine. His symptoms started a few weeks before, occurring in the evenings, often during dinner [130]. He was later diagnosed with late-onset MG AchR positive based on immunological evaluation and neurophysiology. Despite symptomatic treatment, two months later, the patient showed an evolution in generalized MG, requiring hospitalization complicated by pneumonia, requiring ventilator support, and PEG tube insertion. He later recovered. New-onset MGs following COVID-19 vaccination are rarely reported in the literature. A review of vaccine-related adverse events in a large population found only two cases of new-onset MG following the COVID-19 vaccine, with onset within 28 days from vaccination [93]. The two reported cases involved patients in their 70s and occurred 1 to 7 days after administering the second dose of the Pfizer–BioNTech vaccine [93]. Both occurred after the second dose of Pfizer–BioNTech vaccine, one being severe, both male from Israel.

Galassi et al. reported the first case of ocular MG after Oxford–AstraZeneca, onset eight days after the first dose. A 73-year-old male presented with painless left-sided ptosis. He was confirmed MG based on EMG and serum titer of anti-AChR antibodies on day 20 after vaccine injection [131]. To note, he showed high rheumatoid factor titer without signs of joints or arthritic involvement and a recent history of psoriasis.

A large population-based study of more than 32 million people investigated the neurological adverse events associated with the Oxford–AstraZeneca and Pfizer–BioNTech vaccines. They found an increased risk of hospital admission for myasthenic disorders (15–21 days) in those who received the Oxford–AstraZeneca vaccine [13].

9. Acute Disseminated Encephalomyelitis and Weston–Hurst Syndrome

ADEM is a rare acute monophasic demyelinating autoimmune event of CNS potentially occurring after an immunological trigger, such as infections or vaccinations, and sometimes presenting with hyperacute hemorrhagic components (the so-called Weston–Hurst syndrome) [132]. Post-vaccination ADEM seems to occur at any age, with males more affected than females [133]. Several case reports and case series of post-COVID-19 infection and COVID-19 vaccination were described [134]. In contrast with ADEM, Weston–Hurst normally exhibits a very poor prognosis.

In particular, ADEM following Oxford–AstraZeneca all occurred in middle-aged patients, with symptom onset ranging from 10 days after vaccination to a few weeks with progressive MRI and clinical improvement at follow-up [133–136]. Recovery was observed in all patients except for a case occurring in Australia [135] in a 63-year-old man with a medical history of type II diabetes mellitus, ischemic heart disease, and atrial fibrillation, who presented with progressively declining cognition, disorientation, and impaired attention. On day four after admission, he suddenly became poorly responsive. MRI showed numerous bilateral (>20) cerebral white matter lesions with both periventricular and juxtacortical involvement. Stereotactic right frontal craniotomy and open biopsy of a right frontal lesion were performed on day nine, confirming the ADEM diagnosis. He later died on day twenty of admission. Further post-mortem investigations were made, and the cause of death was registered as "ADEM in the setting of recent AstraZeneca COVID-19 vaccination" [135].

Regarding ADEM after mRNA vaccines, except for the case of Kania et al. (describing a 19-year-old female), all other cases occurred in patients in their forties or fifties or even in elderly patients [8,33,137–140]. In all cases, MRI and CSF findings were highly variables. The 19-year-old was injected two weeks prior with Moderna vaccine [141].

Reports of potential vaccine-related adverse effects from the European Medicines Agency's (EMA) EudraVigilance database by November 2021 revealed slightly higher but altogether comparable incidences of 0.07 (Oxford–AstraZeneca) and 0.05 (Janssen) per 100,000 people per year for potential vector-based vaccine-associated ADEM, compared to 0.02 (Pfizer–BioNTech) and 0.04 (Moderna) for mRNA-vaccines [142].

As for Weston–Hurst, Ancu et al. [142] reported three cases with onset within nine days after the first shot of the Oxford–AstraZeneca vaccine. All the patients were treated with steroids, and two with plasma exchange with severe sequelae. One patient, despite treatments, still presented with a vegetative status a few weeks later the onset; a 25-year-old woman developed a medullary thoracic syndrome nine days after the vaccine. Spinal MRI confirmed an autoimmune acute LETM gd+ with a hemorrhagic component (a spinal variant of Weston–Hurst). Brain MRI showed bi-hemispheric white matter lesions with focal contrast enhancement. After treatment, a clinical improvement of the sensory symptoms was observed but persistent paraplegia on a six-week follow-up. Finally, a 55-year-old woman developed progressive nausea, dizziness, and meningism nine days after the first shot. She neurologically deteriorated to a severe spastic tetraparesis and coma. Brain MRI revealed multiple diffuse FLAIR hyperintense and hemorrhagic lesions. She developed hydrocephalus needing an emergency right-sided decompressive hemicraniectomy. The brain cortex biopsy from the affected right temporal lobe revealed perivascular predominantly granulocytic infiltrate and hemorrhages, confirming the diagnosis. The patient died after a few weeks after. This cases series highlighted the heterogeneity in clinical findings and the verity of outcomes in Weston–Hurst syndrome.

10. Autoimmune Encephalitis

Autoimmune encephalitis (AIE) is a rare, recently described group of neurological diseases associated with specific autoantibodies or classified in specific syndromes (e.g., limbic encephalitis). Various subgroups of AIE are distinguished based on autoantibodies, which may lead to clear clinical presentations and different prognoses, particularly concerning paraneoplastic and non-paraneoplastic AIE. Among them, anti-leucine-rich glioma inactivated 1 (LGI1) encephalitis is characterized by cognitive impairment, rapid progressive dementia, psychiatric disorders, faciobrachial dystonic seizures (FBDS), and refractory hyponatremia. Rarely, cerebellitis has been described after vaccination [143,144]. Zlotnik et al. reported a 48-year-old man presenting with severe fatigue developed a few days following his second dose of Pfizer–BioNTech mRNA vaccination and rapidly evolving in progressive cognitive decline and hyponatremia 2.5 weeks after. He was later diagnosed with anti-LGI1 AIE. Other authors reported series of possible AIE according to Grauss Criteria related to prior Oxford–AstraZeneca or Pfizer–BioNTech vaccination. However, none showed specific CSF or serum autoantibodies or hidden neoplastic conditions despite extensive workup [145,146]. A case of acute encephalopathy occurring in a 32-year-old Asian male within a day of receiving the first dose of Moderna was also reported [147]. The second case of acute encephalitis, myoclonus, and Sweet syndrome following the Moderna vaccine occurred in the USA [148]. Kaulen et al. also reported a case of limbic encephalitis where MRI demonstrated bilateral hippocampal hyperintensities, and CSF revealed mild lymphocytic pleocytosis (13/μL), oligoclonal bands type 2, and mild disruption of the blood-brain barrier [122]. In addition, unexplained acute encephalopathy cases have also been described after receiving the COVID-19 vaccine [149,150] and hyperacute reversible encephalopathy related to cytokine storm [151].

Up to June 2021, a total of 79 encephalitis cases were reported with the Oxford–AstraZeneca vaccine (in 99.3 million doses with a resulting incidence of almost 0.08 per 100,000). Only 20 cases in 110.6 million doses with a consequent incidence of nearly 0.02 per 100,000 were reported for Pfizer–BioNTech [145]. However, those reported data included AIE, limbic encephalitis, viral encephalitis, ADEM, Bickerstaff, and other non-infective encephalitides.

11. Aseptic Meningitis

Aseptic meningitis (AM) is an inflammatory disorder of the meninges that can be of iatrogenic origins (e.g., immunoglobulins or NSAID) as a complication of vaccination (against mumps, measles, rubella, and influenza). It has been previously described [152]. However, the etiology remains unclear, although a possible explanation is a reaction to vaccine adjuvants recently implied in an autoimmune/inflammatory syndrome (ASIA) [92]. All reported cases so far occurred after the RNAm vaccine, and the Pfizer–BioNTech vaccine's adjuvant is a nanoparticle-based polyethylene glycol (PEG) stabilizer that has been implied as a trigger of ASIA syndrome in other organs [153]. AM is well recognized as a reaction to certain drugs and is frequently associated with systemic autoimmune disorders. However, Satai's AM was negative for serum anti-PEG antibodies [154]. All cases showed unremarkable brain MRI and good response to steroids. Some also presented with fever and the majority displayed CSF pleocytosis. The first published case was in a middle-aged Japanese nurse who developed a refractory headache, fever, and signs of meningeal involvement one week after vaccination. Although the patient was assuming NSAIDs for a sporadic migraine at home, she was diagnosed with aseptic meningitis based on CSF and peripheral workup. She was initially treated with acyclovir (until viral antibody titers were negative) and later with intravenous methylprednisolone [154]. Subsequently, three other cases were reported in France [155] and Singapore after [156], all after Pfizer–BioNTech.

12. Guillain-Barré Syndrome

GBS is an acquired, inflammatory, acute polyradiculoneuropathy generally clinically characterized by a rapidly progressive ascending flaccid paralysis. It is a rare neurological disorder, with a rate of 1 to 2 per 100,000 person-years [157]. The term GBS includes several variants: the acute inflammatory demyelinating polyradiculoneuropathy (AIDP), which is the most common and primarily expresses demyelinating features; the acute axonal motor neuropathy, less common and characterized by a worse prognosis; the acute motor-sensory axonal polyneuropathy, with both motor and sensory involvement; the Miller Fisher syndrome (MFS), consisting of ophthalmoplegia, ataxia, and areflexia; the Bickerstaff's brain stem encephalitis, which is considered a variant of the MFS and characterized by altered consciousness, paradoxical hyperreflexia, ataxia, and ophthalmoparesis; other less common presentations, include the pharyngeal-cervical-brachial motor variant, the pure facial diplegia and the pandysautonomic variant [158,159]. The pathogenesis of GBS is primarily immuno-mediated by an immune response triggered by preceding infections or vaccinations throughout a cross-reactivity involving the axonal or myelin constituents of the peripheral nerves [160].

The relationship between COVID-19 and GBS is highly complex since COVID-19 infection can trigger GBS [161], and COVID-19 vaccination has been associated with the development of GBS, irrespective of the vaccine used [162].

Most cases of GBS are described following the first dose of any vaccination [163–165]. The interval between vaccination and GBS onset should be within six weeks [164]. GBS after COVID-19 vaccination may occur in an extensive range of ages between 20 and 90 years [166], and the onset is usually within the second week after vaccination [166], ranging from three to 30 days [122,164]. Neurophysiological findings are often typical, with the demyelinating forms of AIDP as the most described [164,166,167]. Cranial involvement is not uncommon and facial diplegia is frequent [164,168–171]. Specifically, facial palsy as a feature of GBS at a higher rate than expected seems to be more frequent after adenovirus-vectored vaccines [167,172]. Allen and colleagues presented four cases of bifacial weakness without any other typical sign, thus not showing areflexia, objective sensorimotor signals in the limbs, or dysautonomia, all showing albumin-cytological dissociation [169]. In other case series, albumin-cytologic dissociation has been detected between 75 and 90% [164,166]. In the work of Kim and colleagues, a relatively low rate of Anti-ganglioside antibodies has been reported [164].

Prognosis may be variable. Respiratory failure was described in 30% of patients in the Kim and colleagues report and 85% of patients described by Maramattom and colleagues [168]. Treatment generally includes intravenous immunoglobulin and, in fewer patients, plasmapheresis [162,166,167]. A prompt treatment with intravenous immunoglobulin followed by early physical therapy usually induces improvement of the motor and sensory deficits [122,173], but recovery may be partial [162]. Even if recurrency of GBS after COVID-19 vaccines in patients with a history of a previous GBS has also been reported [163], a history of GBS does not increase the risk of a relapse. The Israelian study of Shapiro and colleagues on a cohort of about seven hundred patients with a previous diagnosis of GBS showed the recurrence of GBS only in one patient that was successfully treated with plasmapheresis with residual minor proximal weakness [174].

MFS is also associated with COVID-19 infection and after COVID-19 vaccination [175,176]. Ophthalmoplegia and diplopia represent the most common features in the described cases, followed by ataxia [176]. Most patients were treated with intravenous immunoglobulins and showed a favorable prognosis, with recovery within four-to-six weeks [177–180].

In conclusion, GBS may follow COVID-19 vaccination with any vaccine. However, the reporting rate of GBS after the Janssen vaccine analyzed from the United States Vaccine Adverse Event Reporting System (VAERS) was higher than after the Pfizer–BioNTech and the Moderna vaccines [181]. In the VAERS, a national passive surveillance system for monitoring vaccine safety, 130 cases of GBS were reported from February 2021 to July 2021, with an absolute rate increase of 6.36 per 100,000 person-years [182]. For this reason, in December 2021, the Advisory Committee on Immunization Practices recommended mRNA vaccines over the Janssen vaccine due to the latter choice's lower benefit/risk balance [183]. Similarly, the analysis of the English National Immunisation Database of COVID-19 vaccination between 1 December 2020 and 31 May 2021 showed a rate of GBS following the Oxford–AstraZeneca significantly higher than the background rates within 28 days of the first dose [13].

A recent cohort study analyzing data from the United States Vaccine Safety Datalink assessed the incidence of GBS following COVID-19 vaccination with the Pfizer–BioNTech, the Moderna, and the Janssen vaccine. The analysis involved more than ten million participants. It confirmed a higher rate of GBS after the Janssen vaccine than the mRNA vaccines, reaching an unadjusted incidence rate of 32.4 GBS per 100,000 person-year [184]. Post-vaccination surveillance is still ongoing, and these results impose caution but need further confirmation.

13. Bell's Palsy and Cranial Neuropathies

Peripheral facial nerve (Bell) palsy occurrence has been associated with physical stress, pregnancy, cancer, infections, and vaccination [185]. Since the COVID-19 vaccination trials, this neurological condition has been reported as a possible adverse effect. In the Pfizer–BioNTech clinical trial, four cases of Bell's palsy were described in the vaccine group and no cases in the placebo arm. Similarly, in the Moderna trial, three cases of Bell's palsy were observed in the vaccine group and one in the placebo arm [186]. Since then, great attention has been paid to this aspect. In several studies, Bell's palsy has been observed after COVID-19 vaccination at a frequency higher than expected [186,187]. In addition, compared with other viral vaccination, facial paralysis was reported significantly more frequently after COVID-19 vaccination than other vaccines, with a higher risk for males and older individuals, specifically over 65-year-old [188]. However, other reports observed no association between facial nerve palsy and the vaccination status [189,190], calling for further surveillance in larger cohorts after vaccination.

On the other hand, despite confirming a significant association between the administration of mRNA COVID-19 vaccines and the reporting of Bell's palsy, the analysis of Sato and colleagues on VAERS data from the US FDA suggested that the incidence of facial palsy after COVID-19 vaccination is comparable to that for influenza vaccines [191].

No differences between classical Bell's palsy and the COVID-19-associated vaccine palsy seem to be present, with an excellent response to early corticosteroid therapy [187,189]. Thus, the facial palsy outcome is generally favorable when isolated and not associated with other conditions, which should be considered in the diagnostic workup; in Patone and colleagues' work, 6% of the cohort suffering from Bell's palsy had a concurrent suspect diagnosis of cerebral infarction [13].

Since Bell's palsy is considered a rare complication with a high recovery rate, COVID-19 vaccination benefits outweigh the possible link with its occurrence. In addition, a study reported a higher-than-expected rate of Bell's palsy after SARS-CoV-2 infection, which the vaccine should prevent [192].

Olfactory dysfunction, the most frequent neurological complication after COVID-19, can also be a rare adverse effect of COVID-19 vaccine administration [193]. Lechien and colleagues reported six cases of post-COVID-19 vaccination olfactory and gustatory disturbances, which recovered within seven weeks, and none of the patients reported long-term disorders [194]. A rare case of phantosmia after the COVID-19 vaccine showing MRI evidence of enhancement of the olfactory bulbs and tracts has been described [195].

In isolated cases, mRNA vaccination has been described to be associated with sixth and fourth cranial nerve palsy [196]. Cases of trigeminal neuralgia have also been reported, showing a good response after steroid treatment [197,198]. Post COVID-19 vaccination, otologic complications have been described and include hearing loss, tinnitus, dizziness, and vertigo [199]. Although vertigo was frequently suggested as a common neurological side effect after vaccination, only a few cases of vestibular neuropathy have been reported [200,201].

14. Other Peripheral Nervous System Disorders

Parsonage–Turner syndrome or neuralgic amyotrophy is a rare neurological disorder typically characterized by the abrupt onset of unilateral shoulder pain followed by progressive brachial motor weakness [202]. The etiology of this disorder is unclear, but an inflammation of the brachial plexus is often evident. The syndrome has been associated with the post-surgery, post-traumatic, post-infectious, and post-vaccination state [203]. Several cases of Parsonage–Turner syndrome have been described following the COVID-19 vaccination [204]. The onset of the disorder has been generally reported within ten days from the vaccination, mostly hitting people in the third-to-fifth decade [205]. Neurophysiological and imaging tests sustain the diagnosis, showing altered brachial plexus nerve conduction studies and enlargement of the brachial plexus at the MRI, respectively [206]. As in previously described Parsonage-Turner cases, a good response to high-dose steroids has been reported [207,208]. There is no association with a specific vaccine. The largest Parsonage–Turner case series described twelve patients who developed the syndrome after one of the four major vaccines (five received Pfizer–BioNTech, four Oxford–AstraZeneca, two Janssen, and one had Moderna), generally subsequently to the first dose. Intriguingly the involved arm was homolateral to the site of vaccine injection [209].

Varicella-Zoster Virus (VZV) reactivation after COVID-19 vaccination is reported. Overall, VZV reactivation causing shingles may be due to cell-mediated immunity dysfunction as in aging, diabetes, physical and psychological stress, and, rarely, after vaccination [210]. Few studies described VZV reactivation in patients suffering from autoimmune, such as Sjogren's syndrome, rheumatoid arthritis, systemic lupus erythematosus, and autoimmune inflammatory rheumatic disease, with recovery after antiviral therapy [211,212]. VZV reactivation has also been described in immunocompetent patients [213]. The seven cases described by Psichogiou and colleagues were older than 50 years, without risk factors for the immunosuppressive state, and they all received the Pfizer–BioNTech vaccine. The herpes zoster appeared seven to 20 days after the vaccination, and six out of seven patients recovered with valacyclovir treatment [214]. VZV reactivation is reported mostly after mRNA COVID-19 vaccination, especially with the Pfizer–BioNTech vaccine, and usually after the first dose [215]. However, the link between HZV reactivation and COVID-19 vaccination is not undeniable. In the Israelian observational historical cohort study conducted by Shasha and colleagues, the development of Herpes Zoster after COVID-19 vaccination occurred with a rate of 55.2 per 10,000 person-years vs. 51.5 cases per 10,000 person-years in the control unvaccinated group, thus showing a non-significant increase in the vaccinated population [190].

Few biopsy-proven small fiber neuropathy cases after receiving COVID-19 vaccination have been described. A 57-year-old female presented a subacute onset of intense, distal, burning dysesthesias after the second dose of the Pfizer vaccine, showing a good response to gabapentin symptomatic treatment [216]. A 43-year-old man developed neurosarcoidosis and small fiber neuropathy three days after the first dose of the Pfizer vaccine [93]. The development of autoantibodies targeting neuronal proteins via molecular mimicry has been proposed as the pathogenic mechanism for small fiber neuropathy, which has also been described after the Moderna vaccine [217].

15. Functional Neurological Disorders

The COVID-19 pandemic had a dramatic psychological impact. A higher rate of behavioral and neuropsychiatric disturbances has been described in the general population, healthcare workers, and people suffering from neurological conditions [218–220]. On the other hand, the psychological stress related to mass vaccination may trigger functional or psychogenic disorders [221]. Functional neurological disorders have been described after COVID-19 vaccination and are usually characterized by sudden onset, inconsistency over time, and normal diagnostic workup [222]. Anxiety-related events, feeling syncope, and dizziness are prevalent [223]. Functional neurological manifestations may be highly variable and include motor and sensory deficits mimicking strokes [224], episodic loss of consciousness and pseudo crisis [225], and bizarre movement disorders [226]. Multiple functional neurological disturbances, including bilateral facial palsy, hemiparesis, and facial hypoesthesia, have been described in a man as occurring after every single dose of vaccine [227]. The occurrence of psychogenic neurological disorders is a particularly delicate theme since they can devastate public opinion [10], feeding the debate against the COVID-19 mass vaccination and favoring the hesitancy phenomenon.

16. Discussion

In the vaccine human history, all kinds of vaccines have been associated with the possible appearance of neurological complications. COVID-19 vaccines are not free from neurological side effects, which have been reported since the first months of mass vaccination.

Four major types of vaccines are at the moment available for the COVID-19 vaccines: DNA-based vaccines, mRNA-based vaccines, protein-based vaccines, and inactivated viruses. DNA-based vaccines use viral vectors to introduce the DNA coding for the spike protein. The mRNA vaccines use lipid nanoparticles to introduce mRNA into cells. Protein-based vaccines use the spike protein or its fragments. Lastly, other vaccines are based on inactivated virus [9].

Regarding the pathogenesis of COVD-19 vaccine-associated adverse events, several proposed mechanisms have been hypothesized based on the type of immunization. The mRNA vaccines play their action by inducing the encoding of viral antigens by the host cells. The elicited immune response stimulates CD4 and CD8 lymphocytes, eventually forming memory B cells. Viral vector vaccines are based on administering non-pathogenic viruses carrying antigen genes of the target (SARS-CoV-2) virus. Thus, the expression of the antigenic protein of the target virus induces the immune response [17,228]. Vaccines can cause general adverse events, which are strictly linked to the specific characteristics of the vaccine, and include local pain and edema, malaise, and tiredness. In addition, systemic and more severe events can be triggered by the vaccine-related immune response or associated with allergic reactions [17]. Human cells' spike protein expression might trigger an inflammatory reaction leading to neurological autoimmune complications [82,85,89]. On the other hand, severe allergic reactions are rare but reported events since all vaccines (and the vaccine components) may induce anaphylaxis in some people [228].

Any patient or healthcare provider can report vaccine side effects through the Centers for Disease Control VAERS. However, a VAERS database limitation is that it is based on passive surveillance, therefore possibly reporting bias and errors [9]. Real-world data provided reports on a broad spectrum of severe neurological complications following COVID-19 vaccination. However, the real-world evidence is reassuring [229]. Large-scale epidemiological studies are needed to appropriately investigate the effective frequency of neurological complications after the COVID-19 vaccine.

Our review discusses the neurological side effects of COVID-19 vaccinations that are usually mild, short-duration, and self-limiting [230]. Severe unexpected post-vaccination complications that might occur due to molecular mimicry and subsequent neuronal damage are rare but possible events. Most severe neurological complications are reported in isolated case reports or small case series; therefore, a causal association between these adverse events is controversial. Establishing a clear relationship between the COVID-19 vaccine and autoimmune events is difficult. It is mandatory to have ongoing surveillance and reporting of adverse events associated with COVID-19 vaccines to ensure transparency concerning potential risks to patients.

The fear of neurological complications fed the "vaccine hesitancy" [10]. COVID-19 vaccinations are usually well-tolerated, and the benefit for the global population outweighs the relatively rare mild-to-moderate side effects. In addition, the SARS-CoV-2 infection seems to be associated with an increased risk of neurological complications, which exceeds that of all the COVID-19 vaccines [192]. The risk of all neurological complications in the month after a SARS-CoV-2 infection is substantially higher than the risk of neurological side effects after COVID-19 vaccination [192].

The theme of vaccination remains of utmost importance since recent evidence confirmed the efficacy of the vaccine in preventing the COVID-19 infection but also suggests the waning of the immunity over time [231]. An important communication challenge remains, especially when considering individuals developing neurological complications and the impact of such effects on public opinion. Clinicians should discuss this occurrence with the patients and their families, especially prospecting new future pandemic waves. Vaccines are considered safe and effective drugs, but adverse effects are inevitable, especially during mass immunization, which allowed us to crush the pandemic.

Author Contributions: Conceptualization, G.T. and C.C.; methodology, G.T.; software, A.B.; investigation, G.T., E.V. and A.N.; resources, C.C.; data curation, G.T. and E.V.; writing—original draft preparation, G.T., E.V. and A.N.; writing—review and editing, G.T. and C.C.; visualization, C.C.; supervision, C.C.; funding acquisition, C.C. All authors have read and agreed to the published version of the manuscript.

Funding: This research received no external funding.

Institutional Review Board Statement: Not applicable.

Informed Consent Statement: Not applicable.

Data Availability Statement: Not applicable.

Acknowledgments: The graphical abstract was created with Biorender.com (Agreement number DJ24B0VW2E). Publication and Licensing Rights are available upon request to the Author (E.V.).

Conflicts of Interest: The authors declare no conflict of interest.

References

1. Platto, S.; Wang, Y.; Zhou, J.; Carafoli, E. History of the COVID-19 pandemic: Origin, explosion, worldwide spreading. *Biochem. Biophys. Res. Commun.* **2021**, *538*, 14–23. [CrossRef] [PubMed]
2. Shahcheraghi, S.H.; Ayatollahi, J.; Aljabali, A.A.; Shastri, M.D.; Shukla, S.D.; Chellappan, D.K.; Jha, N.K.; Anand, K.; Katari, N.K.; Mehta, M.; et al. An overview of vaccine development for COVID-19. *Ther. Deliv.* **2021**, *12*, 235–244. [CrossRef] [PubMed]
3. Francis, A.I.; Ghany, S.; Gilkes, T.; Umakanthan, S. Review of COVID-19 vaccine subtypes, efficacy and geographical distributions. *Postgrad. Med. J.* **2022**, *98*, 389–394. [CrossRef] [PubMed]
4. Pavord, S.; Scully, M.; Hunt, B.J.; Lester, W.; Bagot, C.; Craven, B.; Rampotas, A.; Ambler, G.; Makris, M. Clinical Features of Vaccine-Induced Immune Thrombocytopenia and Thrombosis. *N. Engl. J. Med.* **2021**, *385*, 1680–1689. [CrossRef]
5. Assiri, S.A.; Althaqafi, R.M.M.; Alswat, K.; Alghamdi, A.A.; Alomairi, N.E.; Nemenqani, D.M.; Ibrahim, Z.S.; Elkady, A. Post COVID-19 Vaccination-Associated Neurological Complications. *Neuropsychiatr. Dis. Treat.* **2022**, *18*, 137–154. [CrossRef]
6. Sallam, M.; Al-Sanafi, M.; Sallam, M. A Global Map of COVID-19 Vaccine Acceptance Rates per Country: An Updated Concise Narrative Review. *J. Multidiscip. Healthc.* **2022**, *15*, 21–45. [CrossRef]
7. García-Grimshaw, M.; Ceballos-Liceaga, S.E.; Hernández-Vanegas, L.E.; Núñez, I.; Hernández-Valdivia, N.; Carrillo-García, D.A.; Michel-Chávez, A.; Galnares-Olalde, J.A.; Carbajal-Sandoval, G.; Del Mar Saniger-Alba, M.; et al. Neurologic adverse events among 704,003 first-dose recipients of the BNT162b2 mRNA COVID-19 vaccine in Mexico: A nationwide descriptive study. *Clin. Immunol.* **2021**, *229*, 108786. [CrossRef]
8. Garg, R.K.; Paliwal, V.K. Spectrum of neurological complications following COVID-19 vaccination. *Neurol. Sci.* **2022**, *43*, 3–40. [CrossRef]
9. Goss, A.L.; Samudralwar, R.D.; Das, R.R.; Nath, A. ANA Investigates: Neurological Complications of COVID-19 Vaccines. *Ann. Neurol.* **2021**, *89*, 856–857. [CrossRef]
10. Kim, D.D.; Kung, C.S.; Perez, D.L. Helping the Public Understand Adverse Events Associated With COVID-19 Vaccinations. *JAMA Neurol.* **2021**, *78*, 789. [CrossRef]
11. Lee, Y.W.; Lim, S.Y.; Lee, J.H.; Lim, J.S.; Kim, M.; Kwon, S.; Joo, J.; Kwak, S.H.; Kim, E.O.; Jung, J.; et al. Adverse Reactions of the Second Dose of the BNT162b2 mRNA COVID-19 Vaccine in Healthcare Workers in Korea. *J. Korean Med. Sci.* **2021**, *36*, e153. [CrossRef] [PubMed]
12. Maruyama, A.; Sawa, T.; Teramukai, S.; Katoh, N. Adverse reactions to the first and second doses of Pfizer-BioNTech COVID-19 vaccine among healthcare workers. *J. Infect. Chemother.* **2022**, *28*, 934–942. [CrossRef] [PubMed]
13. Patone, M.; Handunnetthi, L.; Saatci, D.; Pan, J.; Katikireddi, S.V.; Razvi, S.; Hunt, D.; Mei, X.W.; Dixon, S.; Zaccardi, F.; et al. Neurological complications after first dose of COVID-19 vaccines and SARS-CoV-2 infection. *Nat. Med.* **2021**, *27*, 2144–2153. [CrossRef] [PubMed]
14. Klugar, M.; Riad, A.; Mekhemar, M.; Conrad, J.; Buchbender, M.; Howaldt, H.P.; Attia, S. Side Effects of mRNA-Based and Viral Vector-Based COVID-19 Vaccines among German Healthcare Workers. *Biology* **2021**, *10*, 752. [CrossRef]
15. Lounis, M.; Rais, M.A.; Bencherit, D.; Aouissi, H.A.; Oudjedi, A.; Klugarová, J.; Pokorná, A.; Klugar, M.; Riad, A. Side Effects of COVID-19 Inactivated Virus vs. Adenoviral Vector Vaccines: Experience of Algerian Healthcare Workers. *Front Public Health* **2022**, *10*, 896343. [CrossRef]
16. Menni, C.; Klaser, K.; May, A.; Polidori, L.; Capdevila, J.; Louca, P.; Sudre, C.H.; Nguyen, L.H.; Drew, D.A.; Merino, J.; et al. Vaccine side-effects and SARS-CoV-2 infection after vaccination in users of the COVID Symptom Study app in the UK: A prospective observational study. *Lancet Infect. Dis.* **2021**, *21*, 939–949. [CrossRef]
17. Chen, L.; Cai, X.; Zhao, T.; Han, B.; Xie, M.; Cui, J.; Zhang, J.; Wang, C.; Liu, B.; Lu, Q.; et al. Safety of Global SARS-CoV-2 Vaccines, a Meta-Analysis. *Vaccines* **2022**, *10*, 596. [CrossRef]
18. Bardage, C.; Persson, I.; Ortqvist, A.; Bergman, U.; Ludvigsson, J.F.; Granath, F. Neurological and autoimmune disorders after vaccination against pandemic influenza A (H1N1) with a monovalent adjuvanted vaccine: Population based cohort study in Stockholm, Sweden. *BMJ* **2011**, *343*, d5956. [CrossRef]
19. Mäkelä, A.; Nuorti, J.P.; Peltola, H. Neurologic disorders after measles-mumps-rubella vaccination. *Pediatrics* **2002**, *110*, 957–963. [CrossRef]
20. Zhu, F.C.; Li, Y.H.; Guan, X.H.; Hou, L.H.; Wang, W.J.; Li, J.X.; Wu, S.P.; Wang, B.S.; Wang, Z.; Wang, L.; et al. Safety, tolerability, and immunogenicity of a recombinant adenovirus type-5 vectored COVID-19 vaccine: A dose-escalation, open-label, non-randomised, first-in-human trial. *Lancet* **2020**, *395*, 1845–1854. [CrossRef]
21. Ramasamy, M.N.; Minassian, A.M.; Ewer, K.J.; Flaxman, A.L.; Folegatti, P.M.; Owens, D.R.; Voysey, M.; Aley, P.K.; Angus, B.; Babbage, G.; et al. Safety and immunogenicity of ChAdOx1 nCoV-19 vaccine administered in a prime-boost regimen in young and old adults (COV002): A single-blind, randomised, controlled, phase 2/3 trial. *Lancet* **2021**, *396*, 1979–1993. [CrossRef]
22. Göbel, C.H.; Heinze, A.; Karstedt, S.; Morscheck, M.; Tashiro, L.; Cirkel, A.; Hamid, Q.; Halwani, R.; Temsah, M.H.; Ziemann, M.; et al. Clinical characteristics of headache after vaccination against COVID-19 (coronavirus SARS-CoV-2) with the BNT162b2 mRNA vaccine: A multicentre observational cohort study. *Brain Commun.* **2021**, *3*, fcab169. [CrossRef]

23. Sekiguchi, K.; Watanabe, N.; Miyazaki, N.; Ishizuchi, K.; Iba, C.; Tagashira, Y.; Uno, S.; Shibata, M.; Hasegawa, N.; Takemura, R.; et al. Incidence of headache after COVID-19 vaccination in patients with history of headache: A cross-sectional study. *Cephalalgia* **2022**, *42*, 266–272. [CrossRef]
24. Mattiuzzi, C.; Lippi, G. Headache after COVID-19 vaccination: Updated report from the Italian Medicines Agency database. *Neurol. Sci.* **2021**, *42*, 3531–3532. [CrossRef]
25. García-Azorín, D.; Do, T.P.; Gantenbein, A.R.; Hansen, J.M.; Souza, M.N.P.; Obermann, M.; Pohl, H.; Schankin, C.J.; Schytz, H.W.; Sinclair, A.; et al. Delayed headache after COVID-19 vaccination: A red flag for vaccine induced cerebral venous thrombosis. *J. Headache Pain* **2021**, *22*, 108. [CrossRef]
26. Tsivgoulis, G.; Katsanos, A.H.; Zand, R.; Ishfaq, M.F.; Malik, M.T.; Karapanayiotides, T.; Voumvourakis, K.; Tsiodras, S.; Parissis, J. The association of adult vaccination with the risk of cerebrovascular ischemia: A systematic review and meta-analysis. *J. Neurol. Sci.* **2018**, *386*, 12–18. [CrossRef]
27. Hernández, A.F.; Calina, D.; Poulas, K.; Docea, A.O.; Tsatsakis, A.M. Safety of COVID-19 vaccines administered in the EU: Should we be concerned? *Toxicol. Rep.* **2021**, *8*, 871–879. [CrossRef]
28. De Gregorio, C.; Colarusso, L.; Calcaterra, G.; Bassareo, P.P.; Ieni, A.; Mazzeo, A.T.; Ferrazzo, G.; Noto, A.; Koniari, I.; Mehta, J.L.; et al. Cerebral Venous Sinus Thrombosis following COVID-19 Vaccination: Analysis of 552 Worldwide Cases. *Vaccines* **2022**, *10*, 232. [CrossRef]
29. Sangli, S.; Virani, A.; Cheronis, N.; Vannatter, B.; Minich, C.; Noronha, S.; Bhagavatula, R.; Speredelozzi, D.; Sareen, M.; Kaplan, R.B. Thrombosis With Thrombocytopenia After the Messenger RNA-1273 Vaccine. *Ann. Intern. Med.* **2021**, *174*, 1480–1482. [CrossRef]
30. See, I.; Lale, A.; Marquez, P.; Streiff, M.B.; Wheeler, A.P.; Tepper, N.K.; Woo, E.J.; Broder, K.R.; Edwards, K.M.; Gallego, R.; et al. Case Series of Thrombosis With Thrombocytopenia Syndrome After COVID-19 Vaccination-United States, December 2020 to August 2021. *Ann. Intern. Med.* **2022**, *175*, 513–522. [CrossRef]
31. Pishko, A.M.; Cuker, A. Thrombosis After Vaccination With Messenger RNA-1273: Is This Vaccine-Induced Thrombosis and Thrombocytopenia or Thrombosis With Thrombocytopenia Syndrome? *Ann. Intern. Med.* **2021**, *174*, 1468–1469. [CrossRef]
32. Oldenburg, J.; Klamroth, R.; Langer, F.; Albisetti, M.; von Auer, C.; Ay, C.; Korte, W.; Scharf, R.E.; Pötzsch, B.; Greinacher, A. Diagnosis and Management of Vaccine-Related Thrombosis following AstraZeneca COVID-19 Vaccination: Guidance Statement from the GTH. *Hamostaseologie* **2021**, *41*, 184–189. [CrossRef]
33. Ahmed, I.; Majeed, A.; Powell, R. Heparin induced thrombocytopenia: Diagnosis and management update. *Postgrad. Med. J.* **2007**, *83*, 575–582. [CrossRef]
34. Reilly, R.F. The pathophysiology of immune-mediated heparin-induced thrombocytopenia. *Semin. Dial.* **2003**, *16*, 54–60. [CrossRef]
35. Kelton, J.G.; Smith, J.W.; Warkentin, T.E.; Hayward, C.P.; Denomme, G.A.; Horsewood, P. Immunoglobulin G from patients with heparin-induced thrombocytopenia binds to a complex of heparin and platelet factor 4. *Blood* **1994**, *83*, 3232–3239. [CrossRef]
36. Bakchoul, T.; Greinacher, A. Recent advances in the diagnosis and treatment of heparin-induced thrombocytopenia. *Ther. Adv. Hematol.* **2012**, *3*, 237–251. [CrossRef]
37. Tsilingiris, D.; Vallianou, N.G.; Karampela, I.; Dalamaga, M. Vaccine induced thrombotic thrombocytopenia: The shady chapter of a success story. *Metabol. Open* **2021**, *11*, 100101. [CrossRef]
38. Greinacher, A.; Schönborn, L.; Siegerist, F.; Steil, L.; Palankar, R.; Handtke, S.; Reder, A.; Thiele, T.; Aurich, K.; Methling, K.; et al. Pathogenesis of vaccine-induced immune thrombotic thrombocytopenia (VITT). *Semin. Hematol.* **2022**, *59*, 97–107. [CrossRef] [PubMed]
39. Abrignani, M.G.; Murrone, A.; De Luca, L.; Roncon, L.; Di Lenarda, A.; Valente, S.; Caldarola, P.; Riccio, C.; Oliva, F.; Gulizia, M.M.; et al. COVID-19, Vaccines, and Thrombotic Events: A Narrative Review. *J. Clin. Med.* **2022**, *11*, 948. [CrossRef] [PubMed]
40. Greinacher, A.; Selleng, K.; Palankar, R.; Wesche, J.; Handtke, S.; Wolff, M.; Aurich, K.; Lalk, M.; Methling, K.; Völker, U.; et al. Insights in ChAdOx1 nCoV-19 vaccine-induced immune thrombotic thrombocytopenia. *Blood* **2021**, *138*, 2256–2268. [CrossRef] [PubMed]
41. Franchini, M.; Liumbruno, G.M.; Pezzo, M. COVID-19 vaccine-associated immune thrombosis and thrombocytopenia (VITT): Diagnostic and therapeutic recommendations for a new syndrome. *Eur. J. Haematol.* **2021**, *107*, 173–180. [CrossRef]
42. Cines, D.B.; Bussel, J.B. SARS-CoV-2 Vaccine-Induced Immune Thrombotic Thrombocytopenia. *N. Engl. J. Med.* **2021**, *384*, 2254–2256. [CrossRef]
43. Kim, A.Y.; Woo, W.; Yon, D.K.; Lee, S.W.; Yang, J.W.; Kim, J.H.; Park, S.; Koyanagi, A.; Kim, M.S.; Lee, S.; et al. Thrombosis patterns and clinical outcome of COVID-19 vaccine-induced immune thrombotic thrombocytopenia: A Systematic Review and Meta-Analysis. *Int. J. Infect. Dis.* **2022**, *119*, 130–139. [CrossRef]
44. McGonagle, D.; De Marco, G.; Bridgewood, C. Mechanisms of Immunothrombosis in Vaccine-Induced Thrombotic Thrombocytopenia (VITT) Compared to Natural SARS-CoV-2 Infection. *J. Autoimmun.* **2021**, *121*, 102662. [CrossRef]
45. Mirandola, L.; Arena, G.; Pagliaro, M.; Boghi, A.; Naldi, A.; Castellano, D.; Vaccarino, A.; Silengo, D.; Aprà, F.; Cavallo, R.; et al. Massive cerebral venous sinus thrombosis in vaccine-induced immune thrombotic thrombocytopenia after ChAdOx1 nCoV-19 serum: Case report of a successful multidisciplinary approach. *Neurol. Sci.* **2022**, *43*, 1499–1502. [CrossRef]

46. Sharifian-Dorche, M.; Bahmanyar, M.; Sharifian-Dorche, A.; Mohammadi, P.; Nomovi, M.; Mowla, A. Vaccine-induced immune thrombotic thrombocytopenia and cerebral venous sinus thrombosis post COVID-19 vaccination; a systematic review. *J. Neurol. Sci.* **2021**, *428*, 117607. [CrossRef]
47. Kolahchi, Z.; Khanmirzaei, M.; Mowla, A. Acute ischemic stroke and vaccine-induced immune thrombotic thrombocytopenia post COVID-19 vaccination; a systematic review. *J. Neurol. Sci.* **2022**, *439*, 120327. [CrossRef]
48. Greinacher, A.; Langer, F.; Makris, M.; Pai, M.; Pavord, S.; Tran, H.; Warkentin, T.E. Vaccine-induced immune thrombotic thrombocytopenia (VITT): Update on diagnosis and management considering different resources. *J. Thromb. Haemost.* **2022**, *20*, 149–156. [CrossRef]
49. Van de Munckhof, A.; Krzywicka, K.; Aguiar de Sousa, D.; Sánchez van Kammen, M.; Heldner, M.R.; Jood, K.; Lindgren, E.; Tatlisumak, T.; Putaala, J.; Kremer Hovinga, J.A.; et al. Declining mortality of cerebral venous sinus thrombosis with thrombocytopenia after SARS-CoV-2 vaccination. *Eur. J. Neurol.* **2022**, *29*, 339–344. [CrossRef]
50. Nazy, I.; Sachs, U.J.; Arnold, D.M.; McKenzie, S.E.; Choi, P.; Althaus, K.; Ahlen, M.T.; Sharma, R.; Grace, R.F.; Bakchoul, T. Recommendations for the clinical and laboratory diagnosis of VITT against COVID-19: Communication from the ISTH SSC Subcommittee on Platelet Immunology. *J. Thromb. Haemost.* **2021**, *19*, 1585–1588. [CrossRef]
51. Aleem, A.; Nadeem, A.J. Coronavirus (COVID-19) Vaccine-Induced Immune Thrombotic Thrombocytopenia (VITT). In *StatPearls*; StatPearls Publishing LLC.: Treasure Island, FL, USA, 2022.
52. Zazzeron, L.; Rosovsky, R.P.; Bittner, E.A.; Chang, M.G. Comparison of Published Guidelines for the Diagnosis and the Management of Vaccine-Induced Immune Thrombotic Thrombocytopenia. *Crit. Care Explor.* **2021**, *3*, e0519. [CrossRef]
53. Gresele, P.; Marietta, M.; Ageno, W.; Marcucci, R.; Contino, L.; Donadini, M.P.; Russo, L.; Tiscia, G.L.; Palareti, G.; Tripodi, A.; et al. Management of cerebral and splanchnic vein thrombosis associated with thrombocytopenia in subjects previously vaccinated with Vaxzevria (AstraZeneca): A position statement from the Italian Society for the Study of Haemostasis and Thrombosis (SISET). *Blood Transfus.* **2021**, *19*, 281–283. [CrossRef]
54. Favaloro, E.J.; Pasalic, L.; Lippi, G. Review and evolution of guidelines for diagnosis of COVID-19 vaccine induced thrombotic thrombocytopenia (VITT). *Clin. Chem. Lab. Med.* **2022**, *60*, 7–17. [CrossRef]
55. Gurjar, H.; Dhallu, M.; Lvovsky, D.; Sadullah, S.; Chilimuri, S. A Rare Case of Coronavirus Disease 2019 Vaccine-Associated Cerebral Venous Sinus Thrombosis Treated with Mechanical Thrombectomy. *Am. J. Case Rep.* **2022**, *23*, e935355. [CrossRef]
56. Chew, H.S.; Al-Ali, S.; Butler, B.; Rajapakse, D.; Nader, K.; Chavda, S.; Lamin, S. Mechanical Thrombectomy for Treatment of Cerebral Venous Sinus Thrombosis in Vaccine-Induced Immune Thrombotic Thrombocytopenia. *AJNR Am. J. Neuroradiol.* **2022**, *43*, 98–101. [CrossRef]
57. Warkentin, T.E. Vaccine-induced immune thrombotic thrombocytopenia (VITT). *Semin. Hematol.* **2022**, *59*, 57–58. [CrossRef]
58. Ikenberg, B.; Demleitner, A.F.; Thiele, T.; Wiestler, B.; Götze, K.; Mößmer, G.; Lingor, P. Cerebral venous sinus thrombosis after ChAdOx1 nCov-19 vaccination with a misleading first cerebral MRI scan. *Stroke Vasc. Neurol.* **2021**, *6*, 668–670. [CrossRef]
59. Dias, L.; Soares-Dos-Reis, R.; Meira, J.; Ferrão, D.; Soares, P.R.; Pastor, A.; Gama, G.; Fonseca, L.; Fagundes, V.; Carvalho, M. Cerebral Venous Thrombosis after BNT162b2 mRNA SARS-CoV-2 vaccine. *J. Stroke Cerebrovasc. Dis.* **2021**, *30*, 105906. [CrossRef]
60. Dutta, A.; Ghosh, R.; Bhattacharya, D.; Bhat, S.; Ray, A.; Pandit, A.; Das, S.; Dubey, S. Anti-PF4 antibody negative cerebral venous sinus thrombosis without thrombocytopenia following immunization with COVID-19 vaccine in an elderly non-comorbid Indian male, managed with conventional heparin-warfarin based anticoagulation. *Diabetes Metab. Syndr.* **2021**, *15*, 102184. [CrossRef] [PubMed]
61. Di Pietro, M.; Dono, F.; Consoli, S.; Evangelista, G.; Pozzilli, V.; Calisi, D.; Barbone, F.; Bonanni, L.; Onofrj, M.; De Angelis, M.V.; et al. Cerebral venous thrombosis without thrombocytopenia after a single dose of COVID-19 (Ad26.COV2.S) vaccine injection: A case report. *Neurol. Sci.* **2022**, *43*, 2951–2956. [CrossRef] [PubMed]
62. Perry, R.J.; Tamborska, A.; Singh, B.; Craven, B.; Marigold, R.; Arthur-Farraj, P.; Yeo, J.M.; Zhang, L.; Hassan-Smith, G.; Jones, M.; et al. Cerebral venous thrombosis after vaccination against COVID-19 in the UK: A multicentre cohort study. *Lancet* **2021**, *398*, 1147–1156. [CrossRef]
63. Palaiodimou, L.; Stefanou, M.I.; de Sousa, D.A.; Coutinho, J.M.; Papadopoulou, M.; Papaevangelou, V.; Vassilakopoulos, T.I.; Tsiodras, S.; Filippou, D.K.; Tsivgoulis, G. Cerebral venous sinus thrombosis in the setting of COVID-19 vaccination: A systematic review and meta-analysis. *J. Neurol.* **2022**, *269*, 3413–3419. [CrossRef]
64. Krzywicka, K.; Heldner, M.R.; Sánchez van Kammen, M.; van Haaps, T.; Hiltunen, S.; Silvis, S.M.; Levi, M.; Kremer Hovinga, J.A.; Jood, K.; Lindgren, E.; et al. Post-SARS-CoV-2-vaccination cerebral venous sinus thrombosis: An analysis of cases notified to the European Medicines Agency. *Eur. J. Neurol.* **2021**, *28*, 3656–3662. [CrossRef]
65. Yoshida, K.; Tanaka, K.; Suto, Y.; Fukuda, H. Repeated Cardioembolic Stroke after COVID-19 mRNA Vaccination: A Case Report. *J. Stroke Cerebrovasc. Dis.* **2022**, *31*, 106233. [CrossRef]
66. Schulz, J.B.; Berlit, P.; Diener, H.C.; Gerloff, C.; Greinacher, A.; Klein, C.; Petzold, G.C.; Piccininni, M.; Poli, S.; Röhrig, R.; et al. COVID-19 Vaccine-Associated Cerebral Venous Thrombosis in Germany. *Ann. Neurol.* **2021**, *90*, 627–639. [CrossRef]
67. Kakovan, M.; Ghorbani Shirkouhi, S.; Zarei, M.; Andalib, S. Stroke Associated with COVID-19 Vaccines. *J. Stroke Cerebrovasc. Dis.* **2022**, *31*, 106440. [CrossRef]
68. Kenda, J.; Lovrič, D.; Škerget, M.; Milivojević, N. Treatment of ChAdOx1 nCoV-19 Vaccine-Induced Immune Thrombotic Thrombocytopenia Related Acute Ischemic Stroke. *J. Stroke Cerebrovasc. Dis.* **2021**, *30*, 106072. [CrossRef]

69. Bayas, A.; Menacher, M.; Christ, M.; Behrens, L.; Rank, A.; Naumann, M. Bilateral superior ophthalmic vein thrombosis, ischaemic stroke, and immune thrombocytopenia after ChAdOx1 nCoV-19 vaccination. *Lancet* **2021**, *397*, e11. [CrossRef]
70. Pang, E.; Ghosh, S.; Chemmanam, T.; Grove, C.; Phillips, T. Cerebral arterial and venous thrombosis due to COVID-19 vaccine-induced immune thrombotic thrombocytopenia. *BMJ Case Rep.* **2022**, *15*, e245445. [CrossRef]
71. Mancuso, M.; Lauretti, D.L.; Cecconi, N.; Santini, M.; Lami, V.; Orlandi, G.; Casagli, S.; Ghetta, A.; Liberti, G.; Elena, B.M.; et al. Arterial intracranial thrombosis as the first manifestation of vaccine-induced immune thrombotic thrombocytopenia (VITT): A case report. *Neurol. Sci.* **2022**, *43*, 2085–2089. [CrossRef]
72. Al-Mayhani, T.; Saber, S.; Stubbs, M.J.; Losseff, N.A.; Perry, R.J.; Simister, R.J.; Gull, D.; Jäger, H.R.; Scully, M.A.; Werring, D.J. Ischaemic stroke as a presenting feature of ChAdOx1 nCoV-19 vaccine-induced immune thrombotic thrombocytopenia. *J. Neurol. Neurosurg. Psychiatry* **2021**, *92*, 1247–1248. [CrossRef] [PubMed]
73. Walter, U.; Fuchs, M.; Grossmann, A.; Walter, M.; Thiele, T.; Storch, A.; Wittstock, M. Adenovirus-Vectored COVID-19 Vaccine-Induced Immune Thrombosis of Carotid Artery: A Case Report. *Neurology* **2021**. [CrossRef] [PubMed]
74. De Michele, M.; Iacobucci, M.; Chistolini, A.; Nicolini, E.; Pulcinelli, F.; Cerbelli, B.; Merenda, E.; Schiavo, O.G.; Sbardella, E.; Berto, I.; et al. Malignant cerebral infarction after ChAdOx1 nCov-19 vaccination: A catastrophic variant of vaccine-induced immune thrombotic thrombocytopenia. *Nat. Commun.* **2021**, *12*, 4663. [CrossRef] [PubMed]
75. Lodigiani, C.; Iapichino, G.; Carenzo, L.; Cecconi, M.; Ferrazzi, P.; Sebastian, T.; Kucher, N.; Studt, J.D.; Sacco, C.; Bertuzzi, A.; et al. Venous and arterial thromboembolic complications in COVID-19 patients admitted to an academic hospital in Milan, Italy. *Thromb. Res.* **2020**, *191*, 9–14. [CrossRef]
76. De Mélo Silva, M.L., Jr.; Lopes, D.P. Large hemorrhagic stroke after ChAdOx1 nCoV-19 vaccination: A case report. *Acta Neurol. Scand.* **2021**, *144*, 717–718. [CrossRef]
77. Finsterer, J.; Korn, M. Aphasia seven days after second dose of an mRNA-based SARS-CoV-2 vaccine. *Brain Hemorrhages* **2021**, *2*, 165–167. [CrossRef]
78. Jabagi, M.J.; Botton, J.; Bertrand, M.; Weill, A.; Farrington, P.; Zureik, M.; Dray-Spira, R. Myocardial Infarction, Stroke, and Pulmonary Embolism After BNT162b2 mRNA COVID-19 Vaccine in People Aged 75 Years or Older. *JAMA* **2022**, *327*, 80–82. [CrossRef]
79. Barda, N.; Dagan, N.; Ben-Shlomo, Y.; Kepten, E.; Waxman, J.; Ohana, R.; Hernán, M.A.; Lipsitch, M.; Kohane, I.; Netzer, D.; et al. Safety of the BNT162b2 mRNA COVID-19 Vaccine in a Nationwide Setting. *N. Engl. J. Med.* **2021**, *385*, 1078–1090. [CrossRef]
80. Klein, N.P.; Lewis, N.; Goddard, K.; Fireman, B.; Zerbo, O.; Hanson, K.E.; Donahue, J.G.; Kharbanda, E.O.; Naleway, A.; Nelson, J.C.; et al. Surveillance for Adverse Events After COVID-19 mRNA Vaccination. *JAMA* **2021**, *326*, 1390–1399. [CrossRef]
81. Mungmunpuntipantip, R.; Wiwanitkit, V. COVID-19, neurovascular thrombotic problem and short summary on blood coagulation disorder: A brief review. *Egypt. J. Neurol. Psychiatry Neurosurg.* **2022**, *58*, 6. [CrossRef]
82. Sormani, M.P.; Schiavetti, I.; Carmisciano, L.; Cordioli, C.; Filippi, M.; Radaelli, M.; Immovilli, P.; Capobianco, M.; De Rossi, N.; Brichetto, G.; et al. COVID-19 Severity in Multiple Sclerosis. *Neurol. —Neuroimmunol. Neuroinflamm.* **2022**, *9*, e1105. [CrossRef]
83. Sormani, M.P.; Salvetti, M.; Labauge, P.; Schiavetti, I.; Zephir, H.; Carmisciano, L.; Bensa, C.; De Rossi, N.; Pelletier, J.; Cordioli, C.; et al. DMTs and COVID-19 severity in MS: A pooled analysis from Italy and France. *Ann. Clin. Transl. Neurol.* **2021**, *8*, 1738–1744. [CrossRef]
84. Coyle, P.K.; Gocke, A.; Vignos, M.; Newsome, S.D. Vaccine Considerations for Multiple Sclerosis in the COVID-19 Era. *Adv. Ther.* **2021**, *38*, 3550–3588. [CrossRef]
85. Sormani, M.P.; Schiavetti, I.; Landi, D.; Carmisciano, L.; De Rossi, N.; Cordioli, C.; Moiola, L.; Radaelli, M.; Immovilli, P.; Capobianco, M.; et al. SARS-CoV-2 serology after COVID-19 in multiple sclerosis: An international cohort study. *Mult. Scler. J.* **2022**, *28*, 1034–1040. [CrossRef]
86. Fujimori, J.; Miyazawa, K.; Nakashima, I. Initial clinical manifestation of multiple sclerosis after immunization with the Pfizer-BioNTech COVID-19 vaccine. *J. Neuroimmunol.* **2021**, *361*, 577755. [CrossRef]
87. Havla, J.; Schultz, Y.; Zimmermann, H.; Hohlfeld, R.; Danek, A.; Kümpfel, T. First manifestation of multiple sclerosis after immunization with the Pfizer-BioNTech COVID-19 vaccine. *J. Neurol.* **2022**, *269*, 55–58. [CrossRef]
88. Ismail, I.I.; Salama, S. A systematic review of cases of CNS demyelination following COVID-19 vaccination. *J. Neuroimmunol.* **2022**, *362*, 577765. [CrossRef]
89. Khayat-Khoei, M.; Bhattacharyya, S.; Katz, J.; Harrison, D.; Tauhid, S.; Bruso, P.; Houtchens, M.K.; Edwards, K.R.; Bakshi, R. COVID-19 mRNA vaccination leading to CNS inflammation: A case series. *J. Neurol.* **2022**, *269*, 1093–1106. [CrossRef]
90. Tagliaferri, A.R.; Horani, G.; Stephens, K.; Michael, P. A rare presentation of undiagnosed multiple sclerosis after the COVID-19 vaccine. *J. Community Hosp. Intern. Med. Perspect.* **2021**, *11*, 772–775. [CrossRef]
91. Toljan, K.; Amin, M.; Kunchok, A.; Ontaneda, D. New diagnosis of multiple sclerosis in the setting of mRNA COVID-19 vaccine exposure. *J. Neuroimmunol.* **2022**, *362*, 577785. [CrossRef]
92. Watad, A.; Quaresma, M.; Bragazzi, N.L.; Cervera, R.; Tervaert, J.W.C.; Amital, H.; Shoenfeld, Y. The autoimmune/inflammatory syndrome induced by adjuvants (ASIA)/Shoenfeld's syndrome: Descriptive analysis of 300 patients from the international ASIA syndrome registry. *Clin. Rheumatol.* **2018**, *37*, 483–493. [CrossRef]
93. Watad, A.; De Marco, G.; Mahajna, H.; Druyan, A.; Eltity, M.; Hijazi, N.; Haddad, A.; Elias, M.; Zisman, D.; Naffaa, M.E.; et al. Immune-Mediated Disease Flares or New-Onset Disease in 27 Subjects Following mRNA/DNA SARS-CoV-2 Vaccination. *Vaccines* **2021**, *9*, 435. [CrossRef]

94. Achiron, A.; Dolev, M.; Menascu, S.; Zohar, D.N.; Dreyer-Alster, S.; Miron, S.; Shirbint, E.; Magalashvili, D.; Flechter, S.; Givon, U.; et al. COVID-19 vaccination in patients with multiple sclerosis: What we have learnt by February 2021. *Mult. Scler.* **2021**, *27*, 864–870. [CrossRef]
95. Lotan, I.; Romanow, G.; Levy, M. Patient-reported safety and tolerability of the COVID-19 vaccines in persons with rare neuroimmunological diseases. *Mult. Scler. Relat. Disord.* **2021**, *55*, 103189. [CrossRef]
96. Lotan, I.; Wilf-Yarkoni, A.; Friedman, Y.; Stiebel-Kalish, H.; Steiner, I.; Hellmann, M.A. Safety of the BNT162b2 COVID-19 vaccine in multiple sclerosis (MS): Early experience from a tertiary MS center in Israel. *Eur. J. Neurol.* **2021**, *28*, 3742–3748. [CrossRef]
97. Virgilio, E.; Vecchio, D.; Vercellino, M.; Naldi, P.; Tesser, F.; Cantello, R.; Cavalla, P.; Comi, C. Paraneoplastic neuromyelitis optica spectrum disorders: A case series. *Neurol. Sci.* **2021**, *42*, 2519–2522. [CrossRef]
98. Vecchio, D.; Virgilio, E.; Naldi, P.; Comi, C.; Cantello, R. MOG-antibody demyelinating diseases: A case of post-partum severe rhombencephalitis and transverse myelitis. *Mult. Scler. Relat. Disord.* **2018**, *21*, 9–10. [CrossRef]
99. Anamnart, C.; Tisavipat, N.; Owattanapanich, W.; Apiwattanakul, M.; Savangned, P.; Prayoonwiwat, N.; Siritho, S.; Rattanathamsakul, N.; Jitprapaikulsan, J. Newly diagnosed neuromyelitis optica spectrum disorders following vaccination: Case report and systematic review. *Mult. Scler. Relat. Disord.* **2022**, *58*, 103414. [CrossRef]
100. Chen, S.; Fan, X.-R.; He, S.; Zhang, J.-W.; Li, S.-J. Watch out for neuromyelitis optica spectrum disorder after inactivated virus vaccination for COVID-19. *Neurol. Sci.* **2021**, *42*, 3537–3539. [CrossRef]
101. Fujikawa, P.; Shah, F.A.; Braford, M.; Patel, K.; Madey, J. Neuromyelitis Optica in a Healthy Female After Severe Acute Respiratory Syndrome Coronavirus 2 mRNA-1273 Vaccine. *Cureus* **2021**, *13*, e17961. [CrossRef]
102. Fernandes, J.; Jaggernauth, S.; Ramnarine, V.; Mohammed, S.R.; Khan, C.; Panday, A. Neurological Conditions Following COVID-19 Vaccinations: Chance or Association? *Cureus* **2022**, *14*, e21919. [CrossRef]
103. Koh, J.S.; Hoe, R.H.M.; Yong, M.H.; Chiew, H.J.; Goh, Y.; Yong, K.P.; Tu, T.M.; Chan, D.W.S.; Tan, B.Y.-Q.; Yeo, L.L.L.; et al. Hospital-based observational study of neurological disorders in patients recently vaccinated with COVID-19 mRNA vaccines. *J. Neurol. Sci.* **2021**, *430*, 120030. [CrossRef] [PubMed]
104. Caliskan, I.; Bulus, E.; Afsar, N.; Altintas, A. A Case With New-Onset Neuromyelitis Optica Spectrum Disorder Following COVID-19 mRNA BNT162b2 Vaccination. *Neurologist* **2022**, *27*, 147–150. [CrossRef] [PubMed]
105. Wright, S. MOG antibody-associated longitudinal extensive myelitis after Oxford-Astra Zeneca's COVID-19 vaccination. *Mult. Scler. J.* **2022**, *28*, 1163. [CrossRef] [PubMed]
106. Dams, L.; Kraemer, M.; Becker, J. MOG-antibody-associated longitudinal extensive myelitis after ChAdOx1 nCoV-19 vaccination. *Mult. Scler. J.* **2022**, *28*, 1159–1162. [CrossRef] [PubMed]
107. Voysey, M.; Clemens, S.A.C.; Madhi, S.A.; Weckx, L.Y.; Folegatti, P.M.; Aley, P.K.; Angus, B.; Baillie, V.L.; Barnabas, S.L.; Bhorat, Q.E.; et al. Safety and efficacy of the ChAdOx1 nCoV-19 vaccine (AZD1222) against SARS-CoV-2: An interim analysis of four randomised controlled trials in Brazil, South Africa, and the UK. *Lancet* **2021**, *397*, 99–111. [CrossRef]
108. Román, G.C.; Gracia, F.; Torres, A.; Palacios, A.; Gracia, K.; Harris, D. Acute Transverse Myelitis (ATM):Clinical Review of 43 Patients With COVID-19-Associated ATM and 3 Post-Vaccination ATM Serious Adverse Events With the ChAdOx1 nCoV-19 Vaccine (AZD1222). *Front. Immunol.* **2021**, *12*, 653786. [CrossRef]
109. Pagenkopf, C.; Südmeyer, M. A case of longitudinally extensive transverse myelitis following vaccination against COVID-19. *J. Neuroimmunol.* **2021**, *358*, 577606. [CrossRef]
110. Tan, W.Y.; Yusof Khan, A.H.K.; Mohd Yaakob, M.N.; Abdul Rashid, A.M.; Loh, W.C.; Baharin, J.; Ibrahim, A.; Ismail, M.R.; Inche Mat, L.N.; Wan Sulaiman, W.A.; et al. Longitudinal extensive transverse myelitis following ChAdOx1 nCOV-19 vaccine: A case report. *BMC Neurol.* **2021**, *21*, 1–4. [CrossRef]
111. Malhotra, H.S.; Gupta, P.; Prabhu, V.; Kumar Garg, R.; Dandu, H.; Agarwal, V. COVID-19 vaccination-associated myelitis. *QJM* **2021**, *114*, 591–593. [CrossRef]
112. Vegezzi, E.; Ravaglia, S.; Buongarzone, G.; Bini, P.; Diamanti, L.; Gastaldi, M.; Prunetti, P.; Rognone, E.; Marchioni, E. Acute myelitis and ChAdOx1 nCoV-19 vaccine: Casual or causal association? *J. Neuroimmunol.* **2021**, *359*, 577686. [CrossRef]
113. Hsiao, Y.-T.; Tsai, M.-J.; Chen, Y.-H.; Hsu, C.-F. Acute Transverse Myelitis after COVID-19 Vaccination. *Medicina* **2021**, *57*, 1010. [CrossRef]
114. Notghi, A.A.; Atley, J.; Silva, M. Lessons of the month 1: Longitudinal extensive transverse myelitis following AstraZeneca COVID-19 vaccination. *Clin. Med.* **2021**, *21*, e535–e538. [CrossRef]
115. Maroufi, S.F.; Naderi Behdani, F.; Rezania, F.; Tanhapour Khotbehsara, S.; Mirzaasgari, Z. Longitudinally extensive transverse myelitis after COVID-19 vaccination: Case report and review of literature. *Hum. Vaccin. Immunother.* **2022**, *18*, 2040239. [CrossRef]
116. Wraith, D.C.; Goldman, M.; Lambert, P.H. Vaccination and autoimmune disease: What is the evidence? *Lancet* **2003**, *362*, 1659–1666. [CrossRef]
117. Polack, F.P.; Thomas, S.J.; Kitchin, N.; Absalon, J.; Gurtman, A.; Lockhart, S.; Perez, J.L.; Pérez Marc, G.; Moreira, E.D.; Zerbini, C.; et al. Safety and Efficacy of the BNT162b2 mRNA COVID-19 Vaccine. *N. Engl. J. Med.* **2020**, *383*, 2603–2615. [CrossRef]
118. Alabkal, J.; Rebchuk, A.D.; Lyndon, D.; Randhawa, N. Incomplete Subacute Transverse Myelitis Following Vaccination With Pfizer-BioNTech COVID-19 mRNA Vaccine: A Case Report. *Cureus* **2021**, *13*, e20460. [CrossRef]
119. Eom, H.; Kim, S.W.; Kim, M.; Kim, Y.E.; Kim, J.H.; Shin, H.Y.; Lee, H.L. Case Reports of Acute Transverse Myelitis Associated With mRNA Vaccine for COVID-19. *J. Korean Med. Sci.* **2022**, *37*, e52. [CrossRef]

120. Miyaue, N.; Yoshida, A.; Yamanishi, Y.; Tada, S.; Ando, R.; Hosokawa, Y.; Yabe, H.; Nagai, M. A Case of Refractory Longitudinally Extensive Transverse Myelitis after Severe Acute Respiratory Syndrome Coronavirus 2 Vaccination in a Japanese Man. *Intern. Med.* **2022**, *2022*, 8747-21. [CrossRef]
121. McLean, P.; Trefts, L. Transverse myelitis 48 h after the administration of an mRNA COVID 19 vaccine. *Neuroimmunol. Rep.* **2021**, *1*, 100019. [CrossRef]
122. Kaulen, L.D.; Doubrovinskaia, S.; Mooshage, C.; Jordan, B.; Purrucker, J.; Haubner, C.; Seliger, C.; Lorenz, H.M.; Nagel, S.; Wildemann, B.; et al. Neurological autoimmune diseases following vaccinations against SARS-CoV-2: A case series. *Eur. J. Neurol.* **2022**, *29*, 555–563. [CrossRef]
123. Doi, K.; Ohara, Y.; Ouchi, T.; Sasaki, R.; Maki, F.; Mizuno, J. Cervical Transverse Myelitis Following COVID-19 Vaccination. *NMC Case Rep. J.* **2022**, *9*, 145–149. [CrossRef]
124. Nakano, H.; Yamaguchi, K.; Kawabata, K.; Asakawa, M.; Matsumoto, Y. Acute transverse myelitis after BNT162b2 vaccination against COVID-19: Report of a fatal case and review of the literature. *J. Neurol. Sci.* **2022**, *434*, 120102. [CrossRef] [PubMed]
125. Gao, J.-J.; Tseng, H.-P.; Lin, C.-L.; Shiu, J.-S.; Lee, M.-H.; Liu, C.-H. Acute Transverse Myelitis Following COVID-19 Vaccination. *Vaccines* **2021**, *9*, 1008. [CrossRef] [PubMed]
126. Tahir, N.; Koorapati, G.; Prasad, S.; Jeelani, H.M.; Sherchan, R.; Shrestha, J.; Shayuk, M. SARS-CoV-2 Vaccination-Induced Transverse Myelitis. *Cureus* **2021**, *13*, e16624. [CrossRef] [PubMed]
127. García-Estrada, C.; Gómez-Figueroa, E.; Alban, L.; Arias-Cárdenas, A. Optic neuritis after COVID-19 vaccine application. *Clin. Exp. Neuroimmunol.* **2021**, *13*, 72–74. [CrossRef] [PubMed]
128. Barone, V.; Camilli, F.; Crisci, M.; Scandellari, C.; Barboni, P.; Lugaresi, A. Inflammatory optic neuropathy following SARS-CoV-2 mRNA vaccine: Description of two cases. *J. Neurol. Sci.* **2021**, *429*, 118186. [CrossRef]
129. Arnao, V.; Maimone, M.B.; Perini, V.; Giudice, G.L.; Cottone, S. Bilateral optic neuritis after COVID vaccination. *Neurol. Sci.* **2022**, *43*, 2965–2966. [CrossRef]
130. Chavez, A.; Pougnier, C. A Case of COVID-19 Vaccine Associated New Diagnosis Myasthenia Gravis. *J. Prim. Care Community Health* **2021**, *12*, 21501327211051933. [CrossRef]
131. Galassi, G.; Rispoli, V.; Iori, E.; Ariatti, A.; Marchioni, A. Coincidental Onset of Ocular Myasthenia Gravis Following ChAdOx1 n-CoV-19 Vaccine against Severe Acute Respiratory Syndrome Coronavirus 2 (SARS-CoV-2). *Isr. Med. Assoc. J.* **2022**, *24*, 9–10.
132. Wolska-Krawczyk, M. Acute disseminated encephalomyelitis. *Radiologe* **2022**, *62*, 316–321. [CrossRef]
133. Rinaldi, V.; Bellucci, G.; Romano, A.; Bozzao, A.; Salvetti, M. ADEM after ChAdOx1 nCoV-19 vaccine: A case report. *Mult. Scler. J.* **2022**, *28*, 1151–1154. [CrossRef]
134. Manzano, G.S.; McEntire, C.R.S.; Martinez-Lage, M.; Mateen, F.J.; Hutto, S.K. Acute Disseminated Encephalomyelitis and Acute Hemorrhagic Leukoencephalitis Following COVID-19. *Neurol.—Neuroimmunol. Neuroinflamm.* **2021**, *8*, e1080. [CrossRef]
135. Permezel, F.; Borojevic, B.; Lau, S.; De Boer, H.H. Acute disseminated encephalomyelitis (ADEM) following recent Oxford/AstraZeneca COVID-19 vaccination. *Forensic Sci. Med. Pathol.* **2022**, *18*, 74–79. [CrossRef]
136. Al-Quliti, K.; Qureshi, A.; Quadri, M.; Abdulhameed, B.; Alanazi, A.; Alhujeily, R. Acute Demyelinating Encephalomyelitis Post-COVID-19 Vaccination: A Case Report and Literature Review. *Diseases* **2022**, *10*, 13. [CrossRef]
137. Shimizu, M.; Ogaki, K.; Nakamura, R.; Kado, E.; Nakajima, S.; Kurita, N.; Watanabe, M.; Yamashiro, K.; Hattori, N.; Urabe, T. An 88-year-old woman with acute disseminated encephalomyelitis following messenger ribonucleic acid-based COVID-19 vaccination. *eNeurologicalSci* **2021**, *25*, 100381. [CrossRef]
138. Vogrig, A.; Janes, F.; Gigli, G.L.; Curcio, F.; Negro, I.D.; D'Agostini, S.; Fabris, M.; Valente, M. Acute disseminated encephalomyelitis after SARS-CoV-2 vaccination. *Clin. Neurol. Neurosurg.* **2021**, *208*, 106839. [CrossRef]
139. Senda, J.; Ashida, R.; Sugawara, K.; Kawaguchi, K. Acute Meningoencephalitis after COVID-19 Vaccination in an Adult Patient with Rheumatoid Vasculitis. *Intern. Med.* **2022**, *61*, 1609–1612. [CrossRef]
140. Ahmad, H.R.; Timmermans, V.M.; Dakakni, T. Acute Disseminated Encephalomyelitis After SARS-CoV-2 Vaccination. *Am. J. Case Rep.* **2022**, *23*, e936574. [CrossRef]
141. Kania, K.; Ambrosius, W.; Tokarz Kupczyk, E.; Kozubski, W. Acute disseminated encephalomyelitis in a patient vaccinated against SARS-CoV-2. *Ann. Clin. Transl. Neurol.* **2021**, *8*, 2000–2003. [CrossRef]
142. Ancau, M.; Liesche-Starnecker, F.; Niederschweiberer, J.; Krieg, S.M.; Zimmer, C.; Lingg, C.; Kumpfmüller, D.; Ikenberg, B.; Ploner, M.; Hemmer, B.; et al. Case Series: Acute Hemorrhagic Encephalomyelitis After SARS-CoV-2 Vaccination. *Front. Neurol.* **2021**, *12*, 820049. [CrossRef]
143. Huynh, W.; Cordato, D.J.; Kehdi, E.; Masters, L.T.; Dedousis, C. Post-vaccination encephalomyelitis: Literature review and illustrative case. *J. Clin. Neurosci.* **2008**, *15*, 1315–1322. [CrossRef]
144. Lessa, R.; Castillo, M.; Azevedo, R.; Azevedo, F.; Azevedo, H. Neurological complications after H1N1 influenza vaccination: Magnetic resonance imaging findings. *Arq. Neuropsiquiatr.* **2014**, *72*, 496–499. [CrossRef]
145. Zuhorn, F.; Graf, T.; Klingebiel, R.; Schäbitz, W.R.; Rogalewski, A. Postvaccinal Encephalitis after ChAdOx1 nCov-19. *Ann. Neurol.* **2021**, *90*, 506–511. [CrossRef]
146. Sluyts, Y.; Arnst, Y.; Vanhemelryck, T.; De Cauwer, H. COVID-19-booster vaccine-induced encephalitis. *Acta Neurol. Belg.* **2022**, *122*, 579–581. [CrossRef]
147. Al-Mashdali, A.F.; Ata, Y.M.; Sadik, N. Post-COVID-19 vaccine acute hyperactive encephalopathy with dramatic response to methylprednisolone: A case report. *Ann. Med. Surg.* **2021**, *69*, 102803. [CrossRef]

148. Torrealba-Acosta, G.; Martin, J.C.; Huttenbach, Y.; Garcia, C.R.; Sohail, M.R.; Agarwal, S.K.; Wasko, C.; Bershad, E.M.; Hirzallah, M.I. Acute encephalitis, myoclonus and Sweet syndrome after mRNA-1273 vaccine. *BMJ Case Rep.* **2021**, *14*, e243173. [CrossRef]
149. Liu, B.D.; Ugolini, C.; Jha, P. Two Cases of Post-Moderna COVID-19 Vaccine Encephalopathy Associated With Nonconvulsive Status Epilepticus. *Cureus* **2021**, *13*, e16172. [CrossRef]
150. Badrawi, N.; Kumar, N.; Albastaki, U. Post COVID-19 vaccination neuromyelitis optica spectrum disorder: Case report & MRI findings. *Radiol. Case Rep.* **2021**, *16*, 3864–3867. [CrossRef]
151. Baldelli, L.; Amore, G.; Montini, A.; Panzera, I.; Rossi, S.; Cortelli, P.; Guarino, M.; Rinaldi, R.; D'Angelo, R. Hyperacute reversible encephalopathy related to cytokine storm following COVID-19 vaccine. *J. Neuroimmunol.* **2021**, *358*, 577661. [CrossRef]
152. Jolles, S.; Sewell, W.A.; Leighton, C. Drug-induced aseptic meningitis: Diagnosis and management. *Drug Saf.* **2000**, *22*, 215–226. [CrossRef] [PubMed]
153. Vera-Lastra, O.; Ordinola Navarro, A.; Cruz Domiguez, M.P.; Medina, G.; Sánchez Valadez, T.I.; Jara, L.J. Two Cases of Graves' Disease Following SARS-CoV-2 Vaccination: An Autoimmune/Inflammatory Syndrome Induced by Adjuvants. *Thyroid* **2021**, *31*, 1436–1439. [CrossRef] [PubMed]
154. Saito, K.; Shimizu, T.; Suzuki-Inoue, K.; Ishida, T.; Wada, Y. Aseptic meningitis after vaccination of the BNT162b2 mRNA COVID-19 vaccine. *Neurol. Sci.* **2021**, *42*, 4433–4435. [CrossRef] [PubMed]
155. Reis Carneiro, D.; Matos, A.; Morgadinho, A. Steroid-responsive aseptic meningitis after BNT162b2 SARS-CoV-2 vaccine. *Rev. Neurol.* **2022**, *178*, 160–161. [CrossRef] [PubMed]
156. Chan, A.C.; Tan, B.Y.; Goh, Y.; Tan, S.S.; Tambyah, P.A. Aseptic meningitis after BNT-162b2 COVID-19 vaccination. *Brain Behav. Immun. Health* **2022**, *19*, 100406. [CrossRef] [PubMed]
157. Sejvar, J.J.; Baughman, A.L.; Wise, M.; Morgan, O.W. Population incidence of Guillain-Barré syndrome: A systematic review and meta-analysis. *Neuroepidemiology* **2011**, *36*, 123–133. [CrossRef] [PubMed]
158. Malek, E.; Salameh, J. Guillain-Barre Syndrome. *Semin. Neurol.* **2019**, *39*, 589–595. [CrossRef]
159. Dimachkie, M.M.; Barohn, R.J. Guillain-Barré syndrome and variants. *Neurol. Clin.* **2013**, *31*, 491–510. [CrossRef]
160. Israeli, E.; Agmon-Levin, N.; Blank, M.; Chapman, J.; Shoenfeld, Y. Guillain-Barré syndrome–a classical autoimmune disease triggered by infection or vaccination. *Clin. Rev. Allergy Immunol.* **2012**, *42*, 121–130. [CrossRef]
161. Toscano, G.; Palmerini, F.; Ravaglia, S.; Ruiz, L.; Invernizzi, P.; Cuzzoni, M.G.; Franciotta, D.; Baldanti, F.; Daturi, R.; Postorino, P.; et al. Guillain-Barré Syndrome Associated with SARS-CoV-2. *N. Engl. J. Med.* **2020**, *382*, 2574–2576. [CrossRef]
162. Finsterer, J.; Scorza, F.A.; Scorza, C.A. Post SARS-CoV-2 vaccination Guillain-Barre syndrome in 19 patients. *Clinics* **2021**, *76*, e3286. [CrossRef]
163. Finsterer, J. Exacerbating Guillain-Barré Syndrome Eight Days after Vector-Based COVID-19 Vaccination. *Case Rep. Infect. Dis.* **2021**, *2021*, 3619131. [CrossRef]
164. Kim, J.E.; Min, Y.G.; Shin, J.Y.; Kwon, Y.N.; Bae, J.S.; Sung, J.J.; Hong, Y.H. Guillain-Barré Syndrome and Variants Following COVID-19 Vaccination: Report of 13 Cases. *Front. Neurol.* **2021**, *12*, 820723. [CrossRef]
165. Khan, Z.; Ahmad, U.; Ualiyeva, D.; Amissah, O.B.; Khan, A.; Noor, Z.; Zaman, N. Guillain-Barre syndrome: An autoimmune disorder post-COVID-19 vaccination? *Clin. Immunol. Commun.* **2022**, *2*, 1–5. [CrossRef]
166. Shao, S.C.; Wang, C.H.; Chang, K.C.; Hung, M.J.; Chen, H.Y.; Liao, S.C. Guillain-Barré Syndrome Associated with COVID-19 Vaccination. *Emerg. Infect. Dis.* **2021**, *27*, 3175–3178. [CrossRef]
167. Loo, L.K.; Salim, O.; Liang, D.; Goel, A.; Sumangala, S.; Gowda, A.S.; Davies, B.; Rajabally, Y.A. Acute-onset polyradiculoneuropathy after SARS-CoV2 vaccine in the West and North Midlands, United Kingdom. *Muscle Nerve* **2022**, *65*, 233–237. [CrossRef]
168. Maramattom, B.V.; Krishnan, P.; Paul, R.; Padmanabhan, S.; Cherukudal Vishnu Nampoothiri, S.; Syed, A.A.; Mangat, H.S. Guillain-Barré Syndrome following ChAdOx1-S/nCoV-19 Vaccine. *Ann. Neurol.* **2021**, *90*, 312–314. [CrossRef]
169. Allen, C.M.; Ramsamy, S.; Tarr, A.W.; Tighe, P.J.; Irving, W.L.; Tanasescu, R.; Evans, J.R. Guillain-Barré Syndrome Variant Occurring after SARS-CoV-2 Vaccination. *Ann. Neurol.* **2021**, *90*, 315–318. [CrossRef]
170. Bonifacio, G.B.; Patel, D.; Cook, S.; Purcaru, E.; Couzins, M.; Domjan, J.; Ryan, S.; Alareed, A.; Tuohy, O.; Slaght, S.; et al. Bilateral facial weakness with paraesthesia variant of Guillain-Barré syndrome following Vaxzevria COVID-19 vaccine. *J. Neurol. Neurosurg. Psychiatry* **2022**, *93*, 341–342. [CrossRef]
171. Nasuelli, N.A.; De Marchi, F.; Cecchin, M.; De Paoli, I.; Onorato, S.; Pettinaroli, R.; Savoini, G.; Godi, L. A case of acute demyelinating polyradiculoneuropathy with bilateral facial palsy after ChAdOx1 nCoV-19 vaccine. *Neurol. Sci.* **2021**, *42*, 4747–4749. [CrossRef]
172. Pegat, A.; Vogrig, A.; Khouri, C.; Masmoudi, K.; Vial, T.; Bernard, E. Adenovirus COVID-19 Vaccines and Guillain-Barré Syndrome with Facial Paralysis. *Ann. Neurol.* **2022**, *91*, 162–163. [CrossRef] [PubMed]
173. Trimboli, M.; Zoleo, P.; Arabia, G.; Gambardella, A. Guillain-Barré syndrome following BNT162b2 COVID-19 vaccine. *Neurol. Sci.* **2021**, *42*, 4401–4402. [CrossRef] [PubMed]
174. Shapiro Ben David, S.; Potasman, I.; Rahamim-Cohen, D. Rate of Recurrent Guillain-Barré Syndrome After mRNA COVID-19 Vaccine BNT162b2. *JAMA Neurol.* **2021**, *78*, 1409–1411. [CrossRef] [PubMed]
175. Li, Z.; Li, X.; Shen, J.; Chan, M.T.V.; Wu, W.K.K. Miller Fisher syndrome associated with COVID-19: An up-to-date systematic review. *Environ. Sci. Pollut. Res. Int.* **2021**, *28*, 20939–20944. [CrossRef]

176. Siddiqi, A.R.; Khan, T.; Tahir, M.J.; Asghar, M.S.; Islam, M.S.; Yousaf, Z. Miller Fisher syndrome after COVID-19 vaccination: Case report and review of literature. *Medicine* **2022**, *101*, e29333. [CrossRef]
177. Nishiguchi, Y.; Matsuyama, H.; Maeda, K.; Shindo, A.; Tomimoto, H. Miller Fisher syndrome following BNT162b2 mRNA coronavirus 2019 vaccination. *BMC Neurol.* **2021**, *21*, 452. [CrossRef]
178. Abičić, A.; Adamec, I.; Habek, M. Miller Fisher syndrome following Pfizer COVID-19 vaccine. *Neurol. Sci.* **2022**, *43*, 1495–1497. [CrossRef]
179. Dang, Y.L.; Bryson, A. Miller-Fisher Syndrome and Guillain-Barre Syndrome overlap syndrome in a patient post Oxford-AstraZeneca SARS-CoV-2 vaccination. *BMJ Case Rep.* **2021**, *14*, e246701. [CrossRef]
180. Michaelson, N.M.; Lam, T.; Malhotra, A.; Schiff, N.D.; MacGowan, D.J.L. Miller Fisher Syndrome Presenting After a Second Dose of Pfizer-BioNTech Vaccination in a Patient With Resolved COVID-19: A Case Report. *J. Clin. Neuromuscul. Dis.* **2021**, *23*, 113–115. [CrossRef]
181. Rosenblum, H.G.; Hadler, S.C.; Moulia, D.; Shimabukuro, T.T.; Su, J.R.; Tepper, N.K.; Ess, K.C.; Woo, E.J.; Mba-Jonas, A.; Alimchandani, M.; et al. Use of COVID-19 Vaccines After Reports of Adverse Events Among Adult Recipients of Janssen (Johnson & Johnson) and mRNA COVID-19 Vaccines (Pfizer-BioNTech and Moderna): Update from the Advisory Committee on Immunization Practices—United States, July 2021. *MMWR Morb. Mortal. Wkly. Rep.* **2021**, *70*, 1094–1099. [CrossRef]
182. Woo, E.J.; Mba-Jonas, A.; Dimova, R.B.; Alimchandani, M.; Zinderman, C.E.; Nair, N. Association of Receipt of the Ad26.COV2.S COVID-19 Vaccine With Presumptive Guillain-Barré Syndrome, February–July 2021. *JAMA* **2021**, *326*, 1606–1613. [CrossRef]
183. Oliver, S.E.; Wallace, M.; See, I.; Mbaeyi, S.; Godfrey, M.; Hadler, S.C.; Jatlaoui, T.C.; Twentyman, E.; Hughes, M.M.; Rao, A.K.; et al. Use of the Janssen (Johnson & Johnson) COVID-19 Vaccine: Updated Interim Recommendations from the Advisory Committee on Immunization Practices—United States, December 2021. *MMWR Morb. Mortal. Wkly. Rep.* **2022**, *71*, 90–95. [CrossRef]
184. Hanson, K.E.; Goddard, K.; Lewis, N.; Fireman, B.; Myers, T.R.; Bakshi, N.; Weintraub, E.; Donahue, J.G.; Nelson, J.C.; Xu, S.; et al. Incidence of Guillain-Barré Syndrome After COVID-19 Vaccination in the Vaccine Safety Datalink. *JAMA Netw. Open* **2022**, *5*, e228879. [CrossRef]
185. Rath, B.; Gidudu, J.F.; Anyoti, H.; Bollweg, B.; Caubel, P.; Chen, Y.H.; Cornblath, D.; Fernandopulle, R.; Fries, L.; Galama, J.; et al. Facial nerve palsy including Bell's palsy: Case definitions and guidelines for collection, analysis, and presentation of immunisation safety data. *Vaccine* **2017**, *35*, 1972–1983. [CrossRef]
186. Ozonoff, A.; Nanishi, E.; Levy, O. Bell's palsy and SARS-CoV-2 vaccines. *Lancet Infect. Dis.* **2021**, *21*, 450–452. [CrossRef]
187. Shibli, R.; Barnett, O.; Abu-Full, Z.; Gronich, N.; Najjar-Debbiny, R.; Doweck, I.; Rennert, G.; Saliba, W. Association between vaccination with the BNT162b2 mRNA COVID-19 vaccine and Bell's palsy: A population-based study. *Lancet Reg. Health Eur.* **2021**, *11*, 100236. [CrossRef]
188. Noseda, R.; Ripellino, P.; Ghidossi, S.; Bertoli, R.; Ceschi, A. Reporting of Acute Inflammatory Neuropathies with COVID-19 Vaccines: Subgroup Disproportionality Analyses in VigiBase. *Vaccines* **2021**, *9*, 1022. [CrossRef]
189. Shemer, A.; Pras, E.; Einan-Lifshitz, A.; Dubinsky-Pertzov, B.; Hecht, I. Association of COVID-19 Vaccination and Facial Nerve Palsy: A Case-Control Study. *JAMA Otolaryngol. Head Neck Surg.* **2021**, *147*, 739–743. [CrossRef]
190. Shasha, D.; Bareket, R.; Sikron, F.H.; Gertel, O.; Tsamir, J.; Dvir, D.; Mossinson, D.; Heymann, A.D.; Zacay, G. Real-world safety data for the Pfizer BNT162b2 SARS-CoV-2 vaccine: Historical cohort study. *Clin. Microbiol. Infect.* **2022**, *28*, 130–134. [CrossRef]
191. Sato, K.; Mano, T.; Niimi, Y.; Toda, T.; Iwata, A.; Iwatsubo, T. Facial nerve palsy following the administration of COVID-19 mRNA vaccines: Analysis of a self-reporting database. *Int. J. Infect. Dis.* **2021**, *111*, 310–312. [CrossRef]
192. Li, X.; Raventós, B.; Roel, E.; Pistillo, A.; Martinez-Hernandez, E.; Delmestri, A.; Reyes, C.; Strauss, V.; Prieto-Alhambra, D.; Burn, E.; et al. Association between covid-19 vaccination, SARS-CoV-2 infection, and risk of immune mediated neurological events: Population based cohort and self-controlled case series analysis. *BMJ* **2022**, *376*, e068373. [CrossRef]
193. Bersinger, S.; Lagarde, K.; Marlu, R.; Pernod, G.; Payen, J.F. Using Nonheparin Anticoagulant to Treat a Near-Fatal Case With Multiple Venous Thrombotic Lesions During ChAdOx1 nCoV-19 Vaccination-Related Vaccine-Induced Immune Thrombotic Thrombocytopenia. *Crit. Care Med.* **2021**, *49*, e870–e873. [CrossRef] [PubMed]
194. Lechien, J.R.; Diallo, A.O.; Dachy, B.; Le Bon, S.D.; Maniaci, A.; Vaira, L.A.; Saussez, S. COVID-19: Post-vaccine Smell and Taste Disorders: Report of 6 Cases. *Ear Nose Throat J.* **2021**, 1455613211033125. [CrossRef] [PubMed]
195. Keir, G.; Maria, N.I.; Kirsch, C.F.E. Unique Imaging Findings of Neurologic Phantosmia Following Pfizer-BioNtech COVID-19 Vaccination: A Case Report. *Top. Magn. Reson. Imaging* **2021**, *30*, 133–137. [CrossRef] [PubMed]
196. Reyes-Capo, D.P.; Stevens, S.M.; Cavuoto, K.M. Acute abducens nerve palsy following COVID-19 vaccination. *J. AAPOS* **2021**, *25*, 302–303. [CrossRef] [PubMed]
197. Kaya, A.; Kaya, S.Y. A case of trigeminal neuralgia developing after a COVID-19 vaccination. *J. Neurovirol.* **2022**, *28*, 181–182. [CrossRef] [PubMed]
198. Narasimhalu, K.; Lee, W.C.; Salkade, P.R.; De Silva, D.A. Trigeminal and cervical radiculitis after tozinameran vaccination against COVID-19. *BMJ Case Rep.* **2021**, *14*, e242344. [CrossRef] [PubMed]
199. Wichova, H.; Miller, M.E.; Derebery, M.J. Otologic Manifestations After COVID-19 Vaccination: The House Ear Clinic Experience. *Otol. Neurotol.* **2021**, *42*, e1213–e1218. [CrossRef]
200. Kamogashira, T.; Funayama, H.; Asakura, S.; Ishimoto, S. Vestibular Neuritis Following COVID-19 Vaccination: A Retrospective Study. *Cureus* **2022**, *14*, e24277. [CrossRef]

201. Bramer, S.; Jaffe, Y.; Sivagnanaratnam, A. Vestibular neuronitis after COVID-19 vaccination. *BMJ Case Rep.* **2022**, *15*, e247234. [CrossRef]
202. Van Eijk, J.J.; Groothuis, J.T.; Van Alfen, N. Neuralgic amyotrophy: An update on diagnosis, pathophysiology, and treatment. *Muscle Nerve* **2016**, *53*, 337–350. [CrossRef]
203. Van Alfen, N.; van Engelen, B.G. The clinical spectrum of neuralgic amyotrophy in 246 cases. *Brain* **2006**, *129*, 438–450. [CrossRef]
204. Chua, M.M.J.; Hayes, M.T.; Cosgrove, R. Parsonage-Turner syndrome following COVID-19 vaccination and review of the literature. *Surg. Neurol. Int.* **2022**, *13*, 152. [CrossRef]
205. Vitturi, B.K.; Grandis, M.; Beltramini, S.; Orsi, A.; Schenone, A.; Icardi, G.; Durando, P. Parsonage-Turner syndrome following coronavirus disease 2019 immunization with ChAdOx1-S vaccine: A case report and review of the literature. *J. Med. Case Rep.* **2021**, *15*, 589. [CrossRef]
206. Queler, S.C.; Towbin, A.J.; Milani, C.; Whang, J.; Sneag, D.B. Parsonage-Turner Syndrome Following COVID-19 Vaccination: MR Neurography. *Radiology* **2022**, *302*, 84–87. [CrossRef]
207. Mahajan, S.; Zhang, F.; Mahajan, A.; Zimnowodzki, S. Parsonage Turner syndrome after COVID-19 vaccination. *Muscle Nerve* **2021**, *64*, E3–E4. [CrossRef]
208. Diaz-Segarra, N.; Edmond, A.; Gilbert, C.; McKay, O.; Kloepping, C.; Yonclas, P. Painless idiopathic neuralgic amyotrophy after COVID-19 vaccination: A case report. *PM R* **2022**, *14*, 889–891. [CrossRef]
209. Min, Y.G.; Kim, J.E.; Hwang, J.Y.; Shin, J.Y.; Sung, J.J.; Hong, Y.H. Parsonage-Turner syndrome following COVID-19 vaccination. *J. Neurol. Neurosurg. Psychiatry* **2022**. [CrossRef]
210. Andrei, G.; Snoeck, R. Advances and Perspectives in the Management of Varicella-Zoster Virus Infections. *Molecules* **2021**, *26*, 1132. [CrossRef]
211. Furer, V.; Zisman, D.; Kibari, A.; Rimar, D.; Paran, Y.; Elkayam, O. Herpes zoster following BNT162b2 mRNA COVID-19 vaccination in patients with autoimmune inflammatory rheumatic diseases: A case series. *Rheumatology* **2021**, *60*, Si90–Si95. [CrossRef]
212. Lee, T.J.; Lu, C.H.; Hsieh, S.C. Herpes zoster reactivation after mRNA-1273 vaccination in patients with rheumatic diseases. *Ann. Rheum. Dis.* **2022**, *81*, 595–597. [CrossRef]
213. Rodríguez-Jiménez, P.; Chicharro, P.; Cabrera, L.M.; Seguí, M.; Morales-Caballero, Á.; Llamas-Velasco, M.; Sánchez-Pérez, J. Varicella-zoster virus reactivation after SARS-CoV-2 BNT162b2 mRNA vaccination: Report of 5 cases. *JAAD Case Rep.* **2021**, *12*, 58–59. [CrossRef]
214. Psichogiou, M.; Samarkos, M.; Mikos, N.; Hatzakis, A. Reactivation of Varicella Zoster Virus after Vaccination for SARS-CoV-2. *Vaccines* **2021**, *9*, 572. [CrossRef]
215. Katsikas Triantafyllidis, K.; Giannos, P.; Mian, I.T.; Kyrtsonis, G.; Kechagias, K.S. Varicella Zoster Virus Reactivation Following COVID-19 Vaccination: A Systematic Review of Case Reports. *Vaccines* **2021**, *9*, 1013. [CrossRef]
216. Waheed, W.; Carey, M.E.; Tandan, S.R.; Tandan, R. Post COVID-19 vaccine small fiber neuropathy. *Muscle Nerve* **2021**, *64*, E1–E2. [CrossRef]
217. Khokhar, F.; Khan, A.; Hussain, Z.; Yu, J. Small Fiber Neuropathy Associated With the Moderna SARS-CoV-2 Vaccine. *Cureus* **2022**, *14*, e25969. [CrossRef]
218. Taquet, M.; Geddes, J.R.; Husain, M.; Luciano, S.; Harrison, P.J. 6-month neurological and psychiatric outcomes in 236 379 survivors of COVID-19: A retrospective cohort study using electronic health records. *Lancet Psychiatry* **2021**, *8*, 416–427. [CrossRef]
219. Tondo, G.; Aprile, D.; Tesser, F.; Comi, C. Increased Prevalence of Neuropsychiatric Disorders during COVID-19 Pandemic in People Needing a Non-Deferrable Neurological Evaluation. *J. Clin. Med.* **2021**, *10*, 5169. [CrossRef]
220. Naldi, A.; Vallelonga, F.; Di Liberto, A.; Cavallo, R.; Agnesone, M.; Gonella, M.; Sauta, M.D.; Lochner, P.; Tondo, G.; Bragazzi, N.L.; et al. COVID-19 pandemic-related anxiety, distress and burnout: Prevalence and associated factors in healthcare workers of North-West Italy. *BJPsych Open* **2021**, *7*, e27. [CrossRef]
221. Yang, T.U.; Kim, H.J.; Lee, Y.K.; Park, Y.J. Psychogenic illness following vaccination: Exploratory study of mass vaccination against pandemic influenza A (H1N1) in 2009 in South Korea. *Clin. Exp. Vaccine Res.* **2017**, *6*, 31–37. [CrossRef]
222. Fasano, A.; Daniele, A. Functional disorders after COVID-19 vaccine fuel vaccination hesitancy. *J. Neurol. Neurosurg. Psychiatry* **2022**, *93*, 339–340. [CrossRef] [PubMed]
223. Hause, A.M.; Gee, J.; Johnson, T.; Jazwa, A.; Marquez, P.; Miller, E.; Su, J.; Shimabukuro, T.T.; Shay, D.K. Anxiety-Related Adverse Event Clusters After Janssen COVID-19 Vaccination—Five U.S. Mass Vaccination Sites, April 2021. *MMWR Morb. Mortal. Wkly. Rep.* **2021**, *70*, 685–688. [CrossRef] [PubMed]
224. Butler, M.; Coebergh, J.; Safavi, F.; Carson, A.; Hallett, M.; Michael, B.; Pollak, T.A.; Solomon, T.; Stone, J.; Nicholson, T.R. Functional Neurological Disorder After SARS-CoV-2 Vaccines: Two Case Reports and Discussion of Potential Public Health Implications. *J. Neuropsychiatry Clin. Neurosci.* **2021**, *33*, 345–348. [CrossRef]
225. Sanjeev, O.P.; Verma, A.; Mani, V.E.; Singh, R.K. COVID-19 vaccine-related functional neurological disorders in the emergency department. *CJEM* **2022**, *24*, 455–456. [CrossRef] [PubMed]
226. De Souza, A.; Jacques, R.; Mohan, S. Vaccine-Induced Functional Neurological Disorders in the COVID-19 Era. *Can. J. Neurol. Sci.* **2022**, 1–5. [CrossRef] [PubMed]
227. Ercoli, T.; Lutzoni, L.; Orofino, G.; Muroni, A.; Defazio, G. Functional neurological disorder after COVID-19 vaccination. *Neurol. Sci.* **2021**, *42*, 3989–3990. [CrossRef] [PubMed]

228. Bettini, E.; Locci, M. SARS-CoV-2 mRNA Vaccines: Immunological Mechanism and Beyond. *Vaccines* **2021**, *9*, 147. [CrossRef]
229. Pottegård, A.; Klungel, O.H. The neurological safety of covid-19 vaccines. *BMJ* **2022**, *376*, o522. [CrossRef]
230. Finsterer, J. Neurological side effects of SARS-CoV-2 vaccinations. *Acta Neurol. Scand.* **2022**, *145*, 5–9. [CrossRef]
231. Nordström, P.; Ballin, M.; Nordström, A. Risk of infection, hospitalisation, and death up to 9 months after a second dose of COVID-19 vaccine: A retrospective, total population cohort study in Sweden. *Lancet* **2022**, *399*, 814–823. [CrossRef]

life

MDPI

Review

How an Outbreak of COVID-19 Circulated Widely in Nepal: A Chronological Analysis of the National Response to an Unprecedented Pandemic

Basu Dev Pandey [1,*], Mya Myat Ngwe Tun [2], Kishor Pandey [3], Shyam Prakash Dumre [4], Khin Mya Nwe [2], Yogendra Shah [5], Richard Culleton [6], Yuki Takamatsu [2], Anthony Costello [7] and Kouichi Morita [2]

1 Department of Molecular Epidemiology, Institute of Tropical Medicine, Nagasaki University, Nagasaki 852-8523, Japan
2 Department of Virology, Institute of Tropical Medicine, Nagasaki University, Nagasaki 852-8523, Japan; myamyat@tm.nagasaki-u.ac.jp (M.M.N.T.); bb55418851@ms.nagasaki-u.ac.jp (K.M.N.); yukiti@nagasaki-u.ac.jp (Y.T.); moritak@nagasaki-u.ac.jp (K.M.)
3 Central Department of Zoology, Tribhuvan University, Kathmandu 44618, Nepal; kishor.pandey@cdz.tu.edu.np
4 Central Department of Microbiology, Tribhuvan University, Kathmandu 44618, Nepal; shyam.dumre@cdmi.tu.edu.np
5 Seti Provincial Hospital, Kailali 10900, Nepal; yshah@kist.edu.np
6 Division of Molecular Parasitology, Proteo-Science Center, Ehime University, Matsuyama 790-8577, Japan; culleton.richard.oe@ehime-u.ac.jp
7 Institute for Global Health, University College London, London WC1N 1EH, UK; anthony.costello@ucl.ac.uk
* Correspondence: basudevpandey@nagasaki-u.ac.jp; Tel.: +81-95-819-7827

Citation: Pandey, B.D.; Ngwe Tun, M.M.; Pandey, K.; Dumre, S.P.; Nwe, K.M.; Shah, Y.; Culleton, R.; Takamatsu, Y.; Costello, A.; Morita, K. How an Outbreak of COVID-19 Circulated Widely in Nepal: A Chronological Analysis of the National Response to an Unprecedented Pandemic. *Life* **2022**, *12*, 1087. https://doi.org/10.3390/life12071087

Academic Editors: Silvia De Francia and Sarah Allegra

Received: 27 June 2022
Accepted: 19 July 2022
Published: 20 July 2022

Abstract: Coronavirus disease 2019 (COVID-19) is caused by the severe acute respiratory syndrome coronavirus 2 (SARS-CoV-2). The first COVID-19 case was reported in Wuhan, China, in December 2019. In March 2020, the World Health Organization (WHO) declared COVID-19 a global pandemic. The first COVID-19 case in Nepal was reported in January 2020 in a Nepalese man who had returned from Wuhan to Nepal. This study aims to evaluate the government of Nepal's (GoN) response to the COVID-19 pandemic and explore ways to prevent COVID-19 and other pandemic diseases in the future. As of May 2022, a total of 979,140 cases and 11,951 deaths associated with COVID-19 have been reported in Nepal. To prevent the spread of the virus, the GoN initiated various preventive and control measures, including lockdown strategies. The effects of COVID-19 are expected to persist for many years; the best strategies a resource-limited country such as Nepal can implement to control pandemic diseases such as COVID-19 in the pre-vaccine stage are to increase testing, tracing, and isolation capacity.

Keywords: COVID-19; Nepal; pandemic; vaccine

1. Introduction

COVID-19, caused by the severe acute respiratory syndrome coronavirus 2 (SARS-CoV-2), was first reported in Wuhan, China, in December 2019 and is now a global pandemic [1]. Despite substantial efforts to manage and control the pandemic, morbidity, and mortality due to COVID-19 and its variants have led to significant alterations in daily life [2]. Following its identification, SARS-CoV-2 continued to evolve different variants, including the most recent variant of concern, Omicron (B.1.1.529), identified during a sudden surge of cases [3,4]. The World Health Organization (WHO) classified Omicron as a variant of concern on 26 November 2021, adding it to the earlier variants Alpha, Beta, Gamma, and Delta [4,5]. Continuous viral evolution, the lack or limited availability of vaccines, and wave after wave of outbreaks have posed significant challenges to resource-limited countries such as Nepal in responding to the pandemic [6,7].

A Nepalese man, who had returned from Wuhan, China on 13 January 2020, visited the Sukraraj Tropical and Infectious Disease Hospital (STIDH), Kathmandu, complaining of a cough. He was admitted to the hospital, and efforts were made to exclude common infectious diseases prevalent in Nepal. Real-time reverse transcription polymerase chain reaction (RT-PCR) assays for influenza A and B viruses and NS1 antigen rapid tests for dengue viruses, scrub typhus, malaria, and *Brucella* were negative. There was no laboratory facility for SARS-CoV-2 testing in Nepal at the time, so the samples were sent to the WHO laboratory in Hong Kong, where they were confirmed positive for SARS-CoV-2. Full genome analysis showed that the samples were closely related to the strain from Wuhan [4,8]. On 23 January 2020, the Ministry of Health and Population (MoHP) officially declared this as the first case of COVID-19 in Nepal. The second case was detected on 17 March 2020. Shortly afterward, the virus spread rapidly throughout the country [9].

The COVID-19 pandemic remains a devastating crisis for Nepal, partially due to the fact that the country was poorly prepared for this unprecedented situation [10]. This was evident from the awareness and preparedness levels of Nepalese and international health care professionals at the onset of the pandemic [11]. Nepal had never experienced such an event in the past and its limited capacity to respond was detrimental to containment efforts [12]. The country was severely affected, with an ongoing and steady increase in morbidity and mortality of COVID-19 cases more than two years after the first case was diagnosed. As of May 2022, 5,703,008 PCR assays had been performed throughout the country, detecting a total of 979,140 COVID-19 cases. There have been 11,952 deaths until May 2022. The recovery rate was 98.8%, with a case fatality rate of 1.2%. To comprehend how SARS-CoV-2 spread so widely in Nepal and to prepare for future health crises, a clear and accurate understanding of the series of key early epidemiological events including analysis of the concerned authorities' responses to the emergence of COVID-19 must be established. Towards this aim, we present a chronology of key events and actions pertaining to the pandemic from 23 January 2020, when the first case was identified in Nepal, until May 2022, when cases were appearing all over the country has been analyzed.

Based on these findings, we suggest that improving the speed of disease early detection and the implementation of alerts will help the country respond more efficiently to future emergencies of national and global health concern. Establishing the chronology of major events also permits an assessment of the national effort to detect and prevent transboundary diseases and threats to health under the obligations of the WHO's International Health Regulations. Using a conceptual framework for key stages from outbreak to pandemic and lockdown after the second case was confirmed in March 2020, we delineated the series of identified actions to establish how the health systems functioned and pinpoint potential areas and gaps for improvement in the early outbreak alert and response systems [13,14]. Based on these analyses, we propose a set of options for preparing resilient health systems to respond to high-impact pathogens of international concerns with specific features such as those of SARS-CoV-2. These proposals will ultimately support the design and implementation of pandemic preparedness and response frameworks.

2. Materials and Methods

This is a descriptive epidemiological study of COVID-19 in Nepal. Nepal is a landlocked country bordered by India in the east, west, and south and by China in the north and is home to approximately 30 million people. Geographically, Nepal is divided into three regions, the Terai, hills, and mountains. Politically, it is divided into seven provinces, 77 districts, and 753 municipalities.

The COVID-19 data for this study were obtained from official situation reports of the MoHP of the government of Nepal (GoN), which are publicly available. Data were collected from 23 January 2020 to 31 May 2022 (Figure 1). Data on daily COVID-19 cases are available from the MoHP's website. Data were analyzed in Excel 2019 (Microsoft, Los Angles, CA, USA).

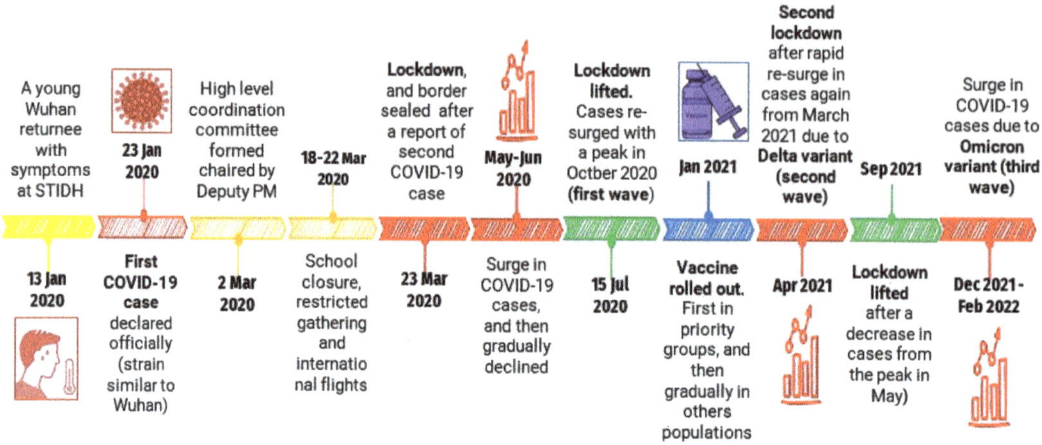

Figure 1. Chronology of key events of COVID-19 in Nepal.

3. Results

3.1. Early Detection of COVID-19 Cases in Nepal

The first SARS-CoV-2 infected case in Nepal was reported on 23 January 2020 in a Nepalese man who had returned to Kathmandu from Wuhan, China (Figure 1). He had symptoms on 3 January 2020, six days before flying back to Nepal. He presented at the outpatient department of Sukraraj Tropical and Infectious Disease Hospital (STIDH), Kathmandu, on 13 January 2020 with a cough; physicians suspected the possibility that the patient was infected with a novel virus that was at the time causing an outbreak of pneumonia in Hunan, China. The patient was immediately admitted to hospital and tested for panel of common infectious diseases using RT-PCR assays for influenza A and B viruses and NS1 antigen rapid tests for dengue viruses, and serological tests for scrub typhus, malaria, and *Brucella*, but all tests were negative. The hospital director called an emergency meeting with the director of the Epidemiology and Disease Control Division (EDCD), the director of the National Public Health Laboratory (NPHL), a representative from the WHO, and the treating physician. As per the meeting decision, throat swabs were collected and sent to the WHO laboratory in Hong Kong for real-time RT-PCR assays for COVID-19 and were found to be positive. On 23 January 2020, the MoHP officially declared the patient Nepal's first COVID-19 case [3]. All 15 staff members working in the special isolation ward of STIDH who had been in close contact with the patient were kept under observation for 14 days. However, none of them showed clinical symptoms during that period. The NPHL's capacity for COVID-19 testing was enhanced by the implementation of using RT-PCR technology for the first time in Nepal.

3.2. The Government's Initial Response to COVID-19

Prompt action was taken by the MoHP to strengthen the country's response to prevent the spread of the virus transmission, mainly through the strengthening of health desks at Tribhuvan International Airport and then at other domestic airports. The ground points of entry on the Nepal–China and the Nepal–India borders were also equipped and strengthened with additional health desks. Travel restrictions were imposed on both sides of these borders [4]. In addition to these initial measures, the GoN sought to repatriate Nepalese nationals from Wuhan, China, resulting in the return of 175 Nepalese citizens. They were quarantined for 14 days and permitted to return to their homes following negative tests. The MoHP activated the Incident Command System (ICS) and Health Emergency Operation Center (HEOC) and initiated coordinated efforts to respond to the virus. The High-Level Coordination Committee (HLCC), led by the Deputy Prime Minister, was formed on 2 March 2020, and subsequently made several crucial decisions intended to limit the spread of COVID-19.

3.3. Detection of a Second Case and Announcement of Lockdown

No cases were reported until 17 March 2020, when a Nepalese person who had returned from France was found to be carrying the virus [4]. There was widespread public concern about the country's diagnostic capabilities as no cases were detected from 23 January 2020 to 17 March 2020. On 18 March 2020, schools were closed, and gatherings of more than 25 people were discouraged. On 22 March 2020, the GoN prohibited all international flights and suspended operations of nonessential businesses and domestic long-distance transportation. All international borders were closed on 23 March 2020. The GoN issued a nationwide lockdown from 24 March 2020 to 21 July 2020, imposing a ban on domestic and international travel and closing borders and nonessential services. There were only two COVID-19 cases and no fatalities recorded at the beginning of the nationwide lockdown.

3.4. Strengthening the System

During the initial phase of the outbreak, PCR tests for SARS-CoV-2 were limited to only at the NPHL, with samples sent from all over the country to Kathmandu. The laboratories were gradually expanded to include the BP Koirala Institute of Health Sciences and public and private laboratories in all seven provinces. Sudarpaschim province borders India, and the provincial government there established seven molecular laboratories. Within a year, starting with one public-sector COVID-19 diagnostic laboratory at the NPHL, the MoHP enhanced a network of more than 100 COVID-19 testing facilities across the country in both the public and private sectors [4]. Antibody-based rapid diagnostic tests (RDTs) were used for screening purposes, particularly at points of entry. However, false-positive and -negative reporting led to discrepancies and the withdrawal of this RDT by the MoHP. On 29 March 2020, the COVID-19 Crisis Management Centre (CCMC), was formed and chaired by the Deputy Prime Minister to serve as the implementing authority of the HLCC.

3.5. Information Management and Risk Communication

After issuing strict lockdown orders, the GoN initiated an effective flow of information to the general public in order to enhance risk awareness and community participation. The MoHP started publishing daily live press briefings in the third week of March. A spokesperson provided daily updates on the situation to ensure a uniform flow of information from the press to the public. The MoHP communicated through various routes, including mobile applications, call centers, hotlines, and social media (Facebook and Viber; MoHP; https://covid19.mohp.gov.np/). The EDCD developed and implemented telephone-based hotlines that received several thousands of calls a day from the public, and these were soon expanded and made available 24 h a day. Educational materials were developed and distributed widely to disseminate messages. On 9 April 2020, the Health Cluster for COVID-19 was activated, headed by the ICS coordinator and co-led by the WHO. With this activation, regular cluster meetings were held every Thursday with health officials at the federal and provincial levels and representatives from private partnerships. Subsequently, health cluster and provincial meetings were held separately to make the discussions more efficient.

3.6. Case Management

The government designated STIDH as the primary hospital for COVID-19 case management, with the Patan Academy of Health Sciences and the Armed Police Forces Hospital as secondary hospitals. Clinical guidelines for case management and standard operating procedures were prepared and distributed by the EDCD. Later, as COVID-19 cases continued to surge, the MoHP designated the Army hospital, the facilities of the Ayurvedic

Hospital Research and Training Center, and Nepal Medical College as COVID-19 treatment facilities and converted Bir Hospital into a COVID-19 special unified hospital. Specific locations were identified to be use as quarantine centers across the country; however, the number of centers was insufficient, and several were mismanaged. As the provision of health services and clinical management are key to saving lives, the MoHP held a series of consultations with the central hub hospitals to prepare for appropriate case management. As the 25 country-wide central hospitals were unable to respond to an international public health emergency such as the COVID-19, the MoHP designated over 150 more hospitals, including private hospitals, as COVID-19 hospitals upon the recommendations of the central and cluster meetings. Hospital services were provided free of charge, and funds were provided to the hospitals according to the number and severity of the cases they managed. Initially, all arrangements for patient admission, including contacting the infected person, arranging an ambulance, finding a bed, and transport to the hospital, were made by an EDCD case-management team. The MoHP, the Curative Services Division led a multi-sectoral and multi-partner team to design and develop a rapid assessment tool to evaluate 12 selected COVID-19 level II hospitals in April 2020. Gaps were identified in multiple areas, including intensive care unit (ICU) capacity, the number of ventilators, infection prevention and control (IPC) measures, logistics, human resources, and training. Training for all health care professionals and non-technical staff, a dedicated IPC budget, and ongoing non-COVID-19 services were recommended.

3.7. Data Management

Initially, when only a few cases were being reported, the NPHL sent information directly to the EDCD, where the data were maintained in Excel. Data received at the EDCD were processed and sent to the MoHP for daily press briefings to the media and for publication. However, as cases began to surge rapidly and several laboratories started COVID-19 testing and reporting data handling became challenging. Therefore, other resources such as Go Data, developed by the Global Outbreak Response Network (GOARN) of the WHO, were utilized. Multiple centers maintained information, including the CCMC and the Home Ministry; these multiple sources led to conflicting information that created discrepancies in data.

In April 2020, COVID-19 hospitals, testing facilities, Health Directorates, Provincial Health Emergency Operations Centers (PHEOC), and the Ministry of Provincial Social Development Ministry all started to observe a surge in cases. This necessitated the development of a uniform reporting system and a template was immediately shared among the stakeholders. However, challenges emerged due to low compliance with the instructions regarding the use of the reporting template and stakeholders' data sharing with the EDCD in multiple formats, such as scanned copies, Excel sheets, or direct emails. Further confusion in data sharing was created because the PHEOCs, the hospitals, and local health facilities reported data to both the EDCD and the HEOC, which led to challenges in verifying the information. In some cases, the provincial and local health agencies hesitated to share data because they believed they did not come under the jurisdiction of the MoHP. The health management information system established an Information Management Unit to ensure the systematic flow of data. The Nepali Army was responsible for managing the deceased and maintaining data pertaining to them, which it provided to the MoHP and the EDCD. However, the data regarding the deceased reported by health facilities to the EDCD differed from the Nepali Army's reports as the definition of death from COVID-19 differed between these institutions.

3.8. Case Investigation and Contact Tracing

On 4 April 2020, the first case of local transmission was confirmed in a 34-year-old women from Kailali District (Figure 2) [4]. A group of people had traveled to New Delhi, India, where they attended a conference, on March the 8th and 9th, and then returned to Nepal via Birgunj, on 11 March 2020. On 17 April, 12 Indian nationals from New Delhi quarantined in a mosque in Bhulke, Udayapur district, tested positive for COVID-19. These people traveled to other parts of the eastern Terai, including Sunsari, Saptari, Rautahat, Parsa, and Udayapur districts, from 30 March to 13 April 2020. The first death due to COVID-19 in Nepal was that of a 29-year-old pregnant woman on 14 May 2020; and by November 2021, 11,496 deaths had been reported [15].

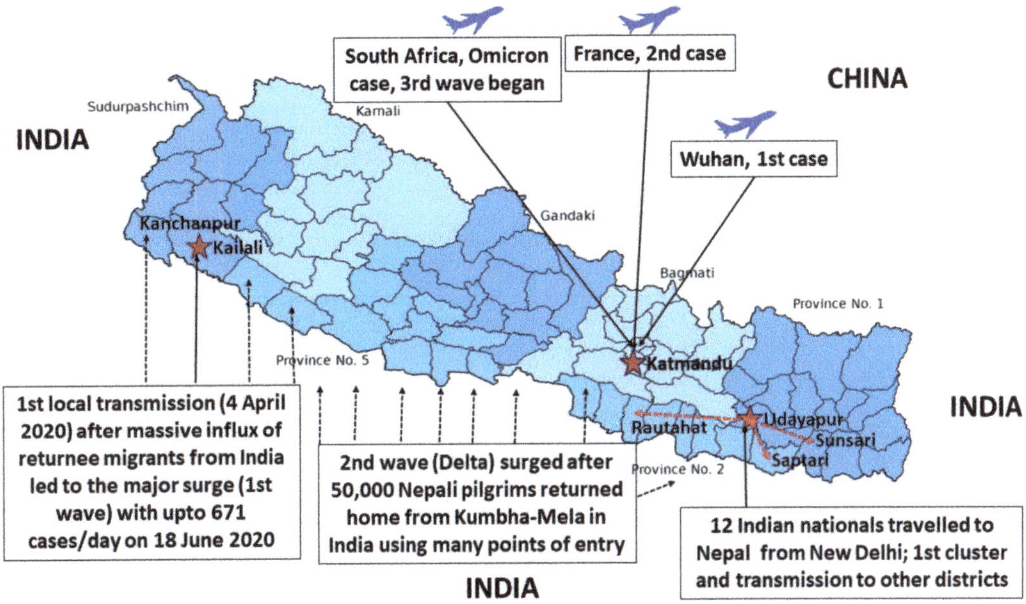

Figure 2. Map showing various surges of corona entering in Nepal.

Initially, the EDCD formed case investigation and contact tracing (CICT) teams, consisting of doctors, public health officers, and laboratory personnel, and these teams were mobilized for each case. However, as case numbers started to increase, the EDCD coordinated with the district health office team and local authorities in Kathmandu Valley for CICT. In August, meetings were held with the HLCC and mayoral forums for the prevention and control of COVID-19. To systematize CICT, measures were initiated with microplanning in meetings with mayors, deputy mayors, district COVID-19 focal persons, municipal health officials, and EDCD officials. These meetings resulted in the EDCD designating health coordinators for municipalities. As the number of cases increased, it became impossible to conduct contact tracing from the central level to the local level. In May 2020, the GoN formulated the Preparedness and Response Plan to prevent and minimize the spread of COVID-19. The plan, which was endorsed by the cabinet, also suggested forming 1075 CICT teams at the local level that would constitute members from the public health, laboratory, nursing, local municipality, administrative staff, and security personnel [16]. The local government was primarily responsible to establish and manage the quarantine centers in each municipality with the support from the government and partial involvement of private organizations as per the guidelines. When the pandemic began to surge in Nepal during March to July 2020, there were 8241 quarantine centers and 238 holding centers across the country [17]. In the beginning, most of the quarantine centers were located at the points of entry along the Indo-Nepal border using school buildings, government facilities and hotels. These centers were converted to the holding centers for the returnees as well as the isolation centers for the symptomatic cases later [18].

3.9. The Initial Surge of the First Wave of COVID-19

The pandemic may have blurred international boundaries and brought much of the world closer together, but less so for Nepal, one of India's neighbors. In fact, Nepal's decision to publish new maps that included areas of dispute with India cooled relations between the countries [19]. The publication by the Minister for Land Management, Cooperatives and Poverty Alleviation of Nepal on 20 May 2020 sparked a dispute between the countries. The dispute was further heated by a high-level parliamentarian's speech that stated that the Indian virus seemed more lethal than the Chinese and Italian viruses. At the same time, toward the end of March, India implemented one of its strictest lockdowns. Thousands of Nepalis working across India, mostly young men, had traveled long distances in the hope of returning to their homes in Nepal [16]. They had been waiting for weeks at the Indo–Nepal border due to travel restrictions imposed by the GoN. The decision to let them enter the country came too late, probably causing a surge of COVID-19 cases in the Sudurpaschim and Madhesh provinces of Nepal. Despite continuous efforts, Nepalese authorities at the central, provincial, and local government have been unable to systematically and efficiently quarantine returnees and migrants from other countries, especially from India. The inflow of COVID-19 cases through the multiple points of entry across Indo–Nepal borders resulted in the first surge, leading to as many as 671 cases in a day on 18 June 2020 (Figure 3). In June and July 2020, there were over 200 daily cases on average, but there were few deaths [20].

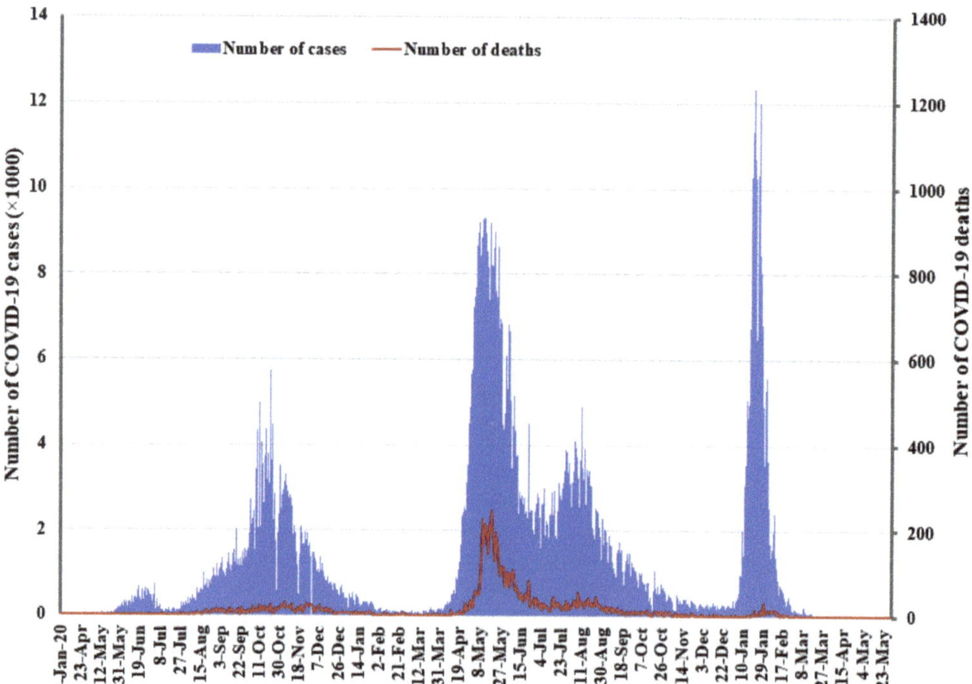

Figure 3. Number of COVID-19 cases and deaths in Nepal.

The GoN made a decision to lift the four-month-long lockdown on 22 July 2020, with a few restrictions in place. Initially, all domestic and international flights were permitted on 17 August 2020; however, it then extended restrictions to 31 August 2020. In July and August 2020, Nepal encountered an unexpected rapid rise in cases, and deaths increased every day. In the first wave, the average number of daily cases rose to 3000, and deaths peaked at fewer than 20 a day. Cases peaked again in October 2020, with 30 deaths on 4 November 2020. The first wave peaked in late October at 5743 cases per day, with a case death rate of less than 1% (13). The number of cases gradually decreased and dropped to a minimum level in January 2021. Although cases reached a baseline level in February of 2021, deaths from earlier infections may have led to a rise in the death rate.

3.10. The Second Wave of COVID-19 in Nepal and the Response

In March 2021, the new Delta variant, labeled as more virulent and infectious by the WHO, was first reported in India [19]. Throughout the spring, hundreds of thousands of people gathered in the streets, engaging in political campaigns to prepare for the May 2021 election and adding to the number of people already congregating for seasonal weddings and festivals. In mid-April, when cases in India were constantly rising, an estimated 50,000 Nepali pilgrims travelled to northern India for Kumbh Mela, a Hindu festive gathering that draws millions of people [19]. While there, many of the pilgrims caught COVID-19, perhaps triggering the second wave in Nepal. The second wave began in April 2021, and cases peaked in May, when the average number of cases per day reached 9317 (on 11 May 2021), and deaths peaked at approximately 246 a day. These numbers coincided with those in India, where the Delta variant contributed to a surge with 414,188 cases recorded on 6 May 2021, setting a global record [19]. On 29 April 2021, the Kathmandu Valley and other parts of Nepal returned to lockdown until 1 September 2021. The existing government facilities were not able to handle the volume of sick patients, and hospitals continued to experience widespread shortages of oxygen and other essential logistic supplies, leading to potentially avoidable losses of life [21]. Cases gradually decreased after June, with an infection rate of 2% in the confirmed cases per day at the end of November 2021. According to the second nationwide seroprevalence survey, between 5 July 2021 to 14 August 2021, over two-thirds of the Nepali population carried antibodies against COVID-19, up from 14.9% in the first seroprevalence survey conducted by the EDCD from 9 October to 20 October 2020 [22,23].

3.11. The Introduction of the Omicron Variant in Nepal

The number of new cases per day began to decline in December 2021 from the peak of the second wave on 11 May 2021. On 6 December 2021, the Omicron (B.1.1.529) variant was observed for the first time in Nepal, merely two weeks after it was first identified in South Africa on 24 November 2021. The appearance of Omicron led to a surge in daily new cases that peaked on 20 January 2022, reaching a record of more than 10,000 cases [17]. The third wave ended after two months (by February 2022). The number of deaths per day (n = 32) was much lower in the third wave compared to those of the second wave (n = 246) [17]. As during previous waves, the long (1800 km), open border between India and Nepal featured inadequate monitoring of compliance with public health protocols, and may have contributed to the size of the third wave. The lessons learned from the first and second waves combined with the introduction of COVID-19 vaccination may have contributed to the management of the third wave in Nepal. Currently, there are few daily new cases and zero deaths reported in the country. Apart from school closures, limitations on mobility, and business operation strategies, the GoN greatly facilitated COVID-19 vaccinations among the population, including in children aged 12–17 years old. S gene target failure or "S gene dropout", producing a false-negative result, has been used as a proxy marker to screen for Omicron, and this option is available in several COVID-19 laboratories throughout Nepal.

3.12. COVID-19 Vaccination

In Nepal, COVID-19 vaccination started on 27 January 2021, with the COVISHIELD vaccine imported from India. As of 18 January 2022, only 40% of Nepal's 30 million people had been fully vaccinated, while approximately 53% had received at least one dose. However, vaccination coverage was quickly extended over the proceeding months.

As of 19 April 2022, 75.9% of the population had received at least one dose of the vaccine, while 66.4% had received two doses. For those within six months of their second dose, 2,474,879 doses of additional vaccine (booster shots) had been administered at the time of writing [24]. No vaccinations have been initiated for children below 12 years of age. Nepal used six different vaccine formulations: Vero Cell, COVISHIELD, AstraZeneca (Japanese and Swedish), Moderna, Pfizer, and Janssen [25].

4. Discussion

This analysis of Nepal's early response to the COVID-19 pandemic from 23 January 2020 to May 2022 highlights crucial lessons and clarifies potential pathways for improving future pandemic preparedness and response. Nepal neighbors China, where the novel coronavirus was first detected on 31 December 2019 [26]. Within a month, on 23 January 2020, Nepal had detected its first case in a young Nepali citizen who had returned from Wuhan, China [3]. Nepal responded quickly and imposed a lockdown after detecting a second COVID-19 case on 24 March 2020; the lockdown lasted almost six months, paralyzing the country [27].

Nepal initiated an early lockdown in order to buy time to prepare the country to respond to the pandemic [28]. This approach to lockdown differed to that applied in other countries, in which restrictions on the population were imposed only after significant numbers of deaths had occurred [29]. Although there is, perhaps, an argument to be made for a universal, WHO-sanctioned protocol for the implementation of lockdown measures at a country level during a pandemic, differences between countries at the political, population and environmental levels make this unrealistic.

The vast majority of early cases in Nepal were in young men who had returned from India; they were usually asymptomatic, and the case fatality rate was initially low at 0.5%, increasing to 1.4% in the second wave [15]. Cases were detected in Kathmandu and a few neighboring districts until March 2020, but quickly thereafter, spread over 77 districts, and a seroprevalence study showed a 14.9% prevalence of antibodies against SARS-CoV-2 by September 2020 [22].

Even in the urban areas such as Kathmandu, over one-quarter of people did not use masks while another 25% did not use them properly during the pandemic. Social distancing was not followed in the public places including in the hospitals, public vehicles and offices [18]. On the other hand, even though there is very high level of knowledge on preventive measures, relatively less practice and no or minimal compliance were observed in areas bordering India [30]. Efforts on enhancing practices of public health measures using face masks, hand washing and maintaining physical distance by the government are key measures to contain the unprecedented transmission.

The first case was confirmed by a laboratory in Hong Kong, and Nepal was unprepared to diagnose a novel disease due to its lack of molecular laboratories. Later, the NPHL installed RT-PCR facilities; more than 100 are now active, but there was a severe shortage of laboratory reagents, logistics, and manpower to operate them during the first wave [31]. At the same time, a lack of face masks and sanitizer created difficulties in implementing public health measures [32]. Unavailability of the face masks, sanitizers, and personal protective equipment (PPEs) was a major issue in the initial stage of the pandemic. However, at least one or more Water Sanitation and Hygiene (WASH) services including hygiene kits, facemask, soap, and sanitizers were provided to 199,252 people and 148,720 returnees in collaboration with partner organizations in the first and second waves of COVID-19 in Nepal [18].

There were few isolation facilities for case management, and STIDH, designated as the primary hospital for COVID-19 management in the beginning, had only five beds with isolation capacity. The health sector emergency response and preparedness plan showed 1595 ICU beds and 840 ventilators available in 194 hospitals [1]. Later, the MoHP designated 111 hospitals to operate COVID-19 clinics and 28 hospitals to treat COVID-19 cases, and Bir Hospital was designated as the COVID-19 Unified Hospital. When cases started to surge, there was no space available in public or private hospitals, leading to potentially avoidable loss of life. Initially, contact tracing was performed by the central government and later by the local governments; the formation of CICT teams made these operations more efficient given the limited manpower and complex logistics. Teams were formed locally with female community health volunteers, and social organizations were also engaged for contact tracing.

COVID-19 vaccination started on 27 January 2021, using the Oxford AstraZeneca vaccine, designated AZD1222 and sold under the brand name COVISHIELD. Although vaccine availability was a major concern when the Serum Institute India stopped supplying COVISHIELD in Nepal until 2022 [33,34], COVID-19 vaccinations continued, and substantial coverage was achieved among different age groups, including children aged 12–17 years old via the introduction of multiple vaccines. Despite this, Nepal has not begun vaccinating children below 12 years, which must begin as early as possible to prevent transmission among the youngest individuals [35]. Properly identifying key shortcomings and gaps and making necessary improvements in the health care delivery system are of the utmost importance to fight the possible next wave or a future, similar health crisis [22].

However, there were gaps at all stages of the response, including insufficient laboratory services, inefficient contact tracing, and poor logistics, coordination, and case management. Overall, Nepal was unprepared to respond to this unprecedented pandemic, and poor management led to potentially avoidable loss of life.

5. Conclusions

The GoN took immediate action, supported by the WHO and international partners, after the first domestic case of COVID-19 was reported. Laboratories were supported to become capable of testing for SARS-CoV-2 in Nepal a week after the first case was reported. The HLCC, led by the Deputy Prime Minister, and the CCMC were formed and made a series of important decisions to break the transmission chain and limit the spread of COVID-19 in Nepal. The MoHP acted immediately to fortify Nepal's efforts to prevent the spread of the COVID-19 virus, such as strengthening the health desks at international and domestic airports and at border crossings between India, China, and Nepal.

To coordinate efforts, the ICS was activated by the MoHP and cluster meetings were initiated. Contract tracing was initially performed by the EDCD via teams that included medical doctors, public health experts, and laboratory staff. This task was later coordinated with the provincial and local governments as the number of cases surged. Efforts to share information with the public and communicate risks were initiated. Hospitals were designated for COVID-19 treatment and specific ambulance services were utilized to transfer patients. Quarantine and isolation centers were created by all three tiers of government; local governments were very proactive regarding quarantine.

The lessons learned from the SARS-CoV-2 pandemic will serve to inform planning for any future outbreak of novel high-impact respiratory pathogens. The national system of Nepal could have responded more efficiently and there is a need for a new framework that provides commitment to prompt detection and fully transparent, harmonized, coordinated, and timely communications and responses.

Author Contributions: Conceptualization, B.D.P., M.M.N.T., K.P., S.P.D. and K.M.; formal analysis, B.D.P., K.P., K.M.N. and Y.T.; methodology, B.D.P., K.P. and Y.S.; supervision, B.D.P., A.C. and K.M.; funding acquisition, B.D.P., M.M.N.T. and K.M.; writing—original draft., B.D.P. and K.P.; writing—review and editing, B.D.P., M.M.N.T., K.P., S.P.D., K.M.N., Y.S., Y.T., A.C., R.C. and K.M. All authors have read and agreed to the published version of the manuscript.

Funding: This research work was supported by the Japan Agency of Medical Research and Development (AMED) under grant number JP22wm0125006 (Japan program for Infectious Diseases Research and Infrastructure).

Institutional Review Board Statement: Not applicable.

Informed Consent Statement: Not applicable.

Data Availability Statement: Not applicable.

Acknowledgments: We thank the Ministry of Health and Population, Nepal and all involved in the COVID-19 response in Nepal and all the members of Department of Virology, Institute of Tropical Medicine, and Nagasaki University.

Conflicts of Interest: The authors declare no conflict of interest.

References

1. Statement on the Second Meeting of the International Health Regulations (2005) Emergency Committee Regarding the Outbreak of Novel Coronavirus (2019-nCoV). 2020. Available online: https://www.who.int/news/item/30-01-2020-statement-on-the-second-meeting-of-the-international-health-regulations-(2005)-emergency-committee-regarding-the-outbreak-of-novel-coronavirus-(2019-ncov) (accessed on 15 November 2021).
2. Woods, J.A.; Hutchinson, N.T.; Powers, S.K.; Roberts, W.O.; Gomez-Cabrera, M.C.; Radak, Z.; Berkes, I.; Boros, A.; Boldogh, I.; Leeuwenburgh, C.; et al. The COVID-19 pandemic and physical activity. *Sports Med. Health Sci.* **2020**, *2*, 55–64. [CrossRef] [PubMed]
3. Bastola, A.; Sah, R.; Rodriguez-Morales, A.J.; Lal, B.K.; Jha, R.; Ojha, H.C.; Shrestha, B.; Chu, D.K.W.; Poon, L.L.M.; Costello, A.; et al. The first 2019 novel coronavirus case in Nepal. *Lancet Infect. Dis.* **2020**, *20*, 279–280. [CrossRef]
4. MOHP. COVID-19. Available online: https://covid19.mohp.gov.np/#/ (accessed on 15 November 2021).
5. Choi, J.Y.; Smith, D.M. SARS-CoV-2 variants of concern. *Yonsei Med. J.* **2021**, *62*, 961. [CrossRef] [PubMed]
6. Gupta, G.P.; Shah, Y.; Pant, D.K.; Lekhak, S.; Shahi, R.; Mandal, A.; Poudel, P.; Dumre, S.P.; Pandey, K.; Pandey, B.D. Preparatory phase for clinical trials of COVID-19 vaccine in Nepal. *Hum. Vaccines Immunother.* **2021**, *17*, 418–419. [CrossRef]
7. Gyanwali, P.; Bista, N.R.; Khadka, M.; Vaidya, A.; Mahato, N.K.; Karn, M.K.; Pant, S.; Ghimire, N.; Pokhrel, A.; Dhimal, M. Assessment of Preparedness of Government of Nepal in COVID Designated Hospitals and Clinics for Pandemic Response. *J. Nepal Health Res. Counc.* **2021**, *19*, 48–54. [CrossRef] [PubMed]
8. Sah, R.; Rodriguez-Morales, A.J.; Jha, R.; Chu, D.K.W.; Gu, H.; Peiris, M.; Bastola, A.; Lal, B.K.; Ojha, H.C.; Rabaan, A.A.; et al. Complete Genome Sequence of a 2019 Novel Coronavirus (SARS-CoV-2) Strain Isolated in Nepal. *Microbiol. Resour. Announc.* **2020**, *9*, e00169-20. [CrossRef] [PubMed]
9. Panthee, B.; Dhungana, S.; Panthee, N.; Paudel, A.; Gyawali, S.; Panthee, S. COVID-19: The current situation in Nepal. *New Microbes New Infect.* **2020**, *37*, 100737. [CrossRef]
10. Adhikari, B.; Devkota, A.; Budhathoki, S.S.; Pinder, R.; Basnet, L.B. COVID-19 crisis in Nepal: A case of systems and governance failure in a low-income country. *J. Glob. Health Econ. Policy* **2021**, *1*, 262996. [CrossRef]
11. Huy, N.T.; Chico, M.R.; Huan, V.T. Awareness and preparedness of healthcare workers against the first wave of the COVID-19 pandemic: A cross-sectional survey across 57 countries. *PLoS ONE* **2021**, *16*, e0258348. [CrossRef]
12. Rai, S.K. Changing trend of infectious diseases in Nepal. In *Advances in Experimental Medicine and Biology*; Springer: Berlin/Heidelberg, Germany, 2018; Volume 1052.
13. Rayamajhee, B.; Pokhrel, A.; Syangtan, G.; Khadka, S.; Lama, B.; Rawal, L.B.; Mehata, S.; Mishra, S.K.; Pokhrel, R.; Yadav, U.N. How Well the Government of Nepal Is Responding to COVID-19? An Experience from a Resource-Limited Country to Confront Unprecedented Pandemic. *Front. Public Health* **2021**, *9*, 597808. [CrossRef]
14. Sharma, K.; Banstola, A.; Parajuli, R.R. Assessment of COVID-19 Pandemic in Nepal: A Lockdown Scenario Analysis. *Front. Public Health* **2021**, *9*, 599280. [CrossRef]
15. Situation Update-84. Coronavirus Disease 2019 (COVID-19) WHO Country Office for Nepal Reporting. 2021. Available online: https://covid19.mohp.gov.np/situation-report (accessed on 10 November 2021).
16. Ministry of Health and Population. *Hospital-Based Interventions Public Health and Social Measures Management and Oversight Health Sector Emergency Response Plan COVID-19 Pandemic*; Ministry of Health and Population: Cairo, Egypt, 2020.
17. Relief Web. Rapid Assessment on Impacts of COVID-19 on Returnee Migrants and Responses of Local Governments of Nepal—Nepal. Available online: https://reliefweb.int/report/nepal/rapid-assessment-impacts-covid-19-returnee-migrants-and-responses-local-governments (accessed on 10 June 2022).
18. NHRC. *Assessment of Compliance with SMS Measures against COVID-19 in Kathmandu Valley 2020, Health Sector Response to COVID-19 Pandemic in Nepal 2022*; NHRC: New Delhi, India, 2022.
19. Kandikattu, H.K.; Yadavalli, C.S.; Venkateshaiah, S.U.; Mishra, A. Vaccine efficacy in mutant SARS-CoV-2 variants. *Int. J. Cell Biol. Physiol.* **2021**, *4*, 1–12.
20. Dhakal, S.; Karki, S. Early Epidemiological Features of COVID-19 in Nepal and Public Health Response. *Front. Med.* **2020**, *7*, 524. [CrossRef]
21. Bhandari, R.; Ellis-Peterson, H. A hopeless situation': Oxygen shortage fuels Nepal's Covid crisis. *The Guardian*. 10 May 2021. Available online: https://www.theguardian.com/world/2021/may/10/hopeless-situation-oxygen-shortage-fuels-nepal-covid-crisis (accessed on 10 November 2021).
22. MoHP. *Enhanced Surveillance on Sero Prevalence of SARS-CoV-2 in General Population of Nepal*; MoHP: Kathmandu, Nepal, 2021.

23. Lama, G.; Shrestha, L.; Karmacharya, N.; Manandhar, R.; Jha, R. Determination of Sero-Prevalence of SARS-CoV-2 antibody among immunized individuals with COVID-19 vaccine in a tertiary center, Nepal. *Int. J. Multidiscip. Res. Growth Eval.* **2022**, *3*, 62–65. [CrossRef]
24. Situation Report-800. Coronavirus Disease 2019 (COVID-19) WHO Country Office for Nepal. 2022. Available online: https://covid19.mohp.gov.np/situation-report (accessed on 20 June 2022).
25. Bhattarai, S.; Dhungana, J. A million-dose success for Nepal: Insights from Nepal's national vaccine deployment plan against COVID-19. *J. Travel Med.* **2021**, *28*, taab027. [CrossRef]
26. Worobey, M. Dissecting the early COVID-19 cases in Wuhan. *Science* **2021**, *374*, 1202–1204. [CrossRef]
27. MyRepublica. Nepal in Second Stage of COVID-19 Transmission. *MyRepublica*. 5 April 2020. Available online: https://myrepublica.nagariknetwork.com/news/nepal-in-second-stage-of-covid-19-transmission/ (accessed on 10 November 2021).
28. Marahatta, S.B.; Mahotra, A.; Aryal, N. How should Nepal apply lock-down exit strategy against rising COVID-19 burden in Nepal? *Appl. Sci. Technol. Ann.* **2020**, *1*, 58–62. [CrossRef]
29. Sharma, A.; Venkatesh, U.; Dong, V.; Raut, A.; Tawfik, G.M.; Ng, S.J.; El Ramly, T.A.; Nam, N.H.; Pavlenko, D.; Alied, M.; et al. Universal checklist for COVID-19 control measures: Consensus via a Delphi method. *Clin. Med. J. R. Coll. Physicians* **2021**, *21*, 25–26. [CrossRef]
30. Sah, P.; Mandal, A.K.; Saud, B.; Yadav, S.K.; Sah, S.K.; Gyawali, M.; Karn, S.K. Knowledge, practices and compliance related to COVID-19 among Nepalese population in Province 2, Nepal. *Dialogues Health* **2022**, *1*, 100002. [CrossRef]
31. Khadka, R.B.; Gyawali, R. Establishing PCR testing in Nepal for COVID-19: Challenges and opportunities. *Kathmandu Univ. Med. J.* **2020**, *18*, 309–312.
32. Mahato, P.; Tamang, P.; Simkhada, P.; Shahi, P.; van Teijlingen, E.; Aryal, N.; Regmi, P. Effects of COVID-19 during lockdown in Nepal. *Eur. J. Med. Sci.* **2020**, *2*, 113–118. [CrossRef]
33. Subedi, D.; Pantha, S.; Subedi, S.; Gautam, A.; Gaire, A.; Sapkota, D.; Sapkota, S.; Kandel, M.; Parajuli, A.; Ghimire, H.; et al. Perceptions towards COVID-19 vaccines and willingness to vaccinate in Nepal. *Vaccines* **2021**, *9*, 1448. [CrossRef]
34. Kansakar, S.; Dumre, S.P.; Raut, A.; Huy, N.T. From lockdown to vaccines: Challenges and response in Nepal during the COVID-19 pandemic. *Lancet Respir. Med.* **2021**, *9*, 694–695. [CrossRef]
35. Sah, R.; Khatiwada, A.P.; Shrestha, S.; Bhuvan, K.C.; Tiwari, R.; Mohapatra, R.K.; Dhama, K.; Rodriguez-Morales, A.J. COVID-19 vaccination campaign in Nepal, emerging UK variant and futuristic vaccination strategies to combat the ongoing pandemic. *Travel Med. Infect. Dis.* **2021**, *41*, 102037. [CrossRef]

life

MDPI

Review

Post-COVID-19 Condition: Where Are We Now?

Paula Boaventura [1,2,3,*], Sofia Macedo [1,2], Filipa Ribeiro [1,2], Sónia Jaconiano [4] and Paula Soares [1,2,3]

1. Institute of Molecular Pathology and Immunology of the University of Porto (IPATIMUP), Rua Júlio Amaral de Carvalho 45, 4200-135 Porto, Portugal; amacedo@ipatimup.pt (S.M.); filipa.ribeiro@i3s.up.pt (F.R.); psoares@ipatimup.pt (P.S.)
2. Instituto de Investigação e Inovação em Saúde (i3S), University of Porto, Rua Alfredo Allen 208, 4200-135 Porto, Portugal
3. Faculty of Medicine of the University of Porto (FMUP), 4200-319 Porto, Portugal
4. School of Architecture, Art and Design (EAAD), University of Minho, 4800-058 Guimarães, Portugal; id5928@alunos.uminho.pt

* Correspondence: mboaventura@ipatimup.pt

Abstract: COVID-19 is currently considered a systemic infection involving multiple systems and causing chronic complications. Compared to other post-viral fatigue syndromes, these complications are wider and more intense. The most frequent symptoms are profound fatigue, dyspnea, sleep difficulties, anxiety or depression, reduced lung capacity, memory/cognitive impairment, and hyposmia/anosmia. Risk factors for this condition are severity of illness, more than five symptoms in the first week of the disease, female sex, older age, the presence of comorbidities, and a weak anti-SARS-CoV-2 antibody response. Different lines of research have attempted to explain these protracted symptoms; chronic persistent inflammation, autonomic nervous system disruption, hypometabolism, and autoimmunity may play a role. Due to thyroid high ACE expression, the key molecular complex SARS-CoV-2 uses to infect the host cells, thyroid may be a target for the coronavirus infection. Thyroid dysfunction after SARS-CoV-2 infection may be a combination of numerous mechanisms, and its role in long-COVID manifestations is not yet established. The proposed mechanisms are a direct effect of SARS-CoV-2 on target cells, an indirect effect of systemic inflammatory immune response, and a dysfunction of the hypothalamic-pituitary-thyroid (HPT) axis leading to decreased serum TSH. Only a few studies have reported the thyroid gland status in the post-COVID-19 condition. The presence of post-COVID symptoms deserves recognition of COVID-19 as a cause of post-viral fatigue syndrome. It is important to recognize the affected individuals at an early stage so we can offer them the most adequate treatments, helping them thrive through the uncertainty of their condition.

Keywords: post-COVID-19 condition; long COVID; SARS-CoV-2; thyroid

Citation: Boaventura, P.; Macedo, S.; Ribeiro, F.; Jaconiano, S.; Soares, P. Post-COVID-19 Condition: Where Are We Now? *Life* **2022**, *12*, 517. https://doi.org/10.3390/life12040517

Academic Editors: Silvia De Francia and Sarah Allegra

Received: 14 March 2022
Accepted: 29 March 2022
Published: 31 March 2022

1. Introduction

COVID-19 was first described as a respiratory disease, but presently it is considered a systemic infection comprising multiple systems and causing chronic complications [1–3] (Figure 1). The pathology results not only from the virus infection but from an aberrant inflammatory host immune response [4]. The immune response has been well described in acute COVID-19 patients, but the lasting consequences of the infection are still not well known [4]. Researchers have been exhaustively surveying the diverse symptoms of long COVID, but until now, no integrated explanation exists for their manifestation [5]. Sykes et al. [6] alerted us to the effects of this poorly known lethal virus, to the societal disruption it has caused, and to the importance it may have in the development of long-lasting physical and mental health symptoms. On the other hand, Sancak and Kilic [1] state that post-COVID-19 condition symptoms can most often be interpreted as somatization; however, the fact that we may not understand them does not mean they are purely psychosomatic [1]. In the study of Xiong et al. [7] in hospitalized patients from Wuhan, a non-infected control group from the general population was used in order to exclude the psychological effects of the long and mandatory isolation period, which caused deconditioning, anxiety, and depression. The authors showed a significant difference between the group of COVID-19 "recovered" patients and the control group, with the latter reporting very few long-term symptoms [7].

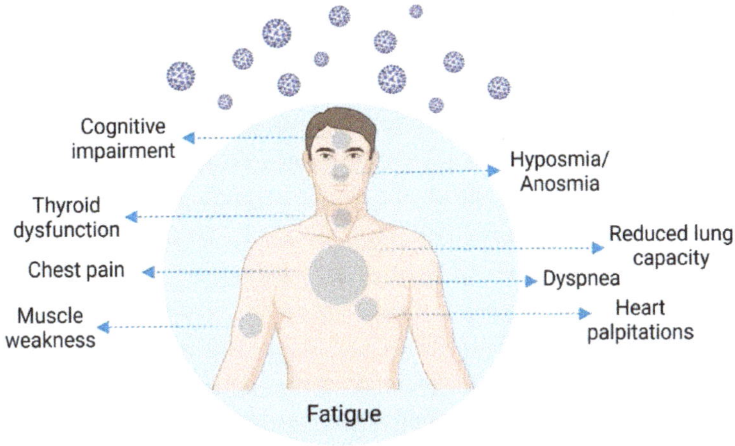

Figure 1. Examples of post-COVID-19 condition chronic complications.

2. Post-COVID-19 Condition Symptomatology and Prevalence

Protracted COVID-19 infection–related symptoms are common [8,9], but the post-COVID-19 condition [10] (previously referred to as long COVID] is a poorly understood aspect of the current pandemic [9,11]. Compared with other post-viral fatigue syndromes, the symptoms are wider and more intense [12]. An exact definition was recently published by the WHO [10]: typically, symptoms with duration ≥2 months [10,13,14] that cannot be explained by an alternative diagnosis are considered post-COVID-19 condition [10]. Post-acute manifestations may be divided into three categories: (1) residual symptoms continuing after recovery from acute infection; (2) organ dysfunction continuing after initial recovery; and (3) new symptoms or syndromes that appear after initial asymptomatic or mild infection [15].

Over several studies, the frequency of post-COVID-19 condition ranged from 4.7 to 80% (n = 25), occurring between 3 to 24 weeks after the acute phase or hospital discharge [16]. Yong [17], in a study on COVID-19 survivors (n = 10), reported that a post-COVID-19 condition persisted for one to six months in 30−80% of patients. Other studies reported a 35% prevalence of residual symptoms in non-hospitalized patients [18], but around 75–87% in hospitalized patients [6,18,19]. In a cohort of patients followed-up for three to nine months after infection, about 30% maintained persistent symptoms [20].

The most frequently reported symptoms, not restricted to severe acute disease [19,21], are profound fatigue [3,6,15,19,22–30] or muscle weakness [6,19,29,31], dyspnea [3,6,24,25,28,30], sleep difficulties [6,19,23,28], anxiety or depression [6,19,29], reduced lung capacity [22,32], memory/cognitive impairment ("brain fog") [23,28,29], hyposmia/anosmia [23], and the inability to fully exercise or work. The most frequent symptom of post-COVID-19 condition is fatigue, which is independent of the acute disease severity or the presence of respiratory problems [33]. A summary of post-COVID-19 condition symptoms, with the frequencies reported and number of patients evaluated, is presented in Table 1. The high variability found between studies is mostly attributable to acute COVID-19 severity, with more frequent symptoms in hospitalized patients compared with patients who suffered from mild or asymptomatic disease.

Table 1. Post-COVID-19 condition symptoms most frequently reported.

Post-COVID-19 Symptoms	Number of Patients Included in the Study	% Patients Suffering from Symptom/References
Fatigue	596, 177, 538, 270, 138, 3065, 134, 242, 115, 143, 96, 1733, 5440, 384, 287	13.1% [34], 13.6% [20], 28.3% [7], 34.8% [25], 39.0% [27,30], 39.6% [6], 41.7% [35], 47% [26], 53.1% [36], 56.3% [28], 63% [19], up to 65% [16], 69% [37], 72.8% [38]
Persistent breathlessness /dyspnea	596, 3065, 287, 270, 96, 138, 143, 384, 134, 5440, 35	6.0% [34], 23.2% [30], 28.2% [38], 34.0% [25], 37.5% [28], 40.0% [27], 43.4% [36], 53.0% [37], 60% [6], up to 61% [16], 80% [23]
Myalgia /muscle weakness	277, 242, 134, 1733	19.6% [25], 35,1% [35], 51.5% [6], 63% [19]
Anxiety	287, 402, 134	38.0% [38], 42% [39], 47.8% [6]
Sleep disturbance	1733, 96, 134, 35, 138	21.1% [35], 26.0% [19], 26.0% [28], 35.1% [6], 40.0% [39], 46% [23], 49% [27]
Joint pain	277, 143, 287	19.6% [25], 27.3% [36], 31.4% [38]
Headache	242, 270, 3065, 287	19.0% [35], 19.8% [25], 23.4% [30], 28.6% [38]
Chest pain	596, 242, 538, 143, 287, 35, 5440	0.8% [34], 10.7% [35], 12.3% [7], 21.7% [36], 28.9% [38], 34.8% [23], up to 89% [16]

To better examine this issue, Gaber et al. [27] looked at the effects of COVID-19 infection in healthcare workers, a population with an expected high level of exposure to the virus. They reported a high incidence of infection and a high prevalence of incapacitating post-COVID-19 symptoms, with fatigue commonly reported [27]. Nonetheless, these health workers were unwilling to either seek medical help or take sick leave, despite their struggle to cope with the symptoms [27].

Taquet et al. [40] found a higher incidence of numerous psychiatric disorders in COVID-19 survivors compared with matched patients with influenza or other respiratory tract infections, in a retrospective cohort study using 236,379 electronic health records. The estimated incidence of a neurological or psychiatric diagnosis in the six months following a COVID-19 diagnosis was 33% (95% CI; 33.17–34.07) [40]. Post-COVID-19 condition presents neurological symptoms similar to chronic fatigue syndrome (CFS) and functional neurological disorder (FND] (except for hypogeusia) [41].

Davis et al. [42] conducted an online survey to characterize post-COVID-19 condition in an international cohort (56 countries), tracing the symptoms over 7 months. They found for 91% of the respondents that the time to recovery exceeded 35 weeks; the most frequent symptoms after six months were fatigue, post-exertional malaise, and cognitive dysfunction [42]. According to the authors, their study represents the largest collection of symptoms recognized in post-COVID-19 condition individuals to date (June 2021). More recent studies have shown that persistent symptoms can be found 12 [43] or up to 15 months after recovery from the acute phase of COVID-19 [44]; symptoms are common both in ambulatory and hospitalized patients [44].

3. Post-COVID-19 Condition Risk Factors

Post-COVID-19 condition is associated with a weak anti-SARS-CoV-2 antibody response [45], severity of illness [19,34,46,47], female sex [3,5,6,18,19,34,35,45], presence of more than five symptoms in the first week of the disease [3,18,48], older age [18], and presence of comorbidities [18]. Concretely, Fernández-de-las-Peñas et al. [49] reported that the most significant risk factor for developing more post-COVID symptoms was the number of symptoms at hospital admission, which supports the idea that a higher symptom burden in the acute phase of the disease is associated with a higher probability of the post-COVID-19 condition.

Early dyspnea, prior psychiatric disorders, and specific biomarkers (e.g., D-dimer, C- reactive protein, and lymphocyte count) have also been reported as risk factors, even though more research is needed to validate them [3]. Peghin et al. [34] suggested that the constantly elevated titers of the serological response against SARS-CoV-2 may constitute an independent risk factor for the post-COVID-19 condition, since the presence of SARS-CoV-2 IgG antibodies is significantly associated with the condition. Contrarily, Seessle et al. [28] reported that patients presenting at least one post-COVID-19 symptom 12 months after infection did not significantly differ in their SARS-CoV-2 antibody levels when compared with patients without symptoms, although their physical and mental quality of life had significantly decreased.

Interestingly, Townsend et al. [26] showed that significant illness persistence after the COVID-19 acute phase of the disease, affecting health perception, ability to return to work, and the existence of lasting fatigue, appears to be unrelated to the severity of the acute phase, though one would expect to see a difference in post-COVID symptoms between hospitalized and non-hospitalized patients; this hypothesis needs to be verified in upcoming studies [50]. In fact, one puzzling feature of post-COVID-19 condition is that it affects COVID-19 patients at all disease severity levels [3], often affecting patients with a mild acute illness [51]. Studies have shown that post-COVID-19 condition affects even mild to moderate cases [3,52,53] and younger adults (or even children) who did not need respiratory support or hospital or intensive care [3]. Post-COVID-19 condition in children is similar to that seen in adults [54], with symptoms such as a fatigue, dyspnea, myalgia, cognitive impairments, headache, palpitations and chest pain [3,55].

In general, it appears that the ratio for post-COVID-19 condition development is 2:1 in women compared with men, but only until around age 60, when the ratio between women and men becomes similar [14].

Post-COVID-19 condition in patients with comorbidities may result from their comorbidity worsening [56].

4. Post-COVID-19 Condition Pathophysiology

Different lines of research are trying to explain these protracted symptoms. A persistent immune activation and/or inflammation may contribute to post-COVID-19 condition, which could explain why many patients with mild COVID-19 disease experience chronic persistent symptoms, involving the cardiovascular, nervous, and respiratory systems [57]. In fact, the persistently elevated inflammatory markers observed in long-COVID patients point towards chronic persistence of inflammation [18,58].

Shuwa et al. [4] observed lasting alterations in the functional potential of CD8+ T cells from recovering COVID-19 patients up to six months following hospital discharge, which may imply a sustained change in cytokine potential, contributing to a constant inflammatory status [4]. Contrarily, B cell changes seem to be largely restored in convalescence [4]. In a more recent study, Glynne et al. [51] reported that CD8+ EM T cells are diminished for up to 400 days following infection, regardless of symptoms, and CD4+ and CD8+ CMT cell PD-1 levels are augmented following COVID-19 (more marked in post-COVID-19 condition). T-cell dysfunction may promote post-COVID-19 condition pathophysiology similarly to what occurs in autoimmune diseases [3]. It remains to be determined if SARS-CoV-2–specific T cells have the capacity to react against self-antigens [57].

Seessle et al. [28] observed several neurocognitive symptoms that were associated with antinuclear antibody titer elevation, pointing to autoimmunity as a cofactor in the etiology of post-COVID-19 neurologic conditions [28]. The autoimmune hypothesis could explain the greater incidence of this condition in women [14,57]. Since thyroid is closely linked to T-cell-mediated autoimmunity, thyroid dysfunction may be important in the pathophysiology of post-COVID-19 condition, as discussed in more detail below [3].

Post-COVID-19 condition has been related to additional characteristics of the innate and adaptive response, involving a weaker initial inflammatory response, with lower baseline levels of C-reactive protein and ferritin [45]. The participation of the immune system in post-COVID-19 condition has been reported in other studies [8,21,57,59,60]. Symptoms such as cognitive dysfunction, persistent fatigue, muscle aches, depression, and other mental health issues are highly associated with an initial immune challenge and/or with a constant dysregulation of the immune system [29,60].

Many neurological anomalies have been described in patients with COVID-19 [41,61], comprising the central and peripheral nervous systems, ranging from mild to fatal, and occurring in patients with severe or asymptomatic SARS-CoV-2 infection [61]. These deferred manifestations may be significant, because they likely affect patients not presenting neurological symptoms in the acute phase [62]. Neurocognitive post-COVID-19 condition symptoms can last for at least one year subsequent to acute infection, diminishing life quality considerably [28].

The involvement of inflammatory cytokines in the etiology of the neuropsychiatric symptoms, reported in current large-scale population-based epidemiological and genetic studies, indicates that these cytokines may have a role in the etiology of the neuropsychiatric symptoms usually observed in patients with post-COVID-19 condition [3,29,60]. This cytokine storm must also be considered as a possible driving factor for the expansion of neuropathies after severe COVID-19 infection, contributing to the chronic pain that appears after acute infection recovery [62]. The augmented cytokine activity, which drives the inflammatory process, disrupts T cell responses, and imposes limitations on neuronal metabolism, may also be an adequate therapeutic target for management and prevention of post-COVID-19 condition [60].

Altered tryptophan absorption and tryptophan-disrupted metabolism have been suggested as key contributors to the enduring symptoms in COVID-19–recovered patients, with numerous studies showing low levels of tryptophan and serotonin in individuals infected with SARS-CoV-2 [63]. Tryptophan is a precursor of melatonin and serotonin, molecules implicated in sleep control and mood disorders, respectively; it is also involved in skeletal muscle mass regulation, a notorious lasting symptom of post-COVID-19 condition [63].

Some symptoms may be related to virus- or immune-mediated disruption of the autonomic nervous system, leading to transient or longstanding orthostatic intolerance syndromes [8,31,64,65]. In orthostatic intolerance, the release of epinephrine and norepinephrine causes pronounced tachycardia, which is experienced as palpitations, breathlessness, fatigue, and chest pain, which are common symptoms of post-COVID-19 condition [8]. Alterations in the autonomic nervous system can promote each of these symptoms, theoretically providing a uniting pathobiology for acute, subacute, and lasting sequelae of the infection, and may also be considered as a target for intervention [31].

Studies have shown that patients with severe symptoms may have more severe autonomic dysfunction when compared with patients presenting mild symptoms, as indicated by the heart rate variability (HRV) analysis [2], which is a reliable non-invasive tool used to evaluate autonomic modulation [2,64]. Patients with severe symptoms presenting amelioration in autonomic parameters also show enhancements in immune and coagulation functions, as well as in cardiac injury biomarkers [2].

Townsend et al. [66] conducted research to assess if fatigue, the most common symptom following infection, was associated with autonomic dysfunction. No association was found with autonomic dysfunction; the authors found an intense association of fatigue with increased anxiety ($p < 0.001$) in patients without pre-existing diagnoses of anxiety [66].

Another potential cause of post-COVID-19 condition could be the SARS-CoV-2 tropism from the olfactory system into the brainstem, and the consequent persistent, low-grade brainstem dysfunction [17]. SARS-CoV-2 may damage the brainstem through viral invasion, inflammation, and vascular activation [17]. Interestingly, functions of the brainstem and post-COVID-19 condition symptoms have a great degree of overlap [17].

SARS-CoV-2 RNA was found in the brain during autopsy of deceased COVID-19 patients in some studies, but in other studies no SARS-CoV-2 materials were found [17]. This suggests that SARS-CoV-2 neurotropism or brain invasion may happen but not in every patient [17]. The presence of SARS-CoV-2 in the central nervous system has not been directly related to the severity of the neuropathological findings, suggesting that neuronal infection may be only one of the pathways through which SARS-CoV-2 could influence brain function and contribute to some of the long-lasting symptoms of post-COVID-19 condition [29].

Hypometabolism has been reported in post-COVID-19 condition patients; specifically, hyposmia/anosmia was associated with cerebellar hypometabolism [23]. In general, areas of hypometabolism comprised the bilateral rectal/orbital gyrus (including the olfactory gyrus], the right temporal lobe (including the amygdala and the hippocampus extending to the right thalamus], the bilateral pons/medulla brainstem, and the bilateral cerebellum [23]. These metabolic groups allowed distinguishing between patients and healthy subjects with a high power of discrimination.

Long-term cardiovascular effects of COVID-19 have been described [67]. Vascular events can happen unpredictably in fit patients with mild or asymptomatic COVID-19 infection, even several weeks after the infection [68]. This means that clinicians should remain attentive for post-infective thrombotic sequelae and carefully manage cardiovascular risk factors in convalescent patients, irrespective of the infection severity and the absence of co-morbidities [68]. In post-COVID-19 condition management it is essential to control blood pressure, lipid levels, and obesity after infection with SARS-CoV-2 [53].

Immunological memory of SARS-CoV-2 is not easy to predict [19]. Neutralizing antibody titers at six-month follow-up are significantly lower compared with the acute phase [8,45]. Contrarily, Sette et al. [22] reported data indicating that T and B cell memory and antibodies probably remain for years in most SARS-CoV-2 infected patients.

As previously mentioned, an additional possibility is that post-COVID-19 condition is caused by an immune system dysfunction that leads the immune system to attack the body, meaning that this condition could be an autoimmune disease [5]. Still, it is precocious to affirm which hypothesis is right and, in fact, it might be the case that each is true in different individuals; preliminary data suggest that post-COVID-19 condition could be various disorders grouped into one [5]. These various disease courses may be traced back to the initial phases of the infection, as shown by the fundamental role of type I IFN responses during the acute phase of SARS-CoV-2 infection [13]. As previously mentioned, the autoimmune hypothesis could explain women's higher susceptibility to this syndrome [14]. Indeed, women present a stronger immune response for genetic and hormonal factors compared with men; this is a double-edged sword, leading to a more severe outcome of acute infection in men, but to more common autoimmune reactions in women [14].

5. Thyroid Involvement in COVID-19

Due to the reported high expression of ACE2, the thyroid may become a target of coronavirus infection, and thyroid involvement in COVID-19 patients has been demonstrated [69]. In fact, SARS-CoV-2 uses ACE2, combined with the transmembrane protease serine 2 (TMPRSS2), as the main molecular complex for the host cell infection [70]. Interestingly, ACE2 and TMPRSS2 expression levels are higher in the thyroid gland than in the lungs [70]. Scappaticcio et al. [70], in their literature review on thyroid dysfunction in COVID-19 patients, presented strong evidence that the thyroid gland and the entire hypothalamic–pituitary–thyroid (HPT) axis may be important targets for SARS-CoV-2 damage.

Coperchini et al. [71] showed that IFN and, to a minor degree TNF-alpha, regularly increase ACE-2 mRNA levels in normal human thyroid primary cultures. As stated by these authors, the increased pro-inflammatory cytokine levels may enable SARS-CoV-2 penetration in the cells through an additional increase of ACE-2 expression and/or account for the diverse grades of severity of the infection [71]. Nevertheless, additional specific studies are needed to validate this hypothesis [71]. Two main mechanisms account for thyroid function alterations in COVID-19 patients: a direct effect of SARS-CoV-2 on target cells and an indirect effect of the systemic inflammatory immune response [70,72–75]. A third hypothesis is that dysfunction of the HPT axis causes centrally a decreased level of serum TSH in the infected patients [74]. Changes in thyroid function tests, mostly defined by a TSH level decrease, were described during the acute phase of the infection [69]. These changes have been associated with either destructive thyroiditis or non-thyroidal illness syndrome (NTIS) [69,76]. NTIS, which is defined by low T3 levels, may be caused by any severe systemic disease [76,77]. It occurs due to the diminished conversion of T4 to T3, which is likely elicited by the same factors that cause a decrease in TSH (increase in cytokines and other inflammatory factors) [77,78]. T3 reduction was observed even in mild COVID-19 disease severity, with increased conversion of T4 to reverse T3 [79].

Since manifestations of post-COVID-19 condition include fatigue, and immune dysregulation is one of the proposed mechanisms involved in the condition development, Lui et al. [80] decided to investigate whether thyroid function and autoimmunity play a role in post-COVID-19 condition. They showed, following-up COVID-19 patients, the spontaneous recovery of most thyroid dysfunction observed in the acute phase of the disease, and that incident thyroid dysfunction was a rare situation. Subgroup analysis revealed that symptom recovery occurred more among patients with positive anti-TPO at the time of re-evaluation, suggesting a potential protective role of anti-TPO in post-COVID-19 condition [80].

Only a few studies have evaluated the thyroid gland condition in the convalescent stage of COVID-19 [81]. Clarke et al. reported that adrenal and thyroid function was maintained $\geq$3 months after COVID-19 diagnosis, even though an important proportion of patients suffered from chronic fatigue [82].

Campi et al. [69] found a temporary situation of low TSH with normal T4 and low T3 levels in patients hospitalized for SARS-CoV-2 infection, which was inversely associated with C-reactive protein, cortisol, and IL-6, and positively associated with normal Tg levels. These authors stated that this temporary change was probably due to the cytokine storm induced by the virus, with a direct or mediated impact on TSH secretion and deiodinase activity, and probably not to a destructive thyroiditis. The THYRCOV study offers early evidence that patients with acute SARS-CoV-2 infection with thyrotoxicosis have statistically significantly higher levels of IL-6 [83]. In a short-term follow-up, Pizzocaro et al. [84] showed a spontaneous normalization of thyroid function in most infected patients with SARS-CoV-2-related thyrotoxicosis. Nevertheless, these authors stated that long-lasting studies are needed, since they found a frequent thyroid hypoecogenicity pattern in the ultrasonographic evaluation of these patients, which may predispose them to late-onset thyroid dysfunction development [84].

Subacute thyroiditis related to COVID-19 typically presents without pain and with thyrotoxicosis, which in some cases is followed by hypothyroidism [73,85]. Subacute thyroiditis was reported in 13 cases (in 10 papers), detected 7 weeks before to 7 weeks after the diagnosis of COVID-19 [76], so only some of these cases could be compatible with post-COVID-19 condition.

Dworakowska et al. [86] stated that clinicians should be aware of subacute thyroiditis likelihood, particularly in the early weeks or months after even mild COVID-19 infection. Subacute thyroiditis might be considered as a late complication of SARS-CoV-2 infection, since it frequently arises a few weeks after the upper respiratory tract infection [75,87]. It may be difficult to promptly diagnose this due to a potential lack of classic symptoms and to shared clinical features between COVID-19 and thyrotoxicosis [88].

Recently, Trimboli et al. [89] conducted a systematic review of subacute thyroiditis in COVID-19 patients, concluding that the size and quality of published data are poor, with only case reports and case series being available. According to the authors, and based on these evidence-based data, subacute thyroiditis cannot yet be considered as a direct or common complication of SARS-CoV-2. Still, this assumption might change in the future, considering the fast worldwide diffusion of SARS-CoV-2 and its variants [89].

Even though clear evidence is missing, infection of the thyrocyte, thyrotroph, and corticotroph may lead to a decrease in T3, T4, TSH, ACTH, and cortisol levels [76]. HPT dysregulation has been considered, at least in part, responsible for hypothyroidism in COVID-19 [74,76]. Low FT3 levels are independently associated with increased mortality [72,76,90] and disease severity [74,91–93] and may be used as a surrogate prognostic biomarker [72,76,90].

Sick euthyroidism is the most common thyroid-related issue in COVID-19 follow-up, especially in patients who were hospitalized or were admitted to intensive care units [73]. Euthyroid sick syndrome is a condition characterized by low serum levels of thyroid hormones in patients with nonthyroidal systemic illness who are clinically euthyroid. These alterations were transitory and recovered during follow-up, although long-term follow-up studies on thyroid function are still needed [73]. Asghar et al. [94], who analysed 54 COVID-19 patients, reported severe COVID-19 patterns in those patients who appeared to have euthyroid sick syndrome. They also reported that the precise clinical importance of a low TSH was uncertain. The authors included a cut-off estimation of TSH decline, predicting disease severity; patients with low TSH levels (<0.996 uIU/mL) showed significantly low survival, whereas patients with sufficient TSH (>0.996 uIU/mL) had a higher cumulative survival proportion [94]. One main limitation of this study was its small sample size. Gong et al. [95] reported that critical illness rates (74.07% vs. 37.40%, $p = 0.001$) and mortality rates (51.85% vs. 22.76%, $p = 0.002$) were significantly higher in the low TSH group compared with a normal TSH group. Zou et al. [96] also reported that euthyroid sick syndrome was significantly associated with the disease severity and inflammatory parameters in COVID-19 patients.

An altered thyroid function is a common situation in COVID-19 patients [72,74,97], especially in critically ill patients [92]. Lui et al. [90] reported that approximately 15% of patients with mild to moderate COVID-19 had thyroid dysfunction. Nevertheless, with the data published so far, it is not possible to assume that thyroid diseases are a risk factor for COVID-19 disease [77,98]. Likewise, a higher occurrence of thyroid disease in patients with COVID-19 has not been observed [77,98]. Although COVID-19 is linked to NTIS, it is not clear if it also raises the risk of developing autoimmune hypothyroidism [98]. It is hypothesized that SARS-CoV-2 might directly influence thyroid morphology and function, leading to an aggravation of a pre-existing autoimmune thyroid disease [98]. Additionally, COVID-19 may worsen autoimmune thyroid disease due to its repercussions on the immune system, which may lead to the development of the cytokine storm [98]. Thyroid autoimmunity, evaluated through the presence of anti-TPO antibodies, was common in COVID-19 patients as compared with pre-pandemic controls [99].

Recent data demonstrate that thyroid hormones have an important role in protecting the lungs from damage, including those related to SARS-CoV-2 infection [98]. The lung is one of many organs that responds to the thyroid hormone, and the T3 receptor is present in alveolar type II cells [98]. T3 increases cell size and number, stimulates surfactant release, and elevates the sodium- and potassium-ATPase pump activity, increasing the cell capacity to translocate fluid and therefore absorb alveolar oedema fluid [98].

Our knowledge of the thyroid patterns of COVID-19 is still incomplete, as is the etiologic view of COVID-19 and thyroid insults [76,100]. To find direct evidence concerning the nature and cause of thyroid SARS-CoV-2 injury, and the full immune response in those patients with thyroid dysfunction, we need a histologic and cytological examination of the thyroid gland in a wide number of patients [74,76]. Poma et al. [101] detected SARS-CoV-2 in a small number of thyroid specimens (9/25, 36%). Currently, there is no clear statement on the importance of SARS-CoV-2-induced apoptosis in the thyroid dysfunction [102], but in the SARS-CoV-2 outbreak, it was shown that apoptosis plays an important role in thyroid injury [102]. Summing up, thyroid dysfunction secondary to SARS-CoV-2 infection is probably a combination of various mechanisms [74], and its role in post-COVID-19 condition is not yet established. Tutal et al. [103], who performed a systematic review of COVID-19 and autoimmune thyroiditis, considered it reasonable to routinely assess thyroid function, both in the acute phase of the infection and during the convalescence, through serum TSH, T4, and T3 evaluation. Contrarily, based on the assumption that thyroid function usually normalizes on follow-up, Pat et al. [104] did not recommend a widespread thyroid function screening.

6. Post-COVID-19 Condition Health Burden and Patient Management

Post-COVID-19 condition (or long COVID) first gained extensive credit among social support groups, and then in scientific and medical communities [3,5,105,106]. It is probably the first illness to be cooperatively identified by patients discovering one another using Twitter and other social media [105]. The term "post-COVID condition" comprises a wide range of organ impairment, and at the moment we do not have enough information to perform a clear diagnosis, to elect a specific treatment, or to indicate a probable prognosis [107]. Some patients may never recover from the illness [52,56], and all age groups are vulnerable [52]. Patients with post-COVID-19 condition are a heterogeneous group, which makes it difficult to advise treatment [108,109]. It is crucial for each patient to find the correct equilibrium between mild activity to avoid deconditioning and not triggering post-exercise malaise [108]. Strategies tackling our levels of stress and/or the stress response, comprising psychosocial intervention, physical exercise, or possibly dietary interventions could be a good approach to counteract some of the negative effects of chronic inflammation [29]. Rebello et al. [110] advanced that physical exercise may counter the neuropsychiatric and endocrine sequelae of post-COVID-19 condition, through the release of circulating factors that mediate the anti-inflammatory response, support brain homeostasis, and increase insulin sensitivity .

Management of post-acute COVID-19 syndrome requires a comprehensive team, including physicians of various specialties (primary care, pulmonology, cardiology, and infectious disease), physiatrists, behavioural health experts, physical and occupational therapists, and social workers, which will address the clinical and psychological aspects of the disease [111].

Although still speculative at the present time, there is a considerable body of literature supporting the anti-stress and anti-inflammatory role of certain seated meditations, yoga asanas, and pranayama practices [112]. The possible benefits of these practices encompass wider neuroimmune systems, which is an advantage since we are facing a systemically dysregulating disease in COVID-19 [112].

Lastly, we may refer to the role of COVID-19 vaccines in post-COVID-19 condition. Although vaccines prevent death and severe illness, it is not yet clear if they may also prevent post-COVID-19 condition [5]. Small studies have shown that AstraZeneca and Pfizer-BioNTech vaccines were associated with overall improvements in post-COVID-19 condition symptoms [113]. Recently, Antonelli et al. [114] found that the odds of having symptoms for 28 days or more after post-vaccination infection were approximately halved by having two doses of the vaccine.

7. Conclusions

It is urgent to better understand this emerging, complex, and puzzling medical condition [16,115]. Post-COVID-19 condition can become a crisis for health systems, which are already facing the challenge of the pandemic [116]. It is essential to be able to better deal with the symptoms of this condition in terms of clinical care, public health, and health resource planning [116]. At the population level, it is necessary to evaluate the burden of post-COVID-19 condition in order to evaluate its impact on the healthcare system and distribute resources in an adequate way [18,48,107,117]. The primary care services, which represent the first approach for patient diagnosis, still have little information or resources to deal with these patients [118]. Patients with post-COVID-19 condition may have a variety of positive and negative healthcare experiences, which can be useful for the creation or adaptation of the healthcare services [119].

The existence of post-COVID symptoms is leading to the recognition of COVID-19 as a cause of post-viral fatigue syndrome, even when the disease acute phase was mild [20,35]. This can help clinicians to organize patient care, namely follow-up visits, rehabilitation, cognitive behavioral therapy, and even simple actions like inclusion of counseling sessions at discharge to diminish patient anxiety about prolonged symptoms [35]. The patients need to be monitored with a systematic protocol, including symptoms of mental and physical health, and specific healthcare programs to support a healthier lifestyle after SARS-CoV-2 infection need to be implemented [120].

There is an urgent need to identify affected individuals early so the most appropriate and efficient treatments may be provided [111,115], helping them to thrive through the uncertainty of their condition [15,121].

Author Contributions: Conceptualization, P.B. and P.S.; investigation, P.B., S.M., F.R. and S.J.; writing—original draft preparation, P.B., S.M., F.R. and S.J.; writing—review and editing, P.B. and P.S.; supervision, P.S.; funding acquisition, P.S. All authors have read and agreed to the published version of the manuscript.

Funding: This work was supported by Portuguese funds through FCT in the framework of a Ph.D. grant to SM (SFRH/BD/137802/2018). This work is part of the project "Cancer Research on Therapy Resistance: From Basic Mechanisms to Novel Targets"—NORTE-01-0145-FEDER-000051, supported by Norte Portugal Regional Operational Programme (NORTE 2020), under the PORTUGAL 2020 Partnership Agreement, through the European Regional Development Fund (ERDF).

Institutional Review Board Statement: Not applicable.

Informed Consent Statement: Not applicable.

Data Availability Statement: Not applicable.

Conflicts of Interest: The authors declare no conflict of interest. The funders had no role in the design of the study; in the collection, analyses, or interpretation of data; in the writing of the manuscript, or in the decision to publish the results.

References

1. Sancak, B.; Kilic, C. A Psychiatrist's Own Experience of Long COVID: Looking beyond the Psychosomatic Perspective. *Psychiatr. Danub.* **2021**, *33*, 250.
2. Pan, Y.; Yu, Z.; Yuan, Y.; Han, J.; Wang, Z.; Chen, H.; Wang, S.; Wang, Z.; Hu, H.; Zhou, L. Alteration of Autonomic Nervous System Is Associated with Severity and Outcomes in Patients with COVID-19. *Front. Physiol.* **2021**, *12*, 630038. [CrossRef] [PubMed]
3. Yong, S.J. Long COVID or post-COVID-19 syndrome: Putative pathophysiology, risk factors, and treatments. *Infect. Dis.* **2021**, *53*, 737–754. [CrossRef] [PubMed]
4. Shuwa, H.A.; Shaw, T.N.; Knight, S.B.; Wemyss, K.; McClure, F.A.; Pearmain, L.; Prise, I.; Jagger, C.; Morgan, D.J.; Khan, S.; et al. Alterations in T and B cell function persist in convalescent COVID-19 patients. *Med* **2021**, *2*, 720–735.e4. [CrossRef]
5. Marshall, M. The four most urgent questions about long COVID. *Nature* **2021**, *594*, 168–170. [CrossRef]
6. Sykes, D.L.; Holdsworth, L.; Jawad, N.; Gunasekera, P.; Morice, A.H.; Crooks, M.G. Post-COVID-19 Symptom Burden: What is Long-COVID and How Should We Manage It? *Lung* **2021**, *199*, 113–119. [CrossRef] [PubMed]
7. Xiong, Q.; Xu, M.; Li, J.; Liu, Y.; Zhang, J.; Xu, Y.; Dong, W. Clinical sequelae of COVID-19 survivors in Wuhan, China: A single-centre longitudinal study. *Clin. Microbiol. Infect.* **2021**, *27*, 89–95. [CrossRef]
8. Dani, M.; Dirksen, A.; Taraborrelli, P.; Torocastro, M.; Panagopoulos, D.; Sutton, R.; Lim, P.B. Autonomic dysfunction in 'long COVID': Rationale, physiology and management strategies. *Clin. Med.* **2021**, *21*, e63–e67. [CrossRef]
9. Mehandru, S.; Merad, M. Pathological sequelae of long-haul COVID. *Nat. Immunol.* **2022**, *23*, 194–202. [CrossRef] [PubMed]
10. WHO. A Clinical Case Definition of Post COVID-19 Condition by a Delphi Consensus, 6 October 2021. Available online: https://www.who.int/publications/i/item/WHO-2019-nCoV-Post_COVID-19_condition-Clinical_case_definition-2021.1 (accessed on 3 March 2022).
11. Shouman, K.; Vanichkachorn, G.; Cheshire, W.P.; Suarez, M.D.; Shelly, S.; Lamotte, G.J.; Sandroni, P.; Benarroch, E.E.; Berini, S.E.; Cutsforth-Gregory, J.K.; et al. Autonomic dysfunction following COVID-19 infection: An early experience. *Clin. Auton. Res.* **2021**, *31*, 385–394. [CrossRef] [PubMed]
12. Jacobs, J.J. Persistent SARS-2 infections contribute to long COVID-19. *Med. Hypotheses* **2021**, *149*, 110538. [CrossRef]
13. Brodin, P. Immune determinants of COVID-19 disease presentation and severity. *Nat. Med.* **2021**, *27*, 28–33. [CrossRef]
14. Ortona, E.; Buonsenso, D.; Carfi, A.; Malorni, W. Long Covid Kids study g. Long COVID: An estrogen-associated autoimmune disease? *Cell Death Discov.* **2021**, *7*, 77. [CrossRef]
15. Amenta, E.M.; Spallone, A.; Rodriguez-Barradas, M.C.; El Sahly, H.M.; Atmar, R.L.; Kulkarni, P.A. Postacute COVID-19: An Overview and Approach to Classification. *Open Forum Infect. Dis.* **2020**, *7*, ofaa509. [CrossRef] [PubMed]
16. Cabrera Martimbianco, A.L.; Pacheco, R.L.; Bagattini, A.M.; Riera, R. Frequency, signs and symptoms, and criteria adopted for long COVID-19: A systematic review. *Int. J. Clin. Pract.* **2021**, *75*, e14357. [CrossRef] [PubMed]
17. Yong, S.J. Persistent Brainstem Dysfunction in Long-COVID: A Hypothesis. *ACS Chem. Neurosci.* **2021**, *12*, 573–580. [CrossRef] [PubMed]

18. Raveendran, A.V.; Jayadevan, R.; Sashidharan, S. Long COVID: An overview. *Diabetes Metab. Syndr.* **2021**, *15*, 869–875. [CrossRef] [PubMed]
19. Huang, C.; Huang, L.; Wang, Y.; Li, X.; Ren, L.; Gu, X.; Kang, L.; Guo, L.; Liu, M.; Zhou, X.; et al. 6-month consequences of COVID-19 in patients discharged from hospital: A cohort study. *Lancet* **2021**, *397*, 220–232. [CrossRef]
20. Logue, J.K.; Franko, N.M.; McCulloch, D.J.; McDonald, D.; Magedson, A.; Wolf, C.R.; Chu, H.Y. Sequelae in Adults at 6 Months After COVID-19 Infection. *JAMA Netw. Open* **2021**, *4*, e210830. [CrossRef] [PubMed]
21. Ramakrishnan, R.K.; Kashour, T.; Hamid, Q.; Halwani, R.; Tleyjeh, I.M. Unraveling the Mystery Surrounding Post-Acute Sequelae of COVID-19. *Front. Immunol.* **2021**, *12*, 686029. [CrossRef]
22. Sette, A.; Crotty, S. Adaptive immunity to SARS-CoV-2 and COVID-19. *Cell* **2021**, *184*, 861–880. [CrossRef]
23. Guedj, E.; Campion, J.Y.; Dudouet, P.; Kaphan, E.; Bregeon, F.; Tissot-Dupont, H.; Guis, S.; Barthelemy, F.; Habert, P.; Ceccaldi, M.; et al. 18F-FDG brain PET hypometabolism in patients with long COVID. *Eur. J. Pediatr.* **2021**, *48*, 2823–2833. [CrossRef]
24. Naeije, R.; Caravita, S. Phenotyping long COVID. *Eur. Respir. J.* **2021**, *58*, 2101763. [CrossRef]
25. Moreno-Pérez, O.; Merino, E.; Leon-Ramirez, J.-M.; Andres, M.; Ramos, J.M.; Arenas-Jiménez, J.; Asensio, S.; Sanchez, R.; Ruiz-Torregrosa, P.; Galan, I.; et al. Post-acute COVID-19 syndrome. Incidence and risk factors: A Mediterranean cohort study. *J. Infect.* **2021**, *82*, 378–383. [CrossRef]
26. Townsend, L.; Dowds, J.; O'Brien, K.; Sheill, G.; Dyer, A.H.; O'Kelly, B.; Hynes, J.P.; Mooney, A.; Dunne, J.; Ni Cheallaigh, C.; et al. Persistent Poor Health after COVID-19 Is Not Associated with Respiratory Complications or Initial Disease Severity. *Ann. Am. Thorac. Soc.* **2021**, *18*, 997–1003. [CrossRef] [PubMed]
27. Gaber, T.A.-Z.K.; Ashish, A.; Unsworth, A. Persistent post-covid symptoms in healthcare workers. *Occup. Med.* **2021**, *71*, 144–146. [CrossRef]
28. Seessle, J.; Waterboer, T.; Hippchen, T.; Simon, J.; Kirchner, M.; Lim, A.; Müller, B.; Merle, U. Persistent symptoms in adult patients one year after COVID-19: A prospective cohort study. *Clin. Infect. Dis.* **2021**, ciab611. [CrossRef]
29. Mondelli, V.; Pariante, C.M. What can neuroimmunology teach us about the symptoms of long-COVID? *Oxf. Open Immunol.* **2021**, *2*, iqab004. [CrossRef]
30. Stephenson, T.; Pereira, S.M.P.; Shafran, R.; de Stavola, B.L.; Rojas, N.; McOwat, K.; Simmons, R.; Zavala, M.; O'Mahoney, L.; Chalder, T.; et al. Physical and mental health 3 months after SARS-CoV-2 infection (long COVID) among adolescents in England (CLoCk): A national matched cohort study. *Lancet Child Adolesc. Health* **2022**, *6*, 230–239. [CrossRef]
31. Becker, R.C. Autonomic dysfunction in SARS-CoV-2 infection acute and long-term implications COVID-19 editor's page series. *J. Thromb. Thrombolysis* **2021**, *52*, 692–707. [CrossRef] [PubMed]
32. Torres-Castro, R.; Vasconcello-Castillo, L.; Alsina-Restoy, X.; Solis-Navarro, L.; Burgos, F.; Puppo, H.; Vilaró, J. Respiratory function in patients post-infection by COVID-19: A systematic review and meta-analysis. *Pulmonology* **2021**, *27*, 328–337. [CrossRef] [PubMed]
33. Bansal, R.; Gubbi, S.; Koch, C.A. COVID-19 and chronic fatigue syndrome: An endocrine perspective. *J. Clin. Transl. Endocrinol.* **2022**, *27*, 100284. [CrossRef]
34. Peghin, M.; Palese, A.; Venturini, M.; De Martino, M.; Gerussi, V.; Graziano, E.; Bontempo, G.; Marrella, F.; Tommasini, A.; Fabris, M.; et al. Post-COVID-19 symptoms 6 months after acute infection among hospitalized and non-hospitalized patients. *Clin. Microbiol. Infect.* **2021**, *27*, 1507–1513. [CrossRef] [PubMed]
35. Kashif, A.; Chaudhry, M.; Fayyaz, T.; Abdullah, M.; Malik, A.; Anwer, J.M.A.; Inam, S.H.A.; Fatima, T.; Iqbal, N.; Shoaib, K. Follow-up of COVID-19 recovered patients with mild disease. *Sci. Rep.* **2021**, *11*, 13414. [CrossRef]
36. Carfì, A.; Bernabei, R.; Landi, F. Persistent Symptoms in Patients After Acute COVID-19. *JAMA* **2020**, *324*, 603–605. [CrossRef] [PubMed]
37. Mandal, S.; Barnett, J.; Brill, S.E.; Brown, J.S.; Denneny, E.K.; Hare, S.S.; Heightman, M.; Hillman, T.E.; Jacob, J.; Jarvis, H.C.; et al. 'Long-COVID': A cross-sectional study of persisting symptoms, biomarker and imaging abnormalities following hospitalisation for COVID-19. *Thorax* **2021**, *76*, 396–398. [CrossRef]
38. Kamal, M.; Abo Omirah, M.; Hussein, A.; Saeed, H. Assessment and characterisation of post-COVID-19 manifestations. *Int. J. Clin. Pract.* **2021**, *75*, e13746. [CrossRef]
39. Mazza, M.G.; De Lorenzo, R.; Conte, C.; Poletti, S.; Vai, B.; Bollettini, I.; Melloni, E.M.T.; Furlan, R.; Ciceri, F.; Rovere-Querini, P.; et al. Anxiety and depression in COVID-19 survivors: Role of inflammatory and clinical predictors. *Brain Behav. Immun.* **2020**, *89*, 594–600. [CrossRef]
40. Taquet, M.; Geddes, J.R.; Husain, M.; Luciano, S.; Harrison, P.J. 6-month neurological and psychiatric outcomes in 236 379 survivors of COVID-19: A retrospective cohort study using electronic health records. *Lancet Psychiatry* **2021**, *8*, 416–427. [CrossRef]
41. Wildwing, T.; Holt, N. The neurological symptoms of COVID-19: A systematic overview of systematic reviews, comparison with other neurological conditions and implications for healthcare services. *Ther. Adv. Chronic Dis.* **2021**, *12*. [CrossRef]
42. Davis, H.E.; Assaf, G.S.; McCorkell, L.; Wei, H.; Low, R.J.; Re'Em, Y.; Redfield, S.; Austin, J.P.; Akrami, A. Characterizing long COVID in an international cohort: 7 months of symptoms and their impact. *eClinicalMedicine* **2021**, *38*, 101019. [CrossRef]
43. Alkodaymi, M.S.; Omrani, O.A.; Fawzy, N.A.; Shaar, B.A.; Almamlouk, R.; Riaz, M.; Obeidat, M.; Obeidat, Y.; Gerberi, D.; Taha, R.M.; et al. Prevalence of post-acute COVID-19 syndrome symptoms at different follow-up periods: A systematic review and meta-analysis. *Clin. Microbiol. Infect.* **2022**. [CrossRef]

44. Duggal, P.; Penson, T.; Manley, H.N.; Vergara, C.; Munday, R.M.; Duchen, D.; Linton, E.A.; Zurn, A.; Kerulz, J.C.; Mehta, S.H.; et al. Post-sequelae symptoms and comorbidities after COVID-19. *J. Med. Virol.* **2022**, *94*, 2060–2066. [CrossRef] [PubMed]
45. Garcia-Abellan, J.; Padilla, S.; Fernandez-Gonzalez, M.; Garcia, J.A.; Agullo, V.; Andreo, M.; Ruiz, S.; Galiana, A.; Gutiérrez, F.; Masiá, M. Antibody Response to SARS-CoV-2 is Associated with Long-term Clinical Outcome in Patients with COVID-19: A Longitudinal Study. *J. Clin. Immunol.* **2021**, *41*, 1490–1501. [CrossRef]
46. Iqbal, F.M.; Lam, K.; Sounderajah, V.; Clarke, J.M.; Ashrafian, H.; Darzi, A. Characteristics and predictors of acute and chronic post-COVID syndrome: A systematic review and meta-analysis. *eClinicalMedicine* **2021**, *36*, 100899. [CrossRef] [PubMed]
47. Cortinovis, M.; Perico, N.; Remuzzi, G. Long-term follow-up of recovered patients with COVID-19. *Lancet* **2021**, *397*, 173–175. [CrossRef]
48. Sudre, C.H.; Murray, B.; Varsavsky, T.; Graham, M.S.; Penfold, R.S.; Bowyer, R.C.; Pujol, J.C.; Klaser, K.; Antonelli, M.; Canas, L.S.; et al. Attributes and predictors of long COVID. *Nat. Med.* **2021**, *27*, 626–631. [CrossRef] [PubMed]
49. Fernandez-de-Las-Penas, C.; Pellicer-Valero, O.J.; Navarro-Pardo, E.; Palacios-Cena, D.; Florencio, L.L.; Guijarro, C.; Martín-Guerrero, J.D. Symptoms Experienced at the Acute Phase of SARS-CoV-2 Infection as Risk Factor of Long-term Post-COVID Symptoms: The LONG-COVID-EXP-CM Multicenter Study. *Int. J. Infect. Dis.* **2022**, *116*, 241–244. [CrossRef]
50. Fernández-De-Las-Peñas, C.; Palacios-Ceña, D.; Gómez-Mayordomo, V.; Cuadrado, M.; Florencio, L. Defining Post-COVID Symptoms (Post-Acute COVID, Long COVID, Persistent Post-COVID): An Integrative Classification. *Int. J. Environ. Res. Public Health* **2021**, *18*, 2621. [CrossRef] [PubMed]
51. Glynne, P.; Tahmasebi, N.; Gant, V.; Gupta, R. Long COVID following mild SARS-CoV-2 infection: Characteristic T cell alterations and response to antihistamines. *J. Investig. Med.* **2022**, *70*, 61–67. [CrossRef]
52. Iyengar, K.P.; Jain, V.K.; Vaishya, R.; Ish, P. Long COVID-19: An emerging pandemic in itself. *Adv. Respir. Med.* **2021**, *89*, 234–236. [CrossRef]
53. Khunti, K.; Davies, M.J.; Kosiborod, M.N.; Nauck, M.A. Long COVID—metabolic risk factors and novel therapeutic management. *Nat. Rev. Endocrinol.* **2021**, *17*, 379–380. [CrossRef] [PubMed]
54. Fainardi, V.; Meoli, A.; Chiopris, G.; Motta, M.; Skenderaj, K.; Grandinetti, R.; Bergomi, A.; Antodaro, F.; Zona, S.; Esposito, S. Long COVID in Children and Adolescents. *Life* **2022**, *12*, 285. [CrossRef]
55. Borch, L.; Holm, M.; Knudsen, M.; Ellermann-Eriksen, S.; Hagstroem, S. Long COVID symptoms and duration in SARS-CoV-2 positive children—A nationwide cohort study. *Eur. J. Pediatr.* **2022**, 1–11. [CrossRef]
56. Yelin, D.; Margalit, I.; Yahav, D.; Runold, M.; Bruchfeld, J. Long COVID-19-it's not over until? *Clin. Microbiol. Infect.* **2021**, *27*, 506–508. [CrossRef]
57. Karlsson, A.C.; Humbert, M.; Buggert, M. The known unknowns of T cell immunity to COVID-19. *Sci. Immunol.* **2020**, *5*, eabe8063. [CrossRef]
58. Doykov, I.; Hallqvist, J.; Gilmour, K.C.; Grandjean, L.; Mills, K.; Heywood, W.E. 'The long tail of COVID-19'—The detection of a prolonged inflammatory response after a SARS-CoV-2 infection in asymptomatic and mildly affected patients. *F1000Research* **2020**, *9*, 1349. [CrossRef] [PubMed]
59. Pasrija, R.; Naime, M. The deregulated immune reaction and cytokines release storm (CRS) in COVID-19 disease. *Int. Immunopharmacol.* **2021**, *90*, 107225. [CrossRef]
60. Kappelmann, N.; Dantzer, R.; Khandaker, G.M. Interleukin-6 as potential mediator of long-term neuropsychiatric symptoms of COVID-19. *Psychoneuroendocrinology* **2021**, *131*, 105295. [CrossRef] [PubMed]
61. Iadecola, C.; Anrather, J.; Kamel, H. Effects of COVID-19 on the Nervous System. *Cell* **2020**, *183*, 16–27.e1. [CrossRef] [PubMed]
62. McFarland, A.J.; Yousuf, M.S.; Shiers, S.; Price, T.J. Neurobiology of SARS-CoV-2 interactions with the peripheral nervous system: Implications for COVID-19 and pain. *Pain Rep.* **2021**, *6*, e885. [CrossRef]
63. Eroğlu, I.; Eroğlu, B.; Güven, G.S. Altered tryptophan absorption and metabolism could underlie long-term symptoms in survivors of coronavirus disease 2019 (COVID-19). *Nutrition* **2021**, *90*, 111308. [CrossRef] [PubMed]
64. Barizien, N.; Le Guen, M.; Russel, S.; Touche, P.; Huang, F.; Vallée, A. Clinical characterization of dysautonomia in long COVID-19 patients. *Sci. Rep.* **2021**, *11*, 14042. [CrossRef]
65. Blitshteyn, S.; Whitelaw, S. Correction to: Postural orthostatic tachycardia syndrome (POTS) and other autonomic disorders after COVID-19 infection: A case series of 20 patients. *Immunol Res.* **2021**, *69*, 205–211. [CrossRef] [PubMed]
66. Townsend, L.; Moloney, D.; Finucane, C.; McCarthy, K.; Bergin, C.; Bannan, C.; Kenny, R.-A. Fatigue following COVID-19 infection is not associated with autonomic dysfunction. *PLoS ONE* **2021**, *16*, e0247280. [CrossRef]
67. Becker, R.C. Anticipating the long-term cardiovascular effects of COVID-19. *J. Thromb. Thrombolysis* **2020**, *50*, 512–524. [CrossRef] [PubMed]
68. Fan, B.E.; Umapathi, T.; Chua, K.; Chia, Y.W.; Wong, S.W.; Tan, G.W.L.; Chandrasekar, S.; Lum, Y.H.; Vasoo, S.; Dalan, R. Delayed catastrophic thrombotic events in young and asymptomatic post COVID-19 patients. *J. Thromb. Thrombolysis* **2021**, *51*, 971–977. [CrossRef]
69. Campi, I.; Bulgarelli, I.; Dubini, A.; Perego, G.B.; Tortorici, E.; Torlasco, C.; Torresani, E.; Rocco, L.; Persani, L.; Fugazzola, L. The spectrum of thyroid function tests during hospitalization for SARS-CoV-2 infection. *Eur. J. Endocrinol.* **2021**, *184*, 699–709. [CrossRef]
70. Scappaticcio, L.; Pitoia, F.; Esposito, K.; Piccardo, A.; Trimboli, P. Impact of COVID-19 on the thyroid gland: An update. *Rev. Endocr. Metab. Disord.* **2020**, *22*, 803–815. [CrossRef]

71. Coperchini, F.; Ricci, G.; Croce, L.; Denegri, M.; Ruggiero, R.; Villani, L.; Magri, F.; Chiovato, L.; Rotondi, M. Modulation of ACE-2 mRNA by inflammatory cytokines in human thyroid cells: A pilot study. *Endocrine* **2021**, *74*, 638–645. [CrossRef]
72. Ruggeri, R.M.; Campennì, A.; Deandreis, D.; Siracusa, M.; Tozzoli, R.; Ovčariček, P.P.; Giovanella, L. SARS-CoV-2-related immune-inflammatory thyroid disorders: Facts and perspectives. *Expert Rev. Clin. Immunol.* **2021**, *17*, 737–759. [CrossRef] [PubMed]
73. Puig-Domingo, M.; Marazuela, M.; Yildiz, B.O.; Giustina, A. COVID-19 and endocrine and metabolic diseases. An updated statement from the European Society of Endocrinology. *Endocrine* **2021**, *72*, 301–316. [CrossRef] [PubMed]
74. Chen, W.; Tian, Y.; Li, Z.; Zhu, J.; Wei, T.; Lei, J. Potential Interaction Between SARS-CoV-2 and Thyroid: A Review. *Endocrinology* **2021**, *162*, bqab004. [CrossRef]
75. Aemaz, U.; Rehman, M.; Farooq, H.; Ali, M.M.; Ebaad, U.; Rehman, M.; Dar, Q.A.; Hussain, A. The Association of Subacute Thyroiditis with COVID-19: A Systematic Review. *SN Compr. Clin. Med.* **2021**, *3*, 1515–1527. [CrossRef] [PubMed]
76. Inaba, H.; Aizawa, T. Coronavirus Disease 2019 and the Thyroid—Progress and Perspectives. *Front. Endocrinol.* **2021**, *12*, 708333. [CrossRef]
77. Gorini, F.; Bianchi, F.; Iervasi, G. COVID-19 and Thyroid: Progress and Prospects. *Int. J. Environ. Res. Public Health* **2020**, *17*, 6630. [CrossRef]
78. Croce, L.; Gangemi, D.; Ancona, G.; Liboà, F.; Bendotti, G.; Minelli, L.; Chiovato, L. The cytokine storm and thyroid hormone changes in COVID-19. *J. Endocrinol. Investig.* **2021**, *44*, 891–904. [CrossRef] [PubMed]
79. Das, L.; Dutta, P.; Walia, R.; Mukherjee, S.; Suri, V.; Puri, G.D.; Mahajan, V.; Malhotra, P.; Chaudhary, S.; Gupta, R.; et al. Spectrum of Endocrine Dysfunction and Association With Disease Severity in Patients With COVID-19: Insights From a Cross-Sectional, Observational Study. *Front. Endocrinol.* **2021**, *12*, 645787. [CrossRef]
80. Lui, D.T.W.; Lee, C.H.; Chow, W.S.; Lee, A.C.H.; Tam, A.R.; Pang, P.; Ho, T.Y.; Fong, C.H.Y.; LAw, C.Y.; Leung, E.K.H.; et al. Long COVID in Patients With Mild to Moderate Disease: Do Thyroid Function and Autoimmunity Play a Role? *Endocr. Pract.* **2021**, *27*, 894–902. [CrossRef]
81. Murugan, A.K.; Alzahrani, A.S. SARS-CoV-2: Emerging Role in the Pathogenesis of Various Thyroid Diseases. *J. Inflamm. Res.* **2021**, *14*, 6191–6221. [CrossRef]
82. Clarke, S.A.; Phylactou, M.; Patel, B.; Mills, E.G.; Muzi, B.; Izzi-Engbeaya, C.; Choudhury, S.; Khoo, B.; Meeran, K.; Comninos, A.N.; et al. Normal Adrenal and Thyroid Function in Patients Who Survive COVID-19 Infection. *J. Clin. Endocrinol. Metab.* **2021**, *106*, 2208–2220. [CrossRef]
83. Lania, A.; Sandri, M.T.; Cellini, M.; Mirani, M.; Lavezzi, E.; Mazziotti, G. Thyrotoxicosis in Patients with COVID-19: The Thyrcov Study. *Eur. J. Endocrinol.* **2020**, *183*, 381–387. [CrossRef]
84. Pizzocaro, A.; Colombo, P.; Vena, W.; Ariano, S.; Magnoni, P.; Reggiani, F.; Favacchio, G.; Mirani, M.; Lavezzi, E.; Voza, A.; et al. Outcome of SARS-CoV-2-related thyrotoxicosis in survivors of COVID-19: A prospective study. *Endocrine* **2021**, *73*, 255–260. [CrossRef]
85. Sandru, F.; Carsote, M.; Petca, R.C.; Gheorghisan-Galateanu, A.A.; Petca, A.; Valea, A.; Dumitrașcu, M.C. COVID-19-related thyroid conditions (Review). *Exp. Ther. Med.* **2021**, *22*, 756. [CrossRef]
86. Dworakowska, D.; Morley, S.; Mulholland, N.; Grossman, A.B. COVID-19-related thyroiditis: A novel disease entity? *Clin. Endocrinol.* **2021**, *95*, 369–377. [CrossRef] [PubMed]
87. Bellastella, G.; Maiorino, M.I.; Esposito, K. Endocrine complications of COVID-19: What happens to the thyroid and adrenal glands? *J. Endocrinol. Investig.* **2020**, *43*, 1169–1170. [CrossRef]
88. Christensen, J.; O'Callaghan, K.; Sinclair, H.; Hawke, K.; Love, A.; Hajkowicz, K.; Stewart, A.G. Risk factors, Treatment and Outcomes of Subacute Thyroiditis Secondary to COVID-19: A Systematic Review. *Intern. Med. J.* **2021**. [CrossRef] [PubMed]
89. Trimboli, P.; Cappelli, C.; Croce, L.; Scappaticcio, L.; Chiovato, L.; Rotondi, M. COVID-19-Associated Subacute Thyroiditis: Evidence-Based Data From a Systematic Review. *Front. Endocrinol.* **2021**, *12*, 707726. [CrossRef] [PubMed]
90. Lui, D.T.W.; Lee, C.H.; Chow, W.S.; Lee, A.C.H.; Tam, A.R.; Fong, C.H.Y.; Law, C.Y.; Leung, E.K.H.; To, K.K.W.; Tan, K.C.B.; et al. Thyroid Dysfunction in Relation to Immune Profile, Disease Status, and Outcome in 191 Patients with COVID-19. *J. Clin. Endocrinol. Metab.* **2021**, *106*, e926–e935. [CrossRef] [PubMed]
91. Schwarz, Y.; Percik, R.; Oberman, B.; Yaffe, D.; Zimlichman, E.; Tirosh, A. Sick Euthyroid Syndrome on Presentation of Patients With COVID-19: A Potential Marker for Disease Severity. *Endocr. Pr.* **2021**, *27*, 101–109. [CrossRef] [PubMed]
92. Güven, M.; Gültekin, H. The prognostic impact of thyroid disorders on the clinical severity of COVID-19: Results of single-centre pandemic hospital. *Int. J. Clin. Pr.* **2021**, *75*, e14129. [CrossRef]
93. Beltrão, F.E.d.L.; Beltrão, D.C.d.A.; Carvalhal, G.; Beltrão, M.F.E.d.L.; Brito, M.A.d.S.; Capistrano, M.K.H.R.; Bastos, I.H.D.A.; Hecht, F.; Daltro, C.H.D.C.; Bianco, A.C.; et al. Thyroid Hormone Levels During Hospital Admission Inform Disease Severity and Mortality in COVID-19 Patients. *Thyroid* **2021**, *31*, 1639–1649. [CrossRef]
94. Asghar, M.S.; Yasmin, F.; Dapke, K.; Phadke, R.; Shah, S.M.I.; Bin Zafar, M.D. Derangements of biochemical markers and thyroid function analysis among COVID-19-positive patients: A developing country single-center experience. *J. Med. Virol.* **2021**, *93*, 5712–5717. [CrossRef]
95. Gong, J.; Wang, D.-K.; Dong, H.; Xia, Q.-S.; Huang, Z.-Y.; Zhao, Y.; Chen, X.; Yuan, F.; Li, J.-B.; Lu, F.-E. Prognostic significance of low TSH concentration in patients with COVID-19 presenting with non-thyroidal illness syndrome. *BMC Endocr. Disord.* **2021**, *21*, 111. [CrossRef] [PubMed]

96. Zou, R.; Wu, C.; Zhang, S.; Wang, G.; Zhang, Q.; Yu, B.; Wu, Y.; Dong, H.; Wu, G.; Wu, S.; et al. Euthyroid Sick Syndrome in Patients With COVID-Front. *Endocrinol.* **2020**, *11*, 566439. [CrossRef]
97. Giovanella, L.; Ruggeri, R.M.; Ovčariček, P.P.; Campenni, A.; Treglia, G.; Deandreis, D. Prevalence of thyroid dysfunction in patients with COVID-19: A systematic review. *Clin. Transl. Imaging* **2021**, *9*, 233–240. [CrossRef] [PubMed]
98. Duntas, L.H.; Jonklaas, J. COVID-19 and Thyroid Diseases: A Bidirectional Impact. *J. Endocr. Soc.* **2021**, *5*, bvab076. [CrossRef] [PubMed]
99. Anaya, J.-M.; Monsalve, D.M.; Rojas, M.; Rodríguez, Y.; Montoya-García, N.; Mancera-Navarro, L.M.; Villadiego-Santana, A.M.; Rodríguez-Leguizamón, G.; Acosta-Ampudia, Y.; Ramírez-Santana, C. Latent rheumatic, thyroid and phospholipid autoimmunity in hospitalized patients with COVID-19. *J. Transl. Autoimmun.* **2021**, *4*, 100091. [CrossRef] [PubMed]
100. Pal, R.; Banerjee, M. COVID-19 and the endocrine system: Exploring the unexplored. *J. Endocrinol. Investig.* **2020**, *43*, 1027–1031. [CrossRef] [PubMed]
101. Poma, A.M.; Bonuccelli, D.; Giannini, R.; Macerola, E.; Vignali, P.; Ugolini, C.; Torregrossa, L.; Proietti, A.; Pistello, M.; Basolo, A.; et al. COVID-19 autopsy cases: Detection of virus in endocrine tissues. *J. Endocrinol. Investig.* **2021**, *45*, 209–214. [CrossRef] [PubMed]
102. Jadali, Z. COVID-19 and Thyroid Infection: Learning the Lessons of the Past. *Acta Endocrinol.* **2020**, *16*, 375–376. [CrossRef] [PubMed]
103. Tutal, E.; Ozaras, R.; Leblebicioglu, H. Systematic review of COVID-19 and autoimmune thyroiditis. *Travel. Med. Infect. Dis.* **2022**, *47*, 102314. [CrossRef]
104. Pal, R.; Joshi, A.; Bhadada, S.K.; Banerjee, M.; Vaikkakara, S.; Mukhopadhyay, S. Endocrine Follow-up During Post-Acute COVID-19: Practical Recommendations Based on Available Clinical Evidence. *Endocr. Pr.* **2022**. [CrossRef]
105. Callard, F.; Perego, E. How and why patients made Long Covid. *Soc. Sci. Med.* **2021**, *268*, 113426. [CrossRef] [PubMed]
106. Roth, P.H.; Gadebusch-Bondio, M. The contested meaning of "long COVID"—Patients, doctors, and the politics of subjective evidence. *Soc. Sci. Med.* **2022**, *292*, 114619. [CrossRef]
107. Sivan, M.; Rayner, C.; Delaney, B. Fresh evidence of the scale and scope of long covid. *BMJ* **2021**, *373*, n853. [CrossRef]
108. Newman, M. Chronic fatigue syndrome and long covid: Moving beyond the controversy. *BMJ* **2021**, *373*, n1559. [CrossRef] [PubMed]
109. Desai, A.D.; Lavelle, M.; Boursiquot, B.C.; Wan, E.Y. Long-term complications of COVID-19. *Am. J. Physiol. Physiol.* **2022**, *322*, C1–C11. [CrossRef]
110. Rebello, C.J.; Axelrod, C.L.; Reynolds, C.F., 3rd; Greenway, F.L.; Kirwan, J.P. Exercise as a Moderator of Persistent Neuroendocrine Symptoms of COVID-19. *Exerc. Sport Sci. Rev.* **2022**, *50*, 65–72. [CrossRef] [PubMed]
111. Chippa, V.; Aleem, A.; Anjum, F. *Post Acute Coronavirus (COVID-19) Syndrome*; StatPearls: Treasure Island, FL, USA, 2022.
112. Bushell, W.; Castle, R.; Williams, M.A.; Brouwer, K.C.; Tanzi, R.E.; Chopra, D.; Mills, P.J. Meditation and Yoga Practices as Potential Adjunctive Treatment of SARS-CoV-2 Infection and COVID-19: A Brief Overview of Key Subjects. *J. Altern. Complement. Med.* **2020**, *26*, 547–556. [CrossRef]
113. Mangge, H.; Kneihsl, M.; Schnedl, W.; Sendlhofer, G.; Curcio, F.; Domenis, R. Immune Responses against SARS-CoV-2-Questions and Experiences. *Biomedicines* **2021**, *9*, 1342. [CrossRef]
114. Antonelli, M.; Penfold, R.S.; Merino, J.; Sudre, C.H.; Molteni, E.; Berry, S.; Canas, L.S.; Graham, M.S.; Klaser, K.; Modat, M.; et al. Risk factors and disease profile of post-vaccination SARS-CoV-2 infection in UK users of the COVID Symptom Study app: A prospective, community-based, nested, case-control study. *Lancet Infect. Dis.* **2021**, *22*, 43–55. [CrossRef]
115. Moghimi, N.; Di Napoli, M.; Biller, J.; Siegler, J.E.; Shekhar, R.; McCullough, L.D.; Harkins, M.S.; Hong, E.; Alaouieh, D.A.; Mansueto, G.; et al. The Neurological Manifestations of Post-Acute Sequelae of SARS-CoV-2 infection. *Curr. Neurol. Neurosci. Rep.* **2021**, *21*, 44. [CrossRef]
116. Murray, T. Unpacking "long COVID". *CMAJ* **2021**, *193*, E318–E319. [CrossRef] [PubMed]
117. Drake, T.M.; Riad, A.M.; Fairfield, C.J.; Egan, C.; Knight, S.R.; Pius, R.; Hardwick, H.E.; Norman, L.; Shaw, C.A.; McLean, K.A.; et al. Characterisation of in-hospital complications associated with COVID-19 using the ISARIC WHO Clinical Characterisation Protocol UK: A prospective, multicentre cohort study. *Lancet* **2021**, *398*, 223–237. [CrossRef]
118. Hussain, F.A. Facilitating care: A biopsychosocial perspective on long COVID. *Br. J. Gen. Pr.* **2021**, *72*, 30–31. [CrossRef] [PubMed]
119. Macpherson, K.; Cooper, K.; Harbour, J.; Mahal, D.; Miller, C.; Nairn, M. Experiences of living with long COVID and of accessing healthcare services: A qualitative systematic review. *BMJ Open* **2022**, *12*, e050979. [CrossRef]
120. Inciardi, R.M.; Chandra, A. Editorial commentary: Long COVID-19: A tangled web of lungs, heart, mind, and gender. *Trends Cardiovasc. Med.* **2022**, *32*, 18–19. [CrossRef]
121. Higgins, V.; Sohaei, D.; Diamandis, E.P.; Prassas, I. COVID-19: From an acute to chronic disease? Potential long-term health consequences. *Crit. Rev. Clin. Lab. Sci.* **2021**, *58*, 297–310. [CrossRef]

life

MDPI

Review

Asthma and COVID-19 Associations: Focus on IgE-Related Immune Pathology

Chung-Jen Wang [1], Shih-Lung Cheng [1,2] and Sow-Hsong Kuo [1,*]

1 Department of Internal Medicine, Far Eastern Memorial Hospital, New Taipei City 22056, Taiwan; dr.chung.jen.wang@gmail.com (C.-J.W.); shihlungcheng@gmail.com (S.-L.C.)
2 Department of Chemical Engineering and Materials Science, Yuab Ze University, Taoyuan City 32003, Taiwan
* Correspondence: shkuo@ntu.edu.tw

Abstract: Management of patients with asthma during the coronavirus disease 2019 (COVID-19) pandemic is a concern, especially since asthma predisposes patients to respiratory problems. Interestingly, asthma characterized by type 2 inflammation, also known as T-helper type 2-high endotype, displays a cellular and molecular profile that may confer protective effects against COVID-19. The results of experimental and clinical studies have established the actions of immunoglobulin E (IgE) in inducing airway hyperreactivity and weakening an interferon-mediated antiviral response following respiratory viral infection. Robust evidence supports the beneficial effect of the anti-IgE biologic treatment omalizumab on reducing respiratory virus-induced asthma exacerbations and reducing the frequency, duration, and severity of respiratory viral illness in patients with asthma. Indeed, accumulating reports of patients with severe asthma treated with omalizumab during the pandemic have reassuringly shown that continuing omalizumab treatment during COVID-19 is safe, and in fact may help prevent the severe course of COVID-19. Accordingly, guidance issued by the Global Initiative for Asthma recommends that all patients with asthma continue taking their prescribed asthma medications, including biologic therapy, during the COVID-19 pandemic. The impact of biologic treatments on patients with asthma and COVID-19 will be better understood as more evidence emerges.

Keywords: asthma; COVID-19; biologics; IgE; omalizumab

Citation: Wang, C.-J.; Cheng, S.-L.; Kuo, S.-H. Asthma and COVID-19 Associations: Focus on IgE-Related Immune Pathology. *Life* **2022**, *12*, 153. https://doi.org/10.3390/life12020153

Academic Editors: Silvia De Francia and Sarah Allegra

Received: 18 November 2021
Accepted: 19 January 2022
Published: 20 January 2022

1. Introduction

Globally, the coronavirus disease 2019 (COVID-19) pandemic has inflicted enormous health and societal impact, and will likely continue to do so into the foreseeable future. Not only has COVID-19 directly caused morbidity and mortality at historic levels, it has also been attributed to significant consequences in the management of patients with chronic diseases. COVID-19 is caused by the novel severe acute respiratory syndrome coronavirus 2 (SARS-CoV-2), and can lead to respiratory failure and death, similar to novel coronavirus diseases that have occurred in the past, such as the severe acute respiratory syndrome (SARS) and the Middle East respiratory syndrome (MERS) [1]. Understandably, there has been concern that patients with chronic respiratory diseases, such as asthma, may be at an increased risk of poorer outcomes if infected with SARS-CoV-2, which has led to considerations of adjustments to the standard management of these patients during the pandemic. Fortunately, clinical data accumulated thus far have revealed that people with asthma do not seem to suffer a markedly increased risk of SARS-CoV-2 infection or burden from COVID-19 compared to people without asthma [2–5]. However, as the manifestation of COVID-19 clearly shows a high degree of variation among those affected, it can still be reasonably expected that COVID-19 may have a variable impact across patients with different asthma types. Specifically, it has been hypothesized that the cellular and molecular profile of type 2 inflammation confers to a reduced susceptibility to

COVID-19 in patients exhibiting this asthma endotype [6,7]. The immunoglobulin E (IgE) blocking agent omalizumab is a biologic therapeutic agent widely used as an adjunctive treatment in patients with persistent moderate-to-severe asthma, and markers of type 2 inflammation are used as predictors of anti-IgE treatment response [8]. However, the effects of immunomodulation with anti-IgE biologic treatment on COVID-19 have not been fully elucidated. With the aim of improving the understanding of the relationship between asthma, COVID-19, and asthma treatments including anti-IgE biologic therapy, current evidence regarding plausibly linked underlying mechanisms and potential implications in the management of patients with asthma during the COVID-19 pandemic are explored in this review.

2. Asthma Classification into Phenotypes and Endotypes

In all asthma patients, the disease manifests as an obstruction of airflow due to airway hyperresponsiveness, leading to symptoms of wheezing, shortness of breath, cough, and chest tightness. However, recent revelations in pathophysiology of asthma have uncovered underlying heterogeneity, and "asthma" is no longer considered a single disease entity but rather an umbrella term used to describe a collection of various phenotypes and endotypes [9]. While phenotyping categorizes asthma patients according to observable clinical characteristics, endotyping differentiates patients according to disease mechanisms at a molecular level, and thereby serves as an essential basis for an individualized treatment in precision medicine [10]. Traditionally, phenotyping of asthma has grouped patients into the broad categories of atopic/extrinsic asthma and non-atopic/intrinsic asthma based on clinically observable variables such as age at onset, exacerbating factors, concomitant comorbidities, and response to treatment [11–13]. However, experience has revealed that traditional phenotyping does not adequately reflect the diversity of the disease nor account for the varying responses to therapies [9,13]. Urged by the advent of a multitude biologic therapies that target specific inflammatory mediators, a shift toward an endotype-driven asthma treatment paradigm has occurred in recent years.

Based on the landmark study conducted by Wenzel et al. over 20 years ago, the most widely recognized inflammatory endotypes of severe asthma are currently the T-helper type 2 (Th2)-high endotype and Th2-low endotype [12]. The Th2-low endotype is characterized by neutrophilic or paucigranulocytic non-allergic airway inflammation associated with the elevation of the inflammatory mediators interleukin (IL)-1β, IL-6, IL-17, interferon (IFN)-γ, and tumor necrosis factor (TNF)-α [14]. In the Th2-high endotype, Th2 cells generate high levels of IL-4, IL-5, and IL-13, which drive IgE production and the recruitment of eosinophils. More recently, evidence has shown that group 2 innate lymphoid cells (ILC2s) are also producers of Th2-related cytokines in airway tissues and contribute to the initiation and propagation of airway inflammation [15,16]. Concordantly, Th2-high inflammation is also termed type 2 (or T2) inflammation to reflect that both adaptive immunity and innate immunity play important roles in asthma pathophysiology [9]. Although the presence of type 2 inflammation is more accurately determined by an assessment of key cellular effectors via tissue biopsy and induced sputum analysis, these invasive procedures are infeasible in routine clinical practice and are generally reserved for research [10]. Alternatively, measurement of fractional exhaled nitric oxide (FeNO), free and total IgE concentrations in serum, and eosinophil count in blood and sputum have been linked to type 2 cytokine involvement, and these biomarkers are currently utilized to approximately predict responsiveness to type 2 inflammatory pathway-targeted biologic therapies, including omalizumab (anti-IgE agent); mepolizumab, reslizumab, and benralizumab (anti-IL-5/IL-5 receptor agents); and dupilumab (anti-IL-4/IL-13 agent) [8].

3. Asthma, COVID-19, and ACE2 Interrelationship

Compared to individuals without asthma, patients with asthma are known to be at an increased risk for common viral respiratory infections as well as at an increased risk for infection-related complications such as virus-induced asthma exacerbations and viral respiratory infection requiring intensive care [17–19]. The United States Centers for Disease Control and Prevention has cautioned that people with moderate-to-severe asthma may be at an increased risk for SARS-CoV-2 infection and more severe COVID-19 [20]. Interestingly, however, asthma has not been consistently shown to be a clear risk factor for SARS-CoV-2 infection, and in fact, most studies have reported similar if not lower rates of asthma among patients with COVID-19 compared to the general population [21–24]. Asthma has also not been consistently shown to be a risk factor for severe clinical outcomes of COVID-19 [23–26]. A review of 150 studies conducted worldwide found comparable prevalence rates of asthma between patients with COVID-19 who were hospitalized vs. not hospitalized between patients with severe COVID-19 vs. not severe COVID-19, and between patients who died of COVID-19 vs. those who survived [23]. Furthermore, reported rates of hospitalization for COVID-19-related asthma exacerbations and asthma exacerbation during hospitalization for COVID-19 have been low [4,27].

A possible protective effect of a Th2-high asthma endotype against poor clinical outcomes of COVID-19 has been suggested (Figure 1). A nationwide study in Korea showed that among patients with asthma, those with allergic asthma were at a lower risk of COVID-19 morbidity and mortality than patients with non-allergic asthma [24]. According to experimental studies, eosinophils may have a role in promoting respiratory virus clearance and antiviral host defense, which leads to the postulation that asthma patients with type 2 inflammation characterized by an increased number of eosinophils in the airway might be protected against severe COVID-19 outcomes [28,29]. In a retrospective cohort study conducted in Wuhan, China including 59 patients with confirmed COVID-19 and underlying chronic respiratory disease (chronic bronchitis, chronic obstructive pulmonary disease, or asthma), 73% of patients suffering severe COVID-19 had a low blood eosinophil count of less than 0.2×10^9 cells/L compared to 24% of patients who had non-severe COVID-19 ($p < 0.001$) [30]. In a retrospective study including 951 patients with asthma and COVID-19 conducted in the United States, pre-existing eosinophilia defined as a blood eosinophil count of $\geq$150 cells/μL was associated with reduced COVID-19-related hospitalization and mortality [6]. In the same study, among patients who had eosinopenia (absolute eosinophil count of 0 cells/μL) at the time of hospitalization, those in whom the eosinophil count increased to above $\geq$150 cells/μL during admission were significantly less likely to die (mortality rate 9.6%) compared to patients whose eosinophil count remained <150 cells/μL throughout hospitalization (mortality rate 25.8%) (odds ratio [OR], 0.006; 95% confidence interval [CI], 0.0001–0.64; $p = 0.03$). In contrast, Th2-low endotype of asthma is characterized by neutrophilia, which is associated with a neutrophil extracellular trap formation and promotion of tissue injury, as well as an increased level of IL-17, which results in the propagation of proinflammatory cytokines [31].

Rhinoviruses are picoronaviruses that gain entrance into host airway epithelial cells via intracellular adhesion molecule 1 (ICAM-1), a low-density lipoprotein receptor (LDLR), and cadherin-related family member 3 (CDHR3) [32,33]. In patients with chronic allergic asthma, epithelial barrier disruption may lead to increased accessibility of rhinoviruses to CDHR3, which is mainly localized on cell surfaces along intercellular junctions [33]. In contrast, angiotensin-converting enzyme-2 (ACE2) has been identified as the receptor for the SARS-CoV-2 spike protein that provides the entryway for the virus into host cells [34]. Studies have reported a correlation between increased ACE2 expression and increased infectivity of SARS-CoV-2 [35,36]. Compared to airway cells of patients with non-allergic asthma, cells of patients with allergic asthma show lower ACE2 expression [7]. Exposure to the type 2 cytokines IL-4 and IL-13 was shown to reduce ACE2 expression in airway epithelial cells [37]. Furthermore, using a blood eosinophil count as a biomarker for type 2 inflammation, cut-off values of 150 and 300 cells/μL effectively identified patients with differential expression levels of ACE2 [38]. Therefore, evidence suggests that type 2 inflammation characteristic of the Th2-high asthma endotype is associated with lower ACE2 expression in the airway, thus potentially conferring a protective effect against COVID-19.

Conversely, ACE2 expression has been shown to be upregulated by IL-17, which is elevated in the Th2-low asthma endotype [39].

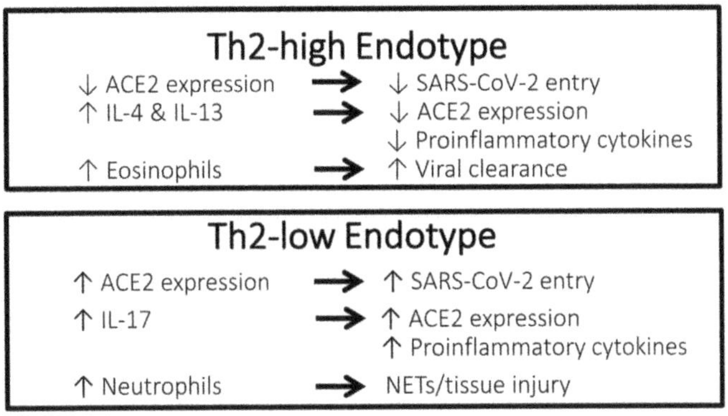

Figure 1. Characteristics of the Th2-high endotype asthma vs. Th2-low endotype asthma that may confer different effects against COVID-19. ACE2, angiotensin-converting enzyme 2 receptor; COVID-19, coronavirus disease 2019; IL, interleukin; NETs, neutrophil extracellular traps; SARS-CoV-2, severe acute respiratory syndrome coronavirus 2; Th2, T-helper type 2.

4. Role of Airway Epithelium in COVID-19 and Asthma

Previously, airway epithelium was considered to simply serve as a mechanical barrier enabling gas exchange. It is now understood that airway epithelium is a complex tissue that performs a multitude of crucial functions, including the mediation of immune mechanisms [40]. The sinonasal airway epithelium is the initial site of SARS-CoV-2 infection and viral replication, and at this stage of infection, ciliated and mucus-secreting goblet cells that express ACE2 are the primary targets [41]. Early stages of COVID-19 are typically associated with relatively mild symptoms, likely related to the dampening of interferon-driven innate immune response to SARS-CoV-2 in nasal and bronchial epithelium [42,43]. Following the initial stage, the disease extends down the respiratory tract to the gas exchange portion of the lung where ACE2-expressing alveolar type II cells become the primary target of viral entry and replication [41]. Whereas in the nasal epithelium, damaged ciliated and secretory cells are replaced by progenitor basal cells that are spared from viral destruction [44], damage to alveolar type II cells results in much more dire consequences. Not only do alveolar type II cells secrete functional surfactant, they are also the progenitor cells for epithelial cells, and their destruction leads to gas exchange dysfunction, alveolar flooding from disrupted epithelium, and initiation of an innate immune response which further propagates alveolar damage due to inflammation [41].

In the context of asthma, the coordinated protective mechanisms of airway epithelium are disturbed differently in Th2-high (type 2 inflammation) and Th2-low (non-type 2 inflammation) asthma endotypes/phenotypes. The dysfunction of ciliated and secretory cells is observed in both asthma types; however, in Th2-high asthma the key early activators released by epithelial cells in response to allergens are IL-25, IL-33, and thymic stromal lymphopoietin (TSLP), whereas in Th2-low asthma, TNF-α, IL-6, IL-8, and IL-1β are the main inflammatory mediators released in response to environmental factors [45]. Importantly, in both asthma types, viruses are among the external factors that can instigate asthma pathogenesis at the epithelium, likely in part due to epigenetic mechanisms by which gene expression is modulated by DNA methylation, histone modulation, or translation modification by microRNAs in response to external stimuli [40,46]. How all of these factors are influenced by COVID-19 requires further study.

5. Therapeutic Management of Asthma Patients during the COVID-19 Pandemic

As asthma has not been clearly shown to be a significant risk factor for SARS-CoV-2 infection or poor outcomes in COVID-19, major modification to established guideline-recommended standard asthma treatment during the COVID-19 pandemic has not been deemed necessary. According to the Global Initiative for Asthma (GINA) guidance about COVID-19 and asthma updated in March 2021, it is important to continue good asthma management in order to maintain optimal symptom control, to reduce the risk of severe exacerbations, and to minimize the need for oral corticosteroids, thus reducing the need to seek urgent medical care and consequential potential exposure to SARS-CoV-2 [47]. In a nationwide health insurance claims data-based study conducted in Korea, which included 218 patients with confirmed COVID-19 and underlying asthma, in univariate analyses, use of short-acting beta agonists was a significant risk factor for intensive care unit admission and use of long-acting beta agonists was a significant protective factor for hospital admission duration, though these factors were no longer significant in multivariate analyses [48]. In terms of the total medical cost burden associated with COVID-19, use of oral short-acting beta agonists in the past year was an independent risk factor for increased cost burden. Furthermore, compared to patients with GINA step 1 asthma (mild asthma not requiring maintenance treatment), patients with GINA step 5 asthma (moderate-to-severe asthma that is difficult to control despite a medium/high dose inhaled corticosteroid plus long-acting beta agonist) required a longer duration of hospitalization for COVID-19 in both univariate and multivariate analyses. These findings support that optimal asthma control with an adequate use of maintenance therapies may improve prognosis in patients with asthma who contract COVID-19; therefore, patients should be advised to continue taking their prescribed asthma medications, including biologic therapy and inhaled or oral corticosteroids, as is recommended by GINA. In addition, all patients should have a written asthma action plan that advises on controller and reliever medication use in case of worsening asthma symptoms. In addition, GINA guidance recommends the avoidance of nebulizer use as a precaution against virus transmission via airborne particles, and that switching to pressurized metered dose inhalers (with a spacer if needed) is preferable. Similarly, avoidance of spirometry during healthcare visits is also recommended as a transmission-based precaution.

6. COVID-19 Vaccination in Patients with Asthma

In addition to maintaining optimal symptom control by continuing all prescribed asthma medications during the pandemic, patients with asthma should be further protected with COVID-19 vaccination. Patients with chronic allergic and atopic diseases, including those treated with biologic agents, have not exhibited increased risk for hypersensitivity reactions after a COVID-19 vaccination [49]. Based on presently understood risks and benefits, GINA guidance recommends that people with asthma undergo COVID-19 vaccination, and the Pfizer/BioNTech and Moderna COVID-19 vaccines were specifically mentioned in the GINA guidance document released in October 2021 [47]. The safety and tolerability of the mRNA COVID-19 Pfizer/BioNTech vaccine in 253 patients with severe asthma were evaluated in a survey study conducted in Italy [50]. According to patient-reported results collected via a vaccination-related adverse events questionnaire, over 80% of patients did not report adverse events after receiving the first and second doses of the vaccine. Among patients who did report experiencing adverse events following vaccination, the reported effects were mostly very common effects such as injection site pain and swelling, weakness, fever, myalgia, arthralgia, and headache, reported by 80% of patients after the first dose and 95% after the second dose, and no patient reported experiencing a rare post-vaccination side effect, including facial asymmetry or severe allergic reaction. In this population of severe asthma patients, 220 (87%) patients were receiving ongoing biologic treatment, and proportions of patients experiencing adverse events following the first and second vaccine doses were comparable across biologic agents (benralizumab 16.9–23.1%, mepolizumab 19.3–20.7%, omalizumab 21.8–22.8%, dupilumab

11.1%). According to GINA guidance, patients with asthma should continue to receive the annual influenza vaccine, with a separation of at least 14 days between COVID-19 and influenza vaccinations. Finally, guidance suggests that biologic therapy and a COVID-19 vaccine should not be given on the same day to allow adverse effects of either to be more easily distinguishable.

7. Impact of Oral and Inhaled Corticosteroids on COVID-19

Compared to patients with asthma of Th2-low endotype, those with Th2-high endotype show higher responsiveness to corticosteroid treatment [51]. As corticosteroids have immunosuppressive effects, the impact of corticosteroids on COVID-19 outcomes may be concerning for many clinicians [52]. Previous studies of corticosteroid treatment during respiratory viral illness, including SARS, have generally shown a lack of effectiveness in reducing morbidity and instead, possible harm [53,54]. Earlier in the pandemic, when COVID-19-specific evidence relating to corticosteroid use was still largely lacking, the World Health Organization (WHO) initially recommended against the use of systemic corticosteroids to treat COVID-19 in its clinical management guidance document released in March 2020 [55]. However, as studies evaluating various treatments, including corticosteroids, in patients with COVID-19 have rapidly accumulated, the WHO changed its stance in a guidance document released in September 2020, which stated a strong recommendation for the use of systemic corticosteroids in the treatment of patients with severe and critical COVID-19 based on the most current evidence [56]. The updated WHO guidance specifically referenced the preliminary report of the now published RECOVERY multicenter, randomized controlled trial, which included 6425 hospitalized patients with COVID-19 who were randomized to receive oral or intravenous dexamethasone 6 mg once daily in addition to usual care or usual care alone [57]. Final results of the RECOVERY study showed that dexamethasone-treated COVID-19 patients had a significantly lower risk of the primary outcome of 28-day mortality compared to those who received usual care alone (age-adjusted rate ratio, 0.83; 95% CI, 0.75–0.93; $p < 0.001$) [57]. However, in the subgroup analysis according to respiratory support at randomization, mortality risk was significantly reduced in patients requiring invasive mechanical ventilation or oxygen only, but was not significantly reduced in patients who did not require respiratory support. In the PRINCIPLE study [58], which is a randomized controlled trial, older non-hospitalized patients with COVID-19 (65 years of older or 50 years or older with comorbidities) who received inhaled budesonide 800 μg twice daily for 14 days in addition to usual care had a significantly shorter time to recovery by approximately 3 days compared to those who received usual care alone.

The burden of COVID-19 among patients with chronic respiratory diseases has not been shown to be increased, and indeed was shown to be decreased in some studies, compared to the general population [59]. This rather unexpected finding has been hypothesized to be attributed to potential protective effects of respiratory disease treatments against SARS-CoV-2 infection and severe disease. Observational studies have shown that inhaled corticosteroid use in patients with asthma was associated with the increased prevalence of non-COVID-19 upper and lower respiratory tract infections [60,61]. In contrast, inhaled corticosteroid treatment has been associated with a reduced expression of ACE2, the entryway of SARS-CoV-2 into host cells, thereby potentially reducing SARS-CoV-2 susceptibility and morbidity [62,63]. Results of an in vitro study have shown that pre-treatment of human nasal and tracheal epithelial cells with budesonide, glycopyrronium, and formoterol inhibited coronavirus HCoV-229E replication and cytokine production [64]. An in vitro study has revealed that the corticosteroids mometasone and ciclesonide suppressed replication of SARS-CoV-2 in a culture medium of infected cells, and that ciclesonide was particularly effective in a concentration-dependent manner [65]. Furthermore, a study that screened a panel of 48 United States Food and Drug Administration-approved drugs identified ciclesonide as an inhibitor of SARS-CoV-2 cytopathic viral activity [66]. In addition to the inhibitory effects on viral replication, treatment with corticosteroids might also reduce the severity of COVID-19 by inhibiting virus-induced cytokine release and dampening the exaggerated inflammatory response responsible for severe symptoms [67]. Overall, the effects of the appropriate use of oral and inhaled corticosteroids is expected to lean toward being beneficial rather than harmful for patients with asthma during the pandemic, and adherence to all maintenance asthma medications to ensure good symptom control and to prevent exacerbation should be emphasized, as is consistently recommended by professional organizations worldwide [68].

8. Role of IgE in the Response to Respiratory Viral Infection

Robust evidence supports the impairment of antiviral response and occurrence of the exacerbation-inducing effect of viral infection in patients with atopic diseases such as asthma [69–73]. Inflammatory mediators released in allergic inflammation result in epithelial barrier disruption and airway remodeling, which may increase susceptibility to respiratory viral infection, and viral infection further induces pro-inflammatory cytokine activity [74,75]. Moreover, atopic asthma with allergic sensitization has been associated with reduced viral clearance, reduced virus-induced IFN responses, and increased viral shedding [18]. Early after respiratory virus infection, myriad cytokines and chemokines activate and attract mast cells, dendritic cells, granulocytes, and monocytes at the infection site [76,77]. Although mast cells are important players in the front-line defense against antigens, the acute hypersensitivity reaction caused by the overactivation of mast cells in highly pathogenic viral infections can be detrimental [77]. In the respiratory tract, mast cell degranulation increases vascular permeability, edema, and mucus production, which lead to airway constriction, congestion, and cough [78–80]. Binding of antigen-specific IgE to the high-affinity IgE receptor (FcεRI) activates mast cells, enhances mast cell survival, and sensitizes mast cells to subsequent encounters with the antigen [81]. Notably, high concentrations of free IgE has been shown to activate mast cells similarly to antigen-bound IgE, thus propagating the immune response even in the absence of antigen stimulation [82]. The anti-IgE therapy omalizumab, which impedes IgE-dependent cellular events by binding to free IgE, prevents binding of IgE to FcεRI, and reduces the expression of FcεRI, was the first biologic therapy to be licensed as an add-on treatment for patients with moderate-to-severe asthma [14].

In the immune response to acute viral infections, plasmacytoid dendritic cells recruited to the respiratory tract produce large quantities of type I IFN, mediated by toll-like receptors, which stimulates T cells and drives antiviral responses (Figure 2) [83]. In a case-control study of patients with allergic asthma, when exposed to the virus, plasmacytoid dendritic cells purified from the whole blood of patients with allergic asthma exhibited significantly reduced expression of IFN-α compared to the plasmacytoid dendritic cells of healthy controls, suggesting a delayed and inefficient antiviral immune response [71]. Expression of FcεRI on plasmacytoid dendritic cells was significantly increased in patients with allergic asthma compared to controls, and both FcεRI expression and serum IgE concentration were significantly inversely correlated with IFN-α secretion upon viral exposure. Importantly, the cross-linking of IgE bound to FcεRI was found to diminish the IFN-α antiviral response of plasmacytoid dendritic cells in a dose-dependent manner. Therefore, findings suggest that elevated serum IgE level and FcεRI expression are associated with an excessive inflammatory response to viral infection as well as a weakened antiviral response. Furthermore, respiratory viruses are known to evade innate type 1 immunity to increase the opportunity for viral replication, which in turn promotes a stronger and more sustained subsequent type 2 inflammatory response [84–86].

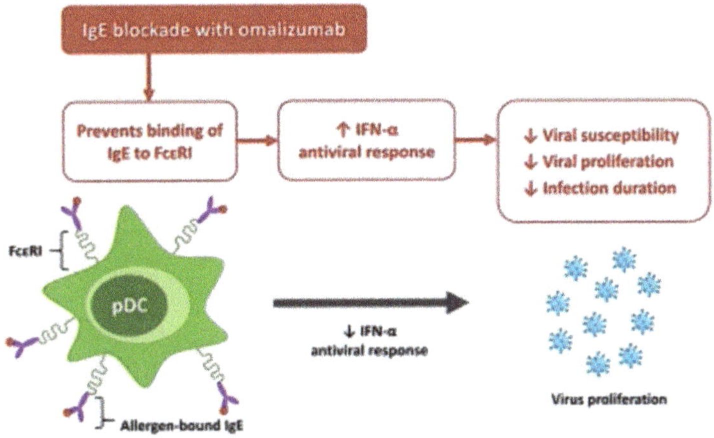

Figure 2. Role of IgE and effects of IgE blockade with omalizumab in antiviral response. Cross-linking of IgE bound to FcεRI on pDCs, which are mainly located in lung interstitium, diminishes IFN-α antiviral response. Anti-IgE therapy omalizumab binds to free IgE, prevents binding of IgE to FcεRI, and reduces the expression of FcεRI, thereby increases IFN-α antiviral response. FcεRI, high-affinity IgE receptor; IFN-α, interferon-α; IgE, immunoglobulin E; and pDC, plasmacytoid dendritic cell.

The critical role of IgE in instigating the inflammatory response to respiratory viral infection has been further supported by the results of multiple randomized trials [87–91]. Notably, studies have demonstrated a reduction in viral infection-induced exacerbations in pediatric asthma patients treated with omalizumab. In the landmark PROSE (Preventative Omalizumab or Step-up Therapy for Severe Fall Exacerbations) study, 478 urban school-aged children with allergic asthma were randomized to receive placebo or omalizumab as an add-on treatment to guideline-based standard asthma care [89,90]. Compared with placebo, treatment with omalizumab significantly decreased the risk of respiratory virus-associated exacerbations in children with severe persistent asthma (OR, 0.35; 95% CI, 0.15–0.85). Ex vivo experiments using nasal mucus samples obtained from participants in this study showed that in the presence of IgE cross-linking, omalizumab significantly increased the IFN-α response to rhinovirus by over three-fold compared to placebo ($p = 0.03$) [90]. Further experiments carried out on the nasal mucus samples showed that omalizumab significantly decreased weekly rhinovirus detection rates (OR, 0.74; 95% CI, 0.60–0.92), indicating a reduced duration of viral infection, and significantly reduced peak viral shedding ($p < 0.04$), suggesting decreased severity of viral illness [89]. These ex vivo results were echoed in the significant reduction in the frequency of rhinovirus illness observed with omalizumab compared with placebo (risk ratio, 0.64; 95% CI, 0.49–0.84). These findings provide strong support that blocking IgE reduces the frequency, duration, and severity of respiratory viral illness in patients with asthma.

9. Anti-IgE Biologic Agent as a Potential Treatment for COVID-19

In the majority of people infected with SARS-CoV-2, the infection follows a mild to moderate self-limiting course that is not much different from the typical course of common respiratory virus infections [92–94]. In a subset of patients, however, SARS-CoV-2 infection evokes a hyperinflammatory immune response characterized by an exaggerated increase in the release of cytokines (IL-2, IL-6, IL-7, IL-10, IL-12, TNF-α, CXCL10, CCL2, CCL3), aptly known as "cytokine storm", and the disproportionate inflammatory activation may lead to acute respiratory distress, multi-organ failure, and possible death [92,95]. Considering the well-established role of IgE in atopic diseases, airway hyperreactivity, and antiviral response, an investigation of anti-IgE as a potential treatment to COVID-19 is warranted. Among the currently available biologic treatments for severe asthma, omalizumab is the only agent with a documented effect in viral respiratory infections [95]. The protective action of omalizumab against viral infections is multifaceted. Omalizumab directly reduces levels of free IgE, inhibits cross-linking of IgE-bound FcεRI, and inhibits mast cell activation that can trigger and propagate the hyperinflammatory cascade (Figure 2) [89]. In addition, omalizumab indirectly reduces the number of FcεRI on basophils, mast cells, and dendritic cells, which helps to prevent the hindrance of an IFN-α-mediated antiviral response [95]. Conversely, as eosinophils have potential antiviral activity, anti-eosinophilic biologic agents might theoretically be less beneficial in COVID-19, though supporting clinical evidence is currently sparse [96]. In a case report of a 41-year-old male patient with severe eosinophilic asthma and a 2-year history of treatment with benralizumab, an anti-IL-5 receptor agent with eosinophil-depleting activity, the patient experienced a mild course of COVID-19 without a marked loss of asthma control [97].

Although still limited in number, reports describing the use of omalizumab in asthma patients during the COVID-19 pandemic are accumulating [98–100]. Lommatzsch et al. reported the results of a 52-year-old German male patient with severe early-onset allergic asthma treated with omalizumab for 6 months who was infected with SARS-CoV-2 [98]. The patient did not experience dyspnea, worsening of asthma, or need for short-acting bronchodilator therapy during the course of infection, and omalizumab treatment was not interrupted. A study conducted in Turkey that surveyed 75 patients with severe asthma treated with omalizumab or mepolizumab revealed that compared to patients who continued biologic treatment, risk of COVID-19 was significantly higher in the 12 patients who interrupted biologic treatment due to a refusal to return to the hospital during the pandemic (relative risk, 2.71; 95% CI, 1.21–6.06) [100]. Another Turkish study showed that among 13 patients with severe asthma treated with omalizumab or mepolizumab who had contracted SARS-CoV-2, five (38.5%) had mild COVID-19, eight (61.5%) had moderate COVID-19, none suffered severe COVID-19 requiring mechanical ventilation or intensive care, and all patients fully recovered [99]. These findings provide reassurance that omalizumab may be safely continued during active COVID-19 infection and may potentially reduce the risk of severe COVID-19 in patients with asthma. Furthermore, maintaining good control of asthma with continued biologic treatment during the pandemic at least reduces the risk of exacerbations requiring medical care and thus reduces the risk of exposure to SARS-CoV-2.

10. Conclusions

Growing evidence reassuringly supports that asthma and asthma treatments do not seem to increase the risk of SARS-CoV-2 infection or severe COVID-19 illness. Indeed, patients with the Th2-high endotype of asthma may be at lower risk for COVID-19 due to increased eosinophil levels and reduced ACE2 expression. Patients should be urged to continue all prescribed asthma medications, including inhaled corticosteroids and biologic treatment, and to receive full COVID-19 vaccination. Th2-high endotype patients treated with omalizumab might be further protected from COVID-19, as the role of IgE in reducing an antiviral response and the beneficial effects of an anti-IgE treatment on immune responses against viral infection are well established. The impact of the different biologic treatments on COVID-19 will be better understood as more evidence emerges. Finally, omalizumab as a potential treatment for COVID-19 warrants investigation.

Author Contributions: Conceptualization, C.-J.W., S.-L.C. and S.-H.K.; methodology, C.-J.W.; software, C.-J.W.; validation, C.-J.W., S.-L.C. and S.-H.K.; formal analysis, C.-J.W.; investigation, C.-J.W.; resources, C.-J.W.; data curation, C.-J.W.; writing—original draft preparation, C.-J.W.; writing—review and editing, S.-L.C. and S.-H.K.; visualization, C.-J.W.; supervision, S.-L.C. and S.-H.K.; project administration, S.-L.C. and S.-H.K. All authors have read and agreed to the published version of the manuscript.

Funding: This research received no external funding.

Institutional Review Board Statement: Not applicable.

Informed Consent Statement: Not applicable.

Data Availability Statement: Not applicable.

Conflicts of Interest: The authors declare no commercial or financial conflict of interest.

References

1. Cao, Y.; Liu, X.; Xiong, L.; Cai, K. Imaging and clinical features of patients with 2019 novel coronavirus SARS-CoV-2: A systematic review and meta-analysis. *J. Med. Virol.* **2020**, *92*, 1449–1459. [CrossRef] [PubMed]
2. Liu, S.; Zhi, Y.; Ying, S. COVID-19 and Asthma: Reflection During the Pandemic. *Clin. Rev. Allergy Immunol.* **2020**, *59*, 78–88. [CrossRef] [PubMed]
3. Broadhurst, R.; Peterson, R.; Wisnivesky, J.P.; Federman, A.; Zimmer, S.M.; Sharma, S.; Wechsler, M.; Holguin, F. Asthma in COVID-19 Hospitalizations: An Overestimated Risk Factor? *Ann. Am. Thorac. Soc.* **2020**, *17*, 1645–1648. [CrossRef]
4. Grandbastien, M.; Piotin, A.; Godet, J.; Abessolo-Amougou, I.; Ederle, C.; Enache, I.; Fraisse, P.; Tu Hoang, T.C.; Kassegne, L.; Labani, A.; et al. SARS-CoV-2 Pneumonia in Hospitalized Asthmatic Patients Did Not Induce Severe Exacerbation. *J. Allergy Clin. Immunol. Pract.* **2020**, *8*, 2600–2607. [CrossRef]
5. Wang, Y.; Chen, J.; Chen, W.; Liu, L.; Dong, M.; Ji, J.; Hu, D.; Zhang, N. Does Asthma Increase the Mortality of Patients with COVID-19?: A Systematic Review and Meta-Analysis. *Int. Arch. Allergy Immunol.* **2021**, *182*, 76–82. [CrossRef] [PubMed]
6. Ferastraoaru, D.; Hudes, G.; Jerschow, E.; Jariwala, S.; Karagic, M.; de Vos, G.; Rosenstreich, D.; Ramesh, M. Eosinophilia in Asthma Patients Is Protective Against Severe COVID-19 Illness. *J. Allergy Clin. Immunol. Pract.* **2021**, *9*, 1152–1162.e3. [CrossRef] [PubMed]
7. Jackson, D.J.; Busse, W.W.; Bacharier, L.B.; Kattan, M.; O'Connor, G.T.; Wood, R.A.; Visness, C.M.; Durham, S.R.; Larson, D.; Esnault, S.; et al. Association of respiratory allergy, asthma, and expression of the SARS-CoV-2 receptor ACE2. *J. Allergy Clin. Immunol.* **2020**, *146*, 203–206.e3. [CrossRef]
8. Busse, W.W. Biological treatments for severe asthma: A major advance in asthma care. *Allergol. Int.* **2019**, *68*, 158–166. [CrossRef]
9. Kuruvilla, M.E.; Lee, F.E.; Lee, G.B. Understanding Asthma Phenotypes, Endotypes, and Mechanisms of Disease. *Clin. Rev. Allergy Immunol.* **2019**, *56*, 219–233. [CrossRef]
10. Busse, W.W.; Kraft, M.; Rabe, K.F.; Deniz, Y.; Rowe, P.J.; Ruddy, M.; Castro, M. Understanding the key issues in the treatment of uncontrolled persistent asthma with type 2 inflammation. *Eur. Respir. J.* **2021**, *58*, 2003393. [CrossRef]
11. Miranda, C.; Busacker, A.; Balzar, S.; Trudeau, J.; Wenzel, S.E. Distinguishing severe asthma phenotypes: Role of age at onset and eosinophilic inflammation. *J. Allergy Clin. Immunol.* **2004**, *113*, 101–108. [CrossRef] [PubMed]
12. Wenzel, S.E.; Schwartz, L.B.; Langmack, E.L.; Halliday, J.L.; Trudeau, J.B.; Gibbs, R.L.; Chu, H.W. Evidence that severe asthma can be divided pathologically into two inflammatory subtypes with distinct physiologic and clinical characteristics. *Am. J. Respir. Crit. Care Med.* **1999**, *160*, 1001–1008. [CrossRef]
13. Svenningsen, S.; Nair, P. Asthma Endotypes and an Overview of Targeted Therapy for Asthma. *Front. Med.* **2017**, *4*, 158. [CrossRef] [PubMed]
14. Chiu, C.J.; Huang, M.T. Asthma in the Precision Medicine Era: Biologics and Probiotics. *Int. J. Mol. Sci.* **2021**, *22*, 4528. [CrossRef] [PubMed]
15. Halim, T.Y.F.; McKenzie, A.N.J. New kids on the block: Group 2 innate lymphoid cells and type 2 inflammation in the lung. *Chest* **2013**, *144*, 1681–1686. [CrossRef]
16. Licona-Limon, P.; Kim, L.K.; Palm, N.W.; Flavell, R.A. TH2, allergy and group 2 innate lymphoid cells. *Nat. Immunol.* **2013**, *14*, 536–542. [CrossRef] [PubMed]
17. James, K.M.; Peebles, R.S., Jr.; Hartert, T.V. Response to infections in patients with asthma and atopic disease: An epiphenomenon or reflection of host susceptibility? *J. Allergy Clin. Immunol.* **2012**, *130*, 343–351. [CrossRef]
18. Jartti, T.; Gern, J.E. Role of viral infections in the development and exacerbation of asthma in children. *J. Allergy Clin. Immunol.* **2017**, *140*, 895–906. [CrossRef] [PubMed]
19. O'Riordan, S.; Barton, M.; Yau, Y.; Read, S.E.; Allen, U.; Tran, D. Risk factors and outcomes among children admitted to hospital with pandemic H1N1 influenza. *CMAJ* **2010**, *182*, 39–44. [CrossRef]
20. Center for Disease Control and Preventions. Coronavirus 19. Available online: https://www.cdc.gov/coronavirus/2019-ncov/need-extra-precautions/asthma.html (accessed on 29 October 2021).
21. Lombardi, C.; Gani, F.; Berti, A.; Comberiati, P.; Peroni, D.; Cottini, M. Asthma and COVID-19: A dangerous liaison? *Asthma Res. Pract.* **2021**, *7*, 9. [CrossRef]
22. Liu, S.; Cao, Y.; Du, T.; Zhi, Y. Prevalence of Comorbid Asthma and Related Outcomes in COVID-19: A Systematic Review and Meta-Analysis. *J. Allergy Clin. Immunol. Pract.* **2021**, *9*, 693–701. [CrossRef]
23. Terry, P.D.; Heidel, R.E.; Dhand, R. Asthma in Adult Patients with COVID-19. Prevalence and Risk of Severe Disease. *Am. J. Respir. Crit. Care Med.* **2021**, *203*, 893–905. [CrossRef] [PubMed]
24. Yang, J.M.; Koh, H.Y.; Moon, S.Y.; Yoo, I.K.; Ha, E.K.; You, S.; Kim, S.Y.; Yon, D.K.; Lee, S.W. Allergic disorders and susceptibility to and severity of COVID-19: A nationwide cohort study. *J. Allergy Clin. Immunol.* **2020**, *146*, 790–798. [CrossRef] [PubMed]

25. Calmes, D.; Graff, S.; Maes, N.; Frix, A.N.; Thys, M.; Bonhomme, O.; Berg, J.; Debruche, M.; Gester, F.; Henket, M.; et al. Asthma and COPD Are Not Risk Factors for ICU Stay and Death in Case of SARS-CoV2 Infection. *J. Allergy Clin. Immunol. Pract.* **2021**, *9*, 160–169. [CrossRef] [PubMed]
26. Zhu, Z.; Hasegawa, K.; Ma, B.; Fujiogi, M.; Camargo, C.A., Jr.; Liang, L. Association of asthma and its genetic predisposition with the risk of severe COVID-19. *J. Allergy Clin. Immunol.* **2020**, *146*, 327–329.e4. [CrossRef] [PubMed]
27. Beurnier, A.; Jutant, E.M.; Jevnikar, M.; Boucly, A.; Pichon, J.; Preda, M.; Frank, M.; Laurent, J.; Richard, C.; Monnet, X.; et al. Characteristics and outcomes of asthmatic patients with COVID-19 pneumonia who require hospitalisation. *Eur. Respir. J.* **2020**, *56*, 2001875. [CrossRef]
28. Carli, G.; Cecchi, L.; Stebbing, J.; Parronchi, P.; Farsi, A. Is asthma protective against COVID-19? *Allergy* **2021**, *76*, 866–868. [CrossRef] [PubMed]
29. Rosenberg, H.F.; Dyer, K.D.; Domachowske, J.B. Respiratory viruses and eosinophils: Exploring the connections. *Antiviral Res.* **2009**, *83*, 1–9. [CrossRef]
30. Chen, D.; Zhang, S.; Feng, Y.; Wu, W.; Chang, C.; Chen, S.; Zhen, G.; Yi, L. Decreased eosinophil counts and elevated lactate dehydrogenase predict severe COVID-19 in patients with underlying chronic airway diseases. *Postgrad. Med. J.* **2021**, 1–8. [CrossRef]
31. Ackermann, M.; Anders, H.J.; Bilyy, R.; Bowlin, G.L.; Daniel, C.; De Lorenzo, R.; Egeblad, M.; Henneck, T.; Hidalgo, A.; Hoffmann, M.; et al. Patients with COVID-19: In the dark-NETs of neutrophils. *Cell Death Differ.* **2021**, *28*, 3125–3139. [CrossRef]
32. Basnet, S.; Palmenberg, A.C.; Gern, J.E. Rhinoviruses and Their Receptors. *Chest* **2019**, *155*, 1018–1025. [CrossRef]
33. Bochkov, Y.A.; Gern, J.E. Rhinoviruses and Their Receptors: Implications for Allergic Disease. *Curr. Allergy Asthma Rep.* **2016**, *16*, 30. [CrossRef] [PubMed]
34. Hoffmann, M.; Kleine-Weber, H.; Schroeder, S.; Kruger, N.; Herrler, T.; Erichsen, S.; Schiergens, T.S.; Herrler, G.; Wu, N.H.; Nitsche, A.; et al. SARS-CoV-2 Cell Entry Depends on ACE2 and TMPRSS2 and Is Blocked by a Clinically Proven Protease Inhibitor. *Cell* **2020**, *181*, 271–280.e8. [CrossRef] [PubMed]
35. Ni, W.; Yang, X.; Yang, D.; Bao, J.; Li, R.; Xiao, Y.; Hou, C.; Wang, H.; Liu, J.; Yang, D.; et al. Role of angiotensin-converting enzyme 2 (ACE2) in COVID-19. *Crit. Care* **2020**, *24*, 422. [CrossRef] [PubMed]
36. Rath, S.; Perikala, V.; Jena, A.B.; Dandapat, J. Factors regulating dynamics of angiotensin-converting enzyme-2 (ACE2), the gateway of SARS-CoV-2: Epigenetic modifications and therapeutic interventions by epidrugs. *Biomed. Pharmacother.* **2021**, *143*, 112095. [CrossRef]
37. Kimura, H.; Francisco, D.; Conway, M.; Martinez, F.D.; Vercelli, D.; Polverino, F.; Billheimer, D.; Kraft, M. Type 2 inflammation modulates ACE2 and TMPRSS2 in airway epithelial cells. *J. Allergy Clin. Immunol.* **2020**, *146*, 80–88.e8. [CrossRef]
38. Camiolo, M.; Gauthier, M.; Kaminski, N.; Ray, A.; Wenzel, S.E. Expression of SARS-CoV-2 receptor ACE2 and coincident host response signature varies by asthma inflammatory phenotype. *J. Allergy Clin. Immunol.* **2020**, *146*, 315–324.e7. [CrossRef]
39. Song, J.; Zeng, M.; Wang, H.; Qin, C.; Hou, H.Y.; Sun, Z.Y.; Xu, S.P.; Wang, G.P.; Guo, C.L.; Deng, Y.K.; et al. Distinct effects of asthma and COPD comorbidity on disease expression and outcome in patients with COVID-19. *Allergy* **2021**, *76*, 483–496. [CrossRef]
40. Alashkar Alhamwe, B.; Miethe, S.; Pogge von Strandmann, E.; Potaczek, D.P.; Garn, H. Epigenetic Regulation of Airway Epithelium Immune Functions in Asthma. *Front. Immunol.* **2020**, *11*, 1747. [CrossRef]
41. Bridges, J.P.; Vladar, E.K.; Huang, H.; Mason, R.J. Respiratory epithelial cell responses to SARS-CoV-2 in COVID-19. *Thorax* **2022**, *77*, 203–209. [CrossRef]
42. Lopez, J.; Mommert, M.; Mouton, W.; Pizzorno, A.; Brengel-Pesce, K.; Mezidi, M.; Villard, M.; Lina, B.; Richard, J.C.; Fassier, J.B.; et al. Early nasal type I IFN immunity against SARS-CoV-2 is compromised in patients with autoantibodies against type I IFNs. *J. Exp. Med.* **2021**, *218*, e20211211. [CrossRef]
43. Pizzorno, A.; Padey, B.; Julien, T.; Trouillet-Assant, S.; Traversier, A.; Errazuriz-Cerda, E.; Fouret, J.; Dubois, J.; Gaymard, A.; Lescure, F.X.; et al. Characterization and Treatment of SARS-CoV-2 in Nasal and Bronchial Human Airway Epithelia. *Cell Rep. Med.* **2020**, *1*, 100059. [CrossRef]
44. Zhu, N.; Wang, W.; Liu, Z.; Liang, C.; Wang, W.; Ye, F.; Huang, B.; Zhao, L.; Wang, H.; Zhou, W.; et al. Morphogenesis and cytopathic effect of SARS-CoV-2 infection in human airway epithelial cells. *Nat. Commun.* **2020**, *11*, 3910. [CrossRef]
45. Potaczek, D.P.; Miethe, S.; Schindler, V.; Alhamdan, F.; Garn, H. Role of airway epithelial cells in the development of different asthma phenotypes. *Cell. Signal.* **2020**, *69*, 109523. [CrossRef]
46. Alashkar Alhamwe, B.; Potaczek, D.P.; Miethe, S.; Alhamdan, F.; Hintz, L.; Magomedov, A.; Garn, H. Extracellular Vesicles and Asthma-More Than Just a Co-Existence. *Int. J. Mol. Sci.* **2021**, *22*, 4984. [CrossRef]
47. Global Initiative for Asthma. GINA Guidance about COVID-19 and Asthma. Updated 30 March 2021. Available online: https://ginasthma.org/wp-content/uploads/2021/03/21_03_30-GINA-COVID-19-and-asthma.pdf (accessed on 29 October 2021).
48. Choi, Y.J.; Park, J.Y.; Lee, H.S.; Suh, J.; Song, J.Y.; Byun, M.K.; Cho, J.H.; Kim, H.J.; Lee, J.H.; Park, J.W.; et al. Effect of asthma and asthma medication on the prognosis of patients with COVID-19. *Eur. Respir. J.* **2021**, *57*, 2002226. [CrossRef]
49. Pfaar, O.; Klimek, L.; Hamelmann, E.; Kleine-Tebbe, J.; Taube, C.; Wagenmann, M.; Werfel, T.; Brehler, R.; Novak, N.; Mulleneisen, N.; et al. COVID-19 vaccination of patients with allergies and type-2 inflammation with concurrent antibody therapy (biologicals)—A Position Paper of the German Society of Allergology and Clinical Immunology (DGAKI) and the German Society for Applied Allergology (AeDA). *Allergol. Select* **2021**, *5*, 140–147. [CrossRef]

50. Caminati, M.; Guarnieri, G.; Batani, V.; Scarpieri, E.; Finocchiaro, A.; Chieco-Bianchi, F.; Senna, G.; Vianello, A. COVID-19 Vaccination in Patients with Severe Asthma on Biologic Treatment: Safety, Tolerability, and Impact on Disease Control. *Vaccines* **2021**, *9*, 853. [CrossRef]
51. Ramadan, A.A.; Gaffin, J.M.; Israel, E.; Phipatanakul, W. Asthma and Corticosteroid Responses in Childhood and Adult Asthma. *Clin. Chest Med.* **2019**, *40*, 163–177. [CrossRef]
52. FakhriRavari, A.; Jin, S.; Kachouei, F.H.; Le, D.; Lopez, M. Systemic corticosteroids for management of COVID-19: Saving lives or causing harm? *Int. J. Immunopathol. Pharmacol.* **2021**, *35*, 20587384211063976. [CrossRef]
53. Russell, C.D.; Millar, J.E.; Baillie, J.K. Clinical evidence does not support corticosteroid treatment for 2019-nCoV lung injury. *Lancet* **2020**, *395*, 473–475. [CrossRef]
54. Stockman, L.J.; Bellamy, R.; Garner, P. SARS: Systematic review of treatment effects. *PLoS Med.* **2006**, *3*, e343. [CrossRef]
55. World Health Organization. Clinical Management of Severe Acute Respiratory Infection (SARI) When COVID-19 Disease is Suspected. Interim Guidance. 13 March 2020. Available online: https://www.who.int/docs/default-source/coronaviruse/clinical-managementof-novel-cov.pdf (accessed on 1 December 2021).
56. World Health Organization. Corticosteroids for COVID-19. Living Guidance, 2 September 2020. Available online: https://www.who.int/publications/i/item/WHO-2019-nCoV-Corticosteroids-2020.1 (accessed on 1 December 2021).
57. Horby, P.; Lim, W.S.; Emberson, J.R.; Mafham, M.; Bell, J.L.; Linsell, L.; Staplin, N.; Brightling, C.; Ustianowski, A.; Elmahi, E.; et al. Dexamethasone in Hospitalized Patients with COVID-19. *N. Engl. J. Med.* **2021**, *384*, 693–704. [CrossRef]
58. Yu, L.M.; Bafadhel, M.; Dorward, J.; Hayward, G.; Saville, B.R.; Gbinigie, O.; Van Hecke, O.; Ogburn, E.; Evans, P.H.; Thomas, N.P.B.; et al. Inhaled budesonide for COVID-19 in people at high risk of complications in the community in the UK (PRINCIPLE): A randomised, controlled, open-label, adaptive platform trial. *Lancet* **2021**, *398*, 843–855. [CrossRef]
59. Halpin, D.M.G.; Faner, R.; Sibila, O.; Badia, J.R.; Agusti, A. Do chronic respiratory diseases or their treatment affect the risk of SARS-CoV-2 infection? *Lancet Respir. Med.* **2020**, *8*, 436–438. [CrossRef]
60. McKeever, T.; Harrison, T.W.; Hubbard, R.; Shaw, D. Inhaled corticosteroids and the risk of pneumonia in people with asthma: A case-control study. *Chest* **2013**, *144*, 1788–1794. [CrossRef]
61. Yang, M.; Chen, H.; Zhang, Y.; Du, Y.; Xu, Y.; Jiang, P.; Xu, Z. Long-term use of inhaled corticosteroids and risk of upper respiratory tract infection in chronic obstructive pulmonary disease: A meta-analysis. *Inhal. Toxicol.* **2017**, *29*, 219–226. [CrossRef]
62. Beyerstedt, S.; Casaro, E.B.; Rangel, E.B. COVID-19: Angiotensin-converting enzyme 2 (ACE2) expression and tissue susceptibility to SARS-CoV-2 infection. *Eur. J. Clin. Microbiol. Infect. Dis.* **2021**, *40*, 905–919. [CrossRef]
63. Peters, M.C.; Sajuthi, S.; Deford, P.; Christenson, S.; Rios, C.L.; Montgomery, M.T.; Woodruff, P.G.; Mauger, D.T.; Erzurum, S.C.; Johansson, M.W.; et al. COVID-19-related Genes in Sputum Cells in Asthma. Relationship to Demographic Features and Corticosteroids. *Am. J. Respir. Crit. Care Med.* **2020**, *202*, 83–90. [CrossRef]
64. Yamaya, M.; Nishimura, H.; Deng, X.; Sugawara, M.; Watanabe, O.; Nomura, K.; Shimotai, Y.; Momma, H.; Ichinose, M.; Kawase, T. Inhibitory effects of glycopyrronium, formoterol, and budesonide on coronavirus HCoV-229E replication and cytokine production by primary cultures of human nasal and tracheal epithelial cells. *Respir. Investig.* **2020**, *58*, 155–168. [CrossRef]
65. Matsuyama, S.; Kawase, M.; Nao, N.; Shirato, K.; Ujike, M.; Kamitani, W.; Shimojima, M.; Fukushi, S. The Inhaled Steroid Ciclesonide Blocks SARS-CoV-2 RNA Replication by Targeting the Viral Replication-Transcription Complex in Cultured Cells. *J. Virol.* **2020**, *95*, e01648-20. [CrossRef]
66. Jeon, S.; Ko, M.; Lee, J.; Choi, I.; Byun, S.Y.; Park, S.; Shum, D.; Kim, S. Identification of Antiviral Drug Candidates against SARS-CoV-2 from FDA-Approved Drugs. *Antimicrob. Agents Chemother.* **2020**, *64*, e00819-20. [CrossRef]
67. Oliver, B.G.; Robinson, P.; Peters, M.; Black, J. Viral infections and asthma: An inflammatory interface? *Eur. Respir. J.* **2014**, *44*, 1666–1681. [CrossRef]
68. Ong, K.Y.; Tan, T.L.; Chan, A.K.W.; Tan, K.L.L.; Koh, M.S. Managing asthma in the COVID-19 pandemic and current recommendations from professional bodies: A review. *J. Asthma* **2021**, *58*, 1536–1543. [CrossRef]
69. Duff, A.L.; Pomeranz, E.S.; Gelber, L.E.; Price, G.W.; Farris, H.; Hayden, F.G.; Platts-Mills, T.A.; Heymann, P.W. Risk factors for acute wheezing in infants and children: Viruses, passive smoke, and IgE antibodies to inhalant allergens. *Pediatrics* **1993**, *92*, 535–540. [CrossRef]
70. Gern, J.E. Virus/Allergen Interaction in Asthma Exacerbation. *Ann. Am. Thorac. Soc.* **2015**, *12* (Suppl. 2), S137–S143. [CrossRef]
71. Gill, M.A.; Bajwa, G.; George, T.A.; Dong, C.C.; Dougherty, I.I.; Jiang, N.; Gan, V.N.; Gruchalla, R.S. Counterregulation between the FcepsilonRI pathway and antiviral responses in human plasmacytoid dendritic cells. *J. Immunol.* **2010**, *184*, 5999–6006. [CrossRef]
72. Lynch, J.P.; Werder, R.B.; Simpson, J.; Loh, Z.; Zhang, V.; Haque, A.; Spann, K.; Sly, P.D.; Mazzone, S.B.; Upham, J.W.; et al. Aeroallergen-induced IL-33 predisposes to respiratory virus-induced asthma by dampening antiviral immunity. *J. Allergy Clin. Immunol.* **2016**, *138*, 1326–1337. [CrossRef]
73. Soto-Quiros, M.; Avila, L.; Platts-Mills, T.A.; Hunt, J.F.; Erdman, D.D.; Carper, H.; Murphy, D.D.; Odio, S.; James, H.R.; Patrie, J.T.; et al. High titers of IgE antibody to dust mite allergen and risk for wheezing among asthmatic children infected with rhinovirus. *J. Allergy Clin. Immunol.* **2012**, *129*, 1499–1505.e5. [CrossRef]
74. Georas, S.N.; Rezaee, F. Epithelial barrier function: At the front line of asthma immunology and allergic airway inflammation. *J. Allergy Clin. Immunol.* **2014**, *134*, 509–520. [CrossRef]

75. Jackson, D.J.; Johnston, S.L. The role of viruses in acute exacerbations of asthma. *J. Allergy Clin. Immunol.* **2010**, *125*, 1178–1187, quiz 1188–1179. [CrossRef]
76. Rossi, G.A.; Colin, A.A. Infantile respiratory syncytial virus and human rhinovirus infections: Respective role in inception and persistence of wheezing. *Eur. Respir. J.* **2015**, *45*, 774–789. [CrossRef]
77. Graham, A.C.; Temple, R.M.; Obar, J.J. Mast cells and influenza a virus: Association with allergic responses and beyond. *Front. Immunol.* **2015**, *6*, 238. [CrossRef]
78. Bradding, P. Allergen immunotherapy and mast cells. *Clin. Exp. Allergy* **1999**, *29*, 1445–1448. [CrossRef]
79. Da Silva, E.Z.; Jamur, M.C.; Oliver, C. Mast cell function: A new vision of an old cell. *J. Histochem. Cytochem.* **2014**, *62*, 698–738. [CrossRef]
80. Hofmann, A.M.; Abraham, S.N. New roles for mast cells in modulating allergic reactions and immunity against pathogens. *Curr. Opin. Immunol.* **2009**, *21*, 679–686. [CrossRef]
81. Galli, S.J.; Tsai, M. IgE and mast cells in allergic disease. *Nat. Med.* **2012**, *18*, 693–704. [CrossRef]
82. Kawakami, T.; Kitaura, J. Mast cell survival and activation by IgE in the absence of antigen: A consideration of the biologic mechanisms and relevance. *J. Immunol.* **2005**, *175*, 4167–4173. [CrossRef]
83. Cella, M.; Facchetti, F.; Lanzavecchia, A.; Colonna, M. Plasmacytoid dendritic cells activated by influenza virus and CD40L drive a potent TH1 polarization. *Nat. Immunol.* **2000**, *1*, 305–310. [CrossRef]
84. Holt, P.G.; Sly, P.D. Viral infections and atopy in asthma pathogenesis: New rationales for asthma prevention and treatment. *Nat. Med.* **2012**, *18*, 726–735. [CrossRef]
85. Subrata, L.S.; Bizzintino, J.; Mamessier, E.; Bosco, A.; McKenna, K.L.; Wikstrom, M.E.; Goldblatt, J.; Sly, P.D.; Hales, B.J.; Thomas, W.R.; et al. Interactions between innate antiviral and atopic immunoinflammatory pathways precipitate and sustain asthma exacerbations in children. *J. Immunol.* **2009**, *183*, 2793–2800. [CrossRef] [PubMed]
86. Kikkert, M. Innate Immune Evasion by Human Respiratory RNA Viruses. *J. Innate Immun.* **2020**, *12*, 4–20. [CrossRef] [PubMed]
87. Busse, W.W.; Morgan, W.J.; Gergen, P.J.; Mitchell, H.E.; Gern, J.E.; Liu, A.H.; Gruchalla, R.S.; Kattan, M.; Teach, S.J.; Pongracic, J.A.; et al. Randomized trial of omalizumab (anti-IgE) for asthma in inner-city children. *N. Engl. J. Med.* **2011**, *364*, 1005–1015. [CrossRef] [PubMed]
88. Heymann, P.W.; Platts-Mills, T.A.E.; Woodfolk, J.A.; Borish, L.; Murphy, D.D.; Carper, H.T.; Conaway, M.R.; Steinke, J.W.; Muehling, L.; Gerald Teague, W.; et al. Understanding the asthmatic response to an experimental rhinovirus infection: Exploring the effects of blocking IgE. *J. Allergy Clin. Immunol.* **2020**, *146*, 545–554. [CrossRef]
89. Esquivel, A.; Busse, W.W.; Calatroni, A.; Togias, A.G.; Grindle, K.G.; Bochkov, Y.A.; Gruchalla, R.S.; Kattan, M.; Kercsmar, C.M.; Khurana Hershey, G.; et al. Effects of Omalizumab on Rhinovirus Infections, Illnesses, and Exacerbations of Asthma. *Am. J. Respir. Crit. Care Med.* **2017**, *196*, 985–992. [CrossRef] [PubMed]
90. Teach, S.J.; Gill, M.A.; Togias, A.; Sorkness, C.A.; Arbes, S.J., Jr.; Calatroni, A.; Wildfire, J.J.; Gergen, P.J.; Cohen, R.T.; Pongracic, J.A.; et al. Preseasonal treatment with either omalizumab or an inhaled corticosteroid boost to prevent fall asthma exacerbations. *J. Allergy Clin. Immunol.* **2015**, *136*, 1476–1485. [CrossRef]
91. Pelaia, C.; Calabrese, C.; Terracciano, R.; de Blasio, F.; Vatrella, A.; Pelaia, G. Omalizumab, the first available antibody for biological treatment of severe asthma: More than a decade of real-life effectiveness. *Ther. Adv. Respir. Dis.* **2018**, *12*, 1753466618810192. [CrossRef]
92. Azkur, A.K.; Akdis, M.; Azkur, D.; Sokolowska, M.; van de Veen, W.; Bruggen, M.C.; O'Mahony, L.; Gao, Y.; Nadeau, K.; Akdis, C.A. Immune response to SARS-CoV-2 and mechanisms of immunopathological changes in COVID-19. *Allergy* **2020**, *75*, 1564–1581. [CrossRef]
93. Huang, C.; Wang, Y.; Li, X.; Ren, L.; Zhao, J.; Hu, Y.; Zhang, L.; Fan, G.; Xu, J.; Gu, X.; et al. Clinical features of patients infected with 2019 novel coronavirus in Wuhan, China. *Lancet* **2020**, *395*, 497–506. [CrossRef]
94. Wang, D.; Hu, B.; Hu, C.; Zhu, F.; Liu, X.; Zhang, J.; Wang, B.; Xiang, H.; Cheng, Z.; Xiong, Y.; et al. Clinical Characteristics of 138 Hospitalized Patients With 2019 Novel Coronavirus-Infected Pneumonia in Wuhan, China. *JAMA* **2020**, *323*, 1061–1069. [CrossRef]
95. Menzella, F.; Ghidoni, G.; Galeone, C.; Capobelli, S.; Scelfo, C.; Facciolongo, N.C. Immunological Aspects Related to Viral Infections in Severe Asthma and the Role of Omalizumab. *Biomedicines* **2021**, *9*, 348. [CrossRef]
96. Renner, A.; Marth, K.; Patocka, K.; Idzko, M.; Pohl, W. COVID-19 in two severe asthmatics receiving benralizumab: Busting the eosinophilia myth. *ERJ Open Res.* **2020**, *6*, 00457–2020. [CrossRef] [PubMed]
97. Renner, A.; Marth, K.; Patocka, K.; Pohl, W. COVID-19 in a severe eosinophilic asthmatic receiving benralizumab—A case study. *J. Asthma* **2021**, *58*, 1270–1272. [CrossRef] [PubMed]
98. Lommatzsch, M.; Stoll, P.; Virchow, J.C. COVID-19 in a patient with severe asthma treated with Omalizumab. *Allergy* **2020**, *75*, 2705–2708. [CrossRef]
99. Aksu, K.; Yesilkaya, S.; Topel, M.; Turkyilmaz, S.; Ercelebi, D.C.; Oncul, A.; Kalkan, I.K.; Ates, H. COVID-19 in a patient with severe asthma using mepolizumab. *Allergy Asthma Proc.* **2021**, *42*, e55–e57. [CrossRef] [PubMed]
100. Tuncay, G.; Cakmak, M.E.; Can Bostan, O.; Kaya, S.B.; Damadoglu, E.; Karakaya, G.; Kalyoncu, A.F. The course of COVID-19 in patients with severe asthma receiving biological treatment. *J. Asthma* **2021**, 1–7. [CrossRef]

life

MDPI

Case Report

The Cumulative Detrimental Effect of COVID-19 Pneumonia in a Patient with Myasthenic Crisis: A Case Report and Overview of the Literature

Georgiana-Cristina Buzatu [1,†], Florin-Teodor Bobirca [2,†], Sebastian Isac [1,3,*], Oana Antonia Mihalache [4], Mihail Cotorogea-Simion [1], Alina Tita [1], Cristian Cobilinschi [5], Maria Daniela Tanasescu [6], Anca Bobirca [7] and Gabriela Droc [1]

1 Department of Anesthesiology and Intensive Care I, 'Fundeni' Clinical Institute, 022328 Bucharest, Romania
2 Department of General Surgery, Dr. Ion Cantacuzino Hospital, 073206 Bucharest, Romania
3 Department of Physiology, Faculty of Medicine, Carol Davila University of Medicine and Pharmacy, 020021 Bucharest, Romania
4 Department of Neurology, Fundeni Clinical Institute, 022328 Bucharest, Romania
5 Department of Anesthesiology and Intensive Care, Clinical Emergency Hospital, 014461 Bucharest, Romania
6 Department of Medical Semiology, Discipline of Internal Medicine I and Nephrology, Faculty of Medicine, Carol Davila University of Medicine and Pharmacy, 020021 Bucharest, Romania
7 Department of Rheumatology, Dr. Ion Cantacuzino Hospital, 073206 Bucharest, Romania
* Correspondence: sebastian.isac@umfcd.ro
† These authors contributed equally to this work.

Simple Summary: A patient with myasthenia gravis, and one very recent acute episode, required ICU admission and immunomodulatory therapy (i.v. immunoglobulins). Shortly after remission of the symptoms, the patient was readmitted to the ICU for COVID-19 pneumonia-induced acute respiratory failure. For the worsening muscle weakness, the patient had to receive five additional sessions of plasma exchange therapy. After receiving remdesivir, plasmapheresis, i.v. dexamethasone, and supportive therapies, the patient was transferred to the neurological ward, which he left 20 days later, after rehabilitation, with no detectable long-term consequences.

Abstract: Background: As the COVID-19 pandemic reached its peak, it became unavoidable that patients with other risk factors for severe pulmonary impairment (such as neuromuscular illnesses) would become afflicted. While the subject of myasthenic crisis secondary to COVID-19 pneumonia represents an interesting topic in the literature, we could not find consistent data that include, as a novel therapeutic approach, both intravenous immunoglobulin and plasma exchange therapy for the treatment of these two concurrent diseases. Case summary: A 69-year-old man with known seropositive generalized myasthenia gravis, hypertension, ischaemic heart disease, NYHA class II-III heart failure, cerebrovascular disease, and recurrent urinary tract infections, was admitted to the ICU for mixed acute respiratory failure, elevated serum lactate and liver function enzymes, and severe thrombocytopenia. A SARS-CoV-2 PCR test was positive, despite a previous COVID-19 pneumonia episode, 10 months prior to the current one. The patient had a recent ICU admission for a myasthenic crisis, which required non-invasive mechanical ventilation and intravenous immunoglobulin therapy. He received supportive therapy, as well as etiological (intravenous remdesivir, plasmapheresis and intravenous dexamethasone). Fifteen days after admission, the patient was transferred to the neurological ward, whence he left 20 days later, with no apparent sequelae. Conclusions: Subsequent intravenous immunoglobulins and plasma exchange therapy appear to be effective and safe in patients with simultaneous acute myasthenic episode and COVID-19 pneumonia.

Keywords: COVID-19 pneumonia; myasthenic crisis; plasmapheresis; immunoglobulin therapy; immunosuppression; case report

Citation: Buzatu, G.-C.; Bobirca, F.-T.; Isac, S.; Mihalache, O.A.; Cotorogea-Simion, M.; Tita, A.; Cobilinschi, C.; Tanasescu, M.D.; Bobirca, A.; Droc, G. The Cumulative Detrimental Effect of COVID-19 Pneumonia in a Patient with Myasthenic Crisis: A Case Report and Overview of the Literature. *Life* **2022**, *12*, 1482. https://doi.org/10.3390/life12101482

Academic Editors: Silvia De Francia and Sarah Allegra

Received: 2 September 2022
Accepted: 22 September 2022
Published: 23 September 2022

1. Introduction

Myasthenia gravis (MG) is a chronic, autoimmune neuromuscular disease, whose pathological trait is the presence of autoantibodies targeting proteins of the neuromuscular junction. The antibodies binding the target proteins, such as the acetylcholine receptor (AChR), the muscle-specific tyrosine kinase (MuSK), the lipoprotein-related protein 4 (LRP4), and other postsynaptic structures, seem to be the cause of the skeletal muscle weakness (ocular, bulbar, limb, respiratory muscles) [1]. There are two major phenotypes: an ocular and a generalized form, with the generalized one representing up to 80–90% of all myasthenia gravis cases [2,3].

The patients with the generalized form are prone to myasthenic crises (MC), which result in the aggravation of the myasthenic deficit, with progressive worsening of muscle weakness. As a result, respiratory failure could occur, requiring intubation and mechanical ventilation [4]. Infectious episodes (including respiratory tract infections) are the most important aggravating factors for MC [5]. The treatment of acute episodes is even more challenging, considering the specific immunosuppressive therapy of these patients [5].

Considering the COVID-19 pandemic, special attention should be given to immunocompromised patients [6,7].

This infection is characterized by an exaggerated inflammatory response which leads, in symptomatic patients, to acute respiratory distress syndrome [ARDS], sepsis, coagulopathy, and, in some cases, death [8]. This exacerbated inflammatory response could be, however, compromised in immunosuppressed patients in MC, with consequent deleterious effects [5].

Recent evidence suggests that COVID-19-associated coagulopathy is a combination of low-grade disseminated intravascular coagulopathy and pulmonary thrombotic microangiopathy, which can lead to pulmonary dysfunction in severely affected patients [9]. Consequently, additional neuromuscular weakness of the upper trunk secondary to MC predisposes patients with COVID-19 pneumonia to a worsened outcome [10].

This case report aims to raise the clinician's awareness regarding this unique detrimental association between parenchymal lung involvement due to COVID-related pneumonia and muscular respiratory weakness secondary to MC. Moreover, highlighting new therapeutic approaches could impact, in the future, the general prognosis of these patients.

2. Case Report

A 69-year old, 75 kg, 175 cm tall Caucasian male was admitted to the intensive care unit (ICU) for mixed respiratory failure due to the onset of a SARS-CoV-2 infection associated with a MC in remission. The patient was not vaccinated against COVID-19. He had recently been admitted to the ICU to undergo intravenous immunoglobulin therapy (IvIG) and non-invasive mechanical ventilation for the aforementioned MC.

His medical records included a seropositive [AChR antibody titer = 25.1 mmol/L] generalized form of MG, with two MC in the previous year, which responded to intravenous immunoglobulin therapy. Repetitive nerve stimulation test result showed a decrement of >10% and the chest CT scan revealed a normal thymus lodge three years prior, when he presented progressive limb weakness, ptosis, and dysphagia. According to the Myasthenia Gravis Foundation of America (MGFA) score, he was classified as 2B and his Myasthenia Gravis Activities of Daily Living (MG-ADL) score was 7.

Additionally, the patient suffered from COVID 19 pneumonia (10 months prior to the current episode), essential hypertension, ischaemic heart disease, New York Heart Association class II–III heart failure, cerebrovascular disease, cataracts, recurring urinary tract infections (one of which was considered to be the triggering factor for the aforementioned myasthenia flare-up).

At admission, the patient was awake and oriented, Glasgow Coma Scale 15 points, complaining of difficulty breathing, with an increased respiratory drive, peripheral blood oxygen saturation = 76% under 10 L/min supplemental oxygen, hemodynamically stable, blood pressure = 125/65 mmHg, sinus tachycardia, heart rate = 105 bpm, with an aggravated myasthenic bulbar deficit, marked difficulty in mobilizing the head and neck, and difficult deglutition. The arterial blood gas sample revealed severe respiratory acidosis, hyperglycemia, anemia, and lactacidemia: pH = 7.12, pCO_2 = 82 mmHg, Na = 137 mmol/L, K = 4.5 mmol/L, Ca = 1.24 mmol/L, glycaemia = 208 mg/dL, lactate = 5.2 mmol/L, serum bicarbonate = 26 mmol/L, base excess = −2.6 mmol/L, haemoglobin = 8.6 g/dL. Moreover, the patient exhibited a moderate hepatic cytolytic syndrome (aspartate aminotransferase = 90 U/L and alanine aminotransferase = 120 U/L), severe thrombocytopenia (platelet coun t = 50,000/µL) and isolated lymphopenia (500/µL), with normal renal function parameters (blood urea nitrogen = 20 mg/dL, creatinine = 0.83 mg/dL). The PCR COVID test turned out positive and confirmed the reinfection. Additionally, we assessed specific prognostic markers for COVID, and the results were as follows: C-reactive protein (1.91 mg/L, upper limit 3 mg/L), procalcitonin (0.59 ng/mL, upper limit 0.07 ng/mL), NT-proBNP (1621 pg/mL), Troponin I (19 ng/mL, cut-off value 29 ng/mL), ferritin (246 µg/L, upper limit 290 µg/L) and interleukin-6 (8.24 pg/mL, upper limit 7 pg/mL). The bacteriological screening at admission turned out negative.

The CT scan revealed mild COVID-19 pneumonia. The results are revealed in Figure 1.

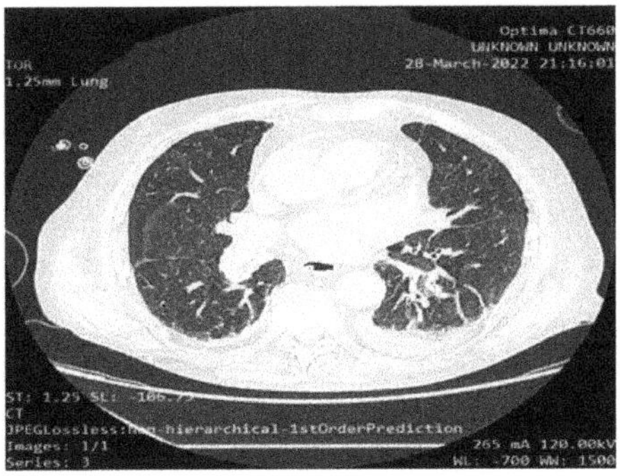

Figure 1. Native CT-scan revealing mild form of COVID-19 pneumonia.

The differential diagnosis included: pulmonary embolism (ruled out based on the chest CT scan), acute heart failure (no suggestive NT-proBNP increase, no peripheral edema, stable blood pressure, no structural changes specific to lung edema), bacterial pneumonia (bacteriological findings were negative, no specific structural changes), other causes for sepsis (negative bacteriological screening and absent clinical signs, with only isolated procalcitonin increase). The above-mentioned results pointed to a severe form of COVID-19 pneumonia in a patient with a relapsed myasthenic crisis.

The treatment strategies involved two main approaches: etiological and supportive treatment. The etiological treatment included: intravenous Remdesivir (Gilead Sciences Ireland UC, Carrigtwohill, Ireland) 200 mg/day and dexamethasone (E.I.P.I.CO MED SRL, Bucharest, Romania) 16 mg/day i.v. for 10 days. Additionally, five distinct sessions of plasma exchange (PLEX), each session with a plasma volume of approximately 4000 mL, were indicated.

Supportive therapy included: (1) non-invasive ventilation (NIV) alternating continuous positive airway pressure (CPAP) with biphasic positive airway pressure (BIPAP), adapted to blood gas analysis and patient comfort; (2) Neostigmine (ZENTIVA S.A., Bucharest, Romania) at an initial dose of 1 mg i.v. every 4 h, followed by 1 mg i.v. every 6 h, in accordance with the local protocol; (3) deep vein thrombosis prophylaxis with Enoxaparin (Sanofi Romania SRL, Bucharest, Romania) 40 mg/day subcutaneously; (4) Antiplatelet therapy with oral Clopidogrel (Sanofi-aventis Groupe, Paris, France) 75 mg/day for mixed primary prophylaxis of cerebral insult and myocardial infarction; (5) Stress ulcer prophylaxis with i.v. Omeprazole (Takeda GmbH, Konstanz, Germany) 40 mg/day, considering the mixed bleeding risks; (6) Beta-blocker therapy with oral Carvedilol (Hexal

AG, Holzkirchen, Germany) 6.25 mg/day for tachycardia; (7) oral Escitalopram (H. Lundbeck A/S, Copenhagen, Denmark) 10 mg/day as antidepressant therapy, alongside with psychological support; and (8) nutritional support via nasojejunal tube.

The patient received 12 days of NIV, followed by three days of oxygen therapy via facial mask with a decremental flow ranging between 4–12 L/min. On the 15th day, the patient was discharged from the ICU. After another 20 days of integrated rehabilitation and specific neurological care, the patient was discharged from the hospital, with no long-term oxygen therapy.

3. Discussion

MC are most often triggered by infections; thus, COVID-19 pneumonia causes respiratory failure in MG patients via two separate mechanisms: firstly, through damage of the lung parenchyma, and secondly, via worsening of myasthenic symptoms [11–13]. Some of the drugs originally used in COVID-19 treatment (such as hydroxychloroquine or azithromycin) can also cause exacerbation of respiratory symptoms in myasthenia patients, either by way of increased acetylcholine receptor antibody production or by directly influencing neuromuscular transmission [14]. Although not the case for our patient, COVID-19-associated ARDS usually warrants the use of sedatives and neuromuscular blocking agents, which make the neurological evaluation of the patients difficult [12]. Moreover, the evolution of COVID-19 pneumonia in these patients is even more challenging to monitor due to the dampened inflammatory response in the context of long-term immunosuppressive therapy [15,16]. Outcomes in patients with concomitant COVID-19 pneumonia and MG are still a topic of debate. Kim et al. [17] reported that such patients have a higher risk of hospital and ICU admission than those without MG (even if suffering from other autoimmune conditions, such as rheumatoid arthritis or systemic lupus erythematosus), but mechanical ventilation requirements were not significantly different, and neither was the risk of death. A retrospective study by Digala et al. [18] noticed that the hospital length of stay was higher in myasthenic patients infected with SARS-CoV-2 than in those without. A systematic review by Abbas et al. [10] suggested that COVID-19 might worsen outcomes in MG cases, but it was not a definitive conclusion, as they lacked the required amount of data.

When managing MG patients with concomitant COVID-19 infection, one study has shown that long-term corticosteroids increase the risk of hospitalization with severe pneumonia, with the effect in direct relation to the dose, while recent rituximab administration increased the odds of death by a factor of over 35 [19].

In patients with MC, where a rapid improvement of muscle strength is required, one could resort to therapies such as IvIG or PLEX [20]. IvIG act by a variety of mechanisms, the most relevant in the context of MG being: (1) binding circulating antibodies; (2) inhibition of B-cell differentiation, migration, and antibody production; and (3) enhanced degradation of the complement system components, thus preventing the formation of membrane attack complexes [21]. PLEX works by separating the cellular and liquid components of blood, discarding the plasma, and replacing it with exogenous fluid (albumin solutions or healthy donor plasma); this way, the circulating antibodies are eliminated, improving the symptoms of MG. When performing a systematic literature review concerning variations in patient outcomes when treated with IvIG or PLEX, Ipe et al. [22] discovered that MC might be more responsive to PLEX than to intravenous immunoglobulin therapy, inducing a quicker, yet shorter-lived remission, leading to shorter ventilation times, with prolonged hospitalization. The literature seems to be remarkably sparse when tackling the topic of combined or consecutive IvIG and PLEX in patients with MC. Normally, an IvIG regimen dictates repeated doses at most every 3–4 weeks. Since our patient had recently received immunoglobulins for a different MC, he had to receive PLEX therapy, which is also proven to help in faster clinical recovery from COVID-19 pneumonia [23].

While the long-term outcomes of MG have been investigated in many studies, which do not cover the pathology addressed here (COVID and MG), we could conclude that the number of relapsing episodes represents an important risk factor for a detrimental long-term prognosis [24]. However, the prognosis of COVID-19 pneumonia in immunosuppressed patients appears to consist of prolonged viral shedding, even at levels lower than conventional assays' detection limits, but not necessarily increased mortality [25–28].

4. Conclusions

COVID-19 pneumonia in patients with MC represents a real therapeutic challenge. The combined IvIG and PLEX at various time points could be beneficial for both pathologies, COVID-19 pneumonia, and myasthenia exacerbation, especially if the timeframe since the previous IvIG therapy is too short. Further clinical studies are needed in order to develop a rational therapeutic approach for the COVID-19 patients with autoimmune neuromuscular disorders.

Author Contributions: G.-C.B., S.I., A.T. and M.C.-S. contributed to manuscript writing and editing; O.A.M., A.B. and F.-T.B. contributed to data collection; C.C., A.B. and M.D.T. contributed to data analysis; F.-T.B., G.D. and S.I. contributed to conceptualization; G.D. contributed to supervision and reviewed the manuscript. All authors have read and agreed to the published version of the manuscript.

Funding: This research received no external funding.

Institutional Review Board Statement: The study was conducted in accordance with the Declaration of Helsinki and approved by the Ethics Committee of Fundeni Clinical Institute (35892/30 June 2022).

Informed Consent Statement: Written informed consent has been obtained from the patient to publish this paper.

Data Availability Statement: Not applicable.

Conflicts of Interest: The authors declare no conflict of interest.

References

1. Silvestri, N.J.; Wolfe, G.I. Myasthenia Gravis. *Semin. Neurol.* **2012**, *32*, 215–226. [CrossRef]
2. Jayam Trouth, A.; Dabi, A.; Solieman, N.; Kurukumbi, M.; Kalyanam, J. Myasthenia Gravis: A Review. *Autoimmune Dis.* **2012**, *2012*, 874680. [CrossRef]
3. Hehir, M.K.; Silvestri, N.J. Generalized Myasthenia Gravis: Classification, Clinical Presentation, Natural History, and Epidemiology. *Neurol. Clin.* **2018**, *36*, 253–260. [CrossRef]
4. Stetefeld, H.; Schroeter, M. SOP Myasthenic Crisis. *Neurol. Res. Pract.* **2019**, *1*, 19. [CrossRef]
5. Wendell, L.C.; Levine, J.M. Myasthenic Crisis. *Neurohospitalist* **2011**, *1*, 16–22. [CrossRef]
6. Punga, D.; Isac, S.; Paraipan, C.; Cotorogea, M.; Stefan, A.; Cobilinschi, C.; Vacaroiu, I.A.; Tulin, R.; Ionescu, D.; Droc, G. Impact of COVID-19 Infection on Liver Transplant Recipients: Does It Make Any Difference? *Cureus* **2022**, *14*, e22687. [CrossRef] [PubMed]
7. Andrei, S.; Isac, S.; Jelea, D.; Martac, C.; Stefan, M.-G.; Cotorogea-Simion, M.; Buzatu, G.-C.; Ingustu, D.; Abdulkareem, I.; Vasilescu, C.; et al. COVID-19 Pandemic Was Associated with Lower Activity but Not Higher Perioperative Mortality in a Large Eastern European Center. *Med. Sci. Monit.* **2022**, *28*, e93809. [CrossRef] [PubMed]
8. Panigada, M.; Bottino, N.; Tagliabue, P.; Grasselli, G.; Novembrino, C.; Chantarangkul, V.; Pesenti, A.; Peyvandi, F.; Tripodi, A. Hypercoagulability of COVID-19 Patients in Intensive Care Unit: A Report of Thromboelastography Findings and Other Parameters of Hemostasis. *J. Thromb. Haemost.* **2020**, *18*, 1738–1742. [CrossRef] [PubMed]
9. Yuki, K.; Fujiogi, M.; Koutsogiannaki, S. COVID-19 Pathophysiology: A Review. *Clin. Immunol.* **2020**, *215*, 108427. [CrossRef] [PubMed]
10. Abbas, A.S.; Hardy, N.; Ghozy, S.; Dibas, M.; Paranjape, G.; Evanson, K.W.; Reierson, N.L.; Cowie, K.; Kamrowski, S.; Schmidt, S.; et al. Characteristics, Treatment, and Outcomes of Myasthenia Gravis in COVID-19 Patients: A Systematic Review. *Clin. Neurol. Neurosurg.* **2022**, *213*, 107140. [CrossRef]
11. Solé, G.; Salort-Campana, E.; Pereon, Y.; Stojkovic, T.; Wahbi, K.; Cintas, P.; Adams, D.; Laforet, P.; Tiffreau, V.; Desguerre, I.; et al. Guidance for the Care of Neuromuscular Patients during the COVID-19 Pandemic Outbreak from the French Rare Health Care for Neuromuscular Diseases Network. *Rev. Neurol.* **2020**, *176*, 507–515. [CrossRef] [PubMed]
12. Anand, P.; Slama, M.C.C.; Kaku, M.; Ong, C.; Cervantes-Arslanian, A.M.; Zhou, L.; David, W.S.; Guidon, A.C. COVID-19 in Patients with Myasthenia Gravis. *Muscle Nerve* **2020**, *62*, 254–258. [CrossRef] [PubMed]
13. Rodrigues, C.L.; Freitas, H.C.; Lima, P.R.O.; Oliveira Junior, P.H.; Fernandes, J.M.A.; D'Almeida, J.A.C.; Nóbrega, P.R. Myasthenia Gravis Exacerbation and Myasthenic Crisis Associated with COVID-19: Case Series and Literature Review. *Neurol. Sci.* **2022**, *43*, 2271–2276. [CrossRef]

14. Sheikh, S.; Alvi, U.; Soliven, B.; Rezania, K. Drugs That Induce or Cause Deterioration of Myasthenia Gravis: An Update. *J. Clin. Med.* **2021**, *10*, 1537. [CrossRef] [PubMed]
15. Raess, N.; Schuetz, P.; Cesana-Nigro, N.; Winzeler, B.; Urwyler, S.A.; Schaedelin, S.; Rodondi, N.; Blum, M.R.; Briel, M.; Mueller, B.; et al. Influence of Prednisone on Inflammatory Biomarkers in Community-Acquired Pneumonia: Secondary Analysis of a Randomized Trial. *J. Clin. Pharmacol.* **2021**, *61*, 1406–1414. [CrossRef] [PubMed]
16. Amanullah, M.M.; Hamid, M.; Hanif, H.M.; Muzaffar, M.; Siddiqui, M.T.; Adhi, F.; Ahmad, K.; Khan, S.; Hasan, Z. Effect of Steroids on Inflammatory Markers and Clinical Parameters in Congenital Open Heart Surgery: A Randomised Controlled Trial. *Cardiol. Young* **2016**, *26*, 506–515. [CrossRef]
17. Kim, Y.; Li, X.; Huang, Y.; Kim, M.; Shaibani, A.; Sheikh, K.; Zhang, G.-Q.; Nguyen, T.P. COVID-19 Outcomes in Myasthenia Gravis Patients: Analysis From Electronic Health Records in the United States. *Front. Neurol.* **2022**, *13*, 802559. [CrossRef] [PubMed]
18. Digala, L.P.; Prasanna, S.; Rao, P.; Qureshi, A.I.; Govindarajan, R. Impact of COVID-19 Infection among Myasthenia Gravis Patients- a Cerner Real-World DataTM Study. *BMC Neurol.* **2022**, *22*, 38. [CrossRef]
19. Jakubíková, M.; Týblová, M.; Tesař, A.; Horáková, M.; Vlažná, D.; Ryšánková, I.; Nováková, I.; Dolečková, K.; Dušek, P.; Piťha, J.; et al. Predictive Factors for a Severe Course of COVID-19 Infection in Myasthenia Gravis Patients with an Overall Impact on Myasthenic Outcome Status and Survival. *Eur. J. Neurol.* **2021**, *28*, 3418–3425. [CrossRef] [PubMed]
20. Županić, S.; Lazibat, I.; Rubinić Majdak, M.; Jeličić, M. Treatment of myasthenia gravis patients with covid-19: Review of the literature. *Acta Clin. Croat.* **2022**, *60*, 496–509. [CrossRef] [PubMed]
21. Jacob, S.; Rajabally, Y.A. Current Proposed Mechanisms of Action of Intravenous Immunoglobulins in Inflammatory Neuropathies. *Curr. Neuropharmacol.* **2009**, *7*, 337–342. [CrossRef] [PubMed]
22. Ipe, T.S.; Davis, A.R.; Raval, J.S. Therapeutic Plasma Exchange in Myasthenia Gravis: A Systematic Literature Review and Meta-Analysis of Comparative Evidence. *Front. Neurol.* **2021**, *12*, 662856. [CrossRef] [PubMed]
23. Faqihi, F.; Alharthy, A.; Abdulaziz, S.; Balhamar, A.; Alomari, A.; AlAseri, Z.; Tamim, H.; Alqahtani, S.A.; Kutsogiannis, D.J.; Brindley, P.G.; et al. Therapeutic Plasma Exchange in Patients with Life-Threatening COVID-19: A Randomised Controlled Clinical Trial. *Int. J. Antimicrob. Agents* **2021**, *57*, 106334. [CrossRef]
24. Nelke, C.; Stascheit, F.; Eckert, C.; Pawlitzki, M.; Schroeter, C.B.; Huntemann, N.; Mergenthaler, P.; Arat, E.; Öztürk, M.; Foell, D.; et al. Independent Risk Factors for Myasthenic Crisis and Disease Exacerbation in a Retrospective Cohort of Myasthenia Gravis Patients. *J. Neuroinflamm.* **2022**, *19*, 89. [CrossRef]
25. Nakajima, Y.; Ogai, A.; Furukawa, K.; Arai, R.; Anan, R.; Nakano, Y.; Kurihara, Y.; Shimizu, H.; Misaki, T.; Okabe, N. Prolonged Viral Shedding of SARS-CoV-2 in an Immunocompromised Patient. *J. Infect. Chemother.* **2021**, *27*, 387–389. [CrossRef] [PubMed]
26. Reuken, P.A.; Stallmach, A.; Pletz, M.W.; Brandt, C.; Andreas, N.; Hahnfeld, S.; Löffler, B.; Baumgart, S.; Kamradt, T.; Bauer, M. Severe Clinical Relapse in an Immunocompromised Host with Persistent SARS-CoV-2 Infection. *Leukemia* **2021**, *35*, 920–923. [CrossRef]
27. Niyonkuru, M.; Pedersen, R.M.; Assing, K.; Andersen, T.E.; Skov, M.N.; Johansen, I.S.; Madsen, L.W. Prolonged Viral Shedding of SARS-CoV-2 in Two Immunocompromised Patients, a Case Report. *Infect. Dis.* **2021**, *21*, 743. [CrossRef]
28. Martínez-Barranco, P.; García-Roa, M.; Trelles-Martínez, R.; Arribalzaga, K.; Velasco, M.; Guijarro, C.; Marcos, J.; Campelo, C.; Acedo-Sanz, J.M.; Villalón, L.; et al. Management of Persistent SARS-CoV-2 Infection in Patients with Follicular Lymphoma. *Acta Haematol.* **2022**, *145*, 384–393. [CrossRef] [PubMed]

life

MDPI

Protocol

Protocol of an Exploratory Single-Arm Study to Evaluate the Safety and Immunogenicity of KD-414 as a Booster Vaccine for SARS-CoV-2 in Healthy Adults (KAPIVARA)

Yuriko Terayama [1], Noriko Tomita [2], Junko Terada-Hirashima [2,*], Yukari Uemura [2], Yosuke Shimizu [2], Junko S. Takeuchi [2], Yuki Takamatsu [3], Kenji Maeda [3], Ayako Mikami [2], Mugen Ujiie [4] and Wataru Sugiura [2]

1 Respiratory Medicine, National Center for Global Health and Medicine, Tokyo 162-8655, Japan; yhorikawa@hosp.ncgm.go.jp

2 Center for Clinical Sciences, National Center for Global Health and Medicine, 1-21-1 Toyama, Shinjuku-ku, Tokyo 162-8655, Japan; ntomita@hosp.ncgm.go.jp (N.T.); yuemura@hosp.ncgm.go.jp (Y.U.); yshimizu@hosp.ncgm.go.jp (Y.S.); jtakeuchi@hosp.ncgm.go.jp (J.S.T.); amikami@hosp.ncgm.go.jp (A.M.); wsugiura@hosp.ncgm.go.jp (W.S.)

3 Department of Refractory Viral Infections, National Center for Global Health and Medicine Research Institute, Tokyo 162-8655, Japan; ytakamatsu@ri.ncgm.go.jp (Y.T.); kmaeda@ri.ncgm.go.jp (K.M.)

4 Disease Control and Prevention Center, National Center for Global Health and Medicine, Tokyo 162-8655, Japan; mgujiie@hosp.ncgm.go.jp

* Correspondence: jterada@hosp.ncgm.go.jp; Tel.: +81-3-3202-7181; Fax: +81-3-5273-6941

Citation: Terayama, Y.; Tomita, N.; Terada-Hirashima, J.; Uemura, Y.; Shimizu, Y.; Takeuchi, J.S.; Takamatsu, Y.; Maeda, K.; Mikami, A.; Ujiie, M.; et al. Protocol of an Exploratory Single-Arm Study to Evaluate the Safety and Immunogenicity of KD-414 as a Booster Vaccine for SARS-CoV-2 in Healthy Adults (KAPIVARA). *Life* **2022**, *12*, 966. https://doi.org/10.3390/life12070966

Academic Editors: Silvia De Francia and Sarah Allegra

Received: 26 May 2022
Accepted: 19 June 2022
Published: 27 June 2022

Abstract: Background: The coronavirus disease 2019 (COVID-19) pandemic is currently ongoing, and there have been significant efforts in the development of COVID-19 vaccines. However, the neutralizing antibody titers in vaccinated individuals are reported to progressively decrease over time. Japanese pharmaceutical companies have published the results of Phase I and II studies on the safety and efficacy of different vaccines. Final clinical trials will be conducted with the aim of practical application by March 2023. To effectively utilize vaccines developed by Japanese companies, the efficacy and safety of a booster dose (i.e., third vaccination) must be evaluated among individuals who have received three doses of different vaccines. Methods: This protocol describes a study that aims to examine the effect of a booster dose of "KD-414", a novel Japanese inactivated vaccine, on antibody titers among participants involved in a previous study. Volunteers in this protocol will be recruited from participants in the previous study and immunized with KD-414 after obtaining consent. The antibody titers, before and after immunization with KD-414, among participants who previously received two doses of the BNT162b2 mRNA vaccine, will be comparatively analyzed. Discussion: The reactogenicity and immunogenicity of seven different COVID-19 vaccines including an inactivated vaccine as a third dose after two doses of ChAdOx1 nCov-19 or BNT162b2, has been tested previously, and found to be superior to control (quadrivalent meningococcal conjugate vaccine) regardless of which vaccine had been received during the initial course. This suggests that many types of third booster doses are efficacious. It is anticipated that this study will provide evidence of the safety and immunogenicity of KD-414 as a booster vaccine, which will have profound public health implications.

Keywords: COVID-19; KD-414; inactivated vaccine; booster

1. Background

The coronavirus disease 2019 (COVID-19) vaccine developed by Pfizer-BioNTech, BNT162b2, was approved for emergency use on 14 February 2021 by the Japanese Ministry of Health, Labor and Welfare. Japan's vaccination program for healthcare workers was started on 17 February 2021. The efficacy of the vaccine is reported to be >90% among individuals who have received two doses of the vaccine [1]. However, the neutralizing antibody titers in vaccinated individuals are reported to progressively decrease over time [2].

The administration of a third booster dose enhanced the neutralizing antibody titers. Based on these findings, the Centers for Disease Control and Prevention (CDC) Advisory Committee of the United States of America recommend a third booster dose for the elderly population (aged $\geq$ 65 years) and patients with comorbidities who are at high risk of severe COVID-19. The US Food and Drug Administration provided emergency use authorization for booster doses to various individuals, including those involved in high-risk occupations. The Japanese Health, Labor, and Welfare Council also acknowledged the need for booster doses. However, no novel COVID-19 vaccine was approved for booster doses until September 2021 and vaccination eligibility for specific booster doses is to be discussed on an ongoing basis.

Japanese pharmaceutical companies have published the results of Phase I and II studies on the safety and efficacy of different vaccines. Final clinical trials will be conducted with the aim of practical application by March 2023. To effectively utilize vaccines developed by Japanese companies, the efficacy and safety of booster doses must be evaluated among individuals who have received two doses of the BNT162b2 mRNA vaccine. It is important to evaluate the effect of combined vaccination, for which primary immunization was examined in detail, at this time, because the opportunity for evaluation in such a setting will be lost when additional immunization becomes widespread. Hence, it will be important to evaluate the effect of combined vaccination, for which primary immunization was examined in detail, at this time.

In March 2021, at public expense, the National Center for Global Health and Medicine (NCGM) conducted an observational study of 100 healthcare volunteers from NCGM staff vaccinated with the BNT162b2 mRNA vaccine (Pfizer-BioNTech). The antibody titers were monitored for nine months (*A survey of vaccine-induced COVID-19 antibody titer to verify temporal change*; NCGM Ethical Review Board Approval Number: NCGM-A-004175-04). An in-depth evaluation of the primary immunization revealed that the highest antibody titers were observed on day 7 post-second dose (day 28 post-first dose), which gradually declined thereafter [3,4].

A previous study has shown that a third dose with CoronaVac, an inactivated vaccine, given eight months after a second dose with CoronaVac, significantly increased neutralizing antibody concentrations against SARS-CoV-2 in healthy adults aged 18 years and older [5]. Another study investigated the reactogenicity and immunogenicity of seven different COVID-19 vaccines, including inactivated vaccine (VLA2001), as a third dose after two doses of ChAdOx1 nCov-19 or BNT162b2, showing that the immunogenicity of homologous or heterologous third dose booster vaccination with all tested vaccines was superior to control (quadrivalent meningococcal conjugate vaccine) regardless of which vaccine had been received during the initial course [6]. This suggests that many types of third booster dose are efficacious and supports the potential use of a booster dose of "KD-414", a novel Japanese inactivated vaccine.

KD-414 is an inactivated vaccine against SARS-CoV-2 being developed by KM-Biologics Co., Ltd. (Kumamoto, Japan). It is an inactivated whole-particle vaccine containing aluminum hydroxide added as an immune aid after growth of SARS-CoV-2 on Vero cells from African green monkey kidney and purification and inactivation of the virus. Currently, Phase I/II clinical trials have been completed in Japan and have shown efficacy of KD-414 in adults and the elderly, and that adverse reactions are comparable to general inactivated vaccines with fewer adverse reactions than for mRNA vaccines. It is thought that KD-414 could provide a new option for those who have difficulty with existing vaccinations or who require higher levels of safety.

In a Phase 1/2 study conducted in JAPAN, only one convalescent fever (39 °C or higher) occurred in 29 subjects with severe adverse reaction after administration of KD-414 containing 10 μg of inactivated coronavirus (SARS-CoV-2). The neutralizing antibody GMT at 28 days after the second vaccination was 12.1 (IU equivalent 44.9–63.4 IU/mL) for adults (aged 20–64 years) and 7.3 (IU equivalent: 27.1–38.3 IU/mL) for the elderly (aged 65 years or older). It approximated an estimated neutralizing antibody titer of 54 IU/mL (95% CI

30–96 IU/mL) required to control the onset of disease [7] In addition, the value was higher than the neutralizing antibody titer for which a 70% efficacy rate of other vaccines could be expected [8].

This protocol describes a study that aims to examine the effect of a booster dose with KD-414 on the antibody titers among participants involved in the previous study [1,2]. Consenting volunteers will be immunized with KD-414. The antibody titers, before and after immunization with KD-414, among participants who had previously received two doses of the BNT162b2 mRNA vaccine will be comparatively analyzed. This trial was registered with the Japan Registry of Clinical Trials (Clinical Trial Plan Number: jRCTs031210388; https://jrct.niph.go.jp/latest-detail/jRCTs031210388; first registration date: 22 October 2021).

2. Methods

Information on drugs used in clinical research (test/comparator products)
Name of the drug (generic and brand name)

(1) Test vaccine: KD-414
(2) Dosage form: Intramuscular injection
(3) Properties: Exhibits uniform white turbidity upon shaking
(4) Storage conditions: Maintained at 2–8 °C under light shielding
(5) Composition: Each vial comprises 0.7 mL of KD-414. Each 0.5 mL of KD-414 contains 10 μg of inactivated coronavirus (SARS-CoV-2) as protein content and aluminum hydroxide as adjuvant.

Dosage administration route and volume
Intramuscular administration with 0.5 mL/inoculation.
Target population (age group, sex, and disease history)
Individuals who have already received two doses of the BNT162b2 mRNA vaccine in the previous study (*"Study on the changes in antibody titers by vaccination with a novel coronavirus vaccine"*) will be enrolled in this study.
Inclusion criteria

(1) Individuals who have received two doses of BNT162b2 mRNA vaccine and have completed the measurement of antibody titer on Day 29 (Day 7 after the second dose) in the previous study [3,4].
(2) Individuals who provide written informed consent, comply with the instructions during the study period, undergo medical examination specified in the study protocol, and can report symptoms.

Exclusion criteria

(1) Individuals who received three doses of the BNT162b2 mRNA vaccine at the time of obtaining informed consent.
(2) Pregnant women, women who are potentially pregnant, women who wish to become pregnant before completing the post-test, and breastfeeding women.
(3) Individuals with fibrodysplasia ossificans progressiva.
(4) Individuals with underlying diseases, such as serious cardiovascular disease, kidney disease, liver disease, blood disease, developmental disorders, respiratory diseases, and diabetes mellitus.
(5) Individuals with convulsion in the past.
(6) Individuals with a previous diagnosis of immunodeficiency and those with relatives with congenital immunodeficiency.
(7) Individuals who may be allergic to the ingredients of KD-414.
(8) Individuals who have participated in another clinical trial within the past four months (120 days) from the date of proposed vaccination with KD-414 and who have received other test substances (excluding placebo), or who are scheduled to participate in another clinical trial during the study period.

(9) Individuals who have received blood transfusions or gamma globulin products within the past three months (90 days) or have received high-dose gamma globulin products (200 or more mg/kg bodyweight) within the past six months (180 days) from the date of proposed vaccination with KD-414.
(10) Patients who have received radiotherapy, immunosuppressive drugs (topical drugs are available), immunosuppressive therapy, antirheumatic drugs, adrenocorticotropic hormones, or corticosteroids that affect immune function within six months (180 days) after the date of vaccination (if the total daily dose of prednisolone was ≥2 mg/kg bodyweight and the treatment duration was ≥14 days, topical drugs may be used).
(11) Individuals immunized with the live vaccine within one month or inactivated vaccine within one week.
(12) Individuals with a history of breakthrough infection who wish to be immunized with KD-414.
(13) Individuals judged by the investigator (or sub-investigator) to be ineligible for the study.

Individuals requiring caution about vaccination

Patients who exhibit the pathological conditions listed below will be carefully monitored through medical examination, and their participation in the study will be considered. The patients will be informed about the aims and endpoints of the study, as well as adverse reactions. Informed consent will be obtained considering their health status and constitution before vaccination.

(1) Individuals with underlying diseases, such as cardiovascular disease, kidney disease, liver disease, hematological disease, developmental disorder, respiratory disease, and diabetes mellitus.
(2) Individuals who have experienced fever within two days of vaccination or who have had suspected allergy symptoms, such as a generalized rash (not applicable if it is confirmed that the causative ingredient is not included in KD-414)
(3) Individuals with convulsions in the past.
(4) Individuals who may be allergic to any of the ingredients of KD-414 (thimerosal).

Criteria for postponement of vaccination and discontinuation of KD-414

Criteria for deferring vaccination of test drugs

Vaccination will be postponed for study subjects who meet any of the following criteria on the scheduled date of vaccination with KD-414. However, if the vaccination deferral criteria are no longer met, vaccination may be performed.

(1) Individuals with fever (above 37.5 °C).
(2) Individuals suffering from serious acute illness.
(3) The investigator or sub-investigator judges the vaccination with KD-414 to be inappropriate.

Criteria for discontinuation

The study participation will be discontinued under the following conditions:

(1) Subjects request withdrawal of consent to participate in the study.
(2) An exclusion criterion is not satisfied.
(3) Compliance with the study protocol is not possible.
(4) The study is discontinued.
(5) The investigator or sub-investigator judged the continuation of the study to be difficult.

Content of clinical research

Type of clinical trial

Open-label, single-arm study.

Randomization and blinding

Not applicable.

Use of pharmaceuticals and medical devices for study subjects

Details of treatment period and observation period (including follow-up)

The schedule of KD-414 inoculation has been approved by the Clinical Research Review Board. The treatment period is one day, followed by a 40-day observation period.

Specific methods (including schedules)

The study will be conducted according to the schedules shown in Tables 1–3 and the recruitment process is shown in Figure 1.

Table 1. Study calendar (KD-414 group).

<table>
<tr><td colspan="2">Visit</td><td></td><td colspan="3">Visit 1
KD-414 inoculation</td><td>Observation period</td><td>Visit 2
Post hoc observation</td><td>Visit 3
Post hoc observation</td></tr>
<tr><td colspan="2">Expiration day (Day) * 1
[Acceptable range]</td><td rowspan="6">Obtaining written consent</td><td>Day 0
Pre</td><td>Inoculation</td><td>After</td><td>Days 1–7</td><td>Day 7
[±1 day]</td><td>Day 40
[±1 day]</td></tr>
<tr><td rowspan="5">Medical institutions</td><td>KD-414 inoculation</td><td></td><td>○</td><td></td><td></td><td></td><td></td></tr>
<tr><td>Examination</td><td>○</td><td></td><td>△ * 2</td><td></td><td>○</td><td>○</td></tr>
<tr><td>Body temperature measurement</td><td>○</td><td></td><td></td><td></td><td></td><td></td></tr>
<tr><td>Volume of blood sampling</td><td>17 mL</td><td></td><td></td><td></td><td>22 mL</td><td>17 mL</td></tr>
<tr><td>Immunogenicity</td><td>○</td><td></td><td></td><td></td><td>○</td><td>○</td></tr>
<tr><td>Home</td><td>Observation of adverse events
(Health diary entries)</td><td></td><td></td><td></td><td></td><td>○</td><td>○</td><td>○</td></tr>
</table>

○: Essential, △: Occurrence of an adverse event. * 1: The day of immunization with KD-414 will be considered day 0. * 2: Performed at 15–30 min post-vaccine inoculation.

Table 2. Study calendar (KD-414 non-vaccinated).

<table>
<tr><td colspan="2">Visit</td><td></td><td>Visit 1</td><td>Visit 2</td><td>Visit 3</td></tr>
<tr><td colspan="2">Expiration Day (Day) * 1</td><td rowspan="4">Obtaining written consent</td><td>Day 0</td><td>Day 7</td><td>Day 40</td></tr>
<tr><td colspan="2">[Acceptable range]</td><td></td><td>[±1 day]</td><td>[±1 day]</td></tr>
<tr><td rowspan="2">Medical institutions</td><td>Volume of blood sampling</td><td></td><td>19 mL</td><td>14 mL</td></tr>
<tr><td>Immunogenicity</td><td></td><td>○</td><td>○</td></tr>
</table>

○: Essential. * 1: The day of KD-414 inoculation is considered day 0.

Table 3. Study calendar (public cost booster).

<table>
<tr><td colspan="2">Visit</td><td>Visit 1</td><td>Observation period</td><td>Visit 2</td><td>Visit 3</td></tr>
<tr><td colspan="2">Expiration Day (Day) * 1
[Acceptable range]</td><td>Day 0</td><td>Day 1–7</td><td>Day 7
[±1 day]</td><td>Day 40
[±1 day]</td></tr>
<tr><td rowspan="5">Medical institutions</td><td>Public cost vaccination</td><td>○</td><td></td><td></td><td></td></tr>
<tr><td>Examination</td><td>○</td><td></td><td>○</td><td>○</td></tr>
<tr><td>Body temperature measurement</td><td>○</td><td></td><td></td><td></td></tr>
<tr><td>Volume of blood sampling</td><td>17 mL</td><td></td><td>22 mL</td><td>17 mL</td></tr>
<tr><td>Immunogenicity</td><td>○</td><td></td><td>○</td><td>○</td></tr>
<tr><td>Home</td><td>Observation of adverse events
(Health diary entries)</td><td>○</td><td>○</td><td>○</td><td>○</td></tr>
</table>

○: Essential. * 1: The day of public cost vaccination date was considered day 0.

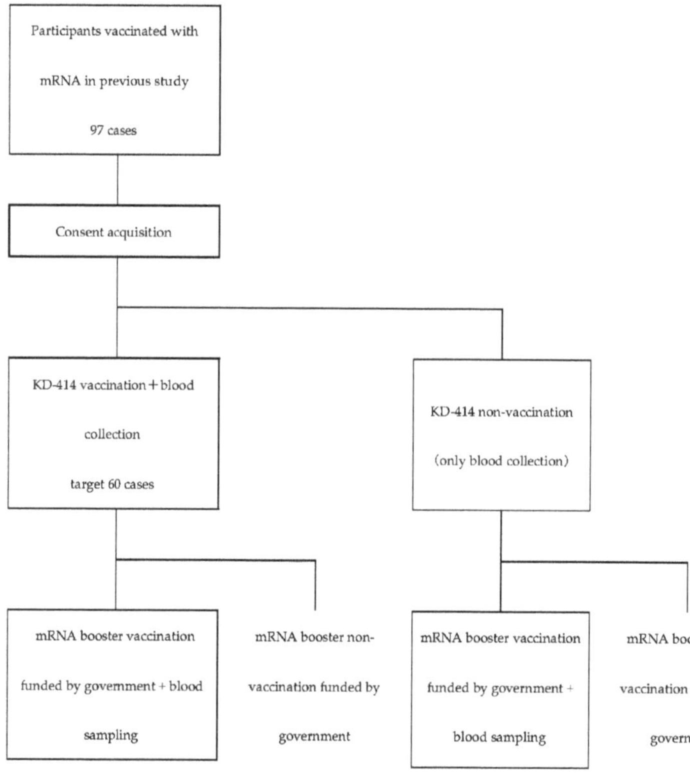

Figure 1. Flowchart of study recruitment process.

If a third dose of BNT162b2 mRNA is received in the future at public expense, the immunogenicity-related parameters and adverse events will be recorded on days 0, 7, and 40 post-vaccination.

Visit 1 (Inclusion of study subjects and KD-414 vaccination/registration)

(1) The investigator or sub-investigator will explain the test and obtain consent in advance. This is described in further detail in the section titled "Method of obtaining consent for clinical research subjects".
(2) The investigator or sub-investigator will confirm the background and eligibility of the study subjects and prepare a registration form.

(hereafter, only in the KD-414 group)

(1) The investigator or sub-investigator will collect the blood for immunogenicity evaluation.
(2) The investigator or sub-investigator will conduct physical examination and body temperature measurement for study subjects and inoculate them with KD-414 if the inoculation is judged to be feasible.
(3) Follow-up observation will be performed for approximately 15–30 min after the inoculation of KD-414. In case of an adverse event, the investigator or sub-investigator will take appropriate measures.
(4) The symptoms should alleviate or stabilize if it is judged necessary to continue to confirm the safety due to the occurrence of adverse events.
(5) Study subjects vaccinated with KD-414 will be requested to complete a health diary

Visits 2 and 3 (post hoc observations)

(1) The investigator or sub-investigator will collect the blood for immunogenicity evaluation.

(2) The investigator or sub-investigator will examine the study subjects and review the entries of the health diary (Visit 2 only).
(3) In case the adverse event caused by KD-414 has not been resolved, the recovery and stabilization of symptoms must be confirmed (only if adverse events occur/continue).

Visits 1–3 (public cost booster)

Investigators will obtain informed consent if booster doses are to be administered to the healthcare workers in the future. Information on vaccinations will be received from the hospital. Personnel who do not receive vaccination in the hospital may be contacted individually.

(1) The investigator or sub-investigator will collect the blood for antibody titration.
(2) Study subjects will receive a publicly funded supplemental coronavirus vaccine during Visit 1.
(3) Study subjects immunized with the vaccine will be requested to complete a health diary.

Survey items

Subject characteristics

Information from previous study (*A survey of vaccine-induced COVID-19 antibody titer to verify temporal change*; NCGM Ethical Review Board Approval Number: NCGM-A-004175-04) should also be used.

(1) Work career (identification of COVID-19 practice engagement and working contacts with patients with COVID-19)
(2) Attributes of study subjects (age, sex, occupation type, etc.)
(3) Diseases and medications that may affect antibody production or the Th1/Th2 ratio (as appropriate)
(4) Adverse reaction information (fever, malaise, etc.)
(5) Medical examination data (anamnesis, antibody-equivalent to SARS-CoV-2) as appropriate
(6) Content of vaccination questionnaire
(7) History of BNT162b2 mRNA (date of receiving the first and second dose of vaccination)
(8) Data from NCGM health survey after vaccination (conducted as a vaccination project)
(9) Data from NCGM staff antibody surveillance on antibodies against SARS-CoV-2 (this study protocol supplemented with data provision for this study was approved by the Ethics Committee of the Center Hospital of the National Center for Global Health and Medicine, Tokyo, Japan.).

Investigation items for efficacy endpoints

(1) COVID-19 disease

Investigation items for safety endpoints

(1) Examination and measurement of body temperature
(2) Health diary entry review

The investigator will request the study subjects to record some measurements in a health diary.

① Subaxillary temperature: The investigator measures and records the body temperature daily for seven days after inoculation with KD-414. The maximum body temperature should be recorded if more than one measurement is performed per day.
② Inoculation site reaction: Study subjects must record the presence or absence of an inoculation site reaction daily for seven days after inoculation. In particular, the major axis should be measured and recorded if redness or swelling is observed at the inoculation site.
③ Subjective symptoms/objective findings: The study subjects should record any subjective symptoms or objective findings.

NOTE: If pyrexia is observed seven days after the inoculation, the body temperature measurement should be continued. The date on which the temperature decreases to <37.5 °C should also be recorded.

Investigation items for immunogenicity endpoints

(1) Timing of blood sampling: Visits 1 (before vaccine inoculation), 2, and 3.
(2) Measurements:

Determination of neutralizing antibody titers against live SARS-CoV-2 viruses (on days 0, 7, and 40 post-inoculation and in the residual sample collected on days 7 and 40 in the previous studies; conducted at the Department of Refractory Viral Infections, National Center for Global Health and Medicine Research Institute, Tokyo, Japan).

① Determination of neutralizing antibodies (on days 0, 7, and 40 post-inoculation) against live SARS-CoV-2 virus (in residual samples collected on days 7 and 40 in previous studies; conducted by the Department of Refractory Viral Infections, National Center for Global Health and Medicine Research Institute, Tokyo, Japan) is shown in Table 4.

Table 4. Evaluation criteria for determination by reagents.

	Outcome	Determination	
		Positive	**Negative**
①	Determination of neutralizing antibodies against live SARS-CoV-2 virus (fold)	≥40	<40
②	Antibody (IgG) titers against the viral nucleocapsid protein (Index or S/C)	≥1.40	<1.40
③	Antibody (IgM) titers against viral spike proteins (Index or S/C)	≥1.00	<1.00
④	Antibody (IgG) titers against viral spike proteins (AU/mL)	≥50.0	≥50.0

Three assays were performed to detect the IgM and IgG antibodies against SARS-CoV-2 spike protein (IgM-S and IgG-S, respectively) and IgG antibodies against SARS-CoV-2 nucleocapsid protein (IgG-N). The presence of IgG-N antibodies can indicate SARS-CoV-2 infection prior to the study and during follow-up, regardless of the BNT162b2 vaccination status. In contrast, the presence of IgM-S and IgG-S indicates previous infection and/or humoral immunity following BNT162b2 vaccination, because BNT162b2 is constructed to express the full-length spike protein.

② Antibody (IgG) titers against the viral nucleocapsid protein

The Abbott ARCHITECT® SARS-CoV-2 anti-N IgG assay based on semi-quantitative CMIA was performed. Determination by reagent is shown in Table 4 (determination on days 0, 7, and 40 post-inoculation)

③ Antibody (IgM) titers against viral spike proteins

The AdviseDx SARS-CoV-2 IgM assay, based on semi-quantitative CMIA, was performed using the Abbott ARCHITECT® according to the manufacturer's instructions. Determination by reagent is shown in Table 4.

④ Antibody (IgG) titers against viral spike proteins

The AdviseDx SARS-CoV-2 IgG II assay was performed using the Abbott ARCHITECT® following the manufacturer's protocol. Determination by reagent is shown in Table 4.

⑤ Flow cytometric analysis of Th1 and Th2 cells (in samples collected on day 7 post-inoculation) (outsourced to SRLs, Tokyo, Japan).

The percentage of Th1 ($IFN\gamma^+$ and $IL\text{-}4^-$) and Th2 ($IFN\gamma^-$ and $IL\text{-}4^+$) cells among $CD4^+$ lymphocytes will be calculated.

⑥ QuantiFERON SARS-CoV-2 RUO test (QIAGEN, Germany; performed on days 0 and 7 post-inoculation)

SARS-CoV-2 specific CD4$^+$ and CD8$^+$ T cell responses were investigated. We quantified T cell-produced interferon-gamma (IFN-γ) in response to SARS-CoV-2 spike peptides using QuantiFERON SARS-CoV-2 RUO (QIAGEN, Germany) according to the manufacturer's instructions.

Volume of blood sampling: 17 mL for Visits 1 and 3 and 22 mL for Visit 2. In total, 56- and 33-mL blood samples were collected from subjects immunized with the study vaccine (KD-414) and unvaccinated subjects, respectively. In total, 56 mL was collected during immunization at public cost.

(1) Treatment method: The blood sample will be allowed to stand at room temperature for 30 min as a guide. After coagulation, the blood samples will be centrifuged at 3000 rpm for 10 min to obtain the serum. The serum will be stored at <-20 °C.
(2) Labeling and transportation methods for storage containers of specimens: Labels will include the name of the study subject, subject identification code, time of blood sampling, and date of blood sampling. The labels will be attached to the sample storage container.

Evaluation and documentation of adverse events

(1) The investigator will record the following items on the case report form regarding adverse events in the study subjects from the date of vaccination to Visit 3 (post hoc observations).
(2) Name of the adverse event
(3) Day of adverse event onset
(4) Severity of adverse event (see the section titled "Severity classification of adverse events").
(5) Seriousness of the adverse event (see the section titled "Reporting of diseases to the certified Clinical Research Review Board (research using unapproved or off-label drugs, etc.);").
(6) Treatment of the adverse event
(7) Outcome (recovery (including remission), recovery but with sequelae, not recovered, death, unknown), and date of outcome
(8) Causal relationship to the study vaccine (see the section titled "Causal relationship between study drugs and adverse events").

Acceptable therapies (including emergency care) before and during the study period

The concomitant use of drugs required for symptomatic treatment is acceptable in the event of an adverse reaction associated with vaccination.

Therapies prohibited before and during the study period

(1) Immunization with other COVID 19 vaccines is prohibited for 40 days post-KD-414 vaccination.
(2) Other vaccines are not allowed for 7 days post-KD-414 vaccination.

Clinical evaluation

The relevant clinical guidelines from the WHO will be followed in relation to the clinical evaluation of the vaccine [9,10].

Evaluation items
Primary endpoint

Neutralizing antibody titers against SARS-CoV-2 on day 7 post-immunization

Rationale: Clinical trials evaluating the efficacy of SARS-CoV-2 vaccine candidates in preventing the development of COVID-19 should be conducted. However, a long study period is required to assess the onset of preventive effects, which is not practical as this is an exploratory study. To investigate the need for a large validation study, the immunogenicity-related parameters, including neutralizing antibody titers, will be evaluated as an important endpoint in this study. The primary assessment time point in this study is day 7 post-immunization as the highest titer was observed at this time point in the previous study.

Secondary endpoint

<Immunogenicity>

(1) Titers of neutralizing antibodies against SARS-CoV-2 on day 40 post-inoculation
(2) IgG antibodies against SARS-CoV-2 spike protein (IgG-S) on days 7 and 40 post-inoculation
(3) IgG antibodies against SARS-CoV-2 nucleocapsid protein (IgG-N) on days 7 and 40 post-inoculation
(4) IgM antibodies against SARS-CoV-2 spike protein (IgM-S) on days 7 and 40 post-inoculation
(5) Th1 and Th2 responses against SARS-CoV-2 on day 7 post-inoculation
(6) QuantiFERON SARS-CoV-2 RUO test on days 7 and 40 post-inoculation

<Efficacy>

(1) COVID-19 incidence till day 40 post-inoculation

<Safety>

(1) All adverse events occurring between days 0 and 40 post-vaccination, deaths due to adverse events, serious adverse events other than death, significant adverse events, and severe (Grade 3 or higher) adverse events
(2) Physician-reported outcomes
   ① Specific local adverse events *1
   ② Specific systemic adverse events *2
   ③ Unspecified adverse events *3
(3) Maximum body temperature from day 0 to day 6 post-inoculation
(4) Subject-reported outcomes

*1 Specific local adverse events: erythema, swelling, induration, and pain at the injection site between day 0 and day 6 post-inoculation.

*2 Specific systemic adverse events: fever (body temperature of ≥37.5 °C, if present), headache, malaise, nausea, and myalgia between day 0 and day 6 post-inoculation

*3 Unspecified adverse events: Adverse events that do not fall into the category of specified adverse events between day 0 and day 6 post-inoculation are classified as unspecified adverse events. Erythema, swelling, induration, and pain at the injection site, pyrexia, headache, malaise, nausea, and myalgia at day 7 post-inoculation are defined as non-specified adverse events.

Statistical analysis items
Analysis sets
Study protocol-adapted analysis set (per-protocol set; PPS)

Among the enrolled participants, subjects associated with major protocol failure and those in whom blood sampling for titration on day 7 post-immunization cannot be performed are excluded.

Safety set

All patients who received KD-414 will be included.

Target sample size and rationale

Target number of subjects receiving KD-414 to evaluate the main hypothesis: 60 subjects
Target sample size for this study: 97 patients

Rationale: Determining if the antibody titer after the booster dose (KD-414) was lower than that after the second dose of BNT162b2 mRNA. The IgG-S antibody titer after the administration of the second dose of BNT162b2 mRNA vaccine will be investigated because the data for neutralizing antibody titers evaluated using the assay used in this study are not available. The titers of IgG-S and neutralizing antibodies are reported to be highly correlated. We assume that the titer of IgG-S after immunization with KD-414 is similar to that after immunization with the second dose of BNT162b2 mRNA in this study population (4.25 (standard deviation; SD 0.35) on a $\log_{10}$ scale). The World Health Organization guidelines on vaccine development set a geometric mean titer (GMT) ratio of 0.67 as the threshold for testing non-inferiority. In the case that the lower limit of the 95% confidence interval for the GMT of IgG-S for the KD-414 to that of Pfizer vaccine is >0.67, the antibody titer after the booster of KD-414 is not inferior to the second dose of BNT162b2 mRNA. Based on these assumptions, the required sample size for 90% power would be 45 for an SD of 0.35, whereas it would be 58 for an SD of 0.4 (Table 5).

Table 5. Power calculations.

	Detection Power 80%	Detection Power 85%	Detection Power 90%
SD 0.35	34	39	45
SD 0.40	44	50	58

The main objective of this study is to obtain consent for immunization with KD-414 from 97 patients with titer test results up to the second dose. Although the rate of informed consent is unknown, the target sample size will be 60 to determine 58 post-vaccination titer values after immunization with KD-414. However, even if the number of subjects enrolled is below 60, a certain power will be assured.

Additionally, patients who do not agree to receive KD-414 (consented to participate in this study) will also be enrolled to compare the antibody titers of non-vaccinated subjects with KD-414 with those of vaccinated subjects. The target sample size for non-vaccinated and vaccinated cases combined will be 97 patients who participated in the previous study.

Criteria for discontinuation of clinical studies

(1) Ethical or medical reasons (including those to ensure the safety of study subjects)
(2) Significant or continued non-compliance determined by the investigator or research institution
(3) The Clinical Research Review Board's decision to terminate or suspend the clinical trial during the continuation review of the ongoing trial
(4) In cases where study subjects are unlikely to satisfy the inclusion criteria
(5) The scientific justification for conducting this study is not satisfactory
(6) The research physician has abandoned the study

Handling of cases

The principal investigator and the individual responsible for statistical analysis will decide on the handling of the enrolled cases after discussion. In case of a problem, the investigator and the statistical analysis manager should discuss and decide on the handling of the case. The details of the case-handling decisions will be retained in the record.

Data handling

In case of any doubt regarding the handling of data during data collection or analysis, the principal investigator and the statistical analysis manager will decide how to proceed after discussion. No imputation is performed for the missing data. Details, including handling outliers, are specified in the statistical analysis plan.

Statistical analysis items and analysis plan

The original version of the analysis plan will be generated prior to data fixation.

Analysis of study subject background

Background factors will be tabulated for PPS and KD-414-vaccinated and non-vaccinated cases.

Primary endpoint analysis

- Primary analysis

The geometric mean of the neutralizing antibody titers will be calculated for PPS cases on day 7 post-inoculation. Meanwhile, the geometric mean ratio of the neutralizing antibody titer at day 7 post-BNT162b2 mRNA vaccination to the geometric mean value and the 95% confidence interval calculated in the previous study ("*A survey of vaccine-induced COVID-19 antibody titer to verify temporal change*": NCGM Ethical Review Board Approval Number: NCGM-A-004175-04) will be used.

- Secondary analysis for the primary endpoint
    (1) The geometric mean of the neutralizing antibody titer on day 7 against day 0 will be calculated for PPS cases. Additionally, the geometric mean of neutralizing antibody titers in non-vaccinated subjects on day 40 will be calculated. The geometric mean ratio of neutralizing antibody titers in vaccinated subjects to that in non-vaccinated subjects among PPS cases will also be calculated.
    (2) The percentage of cases in which the geometric mean ratio of neutralizing titers on day 7 to those on day 0 was $\geq$4 will be calculated.
    (3) Subgroup analyses will be performed on the following items for test drug-vaccinated and non-vaccinated cases among PPS cases:

- Work history (presence or absence of work-related contacts)
- Sex (male/female)
- Age (below or above the median)
- Medical history
- Interval between the third dose and the second dose of the BNT162b2 mRNA vaccine.
- By variant
    (4) Among PPS cases, the geometric mean neutralizing antibody titer on day 7 will be calculated for unvaccinated cases. Additionally, if the non-vaccinated subject received the third dose of other vaccines, the geometric mean of neutralizing antibodies on day 7 after the third dose of other vaccines will be calculated.
    (5) Among PPS cases, the geometric mean neutralizing antibody titers on day 7 will be calculated for unvaccinated and vaccinated cases and compared using the Wilcoxon rank-sum test. Additionally, the geometric mean ratio of neutralizing antibodies on day 7 post-third dose of other vaccines against the vaccinated subject and the 95% confidence interval will be calculated for non-vaccinated subjects.

Analysis of the secondary variables

(1) Neutralizing antibody titers against SARS-CoV-2 on day 40

The geometric mean of the neutralizing titers among PPS cases on day 40 will be calculated. Additionally, the geometric mean ratio of the neutralizing titer on day 40-post immunization with the second dose of BNT162b2 mRNA to the geometric mean value and the 95% confidence interval calculated in the previous study ("*Study on the changes in antibody titers by vaccination with a novel coronavirus vaccine*") will be used; the geometric mean of the neutralizing titer on day 40 and day 0 will be calculated for the test drug-inoculated cases among the PSS cases.

Additionally, the geometric mean of neutralizing antibody titers in non-vaccinated subjects on day 40 will be calculated to compare with that in KD-414 inoculated subjects.

(2) IgG-S on days 7 and 40 post-inoculation
(3) IgG-N on days 7 and 40 post-inoculation
(4) IgM-S on days 7 and 40 post-inoculation

(5) Th1 and Th2 responses against SARS-CoV-2 on day 7 post-inoculation
(6) QuantiFERON SARS-CoV-2 test on days 7 and 40 post-inoculation

Similar analyses as in primary endpoint analysis and analysis of the secondary variables (1) will be performed for the above six items. However, for the Th1 and Th2 endpoints, the analysis will be performed on data collected on day 7 post-inoculation.

(7) Among the PPS cases, the number and percentage of COVID-19 cases in the vaccinated and unvaccinated groups will be calculated.

Analysis of safety endpoints

The following safety endpoints will be summarized for the safety analysis set:

(1) Compilation of the incidence of all adverse events occurring between day 0 and day 40, deaths due to adverse events, serious adverse events other than death, significant adverse events, high-severity (Grade 3 or higher) adverse events, and causal relationship to study drug
(2) Physician-reported outcomes
    - Incidence, severity, days to onset, duration, incidence by dose, and causal relationship to study drugs of specified local adverse events
    - Incidence, severity, days to onset, duration, incidence rate by dose, and causal relationship to study drugs of specified systemic adverse events
    - Incidence, severity, days to onset, duration, incidence by dose, and causal relationship to study drugs of unspecified adverse events
(3) Calculation of summary statistics for maximum body temperature from day 0 to day 6 post-inoculation for each test drug
(4) Subject-reported outcomes

The following analyses will be conducted for the outcome measures in the health observation diary completed by the study subjects.

- The number and percentage of patients with symptoms of redness, swelling, induration, and pain during the 7-day period. A summary statistic is calculated for the maximum value in the case with the maximum diameter reported for redness and induration.
- In case of subjective symptoms and objective findings, the number and percentage of subjects with symptoms in the 7-day period will be calculated. In addition, a contingency table will be generated for the most common symptoms throughout the 7-day period.
- The number and percentage of subjects who used antipyretics will be calculated.

Interim analysis

No interim analysis will be performed in this study.

Procedures for changing the original analysis plan

If there are changes from the original statistical analysis plan, the study plan or statistical analysis plan should be revised and explained in the clinical study report.

Final analysis

Analysis will be performed after fixation of the case for which data are available. The statistical analysis manager will produce the analysis report and submit it to the principal investigator.

Once the data until day 40 are fixed, they will be analyzed. When a booster vaccination at public expense is implemented, data until day 40 will be collected again and analyzed after data fixation.

Adverse events and diseases
Definition of adverse events

An adverse event is an unfavorable symptom, sign, disease, or laboratory abnormality in the study subject, which may or may not be related to the test vaccine. Adverse events are defined as those occurring on or after day 0 (day of vaccine administration).

Definition of disease

The adverse events suspected to be attributable to the test vaccine or procedure are classified as "diseases." The disease should be related to the drugs or procedure used in the study. The investigator or sub-investigator will determine the causal relationship based on the evaluation shown in "Causal relationship between KD-414 and adverse events", below.

Defining serious adverse events

Serious adverse events are defined by the investigator or sub-investigator (hereafter referred to as the "investigator" or "sub-investigator") based on the following criteria:

1. Leading to death
2. Requiring admission to a medical institution or prolongation of hospital stay for treatment
3. Leading to disability
4. Serious according to 1–3
5. Congenital disease or abnormality in later generations

Additionally, adverse events that are non-fatal, non-life-threatening, and do not require hospitalization may be considered "serious". Serious adverse events are those that are judged to be important medical events that may put the subject at risk or require treatment or surgery to avoid the occurrence of the event. However, hospitalization for examination purposes and prolongation of hospitalization for examination purposes scheduled before participation in the study are not considered adverse events.

Measures to be taken in case of adverse events

(1) In case of an adverse event, the investigator will consider necessary medical treatment to ensure the safety of the subject
(2) The investigator should inform the subject when medical treatment is required.
(3) The investigator will confirm the resolution or stabilization of adverse events.

Causal relationship between KD-414 and adverse events

The causal relationship between all adverse events and KD-414 will be determined by the physicians enrolled in the study. The judgment should be based not only on the temporal relationship with the inoculation of KD-414 but also on the underlying disease course, complications, concomitant medications, research procedures, accidents, and other external factors. Causality will be judged and recorded according to the following criteria:

It is reasonable or possibly reasonable to attribute the adverse events to the research procedures or KD-414 inoculation.

Undeniable causal relationship: the decision should be made according to the following criteria irrespective of whether adverse events are known to occur in the clinical study or during vaccination:

- ➢ There is a temporal relationship between studies.
- ➢ No other cause is shown, and a causal relationship to KD-414 cannot be ruled out.

- Not related; judged according to the following criteria:

- ➢ Not reasonable to attribute the adverse events to research procedures or KD-414 inoculation
- ➢ Show no temporal relationship
- ➢ Can indicate other causes

Assessment of adverse events
Evaluation of adverse events

The presence or absence of abnormal findings will be determined at each examination during the period from post-vaccination to Visit 2 (post-hoc observations). The details of the event judged as "abnormal" will be recorded in the Case Report Form as an adverse event.

To determine adverse events at the inoculation site

The details of the local reaction observed during the period from post-vaccination to Visit 3 (post-hoc observations) will be recorded in the Case Report Form as an adverse event.

Evaluation of adverse events listed in the health diary

The contents of the health diary during the period from post-vaccination to Visit 3 (post-hoc observation) will be reviewed and the adverse events will be assessed based on medical judgment.

Severity classification of adverse events

The investigator or sub-investigator will determine and record the severity (severity) of each adverse event. Specific local adverse events and specified systemic adverse events will be judged using the Guidance for Industry Toxicity grading scale for healthy adult and adolescent volunteers enrolled in preventive vaccine clinical trials [11] and recorded in grades 0–4. Unspecified adverse events are categorized according to their impact on daily activities as Grade 1 (those that do not interfere with daily activities), Grade 2 (those that interfere with daily activities), Grade 3 (those that prevent performance of daily activities), and Grade 4 (potentially life-threatening).

Severity classification of specified local adverse events (inoculation site)

The investigator or sub-investigator will determine the severity of local adverse events based on the "Classification of severity of local reactions (inoculation site)" definition in Table 6.

Table 6. Classification of severity of local reactions (inoculation site).

At the Injection Site; Local Reaction	Mild (Grade 0)	Mild (Grade 1)	Moderate (Grade 2)	Severe (Grade 3)	Life Threatening; to Be Feared (Grade 4)
Pain	-	Prevent activity	Repeated use of non-narcotic analgesics; >24 h or interfere with daily activity	Use of narcotic analgesics; interferes with daily activities	Emergency room visits or hospitalization
Erythema/redness *	<2.5 cm	2.5–5 cm	5.1–10 cm	>10 cm	Necrotizing or exfoliative dermatitis
Indurated **	<2.5 cm	2.5–5 cm; does not interfere with activity	≥5.1–10 cm or interferes with activity	>10 cm daily; interferes with daily activities	Necrosis
Swelling **	<2.5 cm	2.5–5 cm; does not interfere with daily activity	≥5.1–10 cm or more or interferes with daily activity	>10 cm daily; interferes with daily activities	Necrosis

* The measured local response is graded by its maximum diameter, and measurements are recorded. ** They are evaluated and graded using the Functional Scale, as well as measured values.

Severity classification of specific systemic adverse events

The investigator will determine the severity of specific systemic adverse events for the subject based on Tables 7 and 8.

Table 7. Severity classification of systemic reactions (1).

Vital Sign *	Mild (Grade 0)	Mild (Grade 1)	Moderate (Grade 2)	Severe (Grade 3)	Life-Threatening; to Be Feared (Grade 4)
Fever (°C) **	37.5–37.9	38.0–38.4	38.5–38.9	39.0–40	>40

* All vital signs should be measured at rest. ** Oral temperature; no recent hot or cold drinks or smoking.

Table 8. Severity classification of systemic reactions (2).

Whole-Body	Mild (Grade 1)	Moderate (Grade 2)	Severe (Grade 3)	Life-Threatening (Grade 4)
Nausea and vomiting	No interference with activity or vomiting 1–2 times within 24 h	Any interference with activity or vomiting ≥2 times within 24 h	Interferes with daily activities; requires intravenous fluids in the outpatient setting	Emergency room (ER) visits or hospitalization for hypotensive shock
Headache	No interference with activity	Use of non-narcotic analgesics for more than 24 h, causing some interference with activity	IMPORTANCE: Use of narcotic analgesics Interferes with daily activities	ER visits or hospitalization
Fatigue (Malaise)	No interference with activity	Cause some interference with activity	IMPORTANCE: Impairing daily activities	ER visits or hospitalization
Myalgia	No interference with activity	Cause some interference with activity	IMPORTANCE: Impairing daily activities	ER visits or hospitalization

Predictive judgment

The predictability of adverse events will be determined based on the description in the Investigator's Brochure. Adverse events not listed in the Investigator's Brochure and those listed but with inconsistent severity or frequency are considered unknown adverse events.

Procedures for collecting, recording, and reporting information on diseases

The investigator or sub-investigator will record all adverse events in study subjects during the study period to determine a causal relationship.

In the event of an illness (a serious adverse event) in the study subject, the NCGM will promptly report the event to the supervisor of the participating medical organization and NCGM Clinical Research Safety Management Office irrespective of its relationship with the administered vaccine. NCGM reports the event (a disease) to the NCGM Clinical Research Review Board. Adverse events suspected to be attributable to the conduct of clinical research are classified as "diseases." Events attributable to the study should be related to the drug used in the research or the research procedure. Detailed reporting procedures shall be according to the "Procedures for response to adverse events/diseases."

Duration of adverse event information collection: 40 Days

Reporting of diseases to the certified Clinical Research Review Board (research using unapproved or off-label drugs, etc.; Table 9)

Table 9. Reporting of diseases to the certified Clinical Research Review Board.

Studies Classification	Predictability	Severity of Illness	Reporting Deadline
Clinical research using unapproved or off-label drugs, etc.	Not possible	Death	7 days
		Diseases that may lead to death	
	Possible	Death	15 days
		Diseases that may lead to death	
	Not possible	Diseases requiring admission to a medical institution or prolonged hospital stay for treatment	15 days
		Disorder	
		Diseases that may lead to disability	
		Serious illnesses accordance to the above and illnesses that may lead to death	
		Any congenital disease or anomaly in the offspring of a treated patient.	
Occurrence of diseases suspected to be attributable to the conduct of clinical research (other than those reported above)			Periodic Report

① The investigator should report a disease that may lead to death to the NCGM Institutional Review Board, the MHLW, and the Pharmaceuticals and Medical Devices Agency (PMDA) after reporting it to the supervisor of the participating medical institution and NCGM Clinical Research Safety Management Office within 7 days. Forecasts should be reported to the NCGM Clinical Research Review Board after reporting them to the administrators and NCGM Clinical Research Safety Management Office of the participating institution within 15 days.

Additionally, the following diseases that cannot be predicted are reported to the supervisor of the participating medical organization and NCGM Clinical Research Safety Management Office within 15 days before reporting them to the NCGM Clinical Research Review Board, the MHLW, and PMDA:

② Diseases requiring admission to a medical institution or prolonged hospital stay for treatment
③ Diseases that may lead to disability
④ Serious diseases according to ①–③.
⑤ Any congenital disease or anomaly in the offspring of a treated patient.

Furthermore, events that are non-fatal, non-life-threatening, and do not require hospitalization may be considered "serious" if they are judged to be an important medical event that may place the subject at risk, or if treatment or surgery is required to avoid the occurrence of the event. However, hospitalization for examination purposes, and prolongation of hospitalization for examination purposes scheduled before participation in the study, are not considered serious adverse events.

Disease reports to the MHLW will be prepared from disease reports of jRCT (Clinical Research Protocol and Research Outline Disclosure System).

The occurrence of diseases suspected to be attributable to the study procedures (excluding diseases subject to the above reports) will be reported at the time of periodic reporting by the NCGM Review Board.

Observation period of study subjects after the outbreak of disease

Observations of the study subjects after the occurrence of adverse events will be followed up until the alleviation of the adverse event or the investigator (or sub-investigator) judges that no follow-up is required. Detailed follow-up procedures shall be performed according to the "Procedures for response to adverse events/diseases."

Viewing source documents

The investigator and the institution shall ensure direct access to all clinical research-related records, such as source documents during clinical research-related monitoring, auditing, and NCGM Clinical Research Review Board and regulatory review.

Quality control and assurance

Method of monitoring

The investigator will prepare a written procedure for monitoring each study protocol and conduct monitoring as specified in the relevant procedure and study protocol.

Personnel involved in clinical research should not be monitored for tasks directly responsible for such individuals. Personnel involved in monitoring will report the results of such monitoring to the investigator.

Auditing method

The investigator shall prepare a written audit procedure for each study plan and conduct an audit as specified in the relevant procedure and study protocol. Individuals involved in clinical studies must be subjected to audits, whereas those involved in their monitoring should not be audited. Personnel involved in the audit will report the results of the audit to the investigator.

Method of data management

Data management will be performed by the personnel of the JCRAC data center.

Data collection will be performed using the EDC system as an electronic case report form (eCRF) and management tool.

Fixed data will be provided to the analysis manager after data fixation at the data center. Details are specified in the data management plan.

Ethical considerations

Compliance with laws and regulations

This study will be conducted according to the ethical principles, Clinical Trials Act (Act No. 16 of 14 April 2017), related notifications, and the study protocols stipulated in the Declaration of Helsinki. Materials stipulated in the law, including research protocols and protocols, will be discussed and approved by the Certified Clinical Research Review Board. Subsequently, the investigator will complete enrollment in the jRCT with permission from the administrator of the participating institution before initiating the study. The same procedures will be followed if there are any changes in these materials before implementing the amendments. The investigator will ensure that all researchers complete educational training on research ethics and other necessary skills at their respective participating medical institutions and will continue to receive educational training during the study.

Benefits and burdens incurred by clinical research subjects, anticipated disadvantages, as well as measures to minimize them

Participation in this study will determine antibody titers before and after immunization with KD-414. Immunization with KD-414 is expected to elicit antibody responses against SARS-CoV-2 and may increase the likelihood of preventing infection. However, it is unclear whether the antibodies can completely prevent COVID-19.

As KD-414 is not approved and there is no adequate vaccination experience, unknown adverse events may occur. Nonclinical studies and Phase I/II studies conducted with KD-414 have not revealed any safety concerns.

No adverse reactions to KD-414 have been reported. Pfizer comparatively analyzed the frequencies of various adverse events between individuals who received the third booster dose (n = 289) and those who received two doses (n = 2682) (fever, 8.7% vs. 16.4%; malaise, 63.7% vs. 61.5%; headache, 48.4% vs. 54.0%; and myalgia, 39.1% vs. 39.3%). Studies have also reported similar or improved tolerability of vaccinations with different novel coronavirus vaccines.

Based on the results of this study and the prevalence of COVID-19, further booster doses will be considered.

When there is a possibility that research may provide important information on the health of subjects or genetic characteristics that can be inherited by their offspring, as a result of the conduct of the research, the handling of the results of research on study subjects including incidental findings which are findings outside the main objectives of the research).

Not applicable

Handling personal information

Personnel involved in this study (including external parties) should comply with the Act on the Protection of Personal Information (promulgated on 30 May 2003, Law No. 57) and related notices applicable to the protection of personal information of study subjects. Personnel involved in this study should not acquire personal information by false or fraudulent means or reveal the personal information without justifying reasons even after the personnel leave their jobs and strive to protect the personal information and privacy of the study subjects.

Personnel involved in this study should not handle personal information acquired in this study beyond the scope of the informed consent provided by the subjects.

The investigator must identify the purpose of using personal information and maintain accurate and up-to-date personal information within the range necessary to achieve the objectives. Additionally, necessary measures should be taken to prevent the leakage, loss, or damage of personal information and appropriately manage personal information. The methods for these measures shall be specifically stipulated by the Implementation Regulations.

Anonymization methods and management

The information of study subjects should be identified using a novel unique number granted by the personnel involved in the study at the time of enrollment (Clinical Research Subject Identification Code) without the information that will lead to the identification of a particular individual.

The investigator shall prepare a correspondence table for the name and identification code of the study subject and store it appropriately in a lockable location within the participating medical organization.

Handling and storage of records (including data)
If the purpose of use includes the provision of samples or information to another institution.

Research institute: KM Biologics Co., Ltd.

Sample/information: The sera from day 29 (day 7 post-second dose) stored in previous studies and those from day 0, 7, and 40 post-inoculation in this study will be provided to investigate the neutralizing antibodies.

Storage and disposal of samples and information
Retention of records

The investigator will retain the records of this study along with the following documents:

(1) Identification code list for subjects of clinical research, research protocol, protocol, explanation to subjects of clinical research and documents related to their consent, clinical study report, other clinical research law, documents prepared by the investigator according to the regulations for enforcement, and their copies.
(2) Documents related to review opinions received from NCGM Clinical Research Review Board
(3) Monitoring and audit documentation

(4) Source documents
(5) Contract for conducting this study
(6) Prepared documents and records that describe the outline of the drugs used in this study
(7) Other documents required to conduct this study

Retention period and location of records

The investigator will retain the records appropriately in a key archive within the institution for five years from the date of completion of the study (the day the summary of the clinical study report was published to jRCT).

In case these materials are entrusted to an outside contractor and stored, arrangements shall be made in the contract so that they can be periodically monitored and controlled under appropriate conditions during the specified storage period.

Method for sample and information disposal

The personal information must be carefully handled when discarding specimens and information related to this study. The samples used in this study should be discarded appropriately according to the in-house procedures. The information must be discarded by placing the paper medium on a shredder. The electronic recording medium must be discarded in a non-readable state. The file in the personal computer should be completely deleted such that it cannot be reproduced.

Payments and compensation related to the conduct of clinical research
Presence and content of cost and burden reduction expenses for study subjects

KD-414 to be administered in this study will be provided free of charge by companies. The cost of testing will also be covered by the research costs to avoid a cost burden to the subjects. For the research subjects, up to JPY 60,000 will be paid to mitigate the cost burden (JPY 20,000 per blood sampling for KD-414 vaccines and JPY 2000 per blood sampling for non-vaccines). For post-booster vaccination at public expense, up to JPY 6000 will be paid as a burden reduction cost (JPY 2000 per blood sampling).

Presence or absence and content of insurance coverage

To compensate for the loss incurred by the research subject due to health hazards, the research representative physician will have clinical research insurance with the following compensation content. The study subject will be compensated according to the clinical research insurance payment conditions.

- Compensation for death or sequelae disability of study subjects
- Medical expenditures and medical benefits required to treat the health hazards of research subjects

Additionally, the principal investigator and the sub-investigator must also have physician's liability insurance in preparation for health damage to the study subjects resulting from the usual range of medical practices in this study.

Existence and content of compensation other than insurance

In the event of any health hazard to a study subject due to the implementation of this study, the investigator, sub-investigator, and participating medical institution shall take the necessary measures, such as the provision of medical treatment so that the study subject can receive the appropriate diagnosis, treatment, and necessary measures immediately.

Publication of information on clinical research
Method of publication

During this study, jRCT should document the items required by the World Health Organization for publication and other items that contribute to ensuring the transparency of the clinical research process and the selection of the volunteers to participate in clinical research and publish these items. The study will be initiated after jRCT publication.

However, changes will be made to the protocol and the information will be updated according to the progress of the study. If a primary endpoint report or clinical study report is prepared, a summary of the primary endpoint report or clinical study report will also be published. The articles will also be presented.

Content of publications and agreements regarding the timing and results of clinical research with the marketing authorization or marketing authorization of pharmaceuticals that have received grants, if any.

The investigator, statistical analysis manager, and GLIDE office will select the authors after consultation and report them to academic societies and/or journals. Based on the contract with the company, prior notice shall be given to KM Biologics, Ltd. before the announcement of the results. The results shall be publicly announced after due consultation where necessary. However, the company is not involved in the interpretation of the results of this study and cannot refuse to provide public consent without justification. The results will be reported when KM Biologics Co., Ltd. wishes to disclose them.

Duration of clinical research

Planned study period: jRCT publication date to 31 March 2023 (study completion date: date of publication of the summary of the clinical study report in jRCT)

Explanation and obtaining informed consent for subjects in clinical research

The investigator (or sub-investigator) will provide full explanations using documentary information before participation in the study and obtain voluntary consent for participation. The purpose and significance of the clinical research and the methods and duration of the clinical research will be explained.

Preparation of explanatory documents, written consent forms, and consent withdrawal forms

The investigator will prepare a single form of written information, informed consent form, and consent withdrawal form for each study protocol and obtain approval from the NCGM Clinical Research Review Board.

Plain language will be used so that the study subjects can easily understand the contents.

Explanation items

Explanatory matters include the matters provided for in Article 46 of the following regulations.

(1) The name of the specified clinical research to be conducted, the approval received by the manager of the participating medical organization for conducting the specified clinical research, and the implementation plan submitted to the Ministry of Health, Labor and Welfare.
(2) The names and titles of the participating medical organizations and the investigators.
(3) Reasons for selecting subjects for specific clinical studies
(4) Anticipated benefits and disadvantages of conducting specific clinical studies
(5) The option of voluntarily refusing participation in specific clinical research.
(6) Matters concerning withdrawal of consent
(7) Denial of participation in specified clinical research or withdrawal of consent does not result in unfavorable handling.
(8) Method of disclosing information on specific clinical research
(9) The availability or browsing of research protocols and other data related to the conduct of specific clinical research and the methods of obtaining or browsing them in response to the request of the test subjects or their representative (hereafter referred to as the target person of specified clinical research, etc.).
(10) Items related to the protection of personal information of study subjects
(11) Method of storage and disposal of samples.

(12) Status of involvement in specified clinical research as prescribed in the items of Article 21, paragraph (1) of the Ordinance on Specified Clinical Research: Matters concerning a conflict of interest
(13) System for responding to complaints and inquiries
(14) Items related to costs related to the conduct of specified clinical research
(15) Presence and content of other treatment and comparison with expected benefits and adverse benefits from other treatment
(16) Matters related to compensation and provision of medical care for health hazards from conducting specific clinical research
(17) Items to be reviewed by the NCGM Clinical Research Review Board concerning opinions for specified clinical research and other items related to the Accredited Clinical Research Review Board related to such specified clinical research.
(18) Other necessary items related to the conduct of specific clinical research

Method of obtaining consent for clinical research subjects

The investigator or sub-investigator will provide an easy-to-understand explanation on clinical research using an explanatory document with permission from the certified clinical research review board before obtaining written informed consent to participate in the study. The study subjects are allowed to ask questions and provided with time to decide on participation in the study. The investigator or sub-investigator will satisfactorily answer the questions. Additionally, the investigator or sub-investigator will explain that the consent for study participation also includes the consent for direct access to the medical records of the study subject during monitoring, audits, certified clinical research review board meetings, and regulatory surveys. The investigator or sub-investigator and consenting clinical research subjects will sign the informed consent form along with the date. After obtaining the informed consent, the investigator or sub-investigator will retain the original copy of the informed consent form and provide a copy of the informed consent form to the study subjects. The investigator or sub-investigator will document that a copy of the information and informed consent form has been provided to the study subject (original of the consent form, medical record, etc.).

Appointment of a proxy consenter and obtaining consent from the proxy consenter

Not applicable

When minors are included in clinical research

Not applicable

Secondary use of samples and information
The possibility of secondary use of specimens and information or providing them to other research institutions

As the specimens and information of the study participants are invaluable, they may be used for future studies (including human genome/gene analysis studies) that are not identified at the time of receiving consent from the study subjects and may be provided to other research institutions or companies around the country.

When existing data are stored and archived samples are used for new research, a new study protocol, will be provided to the Ethics Review Board and used after approval. Additionally, a public information document is prepared by the opt-out procedure to ensure the opportunity for subjects to refuse to participate in the research. Existing data and stored samples can be provided to researchers or companies in other facilities for use within the scope of informed consent. Measures must be undertaken to prevent the identification of individuals without providing anonymization response tables.

In addition, data and stored samples will be transferred to REBIND at any time to collect information, excluding the remaining sample and personal information after the completion of this study. Specimens and data transferred to REBIND may be transferred to other agencies (including companies).

※REBIND

Repository of Data and Biospecimen of INfectious Disease (REBIND)

A public biobank project was conducted by the Ministry of Health, Labor, and Welfare. See the website for more information on REBIND. <https://rebind.ncgm.go.jp/> (accessed on 18 June 2022).

Procedures for secondary use of samples and information

When archived specimens and existing data are used for new research, a new study protocol will be provided to the Ethics Review Board and used after approval. Additionally, the new protocol should be used only for research contributing to infectious diseases. In case existing data and samples are provided to researchers at other institutions, they shall be used only for research contributing to infectious diseases and provided within the scope of informed consent. Anonymized response tables should not be provided. Measures should be taken to prevent the identification of individuals.

Response to consent withdrawal

In case the study subjects withdraw their consent for participation in the study, the investigator or sub-investigator will consult with them to confirm that they are not reluctant to withdraw consent, confirm the reason for the withdrawal of consent, and explain the examinations and tests specified after consent withdrawal. The subjects will be requested to fill in the consent withdrawal form or the consent withdrawal will be recorded in documents, such as the medical record. In the event the study subjects who have withdrawn consent do not provide permission to use their samples and information, they will be excluded from the analysis.

When the results of the clinical research are already published or when medical devices are implanted in the body and cannot be easily removed, it is not possible to respond to some or all the consent withdrawals from the study subjects. In such cases, the investigator or sub-investigator will explain the reason to study subjects and arrive at an understanding.

Requesting the reasons for withdrawal of consent may lead to the retraction of the offer. Therefore, this should be addressed regardless of whether or not the reason for the offer was provided, except in cases where it interferes with ensuring the safety of the study subjects.

Handling of new information that influences the consent provided by the study subjects

The investigator or sub-investigator should promptly inform the study subjects about new important information and obtain their consent to continue their participation in the research. Additionally, the new information and informed consent form will be revised to obtain approval from the NCGM Clinical Research Review Board.

Revision of documents and informed consent forms (entry of the revised version number and date of preparation)

In case the study-related information and informed consent form are revised, the investigator or sub-investigator will promptly explain the objectives of clinical research to the study subjects using the revised information and informed consent form and obtain written consent from the study subjects for continued participation in the study.

Items required for conducting clinical research

Matters related to conflicts of interest

Names of organizations providing research funding (public research expenditures/companies)

KM Biologics Co., Ltd. (project for emergency improvement of vaccine production systems) (contracted expense).

The test drugs were provided free of charge by KM Biologics Co., Ltd. Additionally, the research fund was used to sign a contract with CTD Co., Ltd. (Tokyo, Japan) for assisting with the administration of this study, and Accerise, Inc. (Tokyo, Japan). for monitoring, auditing, and performing statistical analysis. KM Biologics is not involved in the analysis of the results and will not influence the results or the publication of the findings.

Conflicts of interest management

Before the implementation of this study, the investigator or sub-investigator will prepare and review the conflicts of interest according to the provisions of the NCGM Conflict of Interest Management Committee (including the preparation of the Conflict of Interest Control Standards (Form A), Reports of Affiliated Industries (Form B (if occurring)), Reports of Investigator Conflict of Interest Self-Reports (Form C), and Conflict of Interest Control Plans (Form E)).

The principal investigator will submit the NCGM Conflict of Interest Management Committee review report to the NCGM Clinical Research Review Board for review and approval. Additionally, the principal investigator will continuously monitor, manage, and publish the conflicts of interest throughout the study period.

Studies (such as life-saving clinical research in an emergency setting) for which obtaining consent is difficult

Not applicable.

3. Discussion

The COVID-19 pandemic is currently ongoing, and there have been significant efforts in the development of COVID-19 vaccines. However, the neutralizing antibody titers in vaccinated individuals are reported to progressively decrease over time. Japanese pharmaceutical companies have published the results of Phase I and II studies on the safety and efficacy of different vaccines. Final clinical trials will be conducted with the aim of practical application by March 2023. Since mRNA vaccination is progressing around the world, it is necessary to evaluate the efficacy and safety of booster doses in individuals who have received two doses of mRNA vaccine in order to make effective use of inactivated vaccines in the future. There have been studies evaluating the progression of antibody titers with boosters with inactivated vaccines after initial immunization with inactivated vaccines and with boosters with mRNA vaccines, but few studies have evaluated boosters with inactivated vaccines after initial immunization with mRNA vaccines [12,13]. This is the first study of booster immunization with KD-414 developed in Japan.

This manuscript describes the protocol for a study that aims to examine the effect of a booster dose with KD-144, a novel inactivated Japanese vaccine, on antibody titers. A previous study comparing the efficacy, side effects, and seroconversion of various vaccine formats against SARS-CoV-2, including inactivated, recombinant, mRNA, and nanoparticle-based vaccines suggested that inactivated vaccines have fewer side-effects and similar seroconversion compared to other types of vaccines [14]. Furthermore, heterologous prime-boost vaccine strategies have been shown to result in improved immunogenicity, reactogenicity, safety, effectiveness and flexibility, as well as to mitigate against intermittent supply shortages [15,16].

There are some limitations to this protocol. First, it is designed as a single-arm study with no randomization. Second, the follow-up period will be relatively short (40 days) because the COVID-19 vaccination at public expense commences 42 days after KD-414 vaccination. Third, the definition of an 'appropriate' amount of KD-414 vaccination is unclear; the dose used in this study was determined based on the dose of the first and second shots of KD-414 in Phase I and II clinical trials.

Author Contributions: Y.U. and Y.S. analyzed the data and performed the statistical analyses; A.M. contributed to the research and development plans; J.T.-H. coordinated, managed the process, reviewing, and editing the manuscript.; M.U. was a major contributor to the administration of this research; J.S.T., Y.T. (Yuki Takamatsu) and K.M. were responsible for the analysis of specimens and contributed to the planning of this study. W.S. supervised the research; Y.T. (Yuriko Terayama) and N.T. were major contributors to writing the manuscript. All authors have read and agreed to the published version of the manuscript.

Funding: Research funding related to the conduct of this study was provided by KM Biological Co., Ltd. (research contracting for emergency improvement project subsidies for vaccine production systems).

Institutional Review Board Statement: This study was approved by the Certified Review Board of the National Center for Global Medicine (Approval number: NCGM-C-004374-05, date of approval: 12 May 2022). Written informed consent will be obtained from the patients for publication of this study.

Informed Consent Statement: Not applicable.

Data Availability Statement: Data sharing is not applicable to this article as no datasets were generated or analyzed during the current study.

Acknowledgments: Editorial support, in the form of medical writing, assembling tables and creating high-resolution images based on authors' detailed directions, collating author comments, and funded by research funding (contracting for emergency improvement project subsidies for vaccine production systems, etc.) provided by KM Biological Co., Ltd.

Conflicts of Interest: This research will be carried out with research funding (contracting for emergency improvement project subsidies for vaccine production systems, etc.) from KM Biological Co., Ltd. KM Biologics is the developer of the test drug (KD-414) that will be used in this research, and KD-414 is provided free of charge by KM Biologics Co., Ltd. In addition, we will conduct monitoring, auditing, and statistical work for Accessible Co., Ltd. using this research funding, and contract with CTD Co., Ltd. to subsidize the secretariat work for this research, but these companies' intentions will not have an undue impact on the research results and presentations. These conflicts of interest will be declared to the NCGM Conflict of Interest Management Committee in advance and will be appropriately managed and publicized based on the Conflict-of-Interest Management Plan reviewed and approved by the Institutional Review Board of the National Center for Global Medicine. In addition, KM Biological Co., Ltd. will not be involved in interpreting the research results, such as analysis of the research data, so that the results of the research will not be biased. Dr. Mugen Ujiie conducts contract research by receiving R&D expenditure from KM Biological Co., Ltd.

Abbreviations

CDC	Centers for Disease Control and Prevention
COVID-19	coronavirus disease 2019
NCGM	National Center for Global Health and Medicine
REBIND	Repository of Data and Biospecimens of INfectious Disease

References

1. Dagan, N.; Barda, N.; Kepten, E.; Miron, O.; Perchik, S.; Katz, M.A.; Hernán, M.A.; Lipsitch, M.; Reis, B.; Balicer, R.D. BNT162b2 mRNA COVID-19 Vaccine in a Nationwide Mass Vaccination Setting. *N. Engl. J. Med.* **2022**, *40*, 1924–1927. [CrossRef] [PubMed]
2. Maeda, K.; Amamo, M.; Uemura, Y.; Tsuchiya, K.; Matsushima, T.; Noda, K.; Shimizu, Y.; Fujiwara, A.; Takamatsu, Y.; Ichikawa, Y.; et al. Correlates of neutralizing/SARS-CoV-2-S1-binding antibody response with adverse effects and immune kinetics in BNT162b2-vaccinated individuals. *medRxiv* **2021**. [CrossRef]
3. Yamamoto, Y.; Fukunaga, A.; Tanaka, A.; Takeuchi, J.S.; Inoue, Y.; Kimura, M.; Maeda, K.; Ueda, G.; Mizoue, T.; Ujiie, M.; et al. Association between reactogenicity and SARS-CoV-2 antibodies after the second dose of the BNT162b2 COVID-19 vaccine. *Vaccine* **2022**, *384*, 1412–1423. [CrossRef] [PubMed]
4. Takeuchi-Shibata, J.; Fukunaga, A.; Yamamoto, Y. SARS-CoV-2 specific T cell and humoral immune responses upon vaccination with 3 BNT162b2. *medRxiv* **2021**. [CrossRef]
5. Zeng, G.; Wu, Q.; Pan, H.; Li, M.; Yang, J.; Wang, L.; Wu, Z.; Jiang, D.; Deng, X.; Chu, K.; et al. Immunogenicity and safety of a third dose of CoronaVac, and immune persistence of a two-dose schedule, in healthy adults: Interim results from two single-centre, double-blind, randomised, placebo-controlled phase 2 clinical trials. *Lancet Infect. Dis.* **2021**, *22*, 483–495. [CrossRef]
6. Bajema, K.L.; Dahl, R.M.; Evener, S.L.; Prill, M.M.; Rodriguez-Barradas, M.C.; Marconi, V.C.; Beenhouwer, D.O.; Holodniy, M.; Lucero-Obusan, C.; Brown, S.T.; et al. Comparative effectiveness and antibody responses to Moderna and Pfizer-BioNTech COVID-19 vaccines among hospitalized veterans—Five Veterans Affairs Medical Centers, United States, February 1–September 30, 2021. *MMWR Morb. Mortal. Wkly. Rep.* **2021**, *70*, 1700–1705. [CrossRef] [PubMed]
7. Khoury, D.S.; Cromer, D.; Reynaldi, A.; Schlub, T.E.; Wheatley, A.K.; Juno, J.A.; Subbarao, K.; Kent, S.J.; Triccas, J.A.; Davenport, M.P. Neutralizing antibody levels are highly predictive of immune protection from symptomatic SARS-CoV-2 infection. *Nat Med.* **2021**, *27*, 1205–1211. [CrossRef] [PubMed]

8. Gilbert, P.B.; Montefiori, D.C.; McDermott, A.; Fong, Y.; Benkeser, D.; Deng, W.; Zhou, H.; Houchens, C.R.; Martins, K.; Jayashankar, L.; et al. Immune Correlates Analysis of the mRNA-1273 COVID-19 Vaccine Efficacy Trial. *medRxiv* **2021**. [CrossRef]
9. World Health Organization. Guidelines on Clinical Evaluation of Vaccines: Regulatory Expectations. 2016. Available online: https://www.who.int/biologicals/BS2287_Clinical_guidelines_final_LINE_NOs_20_July_2016.pdf (accessed on 4 April 2021).
10. World Health Organization. Guidelines on Clinical Evaluation of Vaccines: Regulatory Expectations Replacement of Annex 1 of World Health Organization Technical Report Series, No. 924. 2017. Available online: https://cdn.who.int/media/docs/default-source/prequal/vaccines/who-trs-1004-web-annex-9.pdf?sfvrsn=9c8f4704_2&download=tru (accessed on 4 April 2021).
11. U.S. Department of Health and Human Services Food and Drug Administration Center for Biologics Evaluation and Research. Guidance for Industry. Toxicity Grading Scale for Healthy Adult and Adolescent Volunteers Enrolled in Preventive Vaccine Clinical Trials. 2007. Available online: https://www.fda.gov/media/73679/download (accessed on 4 April 2021).
12. Kanokudom, S.; Assawakosri, S.; Suntronwong, N.; Auphimai, C.; Nilyanimit, P.; Vichaiwattana, P.; Thongmee, T.; Yorsaeng, R.; Srimuan, D.; Thatsanatorn, T.; et al. Safety and Immunogenicity of the Third Booster Dose with Inactivated, Viral Vector, and mRNA COVID-19 Vaccines in Fully Immunized Healthy Adults with Inactivated Vaccine. *Vaccines* **2022**, *10*, 86. [CrossRef] [PubMed]
13. Atmar, R.L.; Lyke, K.E.; Deming, M.E.; Jackson, L.A.; Branche, A.R.; El Sahly, H.M.; Rostad, C.A.; Martin, J.M.; Johnston, C.; Rupp, R.E.; et al. Homologous and Heterologous Covid-19 Booster Vaccinations. *N. Engl. J. Med.* **2022**, *386*, 1046–1057. [CrossRef] [PubMed]
14. Khandker, S.S.; Godman, B.; Jawad, I.; Meghla, B.A.; Tisha, T.A.; Khondoker, M.U.; Haq, A.; Charan, J.; Talukder, A.A.; Azmuda, N.; et al. A Systematic Review on COVID-19 Vaccine Strategies, Their Effectiveness, and Issues. *Vaccines* **2021**, *9*, 1387. [CrossRef] [PubMed]
15. Sapkota, B.; Saud, B.; Shrestha, R.; Al-Fahad, D.; Sah, R.; Shrestha, S.; Rodriguez-Morales, A.J. Heterologous prime–boost strategies for COVID-19 vaccines. *J. Travel Med.* **2022**, *29*, taab191. [CrossRef] [PubMed]
16. Hillus, D.; Schwarz, T.; Tober-Lau, P.; Vanshylla, K.; Hastor, H.; Thibeault, C.; Jentzsch, S.; Helbig, E.T.; Lippert, L.J.; Tscheak, P.; et al. Safety, reactogenicity, and immunogenicity of homologous and heterologous prime-boost immunisation with ChAdOx1 nCoV-19 and BNT162b2: A prospective cohort study. *Lancet Respir. Med.* **2021**, *9*, 1255–1265. [CrossRef]

life

MDPI

Protocol

A Multi-Center, Open-Label, Randomized Controlled Trial to Evaluate the Efficacy of Convalescent Plasma Therapy for Coronavirus Disease 2019: A Trial Protocol (COVIPLA-RCT)

Noriko Tomita [1], Sho Saito [2], Junko Terada-Hirashima [1,*], Ayako Mikami [1], Yukari Uemura [1], Satoshi Kutsuna [2,3], Hidetoshi Nomoto [2], Kyoko Fujisawa [2], Maki Nagashima [2], Mari Terada [1,2], Shinobu Ashida [2], Shinichiro Morioka [2], Masahiro Satake [4], Akira Hangaishi [5], Tomiteru Togano [5], Katsuyuki Shiratori [6], Yuki Takamatsu [7], Kenji Maeda [7], Norio Ohmagari [2], Wataru Sugiura [1] and Hiroaki Mitsuya [7]

1 Center for Clinical Sciences, Center Hospital of the National Center for Global Health and Medicine, Tokyo 162-8655, Japan; ntomita@hosp.ncgm.go.jp (N.T.); amikami@hosp.ncgm.go.jp (A.M.); yuemura@hosp.ncgm.go.jp (Y.U.); materada@hosp.ncgm.go.jp (M.T.); wsugiura@hosp.ncgm.go.jp (W.S.)
2 Disease Control and Prevention Center, Center Hospital of the National Center for Global Health and Medicine, Tokyo 162-8655, Japan; ssaito@hosp.ncgm.go.jp (S.S.); kutsuna@hp-infect.med.osaka-u.ac.jp (S.K.); hnomoto@hosp.ncgm.go.jp (H.N.); kfujisawa@hosp.ncgm.go.jp (K.F.); mnagamatsu@hosp.ncgm.go.jp (M.N.); sashida@hosp.ncgm.go.jp (S.A.); shmorioka@hosp.ncgm.go.jp (S.M.); nohmagari@hosp.ncgm.go.jp (N.O.)
3 Department of Infection Control and Prevention, Osaka University Hospital, Osaka 565-0871, Japan
4 Central Blood Institute, Japanese Red Cross Society, Tokyo 135-8521, Japan; m-satake@jrc.or.jp
5 Department of Hematology, Center Hospital of the National Center for Global Health and Medicine, Tokyo 162-8655, Japan; ahangaishi@hosp.ncgm.go.jp (A.H.); ttogano@hosp.ncgm.go.jp (T.T.)
6 Labotatory Testing Department, Center Hospital of the National Center for Global Health and Medicine, Tokyo 162-8655, Japan; kshiratori@hosp.ncgm.go.jp
7 Department of Refractory Viral Infections, National Center for Global Health and Medicine Research Institute, Tokyo 162-8655, Japan; ytakamatsu@ri.ncgm.go.jp (Y.T.); kmaeda@ri.ncgm.go.jp (K.M.); hmitsuya@hosp.ncgm.go.jp (H.M.)
* Correspondence: jterada@hosp.ncgm.go.jp; Tel.: +81-3-3202-7181; Fax: +81-3-5273-6941

Citation: Tomita, N.; Saito, S.; Terada-Hirashima, J.; Mikami, A.; Uemura, Y.; Kutsuna, S.; Nomoto, H.; Fujisawa, K.; Nagashima, M.; Terada, M.; et al. A Multi-Center, Open-Label, Randomized Controlled Trial to Evaluate the Efficacy of Convalescent Plasma Therapy for Coronavirus Disease 2019: A Trial Protocol (COVIPLA-RCT). *Life* **2022**, *12*, 856. https://doi.org/10.3390/life12060856

Academic Editor: Theodoros Rampias

Received: 16 May 2022
Accepted: 6 June 2022
Published: 8 June 2022

Abstract: Background: Coronavirus disease 2019 is a global public health concern. As of December 2020, the therapeutic agents approved for coronavirus disease 2019 in Japan were limited to two drugs: remdesivir, an antiviral drug, granted a Special Approval for Emergency on 7 May 2020, and dexamethasone, which has an anti-inflammatory effect. The aim of this study is to evaluate the efficacy of convalescent plasma collected from donors who recovered from coronavirus disease 2019. Methods: This is an open-label, randomized controlled trial comprising two groups: a convalescent plasma and a standard-of-care group. Plasma administered to patients with coronavirus disease 2019 randomized in the convalescent plasma group of this trial will be plasma that has been collected and stored in an associated study. Patients with a diagnosis of mild coronavirus disease 2019 will be included in this trial. The efficacy of convalescent plasma transfusion will be evaluated by comparing the convalescent plasma group to the standard-of-care group (without convalescent plasma transfusion) with respect to changes in the viral load and other measures. The primary endpoint will be time-weighted average changes in the SARS-CoV-2 virus load in nasopharyngeal swabs from day 0 to days 3 and 5. It is hypothesized that the intervention should result in a decrease in the viral load in the convalescent plasma group until day 5. This endpoint has been used as a change in viral load has and been used as an index of therapeutic effect in several previous studies. Discussion: The proposed trial has the potential to prevent patients with mild COVID-19 from developing a more severe illness. Several RCTs of convalescent plasma therapy have already been conducted in countries outside of Japan, but no conclusion has been reached with respect to the efficacy of convalescent plasma therapy, which is likely in part because of the heterogeneity of the types of target patients, interventions, and endpoints among trials. Actually, previous clinical trials on plasma therapy have shown inconsistent efficacy and are sometimes ineffective in COVID-19 patients with severe disease, which is due to unmeasured neutralizing antibody titer in the COVID-19 convalescent plasma. To improve this issue, in this study, we measure neutralizing activity of convalescent plasma before administration and provide the plasma with high neutralizing activity to the subjects. It is hoped that this study will further evidence to support the role of convalescent plasma therapy in COVID-19.

Keywords: plasma therapy; coronavirus disease 2019; COVID-19; SARS-CoV-2; viral load

1. Background

Coronavirus disease 2019 (COVID-19) has spread and become a substantial public health concern worldwide. In addition, in Japan, as of 30 May 2022, the number of infected patients has reached 8.78 million or more through the first to sixth waves, and there is a concern over a continuation of the epidemic.

As of December 2020, the therapeutic agents approved for COVID-19 in Japan were limited to two drugs: remdesivir, an antiviral drug, granted a Special Approval for Emergency on 7 May 2020, and dexamethasone, which has an anti-inflammatory effect. In a global, randomized controlled clinical study (ACTT-1) sponsored by the National Institute of Health (NIH), the time to clinical improvement was shorter in the remdesivir group (10 d) by 5 d than in the placebo group (15 d) [1]. However, in an open-label controlled clinical study (SOLIDARITY trial) sponsored by the World Health Organization (WHO), the ratio of the risk of death in the remdesivir group relative to that in the standard of care group was 0.95 [2]. Consequently, the WHO does not recommend the use of remdesivir in its treatment guidelines [3]. In principle, it is indicated for patients with an oxygen saturation of $\leq$94% or those requiring supplemental oxygen, the introduction of extracorporeal membrane oxygenation (ECMO) or invasive mechanical ventilation. In a large, multi-center, randomized, open-label study in the United Kingdom (UK), mortality was lower in patients given dexamethasone than in those given the standard of care. The maximal therapeutic effect was observed in patients who required invasive mechanical ventilation at the time of randomization, in whom mortality was 29.0% in the dexamethasone group and 40.7% in the control group [4].

Convalescent plasma therapy was used in patients with Spanish flu, and a report suggests that its efficacy was demonstrated by an analysis of the patients given the therapy at that time [5]. More than 40 years ago, the therapy was reported to have decreased mortality in a randomized controlled study in patients with Argentine hemorrhagic fever [6]. Recently, convalescent plasma therapy has been used in patients with severe infection, such as avian influenza H5N1 [7] and Ebola disease [8], and those with coronavirus diseases similar to COVID-19, such as severe acute respiratory syndrome (SARS) [9] and Middle East respiratory syndrome (MERS) [10,11].

Several clinical studies of convalescent plasma therapy in patients with COVID-19 have been reported in China and the United States. In the former, a case series from China showed that clinical symptoms rapidly improved, and cures were achieved in all five patients who underwent convalescent plasma transfusion [12]. To date, three randomized controlled trials (RCTs) have been reported. In one RCT in China [13], convalescent plasma therapy was effective in patients with moderate disease who required supplemental oxygen but did not require mechanical ventilation. However, its efficacy was not shown in all patients or those with severe disease requiring mechanical ventilation. This trial was terminated before the planned number of patients was reached because it was conducted in the latter part of the epidemic in China, and the registration of patients decreased sharply in mid-trial. Therefore, it is possible that its efficacy in severe disease could not be shown because of the small sample size. Another RCT conducted on 81 hospitalized patients in Spain [14] showed that survival improved in patients who underwent convalescent plasma transfusion. The transfusion of the convalescent plasma was not effective in RCTs of patients with at least moderate disease in India [15] or Argentina [16]. However, the prevention of severe disease was demonstrated in an RCT where convalescent plasma was administered within 3 d after disease onset in patients with a high risk for severe disease, such as elderly patients and those with underlying diseases [17].

Based on these results, it is likely that convalescent plasma therapy is ineffective in patients with severe disease but most effective when the plasma with a high antibody titer is administered as soon as possible after disease onset. The inconsistent results of clinical studies on plasma therapy could be attributable to the neutralizing antibody titer (i.e., the extent of the potency to inhibit novel coronavirus infection) in the COVID-19 convalescent plasma administered. This value is unmeasured in most studies. Importantly, the intensities of neutralizing antibodies in the COVID-19 convalescent plasma greatly vary among individuals. That is, the neutralizing activity may be high in the plasma of one patient, whereas no neutralizing activity may be observed in the plasma of another. In addition, it has been reported that neutralizing antibodies are often decreased or disappear approximately 1 month after a high level of neutralizing antibodies is observed; a similar finding was confirmed in a study conducted by the Research Institute National Center for Global Health and Medicine [18].

In this trial, subjects will receive plasma with high neutralizing activity after it is determined whether the convalescent plasma from each donor can inhibit infection when susceptible living cells are exposed to infectious novel coronaviruses at the Research Institute National Center for Global Health and Medicine (directed by Hiroaki Mitsuya). Based on the above-mentioned facts, the objective of this trial is to evaluate the efficacy of convalescent plasma with high neutralizing activity collected from donors who have recovered from COVID-19.

As potential risks and adverse events of the convalescent plasma, if a donor is affected by an unknown infection that is currently difficult to screen, a subject who has received plasma may acquire the infection. Although the donors will be interviewed according to the screening procedures used by the Japanese Red Cross Society, the risk of infection in this trial could be higher than that of standard blood transfusion because some screening criteria have been changed in this trial, allowing the acceptance of convalescent plasma from patients with COVID-19 or diabetes mellitus. In addition, adverse reactions, such as hypersensitivity, shock, anaphylaxis, and multi-organ failure, may occur because of immunization and antigen–antibody reactions. If a donor has been pregnant before or has undergone a blood transfusion, human leukocyte antigen (HLA) antibodies could be present in the blood product, which could increase the risk of transfusion-related acute lung injury (TRALI) after administration of the plasma product. However, persons who tested positive for HLA antibodies are excluded from the donors in advance, and this is not expected to increase the risk of TRALI. A study sponsored by the U.S. Food and Drug Administration, in which COVID-19 convalescent plasma was administered to 20,000 hospitalized patients with COVID-19, showed that COVID-19 convalescent plasma transfusion is safe [19].

To minimize the risks of infections and adverse reactions caused by immunization reactions, the collected plasma is tested, including viral and irregular antibody tests, similar to that of typical blood donations. If a donor has been pregnant or has undergone a blood transfusion, an HLA antibody test will be performed to confirm the negative result and minimize the risk of TRALI caused by the treatment. Before transfusion, cross-matching should be performed. Although the plasma is infused, the medical staff should always be with the subject to carefully monitor changes in their condition. The investigator should provide an explanation to the patients in whom the benefits from convalescent plasma therapy outweigh the risks in his/her judgment and should explain the risks in detail at the time of informed consent. Three months after transfusion, the post-transfusion infection tests specified at the National Centre for Global Health and Medicine, including HBV-DNA quantification, HCV core protein, and HIV antibody assays, will be performed.

2. Methods/Design

This trial is an open-label, randomized controlled trial comprising two groups: a convalescent plasma with a standard-care group (convalescent plasma group) and a standard of care group. The plasma administered to patients with COVID-19 randomized in the convalescent plasma group will be that collected and stored in an associated study (study title: A study on the collection of the COVID-19 convalescent plasma and its antibody titer and activity; abbreviation: COVIPLA-D; Institutional Review Board of National Center for Global Health and Medicine; approval No. NCGM-G-003536). Patients with a diagnosis of mild COVID-19 will be included in this trial. The efficacy of convalescent plasma transfusion will be evaluated by comparing the convalescent plasma group to the standard of care group (without convalescent plasma transfusion) with respect to changes in viral load and other measures. This trial was registered with the Japan Registry of Clinical Trials (Clinical Trial Plan Number: jRCTs031200374; https://dbcentre3.jmacct.med.or.jp/jmactr/Default_Eng.aspx; registration date: 24 February 2021).

The latest research protocol is version 2.4, dated 21 October 2021. The principal investigator shall prepare one form of explanation document, consent document, and consent withdrawal form for one research protocol and obtain approval from the NCGM-approved Clinical Research Review Committee. The subject of the clinical research or a surrogate and a witness (a person who attends the consent explanation process for the subject of the clinical research or a surrogate who has the ability to consent but is unable to read the consent document due to visual impairment or other reasons, and who is independent of the conductor of the clinical research) The consent document should be prepared in plain language so that it can be understood by the person who is the subject of the clinical research (a person who is independent of the conductor of the clinical research).

In this multi-center collaborative study, common items other than those specific to each institution (e.g., the name of the principal investigator and contact information for the consultation service) should be described so that the contents of the explanation and consent for the subjects of the clinical research at each institution are consistent.

Persons engaged in this research (including external parties) shall comply with the "Act on the Protection of Personal Information" (promulgated on 30 May 2003, Law No.57) and related notifications that apply to the protection of personal information, etc. of subjects in clinical research. Those engaged in this research shall not obtain personal information through deception or other wrongful means, shall make the utmost efforts to protect the personal information and privacy of clinical research subjects, and shall not divulge personal information obtained in the course of conducting this research without justifiable reason (the same shall apply even after the person involved has left his/her position).

In addition, those involved in this research must not handle personal information obtained in the course of conducting the research beyond the scope for which consent has been obtained in advance from the subjects of the clinical research.

In handling personal information, the principal investigator must specify the purpose of its use to the greatest extent possible and must keep the personal information accurate and up-to-date to the extent necessary to achieve the purpose of use. In addition, measures necessary to prevent leakage, loss, or damage of personal information and to otherwise appropriately manage personal information shall be taken, and the methods of such measures shall be specifically stipulated as implementation rules.

2.1. Trial Inclusion Criteria

- A patient who, or whose legal representative, has provided written consent for the participation of the patient in the trial.
- A hospitalized patient with a confirmed diagnosis of COVID-19 based on a reverse transcription-PCR (RT-PCR) or loop-mediated isothermal amplification (LAMP) assay, antigen test, or other means.
- A patient who met all of the following criteria upon admission to a hospital:
    - Patients who can begin receiving study treatment within 5 d after onset.

- ○ Patients with $SpO_2 \geq 95\%$ on room air.
- ○ Patients aged ≥ 40 years or having at least one of the following underlying diseases: renal impairment, chronic obstructive pulmonary disease, cardiovascular disease, cerebrovascular disorder, malignant tumor, obesity, diabetes mellitus, hypertension, and an immunosuppressive state.

- A patient aged at least 20 years at the time of informed consent.
- A patient who has been infected with SARS-CoV-2 for the first time.
- Our inclusion criteria have been chosen to ensure that the patients are appropriately and ethically included in our study, have a milder form of the disease, are in the appropriate age group, and do not already have SARS-CoV-2 antibodies at the point of enrollment.

2.2. *Trial Exclusion Criteria*

- A patient who is pregnant or breastfeeding.
- A patient who would not undergo a blood transfusion because of their religious beliefs.
- A patient who is participating in an intervention study for the treatment of COVID-19.
- A patient who has been vaccinated against SARS-CoV-2.
- A patient who has already undergone convalescent plasma transfusion.
- A patient with a history of allergy to a blood product.
- A patient with a deficiency in a plasma protein, such as IgA.
- A patient with New York Heart Association class III or IV heart failure.
- A patient whom the principal investigator, investigator, or sub-investigator judged to be ineligible for other reasons.

These exclusion criteria will help ensure the safety of study subjects, that the products used are acceptable to the study subjects, and that the evaluation of the efficacy of our planned intervention is not influenced by other SARS-CoV-2 treatments or vaccination.

2.3. *Intervention*

The intervention in this trial will be injectable COVID-19 convalescent plasma. This product will be the fresh frozen plasma collected by apheresis to remove most of the white blood cells. After being thawed, the liquid appears yellow to yellow-brown, sometimes cloudy because of lipids in the plasma. This product contains an anticoagulant solution (Acid Citrate Dextrose [ACD]-A solution) derived from apheresis. The collected plasma will be aliquoted in a volume of 100 mL; approximately 1.6 to 2.0 g (71 to 78 mEq) of sodium is present in 100 mL [20]. The neutralizing activity of purified IgG from convalescent plasma against SARS-CoV-2 will be evaluated using a previously described in vitro cell-based assay [18]. The neutralizing activity of convalescent plasma will be expressed as total neutralizing units (NU) by purified IgG neutralizing activity, the total amount of human IgG in the plasma, and the total volume of plasma [18]. The activity of the convalescent plasma will be classified into four stages (Supplementary Information S1).

- Grade A: Potent neutralizing activity. The total neutralizing capacity of 200 mL of plasma is $\geq$18,000 NU (Supplementary Information S2).
- Grade B: Moderate neutralizing activity. The total neutralizing capacity of 200 mL of plasma is $\geq$9000 but <18,000 NU.
- Grade C: Mild neutralizing activity. The total neutralizing capacity of 200 mL of plasma is $\geq$4500 but <9000 NU.
- Grade X: Slight neutralizing activity. The total neutralizing capacity of 200 mL of plasma is <4500 NU.

In this trial, there are no required or restricted concomitant drugs or therapies, but other drugs are used as an investigational intervention for COVID-19.

2.4. Randomization and Blinding

Subjects will be randomized to the convalescent plasma group or the standard of care group at a ratio of 1:1 by the randomization system of the Electronic Data Capture (EDC) or a randomization system created separately from the EDC. The randomization program will be registered on the randomization system in advance. Detailed procedures for randomization will be specified separately in the procedure manual (outside the scope of review by the certified review board). The purpose of this trial is to evaluate whether convalescent plasma therapy should be added to usual care, and it will be difficult to conduct blind treatment, considering the invasiveness in regards to the subjects and effects on the standard of care. For clinical improvement as secondary endpoints, blinding could be made possible if independent persons assess it. However, many other parameters will be collected regardless of whether they are blinded or not, and unblinding is expected to have a negligible impact. Therefore, treatment will not be blinded in this trial.

Randomization will be stratified by the following factors:

- Age (≥60 or <60 years old).
- The number of days from the day of onset (set as day 0) until the scheduled day of convalescent plasma transfusion (≤3 or ≥4 d).
- Trial site.

2.5. Primary and Secondary Endpoints

The primary endpoint of the trial will be the time-weighted average changes in the SARS-CoV-2 virus load in nasopharyngeal swabs from day 0 to days 3 and 5. It is hypothesized that the intervention should result in a decrease in the viral load in the convalescent plasma group up to day 5. This endpoint (change in viral load) has been used as an index of therapeutic effect in several previous studies.

The secondary endpoints are shown in Table 1.

Table 1. Secondary endpoints of the trial.

Objective	Endpoint	Rationale for the Endpoint
Prevention of mechanical ventilation or death.	Use of mechanical ventilation or death by days 14 and 28.	Prognosis is an endpoint that is not subjective.
Prevention of death.	Mortality on days 14 and 28.	Prognosis is an endpoint that is not subjective.
Prevention of the need for supplemental oxygen use.	Percentage of subjects who used oxygen on days 3, 5, 7, 14, and 28.	It is an endpoint that is not subjective.
To assess the shortening of the duration of symptoms (the time to clinical improvement).	Clinical improvement is defined as the first day a subject meets one of the three categories on the ordinal scale shown below: • Not requiring hospitalization or supplemental oxygen and not requiring continuation of treatment. • No hospitalization is needed but requires the limitation of activities and/or oxygen therapy at home. • No hospitalization and no limitation of activities.	Clinical improvement is related to efficacy.
To assess clinical improvement on days 3, 5, 7, 14, and 28 in subjects given the convalescent plasma.	Clinical improvement on days 3, 5, 7, 14, and 28 (on an 8-point scale)	Clinical improvement is related to efficacy.
Time to improvement on the National Early Warning Score, UK (NEWS)	Time to discharge from the hospital or the maintenance of NEWS ≤2 for 24 h (whichever occurs first) NEWS on days 3, 5, 7, 14, and 28.	Clinical improvement is related to efficacy.

Table 1. *Cont.*

Objective	Endpoint	Rationale for the Endpoint
Decrease in the viral load in the convalescent plasma group after convalescent plasma transfusion.	Time-weighted average change and the numerical change in the SARS-CoV-2 virus load in nasopharyngeal swabs from day 0 to each day of assessment.	Change in the viral load has been used as the index of the therapeutic effect in many studies.
To assess safety after convalescent plasma transfusion.	Occurrence of adverse events.	It is necessary to evaluate safety.
To screen and identify variants.	Determine if variants are present in nasopharyngeal swab samples on day 0	Variants are related to efficacy because it has been reported that variants may reduce the antiviral activity of neutralizing antibodies.

2.6. Dosage and Administration

After consent is obtained from a subject, the convalescent plasma will be administered intravenously during the period between the day of onset (set as day 0) and day 5. It will be infused intravenously at 40 mL/h for the first 15 min after the start of infusion. If no adverse event is observed during the infusion, it will be continued at 100 mL/h. The infused volume will be dependent upon the following.

As one course of therapy, the infused volume of the plasma is 200 mL for Grade A, 400 mL for Grade B, and 800 mL for Grade C. When the volume of 800 mL is infused, it will be divided into 2 and 400 mL will be infused in a 24 h interval.

Plasma with blood types A, B, and O must be administered to the subjects with the same blood type. However, according to the Guidelines for the Use of Blood Products issued by the Ministry of Health, Labour and Welfare [20], plasma with blood type AB can be administered to subjects with blood types A, B, and O, unless there is plasma with the same blood type.

2.7. Treatment and Observation Periods (Including Follow-Up)

The Grade A and B convalescent plasma will be administered in a day, and the Grade C convalescent plasma will be administered in 2 d. The subjects will be observed according to the schedule shown in Table 2.

Table 2. Study schedule.

Activities	Admission	Day 0 (Day of Transfusion) [i]		Day 1 (1 d after Transfu-sion)	Day 3 (3 d after Transfu-sion)	Day 5 (5 d after Transfu-sion)	Day 7 (7 d after Transfu-sion)	Day 14 (14 d after Transfu-sion)	Day 21 (21 d after Transfu-sion)	Day 28 (28 d after Transfu-sion)	Day 90 (90 d after Transfu-sion)	Discontinuation of the Study
		Before Trans-Fusion	3 h after Start of Trans-Fusion [j]									
Acceptable time window [a]	−3 to 0	Reference day		+1	±1	±1	±1	±3	±3	±3	+30	–
Informed consent	X											
Confirmation of eligibility	X [k]											
Registration and randomization of a subject	X											
Characteristics of a subject	X [k]											
Plasma transfusion [h]		X		X [b]								
Vital signs [d] and clinical condition [c]		X	X [h]	Once daily during hospitalization and days 3, 5, 7, 14, 21, and 28 after discharge								X
Physical findings [d]	X [k]	X	X [h]	X	X	X	X	X	X	X		X
Pregnancy test [l]	X [k]											
Collection of swabs (2 sticks) [m]		X		X	X	X	X	X	X	X		X [e]
Blood test (biochemistry, complete blood count, and coagulation) [m]	X [k]			X	X	X	X	X	X	X		
Blood test (blood type)		X [k]										
Blood test (cross-match) [h]		X										
Blood test (infection screening) [f]		X [k]										
Blood test (post-transfusion infection test) [g,h]											X	
Storage of plasma [m]		X		X	X	X	X	X	X	X		
Storage of serum [m]		X		X	X	X	X	X	X	X		
Radiography (chest X-ray)	X [k]											
Concomitant drugs [c]		X		Every day during hospitalization and days 3, 5, 7, 14, 21, and 28 after discharge								X
Adverse events [c]			X [h]	Every day during hospitalization and days 3, 5, 7, 14, 21, and 28 after discharge							X [h]	X

[a] If the days overlap because of the acceptable time window, it is not allowed to collect the data for 2 d on the same day (e.g., it is not allowed to collect the data for both day 1 and day 3 on day 2). [b] When Grade C plasma is used, it will be infused on day 1. [c] When a subject has been discharged, the data can be collected by phone on or after day 7. [d] When a subject has been discharged, it is allowed to not collect data on or after day 7. [e] This should be performed if a subject discontinues the study on and before day 14. [f] Similar to tests for infectious diseases conducted at registration, HBsAg, HBsAb, HCVAb, HIV-1/2Ab, Syphilis-RPR/TPHA, and HTLV-1 Ab should be measured. [g] Similar to the post-transfusion infection test, HBV-DNA quantification, HCV core protein, and HIV-1/2Ab assays should be performed (the test should also be performed in a subject who has stopped transfusion prematurely). [h] These data will be collected in the convalescent plasma group. [i] The date of transfusion scheduled at the time of randomization is day 0 in both the convalescent plasma group and the standard of care group. [j] When Grade C plasma is used, the data to be collected 3 h after the start of transfusion on day 0 should be collected both on days of first and second transfusions. [k] The results of the tests as the regular practice before informed consent can be used. [l] It is performed in premenopausal female subjects. [m] When a subject has been discharged, it is allowed to not collect samples on or after day 14.

2.8. Observation and Test Parameters

- Characteristics of each subject:
    - Date of birth (age), sex, nationality/race, smoking history, complications, prior medical history, history of the current disease, and pregnancy status (premenopausal female subjects should undergo a pregnancy test).
    - Information on hospitalization (dates of admission and discharge).
    - Body height and weight.
    - Background data related to COVID-19 and overseas travel history.
- Physical findings:
    - Status of supplemental oxygen and the use of mechanical ventilation.
    - Physical conditions will be examined by inspection, palpation, auscultation, and percussion.
- Vital signs:
    - Level of consciousness
    - Body temperature (°C)
    - Blood pressure (mmHg)
    - Pulse rate (beats/min)
    - Respiratory rate (breaths/min)
    - SpO_2 (%)
- Clinical condition: The clinical condition of each subject will be assessed according to the following:
    - Death
    - Hospitalization and the use of invasive mechanical ventilation or ECMO.
    - Hospitalization and the use of noninvasive mechanical ventilation or a high-flow oxygen device.
    - Hospitalization and supplemental oxygen requirement.
    - Not requiring hospitalization or supplemental oxygen but requiring the continuation of treatment (for COVID-19-related or other diseases).
    - Not requiring hospitalization, supplemental oxygen, or the continuation of treatment.
    - Not requiring hospitalization, but requiring the limitation of activities and/or oxygen therapy at home.
    - Not requiring hospitalization or the limitation of activities.
- Laboratory tests (Table 3)
- Imaging
- Plain chest X-ray

Table 3. Laboratory tests performed as part of the trial.

Hematology	Hemoglobin, Hematocrit, White Blood Cell Count with Differential, and Platelet Count
Coagulation	APTT, PT-INR, and D-Dimer
Blood biochemistry	Albumin, AST, ALT, bilirubin, CRP, blood glucose, urea nitrogen, creatinine, LDH, creatine kinase, potassium, and sodium
Infection screening	HBsAg, HBsAb, HCVAb, HIV-1/2Ab, Syphilis-RPR/TPHA, and HTLV-1Ab.
Pregnancy test	Urine or blood (HCG)
Blood type and cross-match	Blood type: A/O/B/AB, Rh +/−; cross-match: compatible/incompatible
SARS-CoV-2 viral load	Nasopharyngeal swabs
Samples for storage	Serum (1.5 mL) and plasma (1.5 mL)
Post-transfusion infection test	HBV-DNA quantification, HCV core protein, and HIV-1/2Ab

The appropriateness of the incorporated participants and the data collected were monitored on a regular basis to ensure quality.

2.9. Study Schedule

The study schedule is shown in Table 2.

2.10. Participation and Follow-Up Periods

The subjects in both the convalescent plasma and standard of care groups will be followed up for up to 28 d to detect any adverse events. In addition, the convalescent plasma group will be examined for post-transfusion infections on day 90. The observation period for the trial will run from the day of registration of the first subject until 31 March 2022 (scheduled).

2.11. Potential Benefits and Risks of the Study Drug

See Supplementary Information S3.

2.12. Sample Size Calculation

The trial was designed to enroll 200 patients. We calculated that this sample size would provide 90% power to detect a between-group log viral load difference of 0.5 (SD 1.1) at the 0.05 (two-sided) level of significance, accounting for several drop-out subjects.

2.13. Statistical Analysis Plan

See Supplementary Information S4.

2.14. Safety Evaluation

See Supplementary Information S5.

2.15. Data Monitoring Committee

The committee was established to neutrally evaluate interim data during the study period and to provide appropriate advice and recommendations to ensure the safety of subjects and the ethical and scientific validity of the study. In fact, it was not conducted due to a lack of progress in incorporation.

In addition, the committee periodically evaluated the following:

1. Changes over time in the treatment regimen in both treatment groups
2. Relationship between the quality of the study drug (Grades A to C) and safety

2.16. Data Management Team

The Director of the Joint Center for Researchers, Associates and Clinicians (JCRAC) Data Center designated a person to be in charge of data management tasks, and the data management staff carried out the tasks specified in the data management plan.

3. Discussion

COVID-19 is a significant public health concern because of its high associated morbidity and mortality. As of December 2020, the therapeutic agents approved for COVID-19 in Japan were limited to two drugs: remdesivir, an antiviral drug, granted a Special Approval for Emergency on 7 May 2020, and dexamethasone, which has an anti-inflammatory effect. Several case reports and clinical studies of convalescent plasma therapy have been reported [21–29], but the results have been conflicting. This could have occurred because the neutralizing antibody titer in the COVID-19 convalescent plasma administered was not measured in most studies, and there appears to be an association between a higher antibody index and severe COVID-19 [30]. Thus, in this trial, subjects will receive plasma with high neutralizing activity after it is determined whether the convalescent plasma from each donor can inhibit infection when susceptible living cells are exposed to infectious novel coronaviruses. The objective of this trial is to evaluate the efficacy of the convalescent plasma collected from donors who recovered from COVID-19.

There have been several important findings in nonclinical and clinical studies of relevance to this study. Purified IgG from convalescent plasma inhibited the cytotoxicity and replication competence of SARS-CoV-2 in an in vitro cell infection system. At a high concentration, it can completely inhibit SARS-CoV-2 infection at the single-cell level [18]. In a hamster model representing an in vivo cell infection system, the administration of convalescent plasma inhibited viral replication in the lung [31]. However, several RCTs of convalescent plasma therapy have already been conducted in countries outside Japan, but no conclusion has been reached with respect to the efficacy of convalescent plasma therapy, which is likely in part because of the heterogeneity of the types of target patients, interventions, and endpoints among trials [32]. Furthermore, it is well known that SARS-CoV-2 viruses have undergone multiple mutations since their appearance in 2019, resulting in changes in virulence that affect disease severity worldwide. Although it was reported that convalescent plasma therapy has led to a significant decrease in mortality rates among SARS-CoV-2 patients, the efficacy of the therapy would be affected by the degree of severity caused by multiple mutations of SARS-CoV-2 variants [25]. Further studies, such as on the early administration of convalescent plasma containing appropriate antibodies, will be needed to demonstrate the efficacy of convalescent plasma therapy.

Limitations of the Trial

As described above, the proposed trial has the potential to prevent patients with mild COVID-19 from developing a more severe illness, but there will be some limitations to this trial. First, for subjects receiving standard treatments for COVID-19 that will not be withheld as part of the trial, it may be difficult to disentangle the effectiveness of convalescent plasma therapy on the outcomes of subjects from one of the standard treatments. Second, although the number of persons vaccinated against SARS-CoV-2 has increased in Japan, the potential for beneficial effects of convalescent plasma therapy for vaccinated subjects remains unknown. In this trial, subjects vaccinated against SARS-CoV-2 will be excluded; therefore, the number of subjects who could take part in the trial will decrease over time. Third, casirivimab/imdevimab and sotrovimab, which are anti-SARS-CoV-2 monoclonal antibody treatments, and molnupiravir, an antiviral drug, were approved in Japan from July to December 2021 to prevent the development of symptomatic COVID-19, approximately 5–10 months after the first patient enrolled in this trial in February 2021. Because the target population groups of casirivimab/imdevimab, sotrovimab, and molnupiravir therapy overlap with that of this trial, this could cause difficulties when enrolling subjects. Taking these limitations into consideration and closely observing the developing trends of novel drugs and therapies, issues regarding subject enrollment should be reconsidered in the future. Nonetheless, improvements shown by our trial may lead to the expansion of the use of convalescent plasma therapy.

Supplementary Materials: The following supporting information can be downloaded at: https://www.mdpi.com/article/10.3390/life12060856/s1. Refs. [19,33–35] are cited in the Supplementary Materials.

Author Contributions: N.T. and J.T.-H. wrote, reviewed and edited main manuscript. J.T.-H. and A.M. administrated the research project. S.S., H.N., K.F., M.N., M.T., S.A., S.M., M.S., A.H., T.T. and K.S. carried out the investigation. N.T., J.T.-H., A.M., Y.U. and W.S. validated the research methods and results. Y.U. performed statistical analyses. S.K., Y.T., K.M. and H.M. suggested critical and essential conceptualization of the research. N.O. and W.S. supervised throughout the research. All authors have read and agreed to the published version of the manuscript.

Funding: Funding related to the conduct of this study was provided by a Health and Labor Sciences Research Grant (Research Project for Promotion of Policies for Emerging and Re-emerging Infectious Diseases and Immunization; Program Grant Number 20HA1006), AMED research grant (Research Project to Promote the Development of Innovative Drugs for Emerging and Reemerging Infectious Diseases; Grant Numbers JP20fk0108502 and JP20fk0108260), and the NCGM Intramural Research Fund (Grant Number 20A2003D). The funders had no role in the design of the study or collection, analysis, or interpretation of data or in writing the manuscript.

Institutional Review Board Statement: The study was conducted in accordance with the Declaration of Helsinki, and approved by the Certified Review Board of National Center for Global Health and Medicine (protocol code 4216 and date of approval 12 March 2021).

Informed Consent Statement: This study was approved by Institutional Review Board of the National Center for Global Medicine (Approval number: NCGM-C-004126-10, date 16 March 2022). Written informed consent for participation will be obtained from each patient or his/her legal representative.

Data Availability Statement: Data are available from the principal investigator on reasonable request.

Acknowledgments: The authors thank Kiyoto Tsuchiya of the AIDS Clinical Center, NCGM and Kouki Matsuda of the Department of Refractory Viral Infections, NCGM for the evaluation of neutralizing activity of convalescent plasmas. Editorial support, in the form of medical writing, assembling tables and creating high-resolution images based on authors' detailed directions, collating author comments, copyediting, fact checking, and referencing, was provided by Editage, Cactus Communications.

Conflicts of Interest: The authors declare that they have no competing interests.

Abbreviations

COVID-19	coronavirus disease 2019
ECMO	extracorporeal membrane oxygenation
EDC	Electronic Data Capture
MERS	Middle East respiratory syndrome.
NIH	National Institute of Health
RCTs	randomized controlled trials
SARS	severe acute respiratory syndrome
WHO	World Health Organization

References

1. Beigel, J.H.; Tomashek, K.M.; Dodd, L.E.; Mehta, A.K.; Zingman, B.S.; Kalil, A.C.; Hohmann, E.; Chu, H.Y.; Luetkemeyer, A.; Kline, S.; et al. Remdesivir for the Treatment of COVID-19—Preliminary report. *N. Engl. J. Med.* **2020**, *383*, 1813–1826. [CrossRef] [PubMed]
2. WHO Solidarity Trial Consortium; Pan, H.; Peto, R.; Henao-Restrepo, A.M.; Preziosi, M.P.; Sathiyamoorthy, V. Repurposed antiviral drugs for COVID-19—Interim WHO solidarity trial results. *N. Engl. J. Med.* **2021**, *384*, 497–511. [PubMed]
3. Agarwal, A.; Rochwerg, B.; Lamontagne, F.; Siemieniuk, R.A.; Agoritsas, T.; Askie, L. A living WHO guideline on drugs for COVID-19. *BMJ* **2020**, *370*, m3379. [CrossRef] [PubMed]
4. Recovery Collaborative Group; Horby, P.; Lim, W.S.; Emberson, J.R.; Mafham, M.; Bell, J.L. Dexamethasone in hospitalized patients with COVID-19. *N. Engl. J. Med.* **2021**, *384*, 693–704. [PubMed]
5. Luke, T.C.; Kilbane, E.M.; Jackson, J.L.; Hoffman, S.L. Meta-analysis: Convalescent blood products for Spanish influenza pneumonia: A future H5N1 treatment? *Ann. Intern. Med.* **2006**, *145*, 599–609. [CrossRef]
6. Maiztegui, J.; Fernandez, N.; De Damilano, A. Efficacy of immune plasma in treatment of argentine hæmorrhagic fever and association between treatment and a late neurological syndrome. *Lancet* **1979**, *314*, 1216–1217. [CrossRef]
7. Zhou, B.; Zhong, N.; Guan, Y. Treatment with Convalescent Plasma for Influenza A (H5N1) Infection. *N. Engl. J. Med.* **2007**, *357*, 1450–1451. [CrossRef]
8. Van Griensven, J.; Edwards, T.; de Lamballerie, X.; Semple, M.G.; Gallian, P.; Baize, S.; Haba, N. Faculty Opinions recommendation of Evaluation of convalescent plasma for ebola virus disease in guinea. *N. Engl. J. Med.* **2016**, *374*, 33–42. [CrossRef]
9. Cheng, Y.; Wong, R.; Soo, Y.O.Y.; Wong, W.S.; Lee, C.K.; Ng, M.H.L.; Chan, P.; Wong, K.C.; Leung, C.B.; Cheng, G. Use of convalescent plasma therapy in SARS patients in Hong Kong. *Eur. J. Clin. Microbiol.* **2004**, *24*, 44–46. [CrossRef]
10. Arabi, Y.M.; Balkhy, H.H.; Hajeer, A.H.; Bouchama, A.; Hayden, F.G.; Al-Omari, A.; Al-Hameed, F.M.; Taha, Y.; Shindo, N.; Whitehead, J.R.; et al. Feasibility, safety, clinical, and laboratory effects of convalescent plasma therapy for patients with Middle East respiratory syndrome coronavirus infection: A study protocol. *Springerplus* **2015**, *4*, 709. [CrossRef]
11. Ko, J.-H.; Seok, H.; Cho, S.Y.; Ha, Y.E.; Baek, J.Y.; Kim, S.H.; Kim, Y.-J.; Park, J.K.; Chung, C.R.; Kang, E.-S.; et al. Challenges of Convalescent Plasma Infusion Therapy in Middle East Respiratory Coronavirus Infection: A Single Centre Experience. *Antivir. Ther.* **2017**, *23*, 617–622. [CrossRef] [PubMed]
12. Shen, C.; Wang, Z.; Zhao, F.; Yang, Y.; Li, J.; Yuan, J.; Wang, F.; Li, D.; Yang, M.; Xing, L.; et al. Treatment of 5 Critically Ill Patients with COVID-19 with Convalescent Plasma. *JAMA* **2020**, *323*, 1582. [CrossRef] [PubMed]
13. Li, L.; Zhang, W.; Hu, Y.; Tong, X.; Zheng, S.; Yang, J. Effect of convalescent plasma therapy on time to clinical improvement in patients with severe and life-threatening COVID-19: A randomized clinical trial. *JAMA* **2020**, *324*, 460–470. [CrossRef] [PubMed]

14. Avendano-Sola, C.; Ramos-Martinez, A.; Munez-Rubio, E.; Ruiz-Antorán, B.; Malo, d.M.R.; Torres, F. Convalescent plasma for COVID-19: A multicenter, randomized clinical trial. *medRxiv* **2020**, arXiv:2020.08.26.20182444.
15. Agarwal, A.; Mukherjee, A.; Kumar, G.; Chatterjee, P.; Bhatnagar, T.; Malhotra, P.; PLACID Trial Collaborators. Convalescent plasma in the management of moderate COVID-19 in adults in India: Open label phase II multicentre randomised controlled trial (PLACID Trial). *BMJ* **2020**, *371*, m3939. [CrossRef]
16. Simonovich, V.A.; Pratx, L.D.B.; Scibona, P.; Beruto, M.V.; Vallone, M.G.; Vázquez, C.; Savoy, N.; Giunta, D.H.; Pérez, L.G.; Sánchez, M.D.L.; et al. A Randomized Trial of Convalescent Plasma in COVID-19 Severe Pneumonia. *N. Engl. J. Med.* **2021**, *384*, 619–629. [CrossRef]
17. Libster, R.; Marc, G.P.; Wappner, D.; Coviello, S.; Bianchi, A.; Braem, V. Prevention of severe COVID-19 in the elderly by early high-titer plasma. *medRxiv* **2020**, arXiv:2020.11.20.20234013.
18. Maeda, K.; Higashi-Kuwata, N.; Kinoshita, N.; Kutsuna, S.; Tsuchiya, K.; Hattori, S.I. Neutralization of SARS-CoV-2 with IgG from COVID-19-convalescent plasma. *Sci. Rep.* **2021**, *11*, 5563. [CrossRef]
19. Joyner, M.J.; Bruno, K.A.; Klassen, S.A.; Kunze, K.L.; Johnson, P.W.; Lesser, E.R. Safety update: COVID-19 convalescent plasma in 20,000 hospitalized patients. *Mayo Clin. Proc.* **2020**, *95*, 1888–1897. [CrossRef]
20. PSEHB Notification No. 0331-15 Dated March 31, 2017, Revision of the Guidelines for the Use of Blood Products, Attachment: Guidelines for the Use of Blood Products. Pharmaceutical Safety and Environmental Health Bureau, Ministry of Health, Labour and Welfare. Available online: https://www.mhlw.go.jp/file/06-Seisakujouhou-11120000-Iyakushokuhinkyoku/0000161115.pdf (accessed on 6 June 2022).
21. Meher, B.R.; Padhy, B.M.; Das, S.; Mohanty, R.R.; Agrawal, K. Effectiveness of convalescent plasma therapy in the treatment of moderate to severe COVID 19 patients: A systematic review and meta-analysis. *J. Assoc. Physicians India* **2020**, *68*, 35–43.
22. Eren, E.; Ulu-Kılıç, A.; Korkmaz, S.; Yıldız, M.; Yüksel, R.C.; Kılınç-Toker, A. Retrospective analysis on efficacy of convalescent plasma in acute respiratory distress syndrome due to COVID-19. *Sao Paulo Med. J.* **2022**, *140*, 12–16. [CrossRef] [PubMed]
23. Bošnjak, B.; Odak, I.; Ritter, C.; Stahl, K.; Graalmann, T.; Steinbrück, L. Case report: Convalescent plasma therapy induced Anti-SARS-CoV-2 T cell expansion, NK cell maturation and virus clearance in a B cell deficient patient after CD19 CAR T cell therapy. *Front. Immunol.* **2021**, *12*, 721738. [CrossRef] [PubMed]
24. Starshinova, A.; Malkova, A.; Zinchenko, U.; Kudlay, D.; Glushkova, A.; Dovgalyk, I. Efficacy of different types of therapy for COVID-19: A comprehensive review. *Life* **2021**, *11*, 753. [CrossRef] [PubMed]
25. Moubarak, M.; Kasozi, K.I.; Hetta, H.F.; Shaheen, H.M.; Rauf, A.; Al-Kuraishy, H.M. The rise of SARS-CoV-2 variants and the role of convalescent plasma therapy for management of infections. *Life* **2021**, *11*, 734. [CrossRef]
26. Cho, Y.; Sohn, Y.; Hyun, J.; Baek, Y.; Kim, M.; Kim, J. Effectiveness of convalescent plasma therapy in severe or critically ill COVID-19 patients: A retrospective cohort study. *Yonsei Med. J.* **2021**, *62*, 799–805. [CrossRef]
27. Huang, L.; Zhang, C.; Zhou, X.; Zhao, Z.; Wang, W.; Leng, W. Convalescent plasma is of limited clinical benefit in critically ill patients with coronavirus disease-2019: A cohort study. *J. Transl. Med.* **2021**, *19*, 365. [CrossRef]
28. Nusshag, C.; Morath, C.; Speer, C.; Kaelble, F.; Zeier, M.; Boxberger, M. Plasma exchange in patients with severe coronavirus disease 2019: A single-center experience. *Crit. Care Explor.* **2021**, *3*, e0517. [CrossRef]
29. Parikh, P.M.; Sood, T.; Agarwala, V.; Kate, S. Has convalescent plasma therapy hesitancy increased COVID-19 mortality? *South Asian J. Cancer* **2021**, *10*, 42–45. [CrossRef]
30. Bonetti, T.C.S.; Latini, F.R.M.; Invitti, A.L.; Fonseca, M.C.M.; Scorza, F.A.; Saldanha, M.G. Antibody indexes in COVID-19 convalescent plasma donors: Unanswered questions. *Clinics* **2021**, *76*, e2818.
31. Takamatsu, Y.; Imai, M.; Maeda, K.; Nakajima, N.; Higashi-Kuwata, N.; Iwatsuki-Horimoto, K. Highly Neutralizing COVID-19 Convalescent Plasmas Potently Block SARS-CoV-2 Replication and Pneumonia in Syrian Hamsters. *J. Virol.* **2022**, *96*, e0155121. [CrossRef]
32. Piechotta, V.; Iannizzi, C.; Chai, K.L.; Valk, S.J.; Kimber, C.; Dorando, E. Convalescent plasma or hyperimmune immunoglobulin for people with COVID-19: A living systematic review. *Cochrane Database Syst. Rev.* **2021**, *5*, CD013600. [PubMed]
33. Xu, T.; Chen, C.; Zhu, Z.; Cui, M.; Chen, C.; Dai, H.; Xue, Y. Clinical features and dynamics of viral load in imported and non-imported patients with COVID-19. *Int. J. Infect. Dis.* **2020**, *94*, 68–71. [CrossRef]
34. Liu, Y.; Yan, L.M.; Wan, L.; Xiang, T.X.; Le, A.; Liu, J.M.; Peiris, M.; Poon, L.L.M.; Zhang, W. Viral dynamics in mild and severe cases of COVID-19. *Lancet Infect. Dis.* **2020**, *20*, 656–657. [CrossRef]
35. Weinreich, D.M.; Sivapalasingam, S.; Norton, T.; Ali, S.; Gao, H.; Bhore, R.; Musser, B.J.; Soo, Y.; Rofail, D.; Im, J.; et al. REGN-COV2, a neutralizing antibody cocktail, in outpatients with COVID-19. *N. Engl. J. Med.* **2021**, *384*, 238–251. [CrossRef] [PubMed]